Environmental Design Guidelines for Low Crested Coastal Structures

Environmental Design Guidelines for Low Crested Coastal Structures

Hans F. Burcharth
Stephen J. Hawkins
Barbara Zanuttigh
Alberto Lamberti

ELSEVIER

AMSTERDAM • BOSTON • HEIDELBERG • LONDON • NEW YORK • OXFORD
PARIS • SAN DIEGO • SAN FRANCISO • SINGAPORE • SYDNEY • TOKYO

Elsevier
The Boulevard, Langford Lane, Kidlington, Oxford OX5 1GB, UK
Radarweg 29, PO Box 211, 1000 AE Amsterdam, The Netherlands

First edition 2007

British Library Cataloguing in Publication Data
A catalogue record for this book is available from the British Library

Library of Congress Cataloging-in-Publication Data
A catalog record for this book is available from the Library of Congress

ISBN–13: 978-0-08-044951-7
ISBN–10: 0-08-044951-4

For information on all Elsevier publications
visit our web site at books.elsevier.com

Printed and bound in the United Kingdom
Transferred to Digital Print 2011

Editors acknowledge the great effort
by Assistant Editors:
Morten Kramer
Paula Moschella

The support of EC through DELOS (EVK3-CT-00041) is gratefully aknowledged.

Contents

Part I
Guidelines

Part II
– *Appendix* –

Part III

– *Tools* –

Prologue

H.F. Burcharth, A. Lamberti

The effect of human activities is primarily local but can extend far away from the location of intervention. This underlines the importance of establishing coastal zone management plans covering large stretches of coastlines.

The interaction of wave climate, beach erosion, beach defence, habitat changes and beach value, which clearly exists based on EC research experiences and particularly on results obtained by DELOS Project (www.delos.unibo.it) for Low Crested Structures (LCSs), suggests the necessity of integrated approaches and thus the relevance of design guidelines covering: structure stability and construction problems, hydro and morphodynamic effects, environmental effects (colonisation of the structure and water quality), societal and economic impacts (recreational benefits, swimming safety, beach quality).

The present guidelines are specifically dedicated to LCSs to provide methodological tools both for the engineering design of structures and for prediction of performance and environmental impacts of such structures. It is anticipated that the guidelines will provide valuable inputs to coastal zone management plans.

The target audience for this set of guidelines is consulting engineers or engineering officers and officials of local authorities dealing with coastal protection schemes. The guidelines are also of relevance in providing a briefing of current best practice for local and national planning authorities, statutory agencies and other stakeholders in the coastal zone. The guidelines have been drafted in a generic way to be appropriate throughout the European Union taking into regard current European Commission policy and directives to promote sustainable development and integrated coastal zone management.

The guidelines are composed of three main parts.

The **first part** (Chapters 1-10) contains the description of the design methodology, from the preliminary identification of design alternatives till the selection of the sustainable scheme and its construction.

The **second part** presents:

- the analysis of the performance of beach defences in DELOS study sites, which were selected to represent a variety of environmental conditions (Chapter 11);
- the application of the proposed methodology to a real prototype case, in order to give a practical example to designers (Chapter 12).

The **third part** contains all the formulae and tools to help engineers (Chapter 13), ecologists (Chapter 14) and economists (Chapter 15) during the design procedure.

These Guidelines are a product of DELOS Consortium; for each section the main authors and their institution are mentioned, whose contact information can be found in the list reported in DELOS Consortium section.

Summary of the DELOS Project

The overall objective of DELOS was to promote effective and environmentally compatible design of low crested structures (LCSs) to defend European shores against erosion and to preserve the littoral environment and economic development of the coast.

Specific objectives and methods were:

- to provide an inventory of existing LCS and a literature based description of their effects;
- to analyse LCS hydrodynamics, stability and effects on beach morphology by surveys on sites, laboratory experiments and numerical modelling;
- to investigate the impacts of LCS on biodiversity and functioning of coastal ecosystems by observations and field experiments;
- to develop a general methodology to quantify benefits to enable implementation of Integrated Coastal Zone Management based on Contingent Valuation methodologies in different European countries;
- to provide local authorities with validated operational guidelines for the design of LCS based on the achieved knowledge of LCS hydrodynamics and stability, water circulation, beach morphology, impacts on coastal assemblages, human perception and related economic effects.

DELOS offered the possibility to achieve these aims through integrated collaboration among engineers, coastal oceanographers, marine ecologists, economists and political institutions, involving 18 partners from 7 European countries and end users.

The work necessary to meet the overall goal of DELOS was grouped in five integrated Research Tasks:

- ➢ **Research Task 1**: to provide an overview of the different types of structure, how effective they are in the different coastal situations, and to identify which parameters may characterise each structure and its effects on the coastal environment.
- ➢ **Research Task 2**: to analyse the hydrodynamics around stability of structure, to provide relationships among water level, discharge and wave characteristics at both sides of the structure, to analyse currents induced by breaking over the structures and their effects on beach morphology, both near to the structure and over the protected beach, up to the swash limit. This shall be done by observation on sites, by laboratory experiments in wave channel and wave basin and by numerical modelling.
- ➢ **Research Task 3**: to identify, quantify and forecast the impacts (perceived as positive or negative) of low-crested breakwaters on the biodiversity and functioning of coastal assemblages of animals and plants at a range of spatial (local, regional and European) and temporal scales (months to years) and in relation to different environmental conditions (including meteorological conditions, tidal range, wave action, human usage, surrounding habitats).
- ➢ **Research Task 4**: to develop a general methodology for Integrated Coastal Zone Management linking economic and environmental components, based on Contingent Valuation values obtained by Contingent Valuation in different countries in Europe and on criteria for transferring them from one country to the other, accounting for the effects of situations specific to each country.
- ➢ **Research Task 5**: to provide guidelines for an environmental design of such structures, based on practical experience, on the most recent scientific results regarding the hydrodynamics around structures and stability of them, water circulation and beach morphology, impacts on coastal assemblages, and accounting for human perception and related economic effects; guidelines will be verified by application to the study sites and selected case studies.
- ➢ **Research Task 6**: to establish communication among partners and with end-users.

Interactions among the Research Tasks is represented in the flow-diagram below.

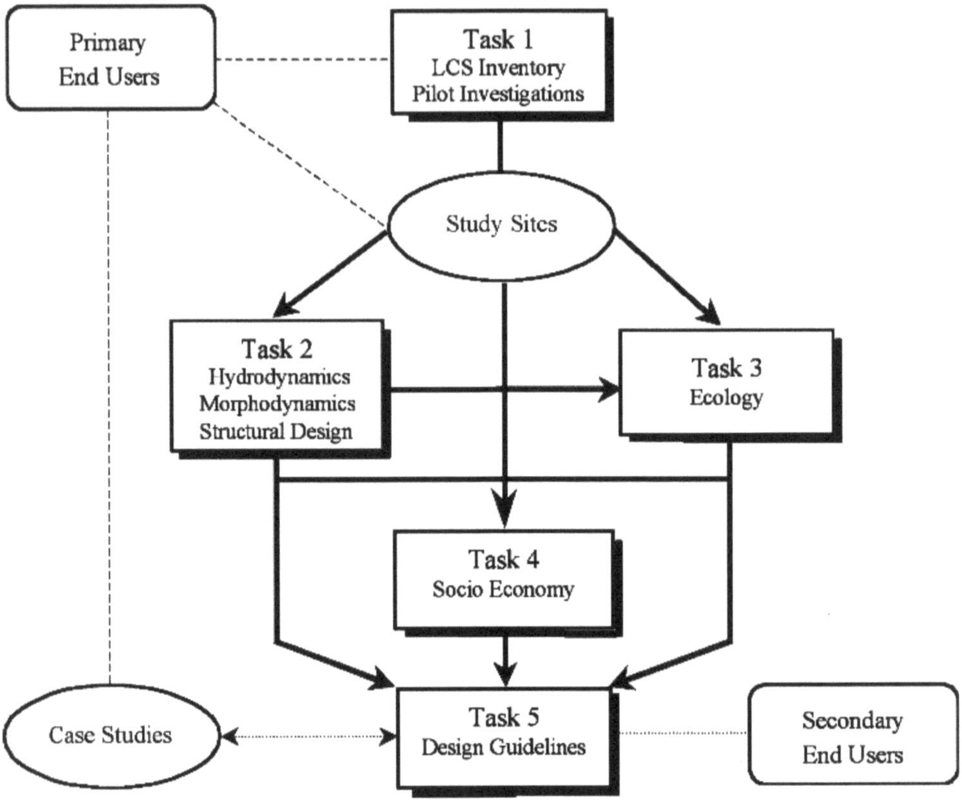

Project results are available at the DELOS web page: **www.delos.unibo.it**.

DELOS Consortium Personnel

Acr.	Institution, Address	Staff			E-Mail
UB	DISTART Idraulica, University of Bologna Viale Risorgimento 2, 40136 Bologna, Italy	Prof.	Lamberti	Alberto	alberto.lamberti@mail.ing.unibo.it
		Dr.	Zanuttigh	Barbara	barbara.zanuttigh@mail.ing.unibo.it
		Dr.	Martinelli	Luca	luca.martinelli@mail.ing.unibo.it
		Dr.	Archetti	Renata	renata.archetti@mail.ing.unibo.it
		PhD St.	Clementi	Emanuela	emanuela.clementi@mail.ing.unibo.it
	Dept. of Economy, University of Bologna Piazza Scaravilli 2, 40126 Bologna, Italy	Prof.	Marzetti	Silva	marzetti@economia.unibo.it
AAU	Dept. of Civil Engineering, Aalborg University, Sohngaardsholmsvej 57, 9000 Aalborg, Denmark	Prof.	Burcharth	Hans F.	burcharth@civil.auc.dk
		Dr.	Kramer	Morten	i5mkr@civil.auc.dk
UCA	University of Cantabria, ETSICCyP, Avd. de los Castros s/n, 39005 Santander, Spain	Prof.	Vidal	Cesar	vidalc@unican.es
		Prof.	Losada	Inigo J.	losadai@unican.es
		Dr.	Garcia	Nicholas	garciani@unican.es
		Dr.	Lara	Javier L.	lopezjav@unican.es
UPC	LIM-UPC, Universidad Polytecnica di Catalunia, Jordi Girona 1-3, 08034 Barcelona, Spain	Prof.	Sánchez-Arcilla	Agustín	agustin.arcilla@upc.es
		Ass. Prof.	Gironella	Xavier	xavi.gironella@upc.es
		Ass. Prof.	Jiménez	José A.	Jose.jimenez@upc.es
		Ass. Prof.	Mösso	Cesar	cesar.mosso@upc.es
		Dr.	Cáceres	Iván	i.caceres@upc.es
		Dr.	Sierra	Joan Pau	joan.pau.sierra@upc.es
		Ph.D. St.	González	Daniel	daniel.gonzalez-marco@upc.es
		Ph.D. St.	Alsina	José M.	jose.alsina@upc.es
UGOT	Göteborgs University, Marine Ecology, Box 461, 405 30 Göteborg, Sweden	Prof.	Åberg	Per	Per.Aberg@marbot.gu.se
		Prof.	Jonsson	Per	pejo@tmbl.gu.se
		Ms.	Granhag	Lena	Lena.Granhag@tmbl.gu.se
		Mr.	Sundelöf	Andreas	Andreas.Sundelof@marbot.gu.se
UTW	Universiteit Twente, Department of Civil Engineering, Transportation Engineering & Management P. O. Box 217, 7500AE, Enschede, The Netherlands	Prof.	van der Veen	Anne	A.vanderVeen@utwente.nl
		Dr.	Polomé	Philippe	polome@ecru.ucl.ac.be
UoS	University of Southampton European Way, SO14 3ZH, Southampton, United Kingdom	Prof.	Collins	Michael	mbc@soc.soton.ac.uk
		Dr.	Paphitis	Doros	doros@soc.soton.ac.uk
		Dr.	Plomaritis	Theocharis	tplo3@soc.soton.ac.uk

Acr.	Institution, Address	Staff			E-Mail
AUTh	Aristotle University Thessaloniki Univ. Campus, 541 24 Thessaloniki, Greece	Prof.	Koutitas	Cristopher	koutitas@civil.auth.gr
		Prof.	Prinos	Panayotis	prinosp@civil.auth.gr
		Prof.	Karambas	Theophanis	karambas@civil.auth.gr
		Dr.	Avgeris	Giannis	iavgeris@civil.auth.gr
FF	Environmental Sciences, University of Ravenna Via S. Alberto 163, 48100 Ravenna, Italy	Prof.	Abbiati	Marco	marco.abbiati@unibo.it
		Prof.	Ceccherelli	Victor U.	victor.ceccherelli@unibo.it
		Dr.	Airoldi	Laura	laura.airoldi2@unibo.it
		Dr.	Bacchiocchi	Francesca	backy@ambra.unibo.it
		Dr	Colangelo	Marina A.	marina.colangelo@unibo.it
		Dr.	Bertasi	Fabio	fabio.bertasi@unibo.it
		Dr.	Bulleri	Fabio	fbulleri@discau.unipi.it
UR3	DSIC-University of Roma TRE, Via Vito Volterra 62, 00146 Rome, Italy	Prof.	Franco	Leopoldo	leofranc@uniroma3.it
		Dr.	Bellotti	Giorgio	bellotti@uniroma3.it
		Dr.	Briganti	Riccardo	briganti@uniroma3.it
		Dr.	Di Risio	Marcello	dirisio@uniroma3.it
ISVA	Technical University Denmark, Bygningst orvet Build. 115, 2800 Kongens Lyngby, Denmark	Prof.	Fredsoe	Jorgen	jf@isva.dtu.dk
		Prof.	Sumer	Mutlu B.	sumer@isva.dtu.dk
DHI	DHI Water & Environment, Agern Allé 5, -2970 Hørsholm, Denmark	Dr.	Zyserman	Julio A.	jaz@dhi.dk
		Dr.	Christensen	Erik D.	edc@dhi.dk
		Dr.	Johnson	Hakeem K.	hkj@dhi.dk
CSIC	CSIC, CEAB C/ Accés a la Cala Sant Francesc 14, 17300 Blanes, Spain	Dr.	Martin	Daniel	dani@ceab.csic.es
		Dr.	Gacia	Esperança	gacia@ceab.csic.es
		Dr.	Macpherson	Enrique	macpherson@ceab.csic.es
		Ms.	Satta	Maria Paola	paoletta@ceab.csic.es
WL-DH	Delft Hydraulics, Rotterdamseweg 185, 2600 MH, Delft 31, The Netherlands	Dr.	De Vries	Mindert	Mindert.deVries@wldelft.nl
INF	Infram ltd, Geomatic Business Park, P.O. box 16, 8316 ZG, Marknesse, The Netherlands	Dr.	van der Meer	Jentsje W.	jentsje.vandermeer@infram.nl
MBA	The Marine Biological Association, The Laboratory, Citadel Hill PL1 2PB, Plymouth, United Kingdom	Prof.	Hawkins	Stephen	sjha@mba.ac.uk
		Dr.	Moschella	Paula	pmos@mba.ac.uk
		Dr.	Frost	Matthew	matfr@mba.ac.uk
	University of Plymouth Drake Circus, PL4 8AA Plymouth, United Kingdom	Dr.	Thompson	Richard	R.C.Thompson@plymouth.ac.uk
BIAU	Aarhus University, Dept. of Marine Ecology Finlands-gade 14, 8200 Aarhus N, Denmark	Ass. Prof.	Cedhagen	Thomas	Cedhagen@biology.au.dk
		Dr.	Dinesen	Grete E.	Grete.Dinesen@biology.au.dk
MOD	MODIMAR, Via Monte Zebio 40 - 00195 Roma, Italy	Prof.	Noli	Alberto	modimar_srl@iol.it
		Ing.	Contini	Paolo	paolo.c@modimar.com

Part I

Guidelines

CHAPTER 1

Definition of LCSs covered by the guidelines

(Burcharth, AAU)

The guidelines cover shore-parallel low crested and submerged structures such as regularly overtopped emergent and submerged detached breakwaters. Whilst LCSs share engineering and ecological features with artificial reefs, these are considered as a separate issue as they are very wide crested, deeply submerged and deployed mainly to enhance fisheries.

The structures reduce the amount of wave energy reaching the shore behind them and as a consequence also influence sediment transport and impose shoreline changes.

LCSs can be constructed as a single structure (Figure 1.1a) or in series (Figure 1.1b). A single structure is used to protect a localized area, whereas a multiple segmented system is designed to protect an extended length of shoreline.

Submerged breakwaters might be constructed as long continuous structures in which case gaps might not be strictly necessary for water exchange. In schemes with emergent breakwaters or slightly submerged structures such gaps might be provided anyway to allow passage of boats. Figure 1.1c shows an example of a scheme consisting of long submerged breakwaters with small gaps between them. Also shown are some submerged terminal groynes forming a cell configuration often used to retain artificial sand fills.

Single structures as shown in Figure 1.1a are usually built in water depths of more than 3-4 metres with the objective of reducing or stopping coastal erosion at a single location and at the same time creating a sheltered area for swimming or mooring of boats. Detached breakwaters in multi-structure schemes are often constructed in very shallow water of few metres water depth close to the shoreline with the single objective of protecting a beach against erosion and flooding of low-lying areas. If built at some distance from the shoreline the objective would most often be a combination of beach protection and creation of a suitable area for recreational usage.

The structures are most commonly constructed of stone material (cf. the cross sections in Figure 1.1). Concrete blocks are used for the armour layers if suitable rock material of sufficient size is not readily available.

Revetments or seawalls are often constructed along the coast as part of LCS-schemes in order to strengthen very exposed coastlines.

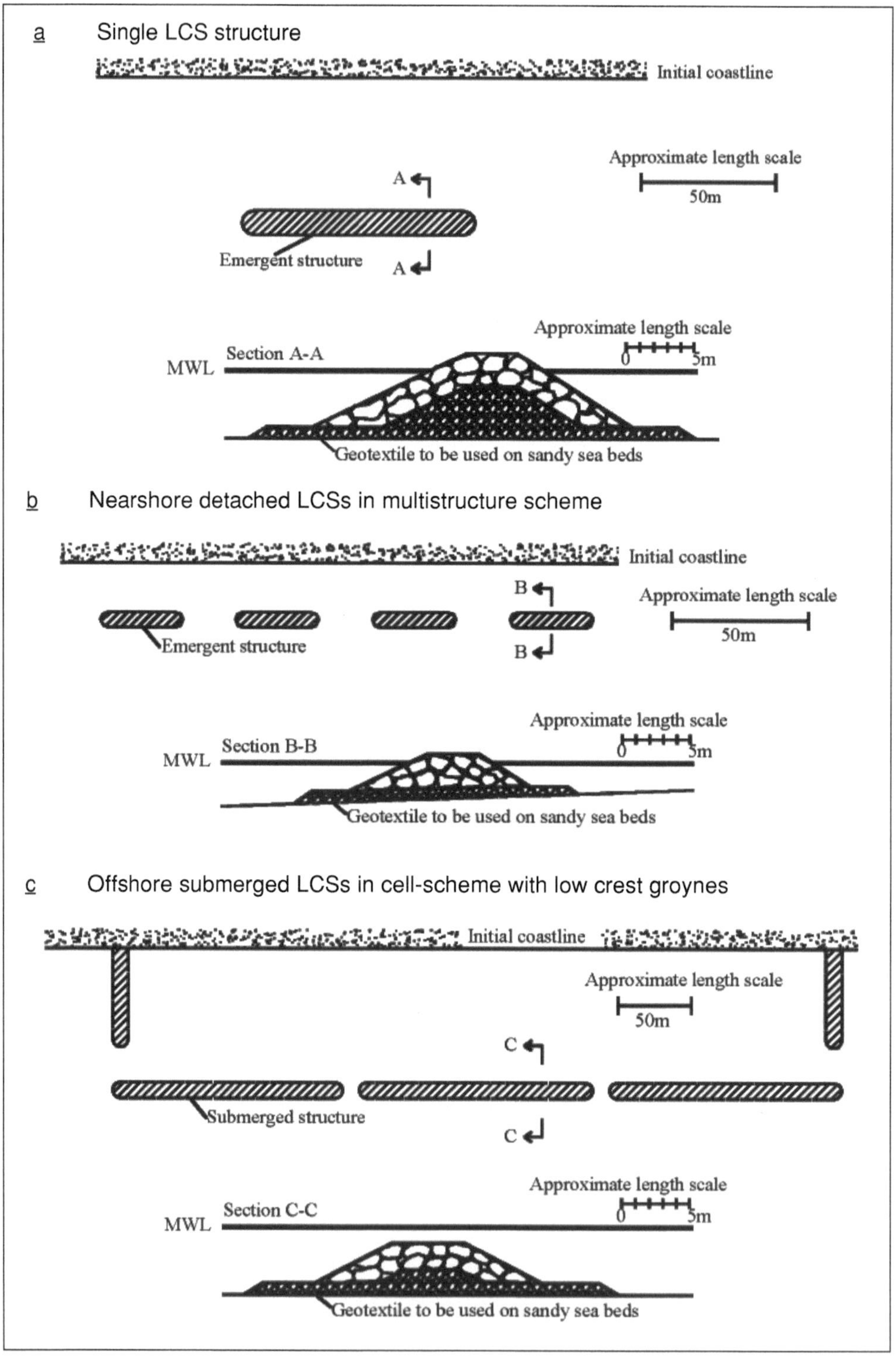

Figure 1.1. Examples of layouts and cross sections of LCSs.

CHAPTER 2

Function of LCSs

2.1. LCSs INTERACTION WITH WAVES, CURRENTS AND SEDIMENT TRANSPORT

(Burcharth, AAU)

When used for beach stabilization the function of LCSs is to reduce wave energy in their lee and thereby reducing the sediment carrying capacity of the waves to the shoreward. They can be designed to reduce or prevent the erosion of an existing beach or a beach fill, or to encourage natural sediment accumulation to form a new beach.

The structures reduce the incoming wave energy across the structure by triggering wave breaking at and on the structure, by partially reflecting the waves and by dissipation related to the wave induced porous flow in the structure. This is illustrated for an emergent structure in Figure 2.1.

Wave breaking accounts for the largest part of the energy reduction, reflection for the second largest part and porous flow for the smallest part. Wave energy is also transmitted

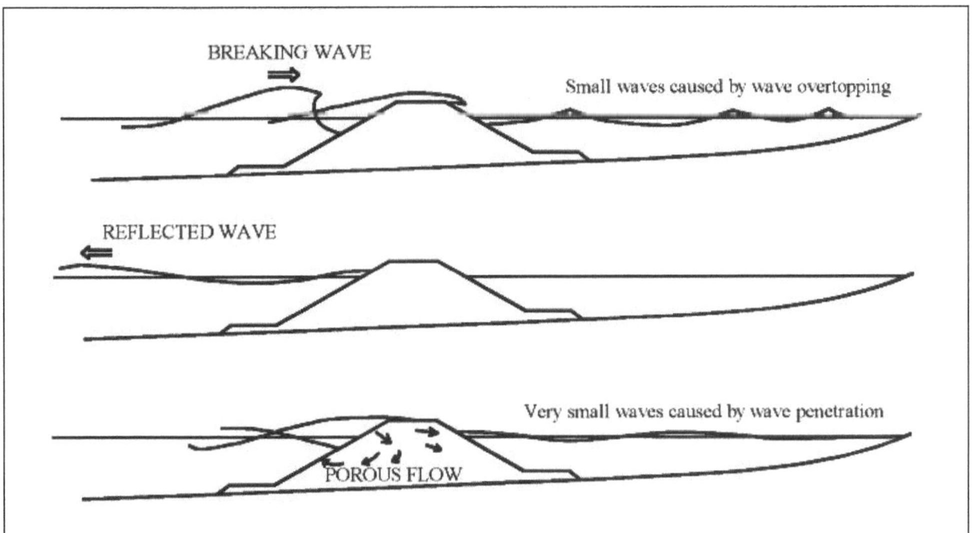

Figure 2.1. Illustration of the sheltering effect of an emergent LCS by reduction in shorewards transmitted wave energy by wave breaking, wave reflection and porous flow.

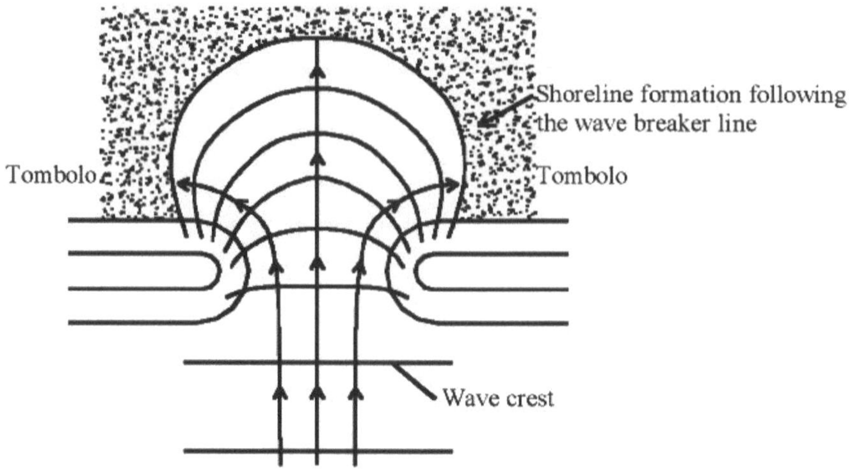

Figure 2.2. Illustration of spreading of wave energy by diffraction and refraction in gap between emergent structures.

horizontally by diffraction and refraction around the heads of the structure into the lee zone as illustrated for an emergent structure in Figure 2.2.

In the case of shorter emergent structures with only limited overtopping the horizontal wave transmission will dominate. The lower the crest level the more dominant will become the wave disturbance caused by overtopping waves. For long submerged structures the wave disturbance is caused almost completely by wave transmission over the crest.

Depending on the sheltering effect of LCSs, more or less littoral material is deposited and retained in the sheltered area behind the structures. If moderately sheltered the sediment will typically appear as a bulge in the beach planform termed as a *salient*. If more protected, the resulting shoreline extends out to the structure thus forming a so-called *tombolo* (cf. Figure 2.3).

The actual morphodynamic changes are to a large extent also determined by currents; not only the tide and storm surge generated currents on the coastal stretch, but indeed by the currents generated locally at and around the structures by wave-structure interaction. Waves passing over a LCS result in a net transport of water across the structure inducing

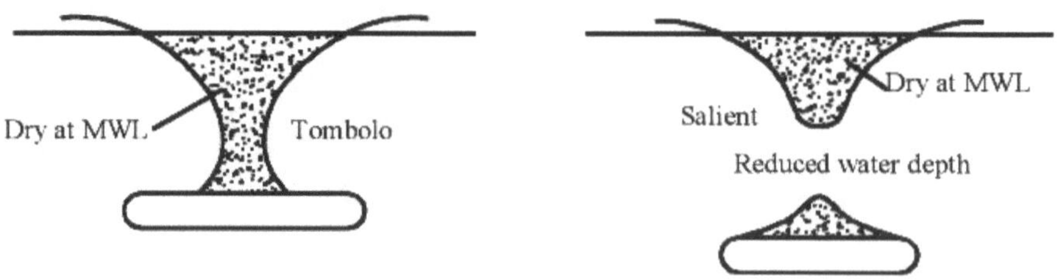

Figure 2.3. Illustration of tombolo and salient.

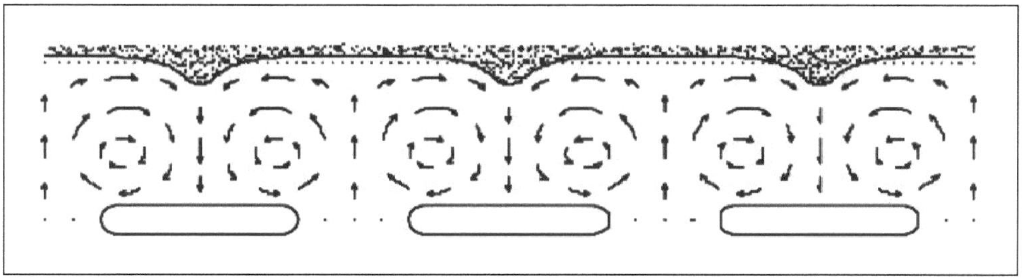

Figure 2.4. Illustration of wave induced currents behind emerged structures without wave transmission across the structures.

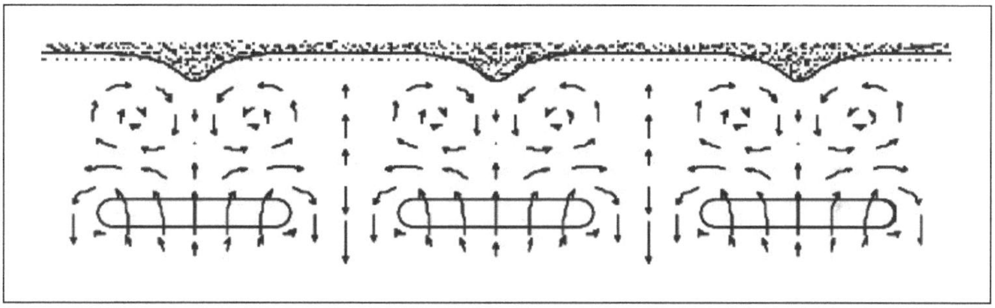

Figure 2.5. Illustration of wave induced currents in case of submerged structures with wave transmission across the crests. Note the very strong outgoing rip-currents in the gaps.

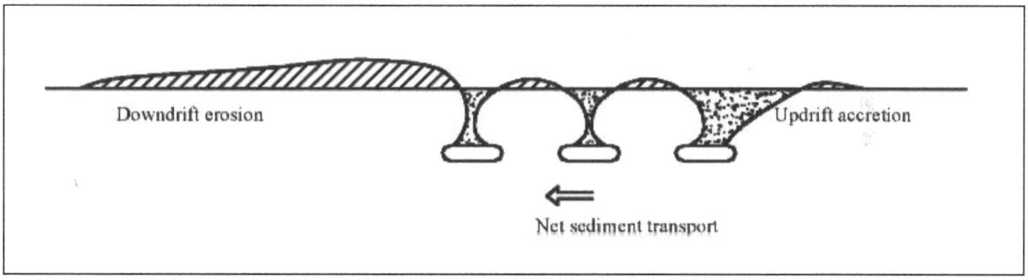

Figure 2.6. Illustration of downdrift erosion and updrift accretion caused by formation of tombolos and salients shoreward of detached breakwaters.

a higher mean water level in the lee of the structure. This creates a seaward net transport of water through the porous structure, but more importantly also horizontal currents and vortices in the lee zone due to head gradients towards the ends of the structures.

The patterns of the currents are different in case of emergent and submerged structures, see Figure 2.4 and 2.5.

The net transport of water into the lee zone causes a water level rise and is balanced mainly by outgoing currents at the heads of the structures. In case of multi-structure schemes these currents will be manifested as concentrated and eroding rip currents in the gaps between the structures (cf. Figure 2.5).

Like other hard structures, LCSs have some drawbacks. Salients or tombolos can interfere with longshore currents and sediment transport and create almost always downdrift erosion on coastlines with one dominating sediment transport direction along the coast (see Figure 2.6). Tombolos have in this respect a stronger negative effect than salients. Moreover, emergent LCSs forming schemes with rather closed cells might result in stagnant water of poor quality. Also the visual impact of emergent structures can be negative at locations of high scenic value.

These factors have resulted in a move towards design of structures with a very low crest or fully submerged.

At a given location and water depth the lower structures are cheaper in material but are less effective in attenuating wave energy than surface-piercing structures. Thus the optimum design will be a balance between these aspects.

Predictions of the actual morphological changes imposed by LCSs, local as well as more distant, are difficult due to the complicated interaction between waves, water levels, currents and sediment transport. These factors change considerably in most places not only over the year but also from year to year. Stable long-term-average beach profiles will not be reached on eroding coasts unless beach nourishment is provided or sufficient natural supply from remote sources is not interrupted.

2.2. ENVIRONMENTAL CONSIDERATIONS AND CONSEQUENCES

(Moschella, MBA; Abbiati, FF; Aberg, UGOT; Airoldi, Bacchiocchi, Bertasi, Bulleri, Ceccherelli, FF; Cedhagen, BIAU; Colangelo, FF; De Vries WL-DH; Dinesen; BIAU; Gacia, CSIC; Granhag, UGOT; Jonsson, UGOT; Macpherson, Martin, Satta, CSIC; Sundelöf, UGOT; Thompson & Hawkins, MBA)

Coastlines are highly dynamic systems subject to geo-morphological processes such as erosion, sediment transport and vertical land movement. These natural processes lead to continuous changes in the coastline that can be affected by human activities.

LCSs, as many man-made constructions in the sea, will have consequences for the natural environment and coastal landscape. These consequences occur at local scale, but may also scale up to the whole coastline. Effects may be site specific, reflecting the complexity, uncertainty and variability of natural systems. Therefore knowledge of environmental context in which coastal defence structures are placed is fundamental to effective design and management of these structures. Although the variability of ecological systems prevents very specific quantitative predictions of impacts, some qualitative trends may be suggested. In particular, the construction of LCSs and other types of hard defence structures results in:

1. the loss of natural sedimentary habitats and associated assemblages of animals and plants. These effects are primarily limited to the immediate vicinity of the structure but can sum up to a significant loss in areas where many LCSs are built; downstream effects can also occur – especially when multiple structure schemes are built along the coast.

2. Effects on surrounding sedimentary habitats as a consequence of the primary objective of the structure itself, which is to reduce wave energy. Such alteration of hydrodynamic regimes directly influences the characteristics of soft sediments (i.e. grain size, content of organic matter, redox conditions) and modifies detrital pathways (Davis *et al.*, 1982). These changes will be most evident in the area between the structure and the shoreline, where water movements will be reduced. This will result in changes in the composition and/or abundance of animals and plants living in and on sedimentary shores and seabeds. Periods

Figure 2.7. Close-up of an LCS in the Adriatic sea, showing the turbidity of the surrounding water and the siltation on the epibiota colonising the structure.

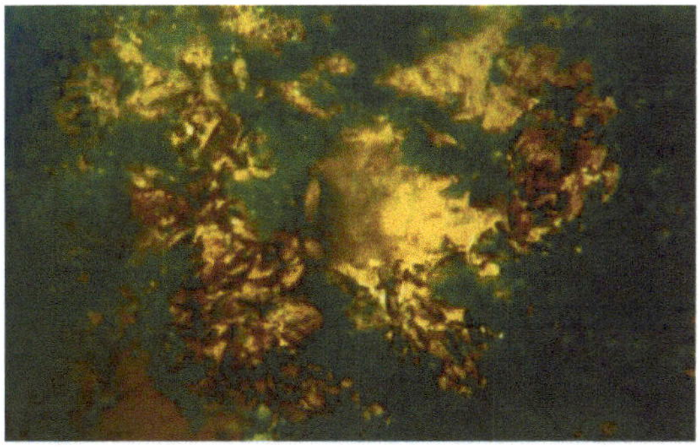

Figure 2.8. Close-up of a submerged rock of an LCS showing mussels and green algae. Deposition of silt is evident on mussels.

with calm weather conditions may further reduce water movement in the protected area leading to stagnant water and degradation of water quality (see Figure 2.7).

3. The introduction of artificial rocky habitats. Similarly to natural rocky reefs, these habitats will be colonised by animal and plants that are typical of rocky coasts such as green algae and mussels (Figure 2.8). On coastlines dominated by sandy shores this will result in the introduction of new species or in an increased abundance of species already present on other types of artificial substrates in the area such as slipways or marinas. These altered distributional patterns cause considerable changes to the identity and/or abundance of species in coastal areas and have important environmental and/or economical consequences. Some of these organisms such as ephemeral green algae may represent a problem for beach

a)

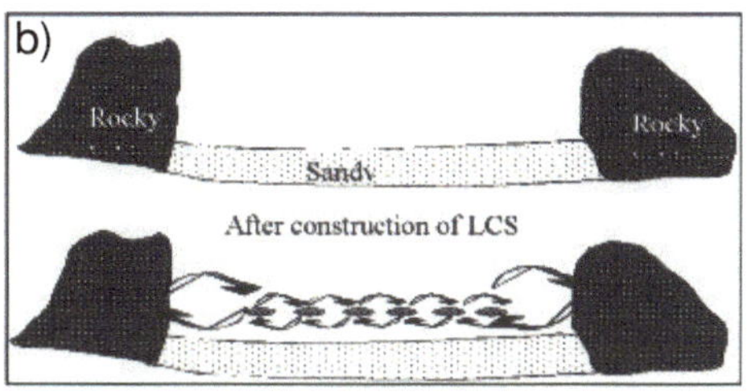

Figure 2.9. Coastal defence structures along the Italian coast of the north Adriatic Sea (left) and a diagram showing multiple LCS acting as stepping stones that facilitate dispersal of species (right).

tourism when turn off the structures and washed up the shore. Conversely, colonisation of LCS by others species such as mussels may be perceived as enhancement of food and/or recreational resources, therefore increasing the socio-economic value of the area.

4. There can be large scale effects. Artificial structures can act as stepping stones that facilitate the dispersal of rocky shore species across habitats that would naturally be unconnected (see Figure 2.9a and b). These structures can facilitate dispersal for many species including the spread of exotic species. Another potential consequence is represented by changes in intrinsic and regional dynamics of many species and communities. An increased connectivity between natural rocky shores can also change the genetic structure within species.

A final consideration is that LCSs are often explicitly or implicitly considered a benefit to coastal sandy areas for their potential to increase local species diversity by allowing settlement of new species that usually live on rocky reefs. The results of DELOS project suggest that although LCSs become colonised by species typical of rocky substrate, their assemblages can differ from those occurring on nearby natural reefs. Diversity is generally lower and assemblages are dominated by ephemeral and early successional species that are more tolerant of disturbance. Primary production does increase as macroalgae only grow on rocky substrata. This can, however, create problems by increasing algal detritus.

In areas lacking of natural rocky shores, extensive sets of LCS in essence completely alter the nature of coastline. A naturally dynamic sedimentary environment is replaced by an impoverished rocky habitat that also interferes with the natural dynamic of geomorphological processes. This should be taken into account when establishing coastal defence plans covering large stretches of coastlines. The design of structures should maximise coastal protection effects but minimise environmental changes by avoiding any unessential engineering.

2.3. SOCIO-ECONOMIC IMPACT OF LCSs

(Van der Veen, UTW)

Economic impacts of LCSs relate to the dynamic behaviour of the coast and thus to protection of land and private and public assets. We might distinguish between mitigating benefits and costs, enhancement benefits, preservation benefits and costs, and indirect benefits and costs. Examples are the reduction of damage due to flooding and erosion, reduction in salinity intrusion, improved navigation, restored recreation opportunities and the preservation of habitats.

CHAPTER 3

Objectives and target effects of LCSs

(Moschella, MBA; Burcharth, AAU)

3.1. PROTECTION OF LAND AND INFRASTRUCTURE BY PREVENTION OR REDUCTION OF COASTAL EROSION

(Moschella & Hawkins, MBA)

Sea level rise, due to global warming, subsidence processes, increased storminess and tidal surges, expose several European coastlines to serious erosion and flooding events. In highly developed coastal areas, erosion and flooding cause conspicuous socio-economic losses in terms of damages to houses, infrastructures such as roads and railways, industries and farmland. The coastal protection provided by LCSs has positive effects on coastal economies. These are:

- protection of recreational beaches against erosion;
- protection of residential properties;
- protection of infrastructures (e.g. roads and railways);
- protection of coastal industries;
- protection of farmlands;
- protection against flooding due to severe storms and surges.

Coastal defences including LCSs must be constructed with due regard to sustainable management of habitats, species and ecosystems and their living natural resource (including goods and services) observing European Directives on habitats, birds, and water plus comply to any national or regional environmental legislation.

An example comes from the Elmer Defence scheme (West Sussex, England), built to protect a low-lying residential area from flooding as a result of severe storms associated with spring tides. Since the construction of the breakwaters in 1993, no flooding events were recorded in that area, causing a significant increase in the property values and a decrease in home insurance premium.

3.2. IMPROVEMENT OF RECREATIONAL CONDITIONS

(Moschella, MBA; Airoldi, FF; Thompson & Hawkins, MBA)

LCSs can stabilize beaches or create wider beaches, improve conditions for swimming as well as beach quality with respect to amenity-friendly beach material such as fine sand. Such development should also observe relevant environmental legislation, guidance and emerging best practice in order to ensure sustainable usage of the coastal zone.

LCSs have significant influence on the recreational conditions for beach users. Some influences are regarded as positive, while others are considered as negative. The influence is either direct due to the physical presence of the structure in the nearshore zone, or indirect due to the consequent effects on the local hydro-morphodynamics (eg. rip currents).

Sea conditions behind LCSs are generally calmer than on open beaches and this can improve bathing conditions, especially for children. The improved safety of bathing and swimming in a calm sheltered zone (probably excluding for boat traffic) is a very positive effect since this is the most common recreational activity taking place in the nearshore area.

However the possible formation of strong rip currents at gaps and/or ends of the LCS shore protection system during rough seas may endanger the safety of bathing.

The presence of organisms that grow on the structures or colonise the sheltered habitats behind LCSs can be a nuisance for beach tourism, leading to expensive beach cleaning or removal of the organisms. Examples of these negative effects on the recreational value of the beach come from the Italian shores of the North Adriatic Sea, where the ephemeral green algae that extensively colonise LCSs (also favoured by local eutrophic conditions) are torn off the structures and washed up the shore, where they decay. In the UK, large amounts of drift algae are trapped on the landward side of the structures and eventually decompose leading to unpleasant smells due to formation of anoxic conditions and increase in number of flies. Further, periods with calm weather conditions may lead to stagnant water and degradation of bathing water quality.

Boating with various craft and surfing may be negatively affected by the presence of the LCS if the crest elevation is not clearly visible, due to the risk of collision. Even more dangerous could be diving into the sea from a boat and hitting on the hard structure.

Conversely, activities like snorkelling or sport fishing can be positively enhanced if the structure provides a new attractive habitat for marine life. If the structure is emergent it favours access for fishermen.

3.3. PROTECT AND MINIMISE IMPACTS ON CULTURAL AND NATURAL HERITAGE OF THE COASTLINE

(Moschella, MBA; Airoldi, FF; Gacia, CSIC; Thompson & Hawkins, MBA)

Coastal erosion and flooding also threaten coastal areas of high ecological value such as intertidal and mud flats, shingle ridges, sand dunes, wetlands, salt marshes, coastal lagoons, maritime cliff grasslands and soft cliffs. These natural habitats are subject to Community interest and many are designated as Special Areas of Conservation (Habitat Directive 92/43/EEC). One of the objectives of the Habitat Directive and the Water Framework Directive is to promote and maintain diversity of natural habitats and their ecosystems and where necessary human intervention can be required to achieve these objectives. Low crested structures can therefore contribute to the protection and maintenance of these coastal habitats providing the following effects:

- protection of habitats with unique geological and geomorphological features;
- protection of habitats that represent nesting sites for protected bird species;
- preservation of endangered or vulnerable species whose survival depends on maintenance of coastal habitats;

- preservation of coastal plant and animal species of scientific interest.

In addition, special features of natural heritage importance (e.g., saline lagoons, saltmarshes, vegetated shingle banks and sand dunes) or special sectors of interest (bird reserves) may be threatened by coastal erosion. Therefore circumstances may occur where a coastal defence structure is proposed to expressly protect endangered habitats or species. For example, in the coastal area between Happisburgh and Winterton-on-sea in Norfolk (East Anglia, UK), a system of LCS and a seawall were built to protect The Broads wetlands.

In Tuscany sea defence structures were built to protect the maritime pine tree forest in the national park of San Rossore endangered by coastal erosion. Sometimes habitats or species protected by conservation legislation such as the vegetated shingles (habitat listed on Annex 1 of the EC Habitats Directive) can indirectly benefit from coastal defence schemes that were built with the only purpose of protecting properties. For example, the Elmer defence schemes in West Sussex (South of England) also protects the vegetated shingle ridge which host species of special national conservation interest such as little robin *Geranium purpureum*, a rare plant in West Sussex, the toadflax brocade moth *Calophasia lunula*, a Biodiversity Action Plan species and many birds which nest in this zone. Elmer defence scheme has been designated as an SSSI (Site of Special Scientific Interest). In Poole Bay (south of England), the recently built rock groyne system not only protects residential properties and the beach from erosion but also helps restoration of native vegetated sand dunes. If protected, these will in turn provide additional natural protection against erosion.

LCSs can be used to protect areas of cultural heritage value such as archaeological and historic sites, monuments, churches and buildings threatened by coastal erosion. Non-visible LCSs are probably preferable; if necessary combined with a revetment or a seawall to strengthen the shore. For example, on the Adriatic coast, along the promontory of Conero, a system consisting of LCS and rocks were deployed to protect historic buildings from erosion.

3.4. ENHANCEMENT OF NATURAL LIVING RESOURCES FOR FOOD AND RECREATION

(Moschella, MBA; Airoldi & Bulleri, FF; Thompson & Hawkins, MBA)

Whilst the primary objectives of LCS are to modify hydrodynamic and sedimentary regimes to protect sensitive areas or improve recreational conditions, any LCS that is put in the sea will also become colonised by marine organisms. Such colonisation must be recognised as an important change to the identity and/or abundance of habitats and hence species in coastal areas, and cannot be avoided. It is, however, possible, within the limits set by the primary necessity of engineering performance of LCS, to modify selected design features to enhance growth of selected organisms. Thus features of LCS design can sometimes be used to maximise desired secondary management end points (where perception of desirability or undesirability are intended as value judgement related to societal goals and expectations). Examples of such secondary management end points include:

- provision of suitable habitats to promote living resources for exploitation of food (such as shellfish and fish);
- provision of suitable habitats to promote living resources that are the focus for recreational (such as angling, snorkelling) or educational (such as appreciation of

marine wildlife, «rock-pooling» and ornithology) activities;
- provision of suitable habitats to promote endangered or rare species;
- provision of suitable habitats to promote diverse rocky substrate assemblages for conservation or mitigation purposes.

CHAPTER 4

Outline of design procedure

(Burcharth, AAU; Lamberti UB)

The design procedure is usually divided into a preliminary (or conceptual) design phase and a detailed design phase. The objective of the preliminary design phase is to explore the project feasibility with respect to economy, technical performance, and societal and environmental impacts. This usually involves conceptual design of alternative LCS-schemes. The preferred scheme is then selected for detailed design which basically consists of optimizing the scheme with respect to impacts, structural performance and costs.

Fig. 4.1 shows schematically the design procedure. Each of the blocks is explained in more detail in the following paragraphs and described in the following Chapters.

Initially in the preliminary design the target effects of the LCS-scheme and the legal, physical, environmental, socio-economic and aesthetic constraints must be clarified.

As a basis for both preliminary design and detailed design one has to establish information on historic performance of the beach at the location, on water level variations (tide, storm surge), currents, waves and/or winds, seabed bathymetry, beach topography, sediment characteristics, water quality, and biotic assemblages. Moreover, in both design phases one has to evaluate the hydrodynamic, morphological, ecologic and socio-economic impacts.

The main difference between preliminary design and detailed design is – apart from analyses of alternatives – the more in-depth analyses used in the detailed design, both with respect to environmental background information and performance of the scheme. However, quite often it is necessary also in preliminary design to perform in-depth analyses of some aspects in order to produce a background for a qualified selection among alternatives.

The design of LCSs includes functional design and structural design.

Functional design concerns the impact and performance of the LCS-scheme with respect to coastal protection, improvement of recreational conditions and conservation of natural living resources.

Structural design concerns the resistance of the LCSs to the actions of waves and currents.

It is characteristic for design of coastal protection schemes that prediction of the morphological and ecological impacts are much more difficult than prediction of the performance of the structures themselves. The reasons for this are that the hydrodynamic-morphologic interactions are very complicated, and the related

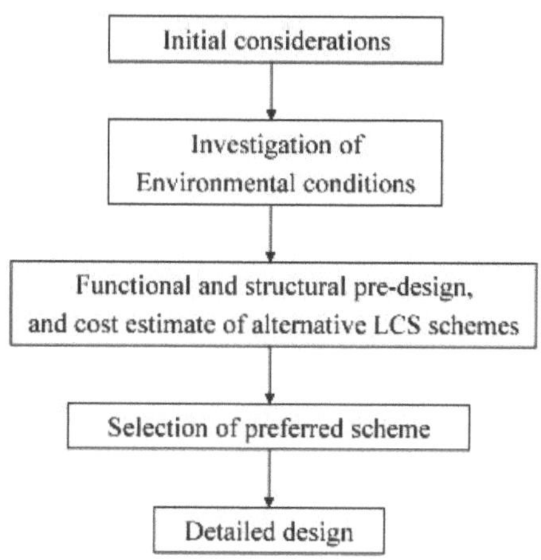

Figure 4.1 Diagram showing the preliminary design procedure.

predictive tools are either indicative simple rules of thumb or complex numerical models. For reliable prediction of the morphological development the latter needs to be run for long-term simulations, not only covering the local areas around the structure but also the sediment cell. To establish the necessary boundary conditions and hydrodynamic input, and to run such models is all together very costly and time consuming. As a consequence they are generally used only for finer tuning of larger schemes. In most cases only more simple numerical models are used locally, and then only for short-term simulations. It follows that the uncertainty related to the long-term prediction of the morphological response will be large.

The tools for structural design are quite reliable formulae for the stability of the various parts of the structures, and/or performance of model tests. The major part of the uncertainty of the structural response is related to the estimation of the design wave climate and, if scour is critical, also to the local currents at the structures.

Because the structure should preserve its shape for the whole project period and because repair cannot take place immediately after damage, it is common practice in structural design to consider the most severe environmental conditions in structure lifetime.

In functional design with respect to impact on beach morphology and ecosystems it is necessary to analyse the long-term effect of all environmental conditions accounting for the variations in intensity and duration that affect the function of the structure.

Most LCSs are located where wave heights are depth limited. As water depth depends both on the water level and the sea bed level, both have to be examined with respect to statistics and variations.

It follows that it is difficult to give more specific guidance with respect to design procedure and selection of design tools. A general statement could be that the marginal costs of further detailed analyses in preliminary and detailed design stages should be compensated by the added value of the certainty of the performance (or reduced risk of failure) of the scheme. Fig. 4.2 outlines a typical optimization procedure of the final design of a LCS scheme where the primary performance factor is the morphological response.

The formal Environmental Impact Assessment (EIA) of important project is usually carried out based on the preliminary project. The imperfect definition at this stage of some parameters should be managed according to a precautionary principle: Evaluate benefits and damages cautiously within the possible scenarios, so that the result of the assessment is not contradicted by any result of the final optimization process. Even if the formal EIA is not carried out, the societal and environmental effects of the scheme shall be evaluated during the final design optimization.

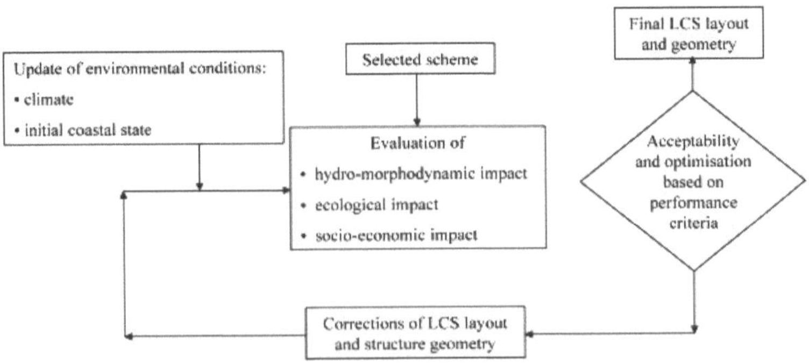

Figure 4.2 Diagram showing the detailed design procedure.

CHAPTER 5

Initial considerations

5.1. CONSIDERATION OF LEGAL, PHYSICAL, ENVIRONMENTAL, SOCIO-ECONOMIC AND AESTHETIC CONSTRAINTS

(Burcharth, AAU; Vidal, UCA; Moschella, MBA; Airoldi, Bulleri, Ceccherelli, Colangelo, FF; Thompson & Hawkins, MBA)

5.1.1. Relevant policy and legislation

Both coastal protection (protection from erosion) and sea defence (defence from inundation) are influenced by EU policy and legislation and by the translation of these at the national level. Other legal issues relate to directives and legislation regarding the procedural steps to obtain the necessary planning permissions and licences for any defence scheme (such as consultation and freedom of access to environmental information).

These approaches and their translation vary across Europe but the overarching EU legislative requirements are the same. Table 5.1 identifies the relevant Directives that will need to be considered when developing proposals for coastal protection and sea defence measures, including LCSs. These directives have been divided into the vertical and horizontal controls impacting on the process. Horizontal directives are the EIA Directive (coastal defence works) and the Strategic Environmental Assessment (SEA) Directive (coastal works to combat erosion and works that alter the coastline). SEA will be required where plans and programmes are from particular sectors or otherwise from those which have significant environmental effects, and set the framework for future development consent of EIA projects (under Directive 85/337/EEC as amended), or any plan which requires an appropriate assessment under the provisions of the Habitats Directive (92/43/EEC).

The SEA Directive had to be translated into national legislation by 21st July 2004. Many of the datasets relevant to implementation of the SEA Directive at the strategic level are also relevant at the individual project level (the EIA Directive level) and will therefore be relevant to individual coast defence project assessments. Sustainability Appraisals (SA), which have been increasingly used at plan and programme level are essentially non-statutory and overlap with many of the requirements of the SEA Directive. Usually SA has a wider remit within the social and economic appraisal than does SEA with its stronger focus on sustainable environment, but SA also has a lower baseline information demand and less analytical approach than SEA.

There are also proposed EU directives and conventions relevant to the development of defences that have been included here since there is already wide adoption of the principles at national level even without the weight of European legislation. A number of the Directives that have influenced the development or that have been active during the development of existing coastal defence structures have since been modified and or amended. These changes have resulted universally in a strengthening of the controls and information requirements to support projects.

Table 5.1. Relevant policies and legislations at international and European level Directives relevant to proposals for coastal protection and sea defence measures.

Directive	*Date*	*Directive No.*
Horizontal		
Environmental Impact Assessment Directive	1985	85/337/EEC amended by Directive 97/11/EC
Strategic Environmental Assessment (SEA) Directive	2001	2001/42/EC
Water Framework Directive	2000	2000/60/EC
Environmental Quality		
Bathing Water Directive	1976 (modified)	76/160/EEC modified 90/656/EEC and 91/692/EEC
Shellfish Waters Directive	1979	79/923/EEC amended by 91/692/EEC
Waste Water Treatment Directive	1991	91/271/EEC
Nitrates Directive for Protection of water against pollution caused by nitrates from agricultural sources	1991	91/676/EEC
Dangerous substances		76/464/EEC amended by Directives 90/656/EEC and 91/692/EEC
Information		
Access to Environmental Information Directive	1990	90/313/EEC replaced by 2003/4/EC
Nature Conservation		
Conservation of Wild Birds	1979	79/409/EEC
Conservation of Natural Habitats and Wild Flora and Fauna (Habitats Directive)	1992	92/43/EEC
Conventions and proposed Directives		
Aarhus Convention on access to information and participation in decision making	2000	Implemented through Directives.
Integrated Coastal Zone Management (ICZM)	2000	Currently a recommendation COM/2000/547
OSPAR Oslo and Paris Convention for the protection of the Marine Environment of the North East Atlantic.	1992	
HELCOM Helsinki Convention for the Protection of the Marine Environment of the Baltic Sea Area.	1974 revised 1992	
Barcelona Convention for the Protection of the Marine Environment and the Coastal Region of the Mediterranean.	1995	
Ramsar Convention (Wetlands of International Importance).	1971	

In addition, there are a number of other international conventions to which the majority of the member states are signatories and are treated alongside the EU legislation. These conventions relate both to horizontal and thematic initiatives.

Two relatively new Directives have a wider role in the strategic assessment of defence projects and for which member states are developing approaches to implementation. Specifically, the Strategic Environmental Assessment Directive and the Water Framework Directive are seen as providing the scope for integrated management of resources, including those on the coast. The Water Framework Directive in particular will provide a new strategic framework for the development of defence plans as part of the overall development of River Basin Management Plans (RBMP) and through these the potential for nationally consistent approaches. Within the UK the RBMPs are likely to act as an overarching framework into which the strategic management of coastal defence will have to be developed. Whether the RBMP can integrate the existing non-statutory approach to Shoreline Management Plans through which strategic defence management is developed is yet to be decided. However, it is likely that any non-statutory plan would be subservient to the objectives developed within any RBMP, which will also cover coastal waters. It is also likely that the objectives of the WFD will influence coastal defence proposals. Defence structures are almost certainly significant modifications to the natural environment and mitigation procedures are therefore likely to be required within LCS scheme to contribute to achieving good ecological status for relevant waterbodies.

The integration of activities along the EU shoreline is also influenced by conventions that target regional seas and consider issues of erosion and water quality. The EU has also considered the requirements for an integrated approach to management of the coastal zone with the adoption of a resolution for the development of an EU strategy for coastal zones (1992). This has lead to the draft strategy for Integrated Coastal Zone Management (ICZM) and a three-year demonstration programme from 1996. The development of ICZM will affect existing legislation and is likely to reinforce the integration of existing Directives and national legislation as well as non-statutory planning guidance.

The development of enhanced integration within spatial planning is also relevant to the coastal zone and the development of the European Spatial Development Perspective (ESDP) offers insight into spatial approaches within integrated coastal zone management planning.

The legislative requirements and policy implementation at member state level for coastal defence planning and management have not been individually assessed here, although it is clear that the approach to Directive implementation and spatial planning differs widely around Europe. In many countries the planning is managed as much by guidance notes and non-statutory plans as they are through legislative provisions.

Many of the member states are also looking more closely at the integration of coastal zone management in advance of any EU ICZM Directive. The complexity of the current administrative and legal system suggests at a national scale (at least in UK) that no EU wide ICZM Directive will be immediately forthcoming. It seems more likely that the ICZM will be implemented through a Council resolution, procedural guidance and best practice.

For example, in England and Wales many of the non-statutory plans focusing on flood and coastal defence would however fall within the assessment of the SEA Directive. These are likely to include Shoreline Management Plans (SMP), Water Level Management Plans

(WLMP), Coastal Habitat Management Plans (CHaMPs). Biodiversity (through Biodiversity Action Plans) will also need to be considered within the scope of defence approaches (DEFRA, 2001). For example, whilst LCSs may develop diverse epibiotic communities, these may not be typical of the area and therefore they may not form appropriate mitigation for significant environmental effects of a flood defence action. However, the development and maintenance of flood and coastal defence may also form integral part of the defence of freshwater sites (e.g. grazing marshes and lagoons) and hence the maintenance of site integrity. The conservation benefits of these LCSs will therefore need careful consideration balancing the environmental losses against the maintenance of biodiversity and potential for enhancement, even where sites are not under international conservation designations.

There are clearly strong overlapping requirements between SEA, EIA, WFD and sectorial Directives. At least there is the potential for the environmental as well as social and economic baseline datasets to be shared between the national implementations of these Directives requirements and also on into non-statutory planning processes – such as shoreline management plans (specifically targeting sea defence and coast protection). Such approaches will help to avoid duplication, provide consistent data and allow national and international status reports to be generated. Further duplication may occur where there is the requirement for multiple assessments (such as where both SEA and Appropriate Assessment under the Habitats Directive would be required). Promotion of the integration of assessments will be important in considering the different objectives of the Directive but also in integrating the findings when applied to coastal planning.

5.1.2. Physical constraints

Physical constraints are mainly given by the bathymetry, the character of neighbouring stretches and by material supply possibilities.

In case of a steep seabed it will be expensive to place the structures at some distance to the shore.

Sedimentary neighbouring coasts vulnerable to erosion cause serious constraints with respect to the tolerable impact of the LCS-scheme on the coastal development. Down-drift erosion is the most serious problem in this respect.

The use of natural rock as building material depends on the availability, size, quality, quantity and costs for quarrying and transport. If not available then concrete blocks is an alternative solution. The choice of material should, however, take into account environmental constraints and desired ecological effects of LCSs.

5.1.3. Ecological constraints (including ecosystems, natural heritage and living resources)

A variety of constraints should be considered in the design and construction procedures of LCS. Environmental constraints should be clearly identified through the EIA and current practice, following also the requirements of the European Commission Environmental Directive 85/337/EEC. Environmental constraints may include cultural and natural heritage, state and sensitivity of habitats, ecosystems and water quality.

1. Cultural heritage:

- The presence of historic sites.
- The presence of archaeological sites, both land and marine based.
- The presence of listed buildings.

2. Natural heritage:

- The presence of marine and coastal natural heritage areas (NHAs), with designated sites of special interest containing important wildlife habitats, endangered species or unique geological or geomorphological features.
- The presence of special areas of protection and conservation at international (e.g. Ramsar convention), European (e.g. SACs under Habitat Directive), national (e.g. SSSI, and SPAs in the UK, PEIN in Spain) and local (voluntary, statutory or private nature reserves) level.
- The presence of national parks, wildlife sanctuaries and marine protected areas (MPAs).

3. Habitats and associated ecosystems:

- The vulnerability of surrounding habitats and associated biota (benthic fauna, fish, birds). For example, subtidal rocky habitats and boulder fields can be severely affected by alteration of sediment regime and deposition (Airoldi 2003). Similarly seagrass meadows (such as *Posidonia, Zostera, Cymodocea*) are sensitive to changes in sediment and nutrient dynamics (Pergent-Martini *et al.*, 1996; Vermaat *et al.*, 1997).
- The presence of rare or endangered species which could be threatened by the construction of LCS. For example, rare species such as the coarse sand requiring *Branchiostoma lanceolatum* which can be threatened by changes in granulometry (Desprez, 2000).
- The presence of species that are important for the local economy (e.g. *Chamelea gallina, Solen vagina*) and that could be replaced by non-native and not edible species introduced by the new structures.

Indirect effects should be also taken into account, such as the presence of birds that rely on feeding on certain infaunal species in the area affected by LCSs.

4. Water quality:

- The presence of estuaries, as LCSs could affect the distribution and characteristics of sediment and organic load on the coast.
- The presence of source of contaminants such as heavy metals, and persistant organic compounds. LCSs might have a trapping effect, leading to accumulation of these pollutants in finer deposits especially on the landward, sheltered side.
- The eutrophic state and nutrients load. The presence of LCSs leading to greater residence time could trigger macroalgal growth and harmful microalgal blooms including potential toxic species (dinoflagellates) by increasing the eutrophic state of the surrounding waters.

5.1.4. Aesthetic constraints

Coastal defences, especially multiple structure defence schemes, represent one very often significant visual impact on the coastal landscape. This is particularly true for emerging shore-parallel structures that tend to block the view from both land to sea and sea to land. Visual impacts need therefore to be taken in consideration in the choice of LCS layout, design and building material. Spoiling the view from beach and seafront restaurants could also have a negative socio-economic effect, as well as the selection of construction material which is in contrast with the surrounding natural landscape. For example, in most cases rock material is preferred instead of concrete.

Aesthethic constraints include considerations for:

- National Parks or Coastal Reserves of particular landscape or scenic beauty.
- Specially designated Areas of Outstanding Natural Beauty (AONBs).
- Heritage Coasts, primarily designated for the quality of their coastal landscape.
- Historic landscapes, such as coastal monuments or terrestrial archaeological sites.
- Residential houses, hotels and leisure infrastructures on the top of the beach.

5.2. DEFINITION OF THE PRIMARY OBJECTIVES

(Moschella, MBA; Airoldi, FF; Thompson & Hawkins, MBA)

5.2.1. Technical objectives

The engineering objectives for the specific project must be specified, reference is given to Section 3.1.

5.2.2. Environmental objectives

a. Geology-geomorphology

One of the environmental aims of LCSs should be to limit the target changes in the geomorphological processes (e.g. from erosional to accreting beach) to the designated area of influence of these structures. Changes in the sediment transport, usually causing downdrift erosion, should be avoided.

b. Ecology

There are no direct natural heritage benefits which derive from construction of LCSs, except when these structures are built with the clear objective of protecting terrestrial or freshwater ecosystems of high natural value such as freshwater or brackish lagoons, wetlands and saltmarshes. Even in this case there will be concomitant impacts on coastal and marine systems.

Ecological objectives can be incorporated into design to maximise specific management goals. Management goals may include minimising specific impacts on the environment (e.g. minimising changes to the characteristics of surrounding soft-bottom sediments, or spread of exotic species) and/or enhancing specific natural resources (e.g. enhancing species biodiversity for recreational purposes, or recruitment to fisheries).

5.2.3. Socio-economic objectives

The socio-economic objectives constructing LCSs relate to the question «what is it we are protecting?» and secondly, how are we going to protect it? The first question refers to the basic societal need for safety and protection, and consequently economic growth and welfare. However, currently environmental quality aspects of coastal protection receive more and more attention and are being incorporated into a measure of welfare. The second question also refers to an environmental problem: the design of a LCS may disrupt or enhance landscape quality or habitat quality. In conclusion the socio-economic objective of constructing a LCS is one of sustainability.

5.3. CONSIDERATION OF LCSs AS A POSSIBLE CONTRIBUTION TO A FUNCTIONAL AND ECONOMICAL SOLUTION

(Burcharth, AAU)

The most common use of LCSs is in coastal protection schemes. The conventional elements in coastal protection schemes are dikes, seawalls, revetments, groynes, beach nourishment, and shore-parallel breakwaters. The LCSs dealt with in this book belong to the last category. A coastal protection scheme very often contains combinations of some of the mentioned elements. The selection of the optimal scheme has to be based on analyses of a number of possible combinations. It is beyond the scope of the present book to discuss schemes not containing shore parallel breakwaters.

5.4. CONSIDERATION OF PROJECT SERVICE LIFETIME AND STRUCTURE SAFETY CLASSIFICATION

(Moschella, MBA; Burcharth, AAU; Airoldi, FF; Lamberti, UB; Thompson & Hawkins, MBA)

Where LCSs are part of a coastal protection scheme the service lifetime for the structures will be as long as protection is required, provided that the structures are functioning satisfactorily. It can be said that the structure service lifetime should equal to the functional lifetime of the LCS scheme. A 50 years lifetime or more is common for coastal structures. However, due to the dynamic character of many sedimentary coasts it can be foreseen that in some places adjustments to the LCSs have to be made maybe several times within such span of years. This means that the structure lifetime is shorter than the functional lifetime of the LCS-scheme.

It is not important related to design to define a specific service lifetime for the LCSs themselves because LCSs are built close to the shore in shallow water and consequently structurally designed for depth limited waves, the sizes of which will be practically independent of the service lifetime.

Internationally accepted safety classes for coastal structures do not exist. However, LCSs will surely belong to a low safety class as the damage that might occur to the structures will not cause human injury or immediate large economic losses. Moreover, repair can normally be done fairly quickly. However, because maximum waves occur frequently in depth limited conditions and because the extra costs needed for increasing the strength of the structure is very small, the economical optimum corresponds to a very safe structure with marginal probability of damage. More details on safety aspects are given in the section on structural design.

From an environmental viewpoint the project lifetime and required maintenance is one of the most crucial factors affecting composition, abundance and composition of species that colonise the structures themselves. For instance, results of DELOS project have shown that along the Italian coasts of the North Adriatic Sea, frequent maintenance of structures by adding new blocks to the crest has dramatic effects on epibiota. Such frequent and severe disturbance effectively reduces biodiversity to an early stage of succession, with few species compared to those on structures which have not been maintained, and facilitate the development of green ephemeral algae with consequent negative effects on the quality of the beach. On any new LCS it will take time for the biological assemblage to reach a diverse community that is most likely to resemble that of a natural shore. For mature biological communities to develop, LCSs need to be stable and built in such a way that maintenance will be minimal.

Marine life also can influence the lifetime and the functioning of the system, for instance by impact of mussel growth on sediment trapping and porosity. In Mediterranean regions, rock boring organisms such as the date mussel *Lithophaga lithophaga* can in the long-term undermine the integrity and reduce the lifetime of structures. In addition, service lifetime can be limited by impacts in the surrounding areas, for example increased siltation or water quality problems. Safety of structures for navigation should be also considered using current legislation and best practice. The design of structures should also minimise risks for recreational use. These include falling into deep gaps between the rocks, sinking in soft sand and mud forming around the structures, swimming in rip and tidal currents.

5.5. CONSIDERATION OF ENVIRONMENTAL CONTEXT INCLUDING ECOSYSTEM, NATURAL HERITAGE AND NATURAL RESOURCES

(Moschella, MBA; Airoldi, FF; Thompson & Hawkins, MBA)

It is important to be aware that the complexity, uncertainty and diversity of natural ecosystems cause a high degree of spatial variability, and that every system and location may respond differently to the construction of an LCS. Thus while generic suggestions can be made, spatial variability precludes standardised designs but solutions should be site specific. The status, vulnerability and sensitivity and resilience of the coastal ecosystems involved should be carefully assessed prior construction of LCSs. The different compartments of the ecosystems that can be directly and indirectly affected should be considered, including terrestrial and marine habitats.

5.6. SYNTHESIS OF «Go / No Go» DECISION

(Moschella, MBA; De Vries WL-DH; Thompson & Hawkins, MBA)

Initial considerations should function as a preliminary screening phase to address specific issues such as objectives, environmental constraints and socio-economic evaluation. These considerations should then be summarised and integrated to enable decision on whether or not to proceed (Go / No Go) to the environmental assessment, planning and construction of a LCS.

CHAPTER 6

Investigation of environmental conditions

This chapter describes the investigations of environmental conditions recommended for design of LCSs. Instruments and procedures should comply with ISO standards where applicable.

6.1. BATHYMETRY AND TOPOGRAPHY INCLUDING SEASONAL AND LONG-TERM VARIATIONS

(Burcharth, AAU; Martinelli, UB)

The bathymetry, the topography and the coastline must be known at the location of the LCS-scheme.

LCSs are usually placed in the active zone for sediment transport where almost continuous changes in seabed levels take place. Seabed level changes can be characterized as short-term fluctuations if caused by single events like storms; as mid-term variations if caused by seasonal changes in the meteomarine climate; or as long-term variations if caused by climatic changes or changes in the sediment budget along the coastline, for example changes in discharge from rivers, sand mining, etc.

In order to decide the position of LCSs and their foundation level it is necessary to know the expected range of seabed level variations at the actual location of the LCS-scheme, i.e. the observed range of variations before placement of the structures plus the influence on the seabed levels caused by the presence of the structures. A foundation level not higher than the lowest expected seabed level should be chosen.

Historic information on coastline position and seabed bathymetry is often available and should be supplemented by surveys of the actual situation. If no historic information is available it is strongly recommended to carry out bathymetric surveys several times over a year in order to cover seasonal variations and situations after significant storms. The bathymetric surveys can be carried out with cable or echo-sounder. Use of differential GPS installed directly over the sonar is the state of the art, allowing for centimetric precision. Remote sensing techniques do not provide for the moment a bathymetry with sufficient reliability. Older methods, like manual soundings and tide corrections can be used as well. Series of cross shore profiles spaced 15 m to 25 m with few long-shore profiles for cross-checking is sufficient for design purpose.

If the mean sea level is not given at nearby fixpoints the mean sea level should be estimated from measured water surface levels over a sufficiently long period.

6.2. GEOLOGY INCLUDING CHARACTERIZATION OF SURFACE LAYERS (SEDIMENTS)

(Burcharth, AAU; Martinelli, UB)

Information on seabed soil conditions is necessary both for the design of the LCS foundation and for the prediction of the morphological changes caused by the structures.

Settlement and subsidence are critical for the proper function of LCSs because the crest level is one of the most important design parameters. Expected consolidation of the seabed due to the weight of the structure must be estimated from mechanical characterisation of the subsoil. Settlement due to consolidation is a problem only in case of very soft and weak subsoils as the foundation load of LCSs is usually small due to the limited height of the structure.

The levels of more solid soil or rock formations underlying relatively thin loose sedimentary surface layers should be identified in order to investigate the possibility of direct foundation of LCSs on the more solid bed.

Subsidence of parts of LCSs into the seabed sediments will take place only if proper filter layers and scour protection are not provided, or if the sediments are very sensible to liquefaction caused by wave action or earthquakes. Information for the evaluation of such conditions can be obtained by conventional geotechnical surveying techniques and soil characterization methods. The spacing of sampling positions should account for the variability in the soil formations.

For the prediction of morphological changes it is necessary to analyze the seabed as well as the beach surface layer sediments with respect to grain size distribution, mass density and fall velocity. Samples should be taken from several locations covering the whole LCS-scheme and adjacent stretches (sediment cell).

Extraction of liquids or gas from the underground may be responsible of settlement in the coastal zone and should be accounted for in the design of the structure crest levels.

6.3. WATER LEVEL VARIATIONS

(Burcharth, AAU; Lamberti, UB)

Water levels are of outmost importance in structural and functional design of LCS schemes by determining both the maximum wave heights in shallow water (due to depth limitations) and the freeboard of the structures. Together they basically control wave transmission.

Variations in water level are due to astronomical tides, storm surges and climatic changes. Tidal variations follow the cycles of the moon and the sun, and are generally very well predicted at almost all coastal locations by various institutes.

The small uncertainty makes it acceptable to model tides as a deterministic cyclic process.

Storm surges are related to stormy weather which causes the water level to rise due to barometric low pressures, wind stress (wind set-up) and breaking of waves approaching the coast (wave set-up). Storm surge must be regarded as a stochastic variable due to the unpredictability of meteorological variables.

More information on storm surge is given in Subsection 13.1.1.

Sea level rise due to climatic changes is a long-term effect, at the moment predicted with large uncertainty to be in the order of 0.5 m within 100 years. This is significant with respect to consequences for erodible coasts and coastal protection works.

Sea level rise might be modelled as a linear rise with time having a coefficient of variation in the order of 30%.

The relative importance of tides and storm surges varies with location. In general tides will dominate on coasts with relatively steep foreshores facing an ocean (e.g. west coasts of France, Ireland, U.K.), whereas storm surges dominate on shallow water coasts of more confined seas (e.g. coastlines of the Baltic Sea).

The statistics of water levels is needed for the design. For structural design extreme values are needed. For functional design with respect to morphological and ecological impacts the more frequent water levels are needed. The correlation between wave heights and wave periods is important in both cases.

If maximum water levels at or near the actual location have been recorded over many years on a daily or monthly basis, it is possible to fit a statistical distribution from which extreme values as well as frequent values can be extracted corresponding to any return period (exceedence probability). If only annual extreme values are recorded then solely extreme value statistics can be established, see Sub-section 13.1.3 for description of standard methodology. If water level maxima throughout the year in a period of approximately ten years or more are recorded then a Peak Over Threshold (POT) analysis can be used.

If water level records are not available it might be possible to establish an extreme distribution based on synthetic data consisting of hindcasted storm surges and the simultaneous tide given by charged institutes.

For LCS schemes, compared for instance to sea dikes, it is less important to obtain accurate statistics of extreme water levels for the structural design, because structures are frequently overtopped and a high water level will often result in greater protection of the armour layer against wave impacts. Accurate statistics of extremes is however important to assess beach response to storm events.

The joint statistics of water levels and waves are dealt with in Section 6.4.

6.4. WAVE STATISTICS

(Burcharth, AAU; Lamberti & Archetti, UB)

The most important environmental loading parameters for the design of LCS schemes are waves and water levels as they fully determine, together with tidal currents, the hydrodynamic load. As most LCS schemes are built on coasts with limited tidal range, tidal currents are not discussed further in this section.

Because the combined effect of water level and waves determine the impact on structure and morphology, it is necessary to deal with the joint statistics of the two.

Statistics of waves and water levels very seldom exist at the nearshore locations usually selected for LCS schemes. Available information on waves usually relates to deeper water off the coast. However, such information, given as frequencies of wave heights, wave periods and direction of waves, is readily available for almost all locations through hydrographic service institutes.

As wind generated waves are irregular some statistical parameters are used to characterise the sea state. The most important are listed below (see Section 13.2 for other parameters).

- Significant wave height, $H_s = H_{1/3}$, defined as the average of the highest one third of the waves during the peak of the storm usually 1-3 hours long. H_s corresponds closely to the visual estimate of wave height in a sea state.

- Root mean square wave heights, $H_{rms} = \sqrt{\frac{1}{N}\sum_{i=1}^{N} H_i^2}$ where N is the number of waves and H_i is the height of a single wave i.
- A typical wave period, T.
- Wave direction.

H_s and T are used as input in formulae for structure design, overtopping and wave transmission, whereas H_{rms} is often used as input parameter in numerical modelling of morphodynamics. The distribution of wave heights in a sea state with constant H_s follows a specific distribution (Rayleigh) for which reason ratios of wave heights of different exceedence probabilities are always the same, like e.g. the ratio between H_s and H_{rms} ($H_s = 1.416 H_{rms}$). However, the Rayleigh distribution does not apply in shallow water where wave heights are limited due to forced wave breaking when the height exceeds approximately 0.8 times water depth. Consequently also the significant wave height H_s is restricted by the water depth. For example, on a flat sloping sea bed the maximum H_s will be approximately 0.6 times the water depth.

The transformation of waves from deep to shallow water with respect to distribution of heights and to directions is explained in Section 13.2.

Where waves are limited by water depth it is necessary to consider changes in seabed levels in front of the structure together with water level variations. Seabed level changes can be considerable on barred coasts with large longshore sediment transport. Such conditions will modify the otherwise almost full correlation (linear relationship) between design wave heights and water levels. Larger changes in wave period with wave height might cause minor deviations from the linear relationship.

It is important to notice that in shallow water it is not possible to extrapolate wave height statistics without consideration of the physical constraint given by depth limitation of the waves.

Where LCSs are built in deeper water, the joint statistics of waves and water levels must be based either on long term recordings, or synthetic data as described in Section 6.3. The latter could also be composed by real time simulation of storms in accordance with the statistics supplied by hydrographic service institutes combined with real time inclusion of tides (variations are known) and estimated storm surges linked to the height of waves with onshore directions.

6.5. CURRENT STATISTICS INCLUDING TIDAL, BATHYMETRIC AND WAVE GENERATED CURRENTS, RESIDUAL LARGE-SCALE CURRENTS

(*Lamberti & Archetti, UB*)

Currents can be distinguished in offshore currents and littoral currents. Offshore tidal currents have usually a modest velocity, with exception of shallow seas with high tidal range. Offshore wind currents due to storms lasting one or two days have an intensity equal to 2-3% of the wind intensity and deviate about 10-20° from the wind following earth rotation (clockwise in Northern hemisphere). Density currents do not exceed some cm/s. All the currents mentioned above intensify in the vicinity of the coasts.

Tidal currents are very important with respect to sediment transport on littoral coast with high tides. Otherwise in the Mediterranean Sea, currents due to tide are of the order of magnitude of 0.10 m/s, smaller than wave-generated currents. In countries where tidal excursion is large (i.e. UK) these currents are strong and are markedly influenced by the local bathymetry, significantly contributing to the sediment transport processes (see for instance Elmer site in Chapter 11).

Littoral currents develop in the surf zone, forced by momentum released by breaking waves. Their intensity can exceed 1m/s with direction linked to wave obliquity. Their main effect is longshore sediment transport. In general their intensity does not affect directly the stability of LCSs, but they may have to be taken into account with respect to the scour they can cause around LCS heads.

Current measurements can be carried out with current meters (e.g. propellers, acoustics) installed at a fixed position in the study site, or alternatively the movement in time of a mass of water can be recorded by tracers or drifters. In general, current measurements are useful to describe velocity fields and for calibration of hydrodynamic models.

6.6. WIND STATISTICS, SOLAR EXPOSURE AND PRECIPITATION

(*Lamberti & Archetti, UB*)

Winds are measured from fixed stations on land and on ships. Wind data are mainly used as input for estimation of waves. Observed wind data have to be normalized to the wind blowing over the sea at the anemometric standard level (10 m a.m.s.l.).

In absence or to substitute for wind observations, information on atmospheric pressure gradients (isobar maps) can be used for prediction of wind fields over open seas.

Standard analysis of wind data time series provides:

- statistics of wind with respect to velocity and direction (wind rose);
- identification of storms: i.e. of events where a certain wind velocity threshold is exceeded.

Solar exposure, temperature and precipitation are data often available at local environmental offices and can be useful in extreme climate environments. At high latitudes the knowledge of periods with very cold weather can be useful for the estimation of degradation of stones due to frost (Norway, Iceland, Canada etc). The knowledge of precipitation is important where salinity concentration is very high (i.e. Red Sea). Also solar radiation can influence the stone durability in tropical climates. These data are usually given as time-series and statistics.

6.7. SEDIMENT TRANSPORT BY WAVES AND WIND

(Zyserman, DHI)

A detailed understanding of the local sediment transport processes is of large importance when designing LCSs and when assessing the expected impact on sediment transport and coastal morphology of the planned intervention.

This understanding should not be limited to the local area where the structure(s) will be built, but should encompass at least the involved sediment sub-cell or, preferably, the whole sediment cell. The term sediment cell refers to the length of coastline that is relatively self-contained as far as movement of sand and other sediments is concerned, and where interruption of such movements will not have significant effect on neighbouring sediment cells. The boundary of a sediment cell generally coincides with larger estuaries or prominent headlands (Mangor, 2001). By extending the analysis of sediment transport processes to the entire cell, undesired impacts on coastal morphology of the scheme being designed can be avoided.

In order to quantify the sediment transport processes, it is necessary to establish a sediment budget for the investigated coast. Such a budget quantifies the variability of the total longshore drift along the coast and helps in the identification of areas of potential coastal erosion or shoreline advance. Adjustment of beach profile to gradients in cross-shore transport takes place on a significantly shorter time scale than shoreline response, and can thus be left out from this analysis.

Known sources of sediment (e.g. discharge from rivers, nourishment schemes, etc.) and sinks (e.g. sand mining for construction purposes, removal of wind-blown sand from the coastal system, etc.) must be taken into account when the sediment budget is established. The same applies to spatial changes in the characteristics of coastal morphology and sediment properties (granulometry).

In some cases, it is possible to define the sediment budget for a given coast on the basis of recorded long-term changes in shoreline position, e.g. from aerial photographs. However, sediment transport models are frequently used for this purpose, since they also provide useful additional information for the design of LCSs.

Output from transport models will typically include gross and net rates of sediment transport (on a yearly and seasonal basis) and their variation along the coast. Other parameters are the distribution of the transport along the beach profile, the equilibrium alignment of the coastline (which corresponds to zero net transport on a yearly-averaged basis), etc.

Input to the models normally includes information about the local hydrographic conditions (winds, waves and tides), coastal morphology (bathymetry, beach profiles, shoreline position) and sediment characteristics (granulometry, density, etc.).

6.8. SEDIMENT CHARACTERISTICS

(Moschella, MBA; Bertasi, Ceccherelli, Colangelo, FF; Frost, Gacia, Martin, CSIC; Thompson & Hawkins, MBA)

One of the major environmental impacts of coastal defence structures is on the surrounding sediments. Sediment characteristics should be therefore fully investigated. The following sediment descriptors should be considered: geological composition, grain size and other granulometric parameters, redox potential and compactation, organic content, nutrient content and chlorophyll content (to quantify abundance of microphytobenthos). In particular, it is important to quantify sediment features that are more likely to worsen after the construction of LCSs, such as anoxic or organic rich sediments.

6.9. HYDROGRAPHIC PARAMETERS INCLUDING WATER QUALITY

(Moschella, MBA; Airoldi, FF; Thompson & Hawkins, MBA)

Hydrographic parameters include salinity, temperature density and other parameters related to water quality. Water quality refers to the use of a water body for a defined purpose. It is a concept which overlaps with ecological characteristics but is primarily geared to suitability for amenity, recreation, immersion water sports, collection of shellfish or other living resources. The relevant water quality parameters include total suspended solids, clarity (measurable by advanced instrumentation or simple field devices such as Secchi disc), dissolved oxygen and biochemical oxygen demand, nutrients and chlorophyll concentration. In addition, presence of pollutants (e.g. organic compounds, heavy metals) and pathogens (e.g. *Escherichia coli*, total number of streptococci) should be also assessed. These parameters must comply with the European Bathing Water Directive (76/160/EEC) and local legislation.

Aesthetic data on the amount of seaweed detritus and non-biodegradable waste material could also be of relevant importance for water quality (see Section 6.10).

6.10. ECOLOGICAL CONDITIONS (ECOSYSTEMS, HABITAT AND SPECIES)

(Moschella, MBA; Bulleri, Airoldi, FF; Gacia, Martin, CSIC; Frost, Thompson & Hawkins, MBA)

A scoping study (mainly desk based, but supplemented by a site visit) of ecological conditions of the site and coastal cell should be carried out to identify the factors likely to affect the biota and to inform design of environmental impact assessment.

To assess the ecological status of the site and coastal cell both physico-chemical and ecological data should be collected. All the information described in Section 6.1-6.9 (particularly Sections 6.8 and 6.9) is also relevant to the investigation of ecological conditions. The physical and geomorphological information can also be used in Delft biotope prediction model (see Chapter 14) of prior conditions, which need to be verified by site visit and to simulate post-construction impacts.

The following ecological data should be gathered:

- any available information for onshore (maritime) habitats (dunes, lagoons, shingle banks and their vegetation) and associated fauna and flora and geological features likely to be influenced (protected/impacted) including downstream effects.
- Any published information for soft shores in the region (e.g. for UK, Marine Nature Conservation review).
- Any published information for rocky shores in the region (e.g., for UK, Lewis 1964; Marine Nature Conservation Review Mermaid database, MARLIN website).
- Any available information from existing artificial structures (especially jetties, moles, harbour walls, stone groynes, sea walls etc.) in the nearby areas.
- Marine biogeographic province and likely species pool: available from general literature (Lewis 1964; Stephenson & Stephenson 1972; Hawkins & Jones 1992) by broad region

(e.g., Atlantic west coasts; Iberian coasts; French coasts; British and Irish coasts; North Sea coasts; west and east Mediterranean coasts; west and north Baltic coasts). In particular, regional species pool and potential source populations of hard-substrate assemblages.

- Any knowledge on recruitment regimes, for species of particular local interest such as mussels and other shellfish.
- Basic knowledge of the ecology and life histories of soft and hard-substrate species to predict dispersal capability, successional patterns and distribution (e.g. between the landward and seawards sides of the LCSs) of assemblages that will result as a consequence of the construction of LCSs.
- Existing information on pest or nuisance species.
- Identification of exploitable natural resources, including fish, shellfish and crustaceans.
- Distribution of fisheries nursery grounds.

In the absence of relevant information data on i-iv can also be gathered by a site visit. Information on conservation and natural heritage legislation for the site should also be collected (see Sub-section 5.1.3).

The desk-based study should be combined with a rapid field assessment of the site and adjacent coastal areas to verify and integrate the information collected during the desk study (see Chapter 14 for a protocol). The field assessment should include a stretch of coast extending at least 10 km either side of the selected site for the proposed LCS.

CHAPTER 7

Conceptual/pre-design alternatives

A preliminary design has to demonstrate satisfactory functional performance and environmental impact at a level high enough for the objective comparison of several alternatives.

7.1. PROPOSALS FOR LAY-OUT AND CROSS SECTIONS OF POTENTIAL LCS SCHEMES

(Burcharth, AAU)

At the pre-design stage a number of alternatives, all meeting the functional objectives and legislative, environmental and economical restrictions, have to be worked out in such detail that an objective comparison can be performed.

As for lay-out and cross sections no single LCS-scheme geometry can be generally recommended since its performance varies with each coastal site, depending on wave climate and required attenuation, on beach morphology (e.g. slope, grain size), use (recreational bathing, boating, surfing, fishing, etc.) and scope of work. However, some guidance to the initial choice of scheme can be given.

Figure 1.1 (Chapter 1, pag. 4) shows examples of typical lay-outs and cross-sections of three different schemes. At pre-design level the choice with respect to lay-out is more or less shown in this figure.

If the objective is to protect a very limited coastal stretch against severe wave action and at the same time to create a sheltered area for mooring of boats then a single-structure solution is often used with a LCS placed at some distance from the shore in order to have enough space for moorings. The length of the structure is determined by the needed space for moorings and the tolerated wave agitation. The demand for water depth and space usually results in water depths of more than 3-5 metre (LWL) at the structure. The structure will normally be emerging with crest-level high enough to prevent significant wave transmission by overtopping and penetration. Thus the wave agitation in the lee of the structure is mainly caused by diffraction and refraction of waves at the heads of the structure. The tidal range and the water level due to storm surge influences the crest level very much. If of some size the structure will certainly be visible as it emerges several metres above MSL. The high structures are economically built as a multilayer rubble mound breakwater. Figure 1.1.a shows an example. The distance to the shore should be large enough to prevent formation of tombolos and salients of some size as the area for moorings will be reduced and down drift erosion will occur. The problems are, however, difficult to avoid in case of significant sediment transport along the coastline unless the structure is built in deep water.

LCS-schemes with the primary objective of coastal protection and improvement of recreational conditions normally cover a longer stretch of the coastline. Two main types of schemes with dependence on the range of water level variations can be identified.

Schemes with submerging structures or structures with crest levels close to MSL can effectively dampen waves on coastlines with small tidal range and rare storm surge events like in the Mediterranean Sea. Such structures are invisible or only sporadic visible for which reason they can be large (continuous) structures without spoiling the sea view. Distinct openings, often made just as lowering of the crest, can be provided for the access of small vessels. Figure 1.1.c illustrates such a scheme. The net-inflow of water across the structures can generate very strong outflow currents in the openings and their surroundings thus creating scour. Dimensions and number of openings should be determined with due consideration of these problems. The larger the submergence the wider the crest should be in order to reduce transmitted wave energy sufficiently. On the other hand problems with return flow currents will be less. The height of the submerged structures is often so small that a homogeneous rubble mound structure is cheaper than a layered rubble mound structure. Appropriate filter layers and/or geotextiles should be used anyway to prevent penetration of finer materials into coarser materials and vice versa.

The other main type of scheme relates to coasts with frequent larger water level variations, such as coasts with significant tidal range and/or frequent storm surge water level set-up. Relatively high structures with crest elevation well above MSL are necessary in order to reduce the wave action on the coast sufficiently. Such emergent structures are blocking the sea view for which reason large gaps between the structures are required. Creation of pocket beaches (see Figure 2.6) by formation of tombolos or salients (see Figure 2.3) are generally also wanted. This leads to detached shorter structures placed relatively close to the shoreline. The width of the gaps relative to the length of the structures influences the total cost of the scheme significantly, especially in case of high emergent structures. For this reason, and in order to avoid concentrated rip currents, the gap width should be as large as possible considering the necessary protection of the coast.

Land-connection of the longitudinal LCS's by means of groynes is beneficial to avoid strong longshore currents. Moreover, they can provide access to the LCSs and thus serve additional recreational value. However, water movement on the landward side is considerably reduced, often negatively affecting water quality.

Also, by blocking the longshore sediment transport usually serious downdrift erosion problems occur. In this respect formation of salients are less damaging than tombolos as the interference with longshore sediment transport is smaller.

The lower the crest level of the LCSs, the greater the wave transmission, with consequent smaller morphological impacts of the structures. This generally means less protective effect but also less downdrift erosion.

From an environmental viewpoint, LCS design should balance the need for engineering performance in terms of coastal protection with the necessity of minimising impacts on surrounding habitats and associated fauna and flora. For example, if structures are built in such a way that considerable water movement on the landward side is maintained (e.g. by frequent wave overtopping or water penetration through the pores), sediment and water characteristics will be less altered and consequently impacts on the sediment fauna and flora will be limited. Design recommendations for minimising impacts on habitats and ecosystems are provided in Sections 8.3, 8.4.

7.2. PRELIMINARY ESTIMATION OF MORPHOLOGICAL IMPACT BY THE USE OF EMPIRICAL DIAGRAMS, FORMULAE OR EXPERIENCE

(Burcharth, AAU; Vidal, UCA; Zyserman, DHI)

LCSs are mainly located on the submerged beach were they modify the wave field and the wave-driven current patterns. If tides are important, also tidal currents could be altered. The consequences of the altered dynamics can be observed both in the near field (scouring or sedimentation around the LCSs) and far field effects, (changes in the shoreline position).

Focusing on far field effects, the hydrodynamic changes produced by a LCS on the protected beach causes sand accretion in the beach area located on the lee side of the LCS, thus producing a protruding shoreline called a salient (see Figure 2.3, Chapter 2, pag. 6). If the length of the LCS and the distance to the beach is adequate, the salient can reach the structure, forming a tombolo. In very special circumstances, the salient on the beach is accompanied by a second salient in the lee-side of the LCS, forming a double salient. In the case of long more deeply submerged LCSs no salients are formed, see for example Figure 1.1.c.

When LCSs are built on beaches with a dominant direction of longshore transport, care should be taken in the design of LCSs because tombolos act as perpendicular groynes causing the interruption of longshore transport. This interruption causes accretion on the updrift beach and beach erosion on the downdrift side, the same way as in case of groynes. On the other hand, salients allow some bypassing of sand, so the interruption effect is less.

For engineering purposes, there are some empirical approaches that predict the shape of the beach affected by LCSs. Some of these empirical approaches for prediction of the beach profile and the shoreline shape are presented in Sections 13.6 and 13.9.

Initially a number of lay-outs for the structures are sketched on the basis of the target beach planslope and wave transmission, considering also updrift and downdrift effects.

Shoreline response to an offshore LCS is controlled by a number of variables the most important of which are:

- distance offshore, X (from initial coastline);
- distance offshore relative to the width of the surf-zone, X/X'_s;
- length of the structure, L_s;
- length of the gaps between segments, G;
- transmission characteristics of the structure given by $K_t = H_t/H_i$, where H_t and H_i are transmitted and incoming wave heights, respectively;

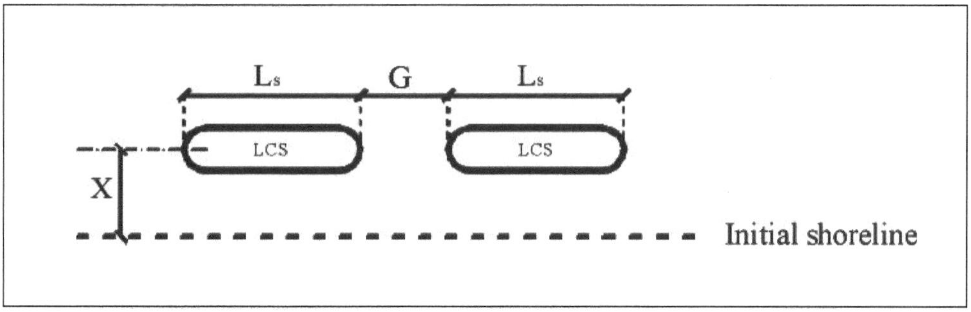

Figure 7.1. Definition of geometrical parameters.

- beach slope and depth at the structure, d_s;
- wave climate (sizes, frequencies, and directions of waves);
- water levels;
- sediment characteristics.

Figure 7.1 shows the definition of the geometrical parameters.

Simple diagrams or rules can give a first indication of the morphological changes imposed by the structures. They all assume the presence of sufficient sediments for the depositions. Example of simple rules are given below (tab. 7.1) for:

- *emergent structures* placed within the littoral drift zone; little or no wave transmission across the structures, i.e. K_t = app. 0.1 to 0.2; shore- parallel structures; almost perpendicular wave approach;

Table 7.1. Conditions for formation of tombolos and salients.

	Emergent structure	
Conditions for formation of		*Reference*
Tombolos	**Salients**	
$L_s/X \geq 1.5$	$1/2 < L_s/X < 2/3$	Dally and Pope (1986)
$L_s/X > 1$	$1/2 < L_s/X < 1$	Herbich (1989)
$L_s/X > 0.9$ to 1	$L_s/X < 0.6$ to 0.7	Mangor (2001)
	Submerged structures	
Conditions for formation of		*Reference*
Tombolos	**Salients**	
$L_s/X > (1.0 \text{ to } 1.5)/(1 - K_t)$	$L_s/X > 1/(1 - K_t)$ $GX/L_s^2 > 0.5(1 - K_t)$	Pilarczyk (2003)

The width of the gap is usually according to Pilarczyk (2003)

$$L \leq G \leq 0.8\, L_s$$

where $L = T\,(g \cdot h)^{0.5}$, T being the wave period and h the water depth at the structure.

Seiji, Uda and Tanaka (1987), referred to Loveless (1999), gave the following conditions for the erosion of the beach behind the gap:

$G/X < 0.8$	no erosion
$0.8 \leq G/X \leq 1.3$	erosion likely
$G/X \geq 1.3$	surely erosion

Simple rules related to reef structures are referred and discussed in Pilarczyk (2003). Tools for more detailed examination of the formation of salients and tombolos behind emerged structures are given in Section 13.9.

The simple rules indicating morphological changes in terms of formation of tombolos and salients cannot give the answer to the main question: can a LCS-scheme, although formation of tombolos or salients will take place, stop the retreat of an otherwise eroding coast? No general answer can be given as it depends on the character of the wave climate, the natural sediment supply and the exposure and erosion rate of the coast. However, for rather exposed coastlines where significant erosion takes place in quite frequent storms it is not possible to stop retreat by means of LCS-schemes unless beach nourishment is applied on regular basis, and/or revetments are installed. However, a LCS-scheme will almost always reduce the erosion rate of the protected stretch like any other reinforcement of the coast. Steepening of the coastal profile seawards of the structures will quite often take place.

All coastal structures sticking out from the coastline cause downdrift erosion and updrift accretion on coastlines with a net-direction of sediment transport. This is also the case for shore parallel structures if they, as is the case for most LCS-schemes, influence the morphology by creating tombolos and salients. Salients, and especially if they are submerged, create less problems than tombolos because total blocking of the longshore sediment transport is avoided. Also, the closer the structures are to the coastline, the less downdrift problems occur.

An approximate prediction of morphological changes to the coast line caused by a LCS-scheme might, at predesign level, be provided by the use of numerical one-line models, cf. Sections 8.1 and 13.10.

The length of LCSs in relation to the width of the gaps together with the crest level and the permeability of the structures determines the water level set-up behind the structures. Generally a large set-up is undesirable as it not only causes reduction of the width of the beach but indeed very strong return currents due to the large pressure gradients. The largest set-up occurs when the structure is impermeable and the crest level is above but close to the still water level, i.e. when the freeboard is small compared to the wave height.

Beach nourishment

Beach nourishment is frequently used together with coastal structures in beach protection and restoration schemes to minimise/counteract the far-field impacts of coastal structures.

Nourishment can be regarded as a natural way of combating coastal erosion by artificially replacing a deficit in the sediment budget over a given stretch of coast with a corresponding volume of sand. The sand used to nourish the coast should have grain size similar or coarser than the native sand.

According to Hanson (2003), approximately 28 million cubic metres of nourishment are placed every year in Europe. The methods and practices applied vary from country to country.

Three nourishment methods can be identified based on the placement of the borrow material along the beach profile (Mangor, 2001): (i) backshore nourishment, (ii) beach nourishment and (iii) shoreface nourishment. In the first case, the upper part of the beach is strengthened by placing nourishment at the backshore or at the foor of dunes. The aim of backshore nourishment is to prevent dune erosion and breaching during storm events. In the case of beach nourishment, sand is supplied to the shore to increase the recreational value and/or to secure the beach against shore erosion by adding sand to the sediment budget. Shoreface nourishment consists of supplying sand to the outer part of the beach profile, usually on the seaward side of a barrier, to strengthen the coastal profile and to add sand to the sediment budget.

Common to all types of nourishment is the fact that, if the cause of erosion is not

Figure 7.2. Salients and tombolos in Pedregalejo artificial beach, Málaga, Spain.

eliminated, the erosion will continue in the nourished sand. This means that nourishment as a stand-alone method for coastal protection will normally require a long-term maintenance effort, based on the definition of the frequency and volumes involved in re-nourishing the coast. Regular re-nourishment requires a permanent and well-functioning organisation, which generally makes nourishment as a stand-alone solution unsuitable for private beaches and small-scale schemes.

The idea of combining beach nourishment and coastal structures is to use the structures to create closed sediment cells in such a way that no significant losses of sediment take place, thus largely reducing or completely eliminating the need for re-nourishment. This might be achieved through shore-normal structures, such as groynes of different shapes or artificial headlands, or by use of shore-parallel structures, typically breakwaters. When shore-parallel structures are used, tombolo formation is usually sought in order to ensure zero sediment transport out of the cell. It is far from always possible to eliminate the need for re-nourishment.

All type of nourishments, especially if regularly repeated, will have serious impacts on habitats and associated biota at both source and destination sites. For example, if sand is extracted from off-shore sites, the seabed will be highly disturbed, leading to significant loss of benthic flora and fauna as well as disturbance to fish. If sand is dredged from harbour bottoms or docks, the risk for contamination of sediments by pollutants and pathogens can be high. This practice may also increase the risk of introducing soft-bottom, non-native species that often occur in harbour areas.

Figure 7.2 illustrates the application of beach nourishment combined with coastal structures to create an artificial beach at Pedragalejo, Málaga, Spain. In this scheme, a detached breakwater has been placed at the centre of the coastal cell to form a salient in order to increase the available length of beach and, thus, its recreational value.

7.3. STRUCTURAL SAFETY OF PREDESIGN

(Burcharth, AAU)

The structural design of LCSs follows the functional design. The outcome of this are the crest level of the structure, the sea bed level at the structure, and the length of the structure (and width of gaps in case of multi-structure schemes). Apart form drawing trunk cross sections and head sections defining the composition of materials to be used obeying filter criteria etc, the structural design consists of determining the size of stone (blocks) in armour, toe and scour protection, see Section 13.11. For this it is necessary to define safety levels if not given in a national standard or design recommendation. If given, they usually relate to larger structures and not to very small structures such as LCSs built close to the foreshore. Typically is safety implemented by definition of a maximum allowable damage, e.g. 5% of the armour blocks displaced, when exposed to the 50-years return period sea state. This implies that a certain return period sea state has to be extracted from the combined information (joint statistics) on water levels and waves. However, as this is very complicated because of several dimensions (water depth, freeboard, wave height, wave direction) it is recommended to establish the statistics on the effect of the various sea states in terms of necessary size of the armour units, and extract from this the size corresponding to the 50-years event.

Economical optimization of rubble mound breakwaters shows very flat minima for lifetime costs as function of armour unit size (Burcharth and Sorensen, 2005). This means that no money is saved by minimizing the armour size, unless at the limit where size of armour units is a supply or a construction problem. If this is not the case and if the waves are depth limited there is no need at predesign level to perform detailed statistical analyses of the sea states as stone size can be based on conservative use of water depth statistics alone. In shallow water there will most often be very small differences between wave heights related to for example the 5-years and the 50-years return period sea states.

If in a standard the demanded safety level is given as a maximum probability P_f of exceedence of a certain damage within service lifetime T_L of the structure, then the structure should as a minimum be designed for a sea state with return period T_R given as

$$T_R = T_L/[-I_n(I - P_f)]$$

The formula expresses the encounter probability which does not include uncertainties related to the parameters and to the formulae. A probabilistic design approach is necessary for the inclusion of these uncertainties, but this is not used for conceptual design of small simple LCS structures.

If no standards or recommendation covering the actual location exist, or if these apply to breakwaters in deeper water, it is recommended to design the main armour of LCSs in shallow water for practically no damage applying a conservative value of wave height, cf. the discussion in Sections 7.5 and 13.11.1. Where toe berms consist of few stones they should also be designed for practically no damage. In case of wide toe berms and scour protection layers consisting of many stones placed in two layers or more, some displacement can normally be tolerated when exposed to the largest depth limited waves.

7.4. IDENTIFICATION OF ENVIRONMENTAL CONDITIONS FOR PREDESIGN

(Burcharth, AAU)

Fundamental understanding of the historic performance of the actual coastal stretch including responses to man made interventions is of outmost importance for drafting of realistic alternatives at predesign level. To obtain such understanding it is necessary to seek historic information and combine it with knowledge about seabed and sediment characteristics, wave climate, water level variations and currents.

The understanding of the morphodynamic processes must cover not only the project area but the whole of the sediment cell. Also, to ensure that the project will not impose unacceptable environmental conditions it is necessary to know the ecological conditions and identify constraints related to conservation and natural heritage.

Chapter 6 describes how the environmental conditions can be investigated. The environmental data needed at predesign level does not need to be very detailed as long as the main characteristics are given. For meteormarine data it means that slightly conservative parameter values are sufficient. This is because calculations related to conceptual designs will normally be deterministic. Stochastic analyses usually await detailed design stages.

The first phase of predesign deals with lay-out and main dimensions of alternative schemes and their tuning to fulfil the set target performances. In most cases the focus is on morphodynamic and recreational performances. The meteormarine input to be used for estimation of the morphodynamic performance of a scheme should reflect the typical conditions at the site including seasonal variations. For this is used simplified time series of combined values of water levels wave height, wave period and direction of waves. The values will typically be chosen to reflect average conditions for each season, but storm conditions might be included as well. Only conditions which cause movement of sediments should be included when defining average conditions. If tidal currents are significant they should be included in a simplified manner.

If there is risk of stagnant water etc. it is important to include time series reflecting also quiet conditions for the study of recreational and environmental performances of the predesign schemes.

7.5. STRUCTURAL DESIGN OF LCSs BASED ON MATERIAL SUPPLY POSSIBILITIES, FORMULAE FOR STABILITY, AND SEMI-EMPIRICAL INFORMATION ON SCOUR

(Kramer & Burcharth, AAU)

In general a LCS consists of the following parts:

- an outer armour layer of large stones or concrete blocks (Sub-section 13.11.1).
- a bedding layer of smaller stones and/or geotextile between the bottom of the structure and the sea bed (Sub-section 13.11.2).
- a toe protection of armour layer stones or smaller stones (Sub-section 13.11.3).

At almost all locations in Europe suitable rock and stone material for LCSs is economically available due to the rather limited costs of long distance shipping materials by barge. However, nearby land-based sources with sufficient quality and sizes of stone and rock materials are also used. Concrete blocks are used only if costs for rock materials are very high.

The fact that finer rock and stone materials generally are cheaper than larger size materials leads to preference for layered designs instead of more homogeneous designs based on very few sizes or classes of materials. In any case, sufficient filter layers must be provided between sandy seabed and the coarser structure materials. Geotextiles are often

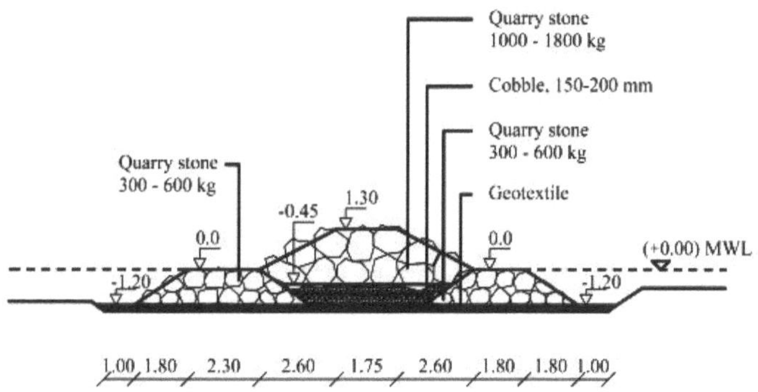

Figure 7.3. Cross-section of breakwaters at Lønstrup, Denmark (Laustrup & Madsen, 1994).

used for this purpose.

For structures of limited height it is not possible to have several layers of different grain/block sizes due to the large size of the armour blocks compared to the total height of the structure. In such cases similar sized blocks will be used for the main body resulting in a very permeable structure as opposed to structures with a core of finer materials. In the case of deeper water there is a choice between homogeneous structures and layered more impermeable structures. The target wave penetration and exchange of water through the structure then determines the type of design.

A toe protection of a certain width must be provided; this is usually made flexible by the use of stone and geotextiles to allow for some sea bed scour close to the structure. Toe protection is necessary both on the front and the rear side of the structure.

Various designs of cross-section composition and shape are possible. A sketch of a characteristic cross-section built to prevent coastal erosion in Denmark is shown in Figure 7.3. The level of the crest is seen to be 1.3 m above MSL indicating that the structure is not low-crested under normal wave conditions. However, storm surge can be around 1.5 m above MSL making the breakwater heavily overtopped. In Figure 7.4 a typical cross-section

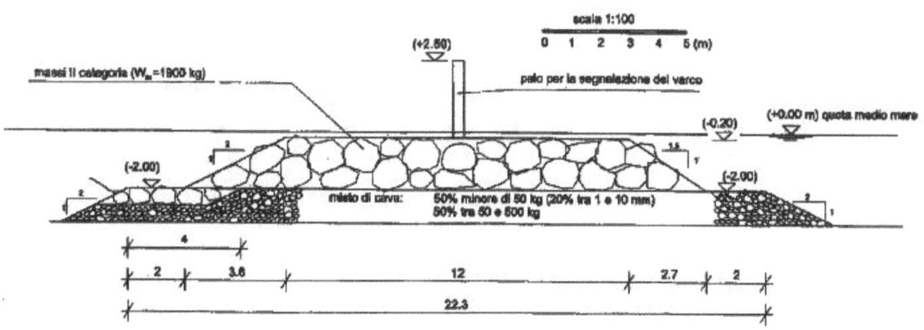

Figure 7.4. Cross-section of a submerged breakwater along Emilia Romagna coast, Italy.

of a submerged breakwater along the Emilia Romagna coast (North Adriatic coast) in Italy is shown.

The cross-section shown in Figure 7.3 is narrow-crested and relatively high compared to the submerged wide-crested breakwater in Figure 7.4. Typically also the leeward side of LCSs are exposed to direct wave action due to overtopping waves and it is therefore necessary to design a toe berm on both sides of the breakwater. If the breakwaters are very high and/or wide, then overtopping will be reduced and the toe berm on the leeward side of the breakwater can be designed using smaller stones.

Stones used in the armour layer of a LCS must be sufficiently large to avoid undesirable displacements caused by the wave action against the structure. As LCSs are built in shallow water the highest waves will often be depth limited. As a consequence the structures will typically be exposed to design waves numerous times during the lifetime. Because damage is cumulative it is important to design such structures with criteria based on a very low damage per storm criteria. Moreover, because narrow-crested breakwaters built in shallow water are only a few stone-sizes high and wide, one stone removed from the edge of the crest will cause a relatively large hole in the cross-section leading to increased wave transmission. Consequently it is recommended to use the limit between the no damage and initiation of damage for the design and to use at the same time a safety factor which compensates for the uncertainties.

For the determination of the armour block size the armour stability formulae given in Sub-section 13.11.1 can be used with a safety factor of 1.1 on the nominal diameter. Generally there are differences in the exposure of armour blocks of the various parts of the structure (heads, trunk crest, trunk seaward and leeward sides). However, for preliminary/conceptual design it is recommended to use the same armour size for the whole structure, corresponding to the most exposed part. The armour stability formulae are in case of depth-limited waves valid only for 1:2 slopes. For LCSs exposed to non-depth limited waves also slopes of 1:1.5 are covered by the formulae. For structures in larger water depth reference can be given to armour stability formulae given in CEM (2003).

Determination of the toe block sizes and scour protection can be based on the formulae given in Sub-section 13.11.3. The extent of the scour protection is given by formulae covering the seaward side of the trunk and the head. The toe berm stability formula can be used for the determination of the size of the scour protection material if the width of the protected area is not too wide. In case of wide areas the stone size should be determined by theory for the transport of granular materials in waves and currents.

Bedding layers and stone filters must fulfil accepted filter criteria, e.g. as given in Sub-section 13.11.2.

7.6. ASSESSMENT OF ENVIRONMENTAL IMPACTS (EIA) AT LOCAL AND REGIONAL SCALE

(Moschella, MBA; De Vries WL-DH; Frost, Thompson, Hawkins, MBA)

An EIA should be performed at this stage to identify and evaluate the potential impacts (or effects) of construction of a LCS in relation to physical, chemical, biological and cultural components of the environment. This should enable environmental issues to be integrated at the planning and decision making phase and hence promote design alternatives that are environmentally sound.

Once relevant ecological information (see scoping study in Section 6.10) has been collected, baseline ecological surveys should be undertaken to identify likely effects of LCS on habitats and species and assess the site sensitivity to impacts. Surveys should be undertaken at both local (near-field) and broader (far-field) scale. Also, they should be spatially and temporally replicated, to allow identification of potential impacts from the background, natural variability of benthic assemblages.

A preliminary field visit should be also carried out prior the detailed survey to define appropriate sampling strategy, for maximising sampling effort and guaranteeing accuracy in the assessment. This can be based either on biotope mapping (e.g. BIOMAR, www.JNCC.gov.uk) or on physical gradients (e.g. height on the shore/bathymetry). The exact format of the survey will depend on the coastal system considered (macrotidal, microtidal), the environmental setting, size and configuration of the LCSs to be built as well as the specific ecological features of the site. Although priority should be given to assessment of physical and biological features of sediments, the nearby rocky shores (if any) and water column should be also characterised in the survey. A protocol indicating general steps to be undertaken in the survey is provided in Chapter 14.

7.7. EVALUATION OF THE SCHEMES BASED ON ECONOMICAL OPTIMISATION

(Martinelli, UB)

The design of the alternatives identified in the preliminary phase should be detailed enough to allow their economic evaluation. These include at least an identification of quantities and methods involved in the building process and the evaluation of the structure performance in time. Both are necessary for the evaluation of the total cost, which is a combination of the initial building cost and of the long term maintenance costs.

Typical construction unit costs for the area where the structure is built may be considered as a starting point.

Maintenance costs are distributed over lifetime; it is suggested to reduce the frequency of maintenance, in order to control possible negative effects on the ecosystem (see Section 8.8). A proper economic life-time should be selected, usuall smaller or equal to the structural lifetime (eg. 20 years), in order to account for the possible change of strategies or environmental conditions. The equivalent initial cost can be obtained by capitalising maintenance costs at present prices using an appropriate interest rate compensated for cost inflation (in Europe it is in the range 2-4%). A lower interest or a longer economic lifetime lead to lower weight of initial costs compared to maintenance costs, but higher initial costs and lifetime costs.

The cost-benefit analysis should be performed considering an area where all the physical and social effects take place, i.e. significantly wider than the intervention area; alternatives shall usually include the «no structure» scenario, and cost and benefits should account for both direct (related to works and beach activities) and indirect economic consequences (e.g. tourism induced effects over the wider area).

7.8. SOCIO-ECONOMIC EVALUATION OF THE SCHEMES

(Zanuttigh, UB)

The construction of different schemes may lead to different visual impact scenarios and to the development of recreational activities that can significantly affect visitor enjoyment and thus beach value.

Schemes including emerged barriers worsen water quality, improve bathing safety especially for children, impose some restrictions to water sports and may have a negative aesthetic impact; groynes are usually welcome from beach visitors for sunbathing, fishing and walking on the crest, if possible; submerged structures can mitigate risk for bathers without degrading water quality and the view from the beach. These effects can increase or decrease the number of people visiting the beach, the time they spend in average on it, the money they are willing to pay for a visit and the money they may spend for recreational activities.

Identification of social effects of design alternatives can be supported by questionnaires and face to face interviews to residents and visitors (see Chapter 15 for details) to determine their evaluation of different beach evolution scenarios and their preferred scheme for recreational purposes.

7.9. INTEGRATION OF TECHNICAL, ECOLOGICAL AND ECONOMIC EVALUATION FOR SELECTION OF THE SUSTAINABLE SCHEME

(Zanuttigh, UB; Burcharth, AAU)

After a preliminary selection of design alternatives, each alternative has to be examined and compared with respect to its technical, socio-economical and environmental performance.

The use of numerical and physical models may help to predict the hydro-morphological consequences of each solution and their suitability to accomplish the design objectives.

Estimated waves and currents allow, for instance, evaluation of the following:

- the inshore wave energy reduction with the consequent level of beach protection;
- the water residence time inside the protected cell to assess water recirculation (and thus also water quality) for ecological purposes;
- the current patterns and intensities, in particular at gaps and roundheads, to verify bathing safety;
- the structure submercenge/emergence due to waves and tide and its frequency, to check the possible dessication of organisms at the structure.

Estimated sediment transport allows, for instance, evaluation of the following:

- the global sand volume balance for the protected cell, in order to estimate if renourishment is necessary and, if it is, its quantity and frequency;
- the formation of local scour that may produce structure instability, in order to redesign a proper toe protection or structure extension;
- the erosive/depositional patterns and their rate to identify the level of disturbance to the assemblages.

The results of analyses and numerical and/or physical modelling have to be judged by different experts and then have to be synthesised defining appropriate indicators such as:

- performance of the scheme for beach protection;
- initial and maintenance costs;
- impact on habitats, species, ecosystem and their living natural resources;
- cultural heritage of the coastline;
- recreational value.

A proper weight has to be assigned to each indicator and a mark for each alternative is derived from the weighted sum of all indicators, providing an objective selection of the «optimum» scheme.

An example of selection of the sustainable scheme starting from several different alternatives is given in details in Chapter 12. Tab. 12.17 shows the selection of the scheme among design alternatives by means of representative weighted indicators; in this case, the intervention is judged based on four main objectives: beach protection, intervention total costs, ecological and social effects; to each objective an equal weight of 1 is assigned and specific indicators within each area are equally weighted; the selected alternative is characterised by the greatest mark, which means a compromise among the judgements achieved for each specific design objective.

CHAPTER 8

Detailed design of preferred scheme

8.1. OPTIMIZATION OF LAY-OUT AND CROSS SECTIONS OF LCSs BASED ON SHORT-TERM AND LONG-TERM MORPHODYNAMIC SIMULATIONS

(González-Marco, Mösso, Sánchez-Arcilla, UPC)

From an engineering[(1)] point of view, the optimization of the lay-out and cross section of LCSs, on the basis of short and long term morphodynamic numerical simulations, should follow these five main steps.

1) Definition of Boundary Conditions for a Refined LCS Design

The optimum structural design (optimization process) must be preceded by a compilation of information/boundary conditions regarding hydrodynamic and morphodynamic pre-existent conditions as a pre-process for numerical modeling. This compilation should include, at least, information regarding average and episodic values of: waves/wind/tide climates, sediment characteristics, sediment transport rates and trends of beach plan and profile dynamics. The accuracy of this pre-existing information will play an important role in the optimization process, since it provides the initial boundary conditions as well as information on the morphodynamic evolution of the affected area. The meta-information of the «transient stages» will also be a useful tool to verify the model performance during this numerical optimization process.

2) Modelling Tools

Depending on the considered temporal and spatial scales as well as the structural/functional parameters to be optimized, it is necessary to make use of different numerical modelling approaches. In this sense, 1-Line morphodynamic models should be used to initially assess structural length, orientation, distance to the coast, functionality of gaps, and other structural parameters within time scales from months to years and spatial scales from hundred meters to kilometers. These models (see e.g. Hanson and Krauss, 1989) have been widely employed to design detached LCSs, mainly emerged. The most important limitation of this kind of models is that they are based on the computation and balance of wave-induced long-shore sediment transport and do not take into account other hydrodynamic processes, which could contribute to sediment transport. This includes the important effect of wave induced currents, overtopping

---

(1) Ecological and socio-economic impacts are out of the scope of these considerations.

and, sometimes, even transmission, amongst others. In this respect, Hanson and Krauss (1990) and later Jimenez and Sanchez-Arcilla (2002) analyzed the influence of wave transmission and LCS freeboard on the shoreline evolution with a 1-Line (1L) model.

However, in order to assess more accurately the morphodynamics associated to these structural parameters at smaller temporal and spatial scales (of about hours to days and meters to hundred meters, respectively) focusing on the effects of mean, storms or extreme conditions, 2-dimensional Depth Averaged (2 DH) morphodynamic simulations should be performed. This type of numerical models must simulate accurately, in a 3D domain, the most important hydro-morphodynamic processes acting around LCSs, both submerged and emerged. This explicitly includes the diffraction and reflection of waves, currents due to waves, wind and tides, turbulence and sediment transport – distinguishing between bed and suspended loads for the different parts of the domain. The morphodynamic evolution results hence as a function of beach state, driving terms and structural geometry. These «coastal area morphodynamic models» allow the modelling of complex hydrodynamic patterns around LCSs, considering the effect of a number of both environmental and design variables (see Figure 8.1) for smaller time and spatial scales in comparison with 1L Models. The applications of 2DH morphodynamic models should be considered within this scope. Figure 8.1 illustrates the most important hydrodynamic fluxes around LCSs which can be simulated by this kind of numerical models within the limits of their application regarding time and spatial scales. Examples of this can be found in Watanabe *et al.* (1986), Zysermann *et al.* (1999), Alsina *et al.* (2003), Alsina (2005) or Sánchez-Arcilla *et al.* (2004, 2005).

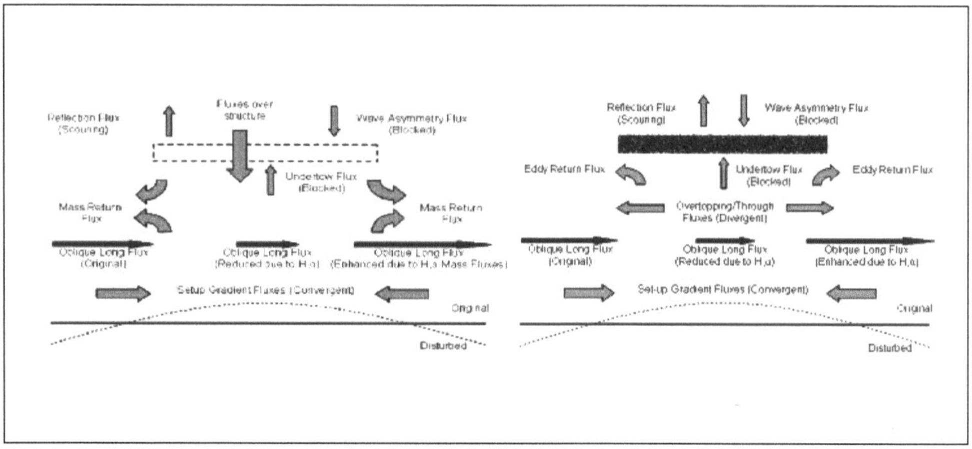

Figure 8.1 Main hydrodynamic fluxes around LCSs for both cases, emerged (right) and submerged (left).

For more complex scenarios, for which it is necessary to take into account additional structural parameters such as freeboard, crest width, permeability, and then more intricate hydrodynamic processes, Quasi 3-Dimensional (Q3D) or 3D morphodynamic simulations are required. These models should deal adequately with the overtopping fluxes and the fluxes through the structure via mass and momentum conservation laws, and provide also the profile dynamics with the presence of the structure in a manner consistent with state-of-art 2-Dimensional Vertical (2DV) profile models. Over the past several years, significant efforts have been dedicated to develop advanced 3D computational fluid dynamics tools, mainly centred on the solution of the three-dimensional Navier-Stokes equations (see e.g. Mayer *et al.* 1998). This level of numerical simulations allows an accurate description of the hydrodynamics acting both around and inside (in case of permeable structures) LCSs. Nowadays the applicability of these models is limited due to the complex process of model calibration, as well as the high computational costs required to run them. For this reason, their use is mainly centred on the solution of very specific problems in small computational regions.

In addition, as a complement to numerical simulations, physical modelling both in flumes and wave tanks should be carried out in order to reduce uncertainties in the hydrodynamic and morphodynamic processes simulated around LCSs.

3) Predictions with Error Bounds

The final objective of numerical simulations must be to improve the knowledge of expected shoreline and beach morphodynamic behaviour (both 2DH and Q3D or 3D) with its corresponding error bounds. These morphological changes will be a function of meteo-oceanographic characteristics (waves, tide, wind, currents), sediment characteristics and structural and geometrical aspects (structure length, orientation and distance to coast, gaps, freeboard, crest width and permeability).

The level of uncertainty of hydro-morphodynamic parameters is well known and described

Table 8.1. Estimated uncertainties intervals for some usual variables in coastal engineering projects (From Soulsby, 1997).

Input Parameter	*Uncertainty*
Density of water, ρ	± 0.2%
Kinematic viscosity of water, υ	± 10%
Sediment density ρ_s	± 2%
Grain diameters, d_{10}, d_{50}, d_{90}, etc.	± 20%
Water depth, h	± 5%
Current speed, $\vec{U}$	± 10%
Current direction	± 10°
Significant wave height, Hs	± 10%
Wave period, T_z	± 10%
Wave direction, θ	± 15°

(see e.g. Soulsby, 1997). The most important error typical values are compiled in Table 1.

These uncertainties, together with those intrinsic to numerical models, have to be taken into account in order to evaluate and interpret numerical results. Then, when making predictions, it is prudent to perform *a priori* a sensitivity analysis of the models in order to estimate differences between prediction methods and errors in the output as a result of the uncertainties in the input parameters. In this respect, in van Rijn *et al.* (2003) there is an intercomparison exercise in which several models (prediction methods) are evaluated for the same scenarios. In the same way, in Mösso (2004) there is an exhaustive sensitivity analysis of a hydromorphodynamic suite of models, in which an extensive number of input parameters has been evaluated.

4) Assessment of Predicted Shoreline and Beach Dynamics

The assessment should be carried out for a full sequence of stages, going from initial to a final, through several transient stages. The predicted shoreline and bottom geometry must be compared with acceptability criteria from three standpoints: 1) Morphodynamics, which is related to the beach physical state, 2) Ecology, which takes into account beach ecological state and 3) Socio-Economy, which represents the relation of the construction and maintenance costs of the structure versus the benefit of the resulting protected beach.

5) Corrections of Lay-Out

In this final step, a re-evaluation of the general state must be done by introducing the corrections resulting from the analysis done within previous steps. It is then necessary to evaluate the convenience of starting an iteration process from step 2 onwards.

8.2. STRUCTURAL DESIGN BY THE USE OF FORMULAE AND MODEL TESTS

(Burcharth, AAU)

Detailed structural design contains a detailed examination of the performance of the various parts of the structure and an economical optimization based on amounts and types of materials, methods of construction, and long-term maintenance.

The formulae for armour stability, toe stability and scour protection, given in Section 13.11, will normally be sufficient for the detailed design for LCSs. In case of design of very large structures reference is given to breakwater design tools, for example as given in the Coastal Engineering Manual (CEM) and the Manual on the use of Rock in Hydraulic Engineering.

If these tools are insufficient, maybe because less uncertainty is wanted, it is necessary to perform hydraulic model tests, cf. Section 13.12.

8.3. STATEMENT OF SOCIO-ENVIRONMENTAL IMPACTS

(Moschella, MBA; Abbiati, Airoldi, Bacchiocchi, Bertasi, Bulleri, Ceccherelli, FF; Cedhagen, BIAU; Colangelo, FF; De Vries WL-DH; Dinesen; BIAU; Aberg & Granhag, UGOT; Jonsson, UGOT; Gacia, Macpherson, Martin & Satta, CSIC; Sundelöf, UGOT; Frost, Thompson & Hawkins, MBA)

LCSs can cause severe impacts on the surrounding environment at both local and regional scale. Soft-sediments are the most affected by LCSs; their presence always induces a disruption in the normal transition of assemblages from deep waters to the shoreline, due to the physical presence of the structure on the sediments as well as to the modification of the hydrodynamic regime. Marked changes in the water characteristics also occur, particularly on the landward side. The construction of LCSs as well as other man-made structures has some implications for rocky-bottom communities as the structures provide new hard substrate for colonisation of species typical of rocky shores that naturally would not be there.

The modifications induced in water circulation patterns, water quality and assemblage types can strongly affect the social enjoyment of protected beaches and consequently beach value and usage.

8.3.1. Impacts on soft-bottoms (habitats and associated biota)

Unavoidable large scale changes in sedimentation patterns of the coastal cell due to the presence of the LCSs may impact not only immediate sea bottoms but also nearby updrift/downdrift areas affected by changed erosion/sedimentation processes with major negative consequences for the associated fauna and flora. The construction of one or more LCSs have two direct consequences: habitat loss and habitat fragmentation. The construction of LCSs leads to loss of sandy areas and the associated infaunal communities.

Where coastlines are defended by a series of LCSs, habitat loss becomes important and can lead to severe disruption of soft-bottoms at large scale. Impacts of LCS on infaunal communities, however, are mainly indirect, through modification of the local hydrodynamics and sediment regime including physical and chemical characteristics of the water column and sediment. Changes to the physical environment are particularly evident on the landward side of the LCS and include reduced water movement, increased scour in proximity of the structures, increase of silt/clay fraction, organic matter and anoxic layer in the sediments, and trapping of coarse material (i.e. pebbles, shells, algal detritus). These modifications of the sedimentary habitat surrounding the structures will in turn affect the associated biota. The main effects are:

- changes in the structure (composition and abundance) of the assemblages. Certain species are more sensitive to changes under the new habitat conditions and can decrease in abundance or in some cases disappear. Others will take advantage of the new environmental conditions and from reduced interspecific competition. As a result, the relative abundance of species in the infaunal assemblages could permanently change as well as diversity being altered.
- In extremely altered conditions the composition of the infaunal community can change completely, leading to replacement of all the local species with others typical of other ecosystems (from an open beach to a lagoon).
- Increased risk of spread of non-native species. The modified habitat can also provide an opportunity for non-native, invasive species to expand their range of distribution.

The presence of soft-sediments vegetation should also be taken into account. Seagrass meadows are important engineering species in the coastal zone providing sediment stability and refugee for associated species. Vegetated soft-bottoms are richer in terms of diversity than unvegetated areas; thus, LCSs should not be built in such areas. This is particularly critical when in the area there are endangered species such as *Posidonia oceanica* in the Mediterranean.

8.3.2. Implications for hard-substrate assemblages

LCSs provide new rocky habitats for colonization by species typical of natural rocky shores. The type of habitat can vary depending on a series of natural factors and processes (see Ecological Tools) but is also influenced by LCSs design features, including the layout of structures and the building material used. Also, the sheltered and exposed side of the structures increases the variety of habitats provided. The main ecological implication is that LCSs can function as «stepping stones» in coastal areas lacking of rocky shores, promoting the expansion of hard bottom species beyond the limits set by the availability of suitable natural habitats. For example, in the UK two species of grazers (*Gibbula umbilicalis* and

Melaraphe neritoides) have extended their distribution along the south east of England by colonizing the LCSs at Elmer. In Italy, the alga *Codium fragile ssp tomentosoides* has much of spread along the north Adriatic coast, colonizing the sheltered side of LCSs. This has serious implications for the identity of rocky shore communities, as the composition and dynamics of assemblages can change considerably after the introduction of non-native species and the detrimental effects of invasive species on native assemblages have already been demonstrated (e.g. *Sargassum muticum*, see review in Rueness, 1989).

8.3.3. Impacts on water quality

Emerged and rarely overtopped structures significantly reduce water movement and mixing on the landward side of the structures, thus oxygen exchange is often minimal and nutrients tend to accumulate. This can lead to hypoxia and increase the risk of algal blooms, particularly in shallow, eutrophic waters such as in the Adriatic Sea. Reduction of water movement on the landward side may also enhance accumulation of algal detritus, leading to anoxic sediments, proliferation of flies and unpleasant odours.

The worsening of water quality, the presence of algae and stagnant enclosed waters will reduce the quality of recreational activities such as swimming and sunbathing.

LCS due to frequent overtopping allow greater water movement and mixing thereby avoiding stagnant conditions. Thus water quality is minimally affected as are recreational activities.

8.3.4. Impacts on safety issues

LCSs partially reduce wave kinetic energy in the protected area and thus increase safety for beach visitors in general. Nevertheless, rip currents at gaps (in case of multiple structures, see Fig. 2.5) and roundheads may occur and be very risky for bathers; moreover, the location of submerged structures has to be marked not to be dangerous for boating and water sports.

8.4. DESIGN MITIGATION MEASURES

(Moschella, MBA; Abbiati, Airoldi, Bacchiocchi, Bertasi, Bulleri, Colangelo & Ceccherelli, FF; Cedhagen, BIAU; De Vries WL-DH; Dinesen; BIAU; Granhag & Jonsson, UGOT; Gacia, Macpherson, Martin & Satta, CSIC; Sundelöf, UGOT; Frost, Thompson & Hawkins, MBA)

LCS are designed to modify hydrodynamics and geo-morphological coastal processes and, inevitably, these changes will have ecological consequences (see Chapter 2). It is therefore important to ensure that adequate measures are considered in the design procedure of LCS to minimise environmental impacts.

The following LCS design features influence the type and magnitude of impacts on the surrounding habitats and associated biota:

a) Extensively defended coastlines

Results of DELOS project have shown that proliferation of LCSs causes broad-scale alteration of the whole coastline, resulting in important changes on habitats and species (see Sub-Section 8.3.2). Along the coasts of the North Adriatic Sea, for example, the proliferation of defence structures has substantially changed the identity and nature of the coastal landscape of this region (see Sub-section 11.4.6 and Chapter 12). Local coastal defence planning should also take into account regional environmental conditions, and avoid any unnecessary overengineering.

b) Spatial arrangement of structures

Spatial arrangement (i.e. location, relative proximity to natural reefs and other artificial structures) of coastal defence structures is of great importance in influencing the type of hard-bottom species that will colonise any novel structure, including the dispersal of invasive species.

c) Distance from the shore

In microtidal systems distance from the shore can be important in determining the degree of impacts on water quality (e.g. sediment suspension, eutrophication, turbidity) on the landward side, especially in shallow waters. In this case, LCSs should not be built too close to the shoreline. In microtidal systems distance from the shore can be important in determining the degree of impacts on water quality (etc., sediment suspension, eutrophication, turbidity) on the landward side, especially in shallow waters.

d) Tombolo and salient formation

Tombolo formation can cause burial of assemblages colonising the lower part of the structures on the landward side. The extent of the zone affected can vary depending on the height of tombolo from the sediment level.

e) Shore connectors, groynes

The addition of perpendicular rock groynes connected or unconnected to the structures significantly decreases water mixing on the landward side, thus worsening impacts on sedimentary habitat and the associated biota and water quality. These additional structures should not be considered in the design of LCSs unless strictly necessary.

f) Length of structures

At a local scale length of structures might affect hydrodynamics, particularly on the landward side. In case of emerged structures, shorter structures should be preferred, as long structures create more sheltered conditions on the landward side to the detriment of water quality and sedimentary habitat. In addition, the very sheltered habitats that are likely to be created by longer structures increase the risk for spread of non-native species such as the invasive species *Codium fragile ssp tomentosoides* along the Adriatic coast.

g) Submerged versus emerged barriers

Height of the structure affects the hydrodynamics at the landward side of the structure. This has important consequences for both soft-bottom and hard-bottom assemblages. Reducing the height of structures allows greater water movement on the landward side thus mitigate impacts on soft-bottom habitats and the water column. Greater water movement also reduces the effects of siltation that negatively affect hard-substrate species. Submerged structures should therefore be preferred, recreational value is lower, however, since the structures can be accessed only by diving or snorkelling as they also minimise aesthetic impacts.

h) Distance between structures

In case of high emerged structures, currents at gaps are usually of low intensity and thus gap width is not a critical design parameter. Conversely, for moderately submerged structures, due to the great velocities that rip currents may reach, wide gaps have to be preferred both for safety issues and ecological reasons. Slower currents will reduce erosion at gaps and hence risk of structure instability and disturbance of colonising organisms.

i) Type of material (see also Section 9.4)

The physical and chemical attributes of materials used to build LCSs will affect the development of the epibiota. In particular, if LCSs are built with materials that are not typical of the area (e.g., granite in an area of limestone bedrock or concrete blocks) this may affect the local distribution of species, providing suitable substrata for species that would normally be rare or absent in the area, including invasive species. For example certain type of smooth geotextiles may be colonised only be ephemeral algae which can represent a nuisance for the local community. Therefore the same or similar stone materials typical of the area should be used. Carbonate rocks used for construction of LCS are softer and are more easily weathered and bioeroded, leading to a more complex topography (crevices, small pits) which enhance colonisation and growth by algae and marine invertebrates.

j) Porosity

Large pores between blocks allow greater water flow through the structures and increase water mixing on the landward side, thus reducing impacts on sediments and water quality (see Sub-Sections 8.3.1 and 8.3.3). In addition, small pores can be easily filled blocked by growth of marine organisms such as mussels and polychaetes (*Sabellaria*), which facilitate sediment trapping thus further reducing porosity.

k) Scouring and abrasion

Scour at the base of the structures causes high level of disturbance to communities, leading to increased mortality, especially for filter feeders such as barnacles and algae. This effect can be minimised by building a berm around the structures, particularly on the seaward side or by providing more refugia such as crevices and holes.

l) Maintenance works

Frequent maintenance of LCSs leads to greater disturbance of epibiotic assemblages. These will remain at a permanent pioneer stage, characterised by abundance of ephemeral green algae (*Ulva* spp.) that are often considered a nuisance for recreational activities. Stability of the structure should be increased to allow development of assemblages and succession of species leading to a more diverse community.

8.5. IDENTIFICATION OF DESIGN OPTIONS THAT MAXIMISE SPECIFIC SECONDARY MANAGEMENT GOALS

(Moschella, MBA; Abbiati, Airoldi, Bacchiocchi, Bertasi, Bulleri, Colangelo & Ceccherelli, FF; Cedhagen, BIAU; Colangelo, FF; De Vries WL-DH; Dinesen; BIAU; Granhag & Jonsson, UGOT; Gacia, Macpherson, Martin & Satta, CSIC; Aberg; Frost, Thompson & Hawkins, MBA)

8.5.1. Tools to maximise recreational activities

Appropriate LCS design can also provide suitable habitat for living resources for exploitation of food (usually non-commercial or recreational) or act as the focus for recreational activities, primarily angling but also snorkelling, appreciation of marine wildlife such as «rock-pooling» and ornithology. In some cases such activities have been an accidental by-product of the building of LCS and other sea defence structures. For example, in some Mediterranean countries such as Italy, shellfish harvesting (mussels, oysters) is a very popular recreational activity on LCS, particularly during summer. In the UK, where the structures can be easily reached at low tide, many people consider LCS as sites of natural interest for observation of marine life. This effect can also have a potential educational value particularly on coastal areas lacking of natural rocky shores. Some recreational activities can, however, compromise the ecological value of the structures. For example, frequent trampling on the rocks and intense mussel harvesting have a negative effect on the diversity and dynamics of epibiotic communities.

8.5.2. Tools to maximise diversity of species (e.g. for recreational or commercial purposes)

Some species are generally perceived as benefits in coastal environments because they represent a resource to exploit for commercial and recreational activities. Other species can also contribute in ameliorating environmental conditions (e.g. bivalves filtering the water, see Allen *et al.*, 1995; Wilkinson *et al.*, 1996).

1) A general rule is that location of structure is one of the most important factors influencing the species that will colonise the structures. Further, for any new LCS introduced into the marine environment it will take time for the biological assemblage to reach a diverse community that is most likely to resemble that of a natural shores. For mature biological communities to develop, LCSs need to be stable and built in such a way that maintenance will be minimal. Unless LCSs meet these criteria, there is little point in introducing additional features to enhance diversity (for example by enhancing complexity), as attempts to repair the structure will result in considerable degradation of developing communities.

2) Surfaces that are complex on different spatial scales enhance settlement of a wide variety of sessile species. Many larvae and algal propagules prefer to settle in small pits or crevices as they provide protection from desiccation, wave exposure and refuges from grazing. The surface of the blocks can be made rougher by chiselling grooves or drilling small pits and deeper holes. The choice of building material can also significantly contribute to increase diversity of microhabitats. Rough or complex surfaces can be easily cast in concrete units, although similar features can be naturally created by weathering and bioerosion when using limestone blocks. Much more time (5-10 years), however, is needed to obtain complex and heterogeneous surfaces on the natural rock.

3) Rock pools can also be incorporated into design of LCSs to increase diversity on blocks located above mean tidal level and to provide suitable habitats for recruitment and settlement of lower shore species and mobile animals such as limpets, winkles (littorinids) and crabs. Artificial rock pools can be created either by pre-cast units or by modification of drainage patterns on the blocks.

4) On macrotidal systems, location of LCS on the shore is also important to determine the number of epibiotic species that will colonise the structure. Structures built lower on the shore will have greater diversity than those built above mean tidal level.

5) Large mobile species (crabs, lobsters, octopuses) need small-medium size (10-20 cm diameter) refuges and the interstices between boulders/blocks provide them. The design should avoid large crevices and cavities where scouring can be exaggerated.

6) Living resources will regenerate if exploited in a sustainable manner. Therefore fishing and shellfish collection may need to be managed. There are a variety of methods (closed seasons, licenses, quotas) to limit these activities. Artificial structures are particularly suitable for management by defining areas open or closed to access to be interspersed along the structures.

8.5.3. Tools for minimising growth of ephemeral green algae

1) Minimising disturbance. The high macroalgal growth on LCS is generally perceived as negative. Along the shores of the North Adriatic, for example, the banks of ephemeral green algae that are torn off the structures and washed up on the shore is a major problem for beach tourism, and leads to major costs to clean the beach. Green ephemeral algae are opportunistic species that flourish on disturbed habitats and they are the first colonisers when a new bare substrate becomes available. Maintenance of LCSs significantly increases disturbance to the epibiotic assemblages, and remove later colonisers. Minimal maintenance should be carried out on LCSs. The stability of the structures should also be ameliorated, in order to minimize translocation and overturning of the blocks, which can provide new substratum for colonisation by early stage colonisers.

2) Increasing recruitment of grazers. Promoting settlement of limpets can be a very useful, cost-effective and environ-mentally sensitive tool for drastically reducing the abundance of nuisance green on LCSs. Settlement of limpets generally occurs in rock pools. Therefore building blocks should be included features such as artificial pools and small pits which retain water during low tide.

Table 8.2. Design parameters for emerged LCS. Reference is given to the scheme in Fig. 12.9.

Water depth (m)	h = 3.0
Crest elevation (m MSL)	Rc = + 1.5
Crest width (m)	B = 4
Shoreward slope	1:2
Seaward slope	1:2
Armour rock weight (kg)	3000-6000
Stones for bedding layer (kg)	0-200
Thickness of bedding layer (m)	1.0

Table 8.3. Design parameters for submerged LCS. Reference is given to the scheme in Fig. 12.7.

Water depth (m)	h = 3.5
Crest elevation (m MSL)	Rc = – 1.5
Crest width (m)	B = 16
Shoreward slope	1:2
Seaward slope	1:2
Armour rock weight (kg)	500-1000
Stones for bedding layer (kg)	0-200
Thickness of bedding layer (m)	0.7

8.6. EVALUATION OF INITIAL AND MAINTENANCE COSTS

(Franco, MOD; Lamberti, UB)

Preliminary analysis of construction costs is carried out as an example for two typical LCS geometries, namely emerged and submerged rubble mounds, assuming unit costs and other typical constraints (wave climate, foreshore slope, sediment characteristics, construction material and technology) of the Italian North East regions.

Table 8.4. Design parameters for gap bed protection.

Water depth (m)	h = 3.5
Crown width (m)	B = 30
Stones for bedding layer (kg)	0-200
Thickness of bedding layer (m)	0.7

Table 8.5. Unit costs for emerged LCS.

Item	*Unit cost*	*Amount*	*Cost*
Armour	40 €/m³	38,50 m³/m	1.540 €/m
Bedding	37 €/m³	28,00 m³/m	1.036 €/m
Geotextile	12 €/m²	34,00 m²/m	408 €/m
Total			**2.984 €/m**

Table 8.6. Unit costs for submerged LCS.

Item	*Unit cost*	*Amount*	*Cost*
Armour	39 €/m³	24,18 m³/m	943 €/m
Bedding	37 €/m³	21,42 m³/m	792 €/m
Geotextile	12 €/m²	38,00 m²/m	456 €/m
Total			**2.191 €/m**

Table 8.7. Unit costs for gap protection among LCS.

Item	*Unit cost*	*Amount*	*Cost*
Bedding	37 €/m³	22,00 m³/m	813 €/m
Geotextile	12 €/m²	38,00 m²/m	456 €/m
Total			**1.269 €/m**

The construction costs include material supply (the material is supposed to be imported from Croatia) and placement with floating equipment.

Geometric-structural characteristics are given in Table 8.2 (emerged LCS), Table 8.3 (submerged LCS), Table 8.4 (gap protection), while corresponding unit costs (per metre length) are given in Tables 8.5-8.6-8.7.

Structure design is provided in Chapter 12, figs. 12.7 and 12.9.

It is obvious that construction costs are proportional to the LCS volume.

Maintenance costs could be determined with reference to the expected damage during LCS lifetime as predicted by stability formulae (see Section 13.11), though the total costs will increase due to the higher mobilization costs of the equipment for a small volume of rock to be placed.

LCS maintenance is relatively expensive and causes disturbance to local ecology and recreational activities and should therefore be reduced to a minimum or avoided with a more conservative and careful design. Significant and rare (every 10 years, once in economic lifetime) maintenance interventions should be preferred to small and frequent ones (twice or more in economic lifetime).

Table 8.8. Summary of potential monitoring activities given the different spatial and temporal scales associated with LCS schemes.

Construction related time scale		*At the structure*	*Scheme-wide*	*At regional scale*
	Pre	Geotechnical survey	Beach profiles (down to low water) • Bathymetric surveys • Historical changes (charts and aerial surveys) • Establish hydrodynamic conditions (winds, waves and currents) • Superficial sediment distribution • Sediment movement, as bedload and suspended load; establish transport pathways	Historical changes (charts and aerial surveys) • General hydrodynamic conditions • Establish regional transport pathways
	Construction	Structural inspections/stability	Suspended sediment measurement • Possible plume dispersion	
	Functional lifetime	• Observation of damage to the structure • Scour measurements around the structure • Wave transformation due to the presence of LCS • Sediment transport in the vicinity of LCS • Current measurements within the porous media of the breakwater	Monitoring, at quarterly intervals • beach profiles (down to low water) • offshore bathymetric surveys • aerial photographs • Periodic measurements of: • superficial sediment distribution • prevailing hydrodynamic conditions • suspended sediment concentrations • bedload movement • Numerical modelling (hydrodynamics, sediment transport and morphodynamics)	Detailed observations of coastal changes on the adjacent coastlines • Hydrodynamic numerical modelling, on a regional scale • Transport pathways

8.7. FORMULATION OF MONITORING PROGRAMMES

(Paphitis, Plomaritis & Collins, UoS; Moschella, Thompson & Hawkins, MBA)

The monitoring programme should incorporate information about beach levels, sediment distributions, tidal information (i.e. tidal currents and levels), wave and wind conditions. The exact techniques used for collection of the data can be decided on the degree of accuracy that each measurement requires and on the monitoring costs. For the case of the beach and intertidal zone the best method is beach profiling that provides both high accuracy and low cost (Serra and Medina, 1997). The spacing between beach profiles (or any beach levelling technique) is very important since it will determine the accuracy of any derived calculation (i.e. sediment budget, beach volume, etc.) (Irish *et al.*, 1997). Where data exist, these can be used for estimating the optimum beach profile sampling interval (Philips, 1985). Beach profiles should extend, in the offshore direction, down to the estimated closure depth for the area. Sediment sampling/analysis should be undertaken following standard techniques (grabs, shallow cores, mechanical sieving, settling towers, microscopy, etc.); care should be taken for the collections of an appropriate number of samples and spatial density for the proper representation of the sedimentary environment. Hydrodynamic information can be collected using various methods (i.e. pressure transducers, current meters, etc.); these will depend upon the required accuracy and frequency of measurements.

When dealing with defence schemes, involving LCSs, the programme for monitoring the structures and assessing the environmental impacts must be comprised of methods and techniques that are referring to different spatial and temporal scales. For an integrated investigation on the performance and impact of the structures, measurements have to be undertaken in the vicinity of individual breakwaters, scheme-wide and on a regional scale (see Table 8.8). Furthermore, especially in the assessment of the impacts, information about the pre-construction environment, together with post-construction information is required. An outline of the methods proposed for the monitoring, is presented in Table 8.8. The different monitoring programmes that can be used will be explored in relation to the timing of the construction.

In the pre-construction period the main task of the monitoring programme should be a desk study; the purpose of this is to identify all the available information which is related to the geological and historical development of the area. Existing monitoring programmes in the area should be evaluated with regards to the collected information. Both on a regional scale and in the area of the future scheme, beach level data and their accuracy should be established. In situations were an ongoing beach level programme is not established by the local authorities, a baseline study must be undertaken before the beginning of the construction works. Superficial sediment samples have to be collected from the area for the determination of seasonal or long-term changes in beach composition and possibly for the identification of sediment transport trends. A combined study of beach profiles and grain parameters can give an indication of beach stability (Mohan and Kana, 1997). Hydrodynamic measurements have to be undertaken to establish the current and wave regime prior to the construction. All the above information can be used to investigate the performance and impact of the proposed scheme by means of numerical and physical models.

During the actual construction of the scheme the monitoring procedures (i.e. beach level, hydrodynamic measurement) may be compromised by the high level of activity in the area. Some construction process necessitates a great amount of excavation work which, in turn, results in unusually high levels of suspended sediment concentrations. In such circumstances the plume development must be monitored. In cases of soft bottom substrate compaction/ subsidence should be monitored, during and after the construction.

The careful monitoring of the early post-construction period is of the utmost importance. Beach level measurements need to be intensified, both in spatial and temporal scales, in order to capture the immediate response of the beach system. Such measurements will also provide information for the sediment budget and the morphodynamic evolution; for this reason an accurate evaluation of the volume changes close to the scheme is important. Irish *et al.* (1997) demonstrated that the error in computing beach volumes from beach profiles is increased with increasing profile spacing. The recommended spacing, in the literature, both for pre and post-construction monitoring seems to be 30 m; in practice 300 m spacing is used from the majority of Local Authorities in their monitoring programmes (Kana and Andrassy, 1995). However, a certain level of flexibility in the spacing of beach profiles was to be adopted, especially in the area of the scheme, as all of the major features of the system (i.e tombolos, salients) have to be monitored. Such flexibility is rather difficult in beach profiling procedures, whereas a 3D beach level measurement, using a total station or kinematic GPS systems, can provide faster beach coverage and better accuracy in the morphological representation. The time interval between successive measurements needs to be more frequent (more often than seasonal measurements), incorporating fast response monitoring after storm events. Offshore bathymetric surveys also have to be undertaken in order to investigate the offshore morphodynamic influence of the scheme. Standard field measurements of sediment distribution, hydrodynamic condition and sediment transport have to be continued as in the pre-construction period. Furthermore, these measurements have to be intensified closer to the LCS for the identification of specific processes taking place (i.e. wave diffraction reflection at the structures, wave energy behind the structures) and the evaluation of their performance. Again the data can be used for the calibration of hydrodynamic and morphodynamic models.

In the vicinity of the breakwater scour measurements at the head and the trunk sections of the structures have to be performed. Although a considerable amount of research has been undertaken in laboratories considering scour development and prediction, field measurements of scour are very rare and difficult. For the long time monitoring of the scour around coastal structures the most common method is the use of scour rods (Dean *et al.*, 1997). Rods are tubes with relatively small diameter and long enough so they can be placed firmly in the study area. A movable disk is placed around the tube on the sand surface and when erosion takes place the disk follows the sands elevation; then the sand is excavated down to the disk and the maximum scour depth is obtained. The disadvantage of this method is that only the maximum scour depth is obtained with no information on the time scale of the process or the shape of the scour hole.

On the regional scale, following construction, the monitoring programme should provide data for the evaluation of significant changes in the adjacent coastlines. These can be done in terms of accretion/erosion and sediment budget calculations. The spatial spacing of beach profiling in the adjacent coastlines should be kept low for a more accurate estimation of sediment volume changes (Irish *et al.*, 1997); such estimations will provide evidence on the probable blockage of longshore sediment transport. For better understanding of the sediment dynamics of the area the regional transport pathways have to be established.

LCSs would be expected to have environmental impacts on short (largely associated with construction) and immediate responses to altered sediment regimes. Thus detailed monitoring needs to be made for 1-2 years. Subsequent ecological effects are likely to be long term and to date have not been measured. Thus programme of biannual survey of sediment infauna (early spring, early autumn) needs to be run around the structure using sample locations selected on the basis of hydrodynamics/sediment modelling. Particular attention should be given to the sampling at various distances to the seaward/landward side of the structure, at least two control areas outside the influence of the structure (ideally on either side). Samples should also be located at the round heads (simple structures) or gaps (multiple structures). At the end of the 2nd year the number of station can be minimised on the basis of experience. Within the sediments granulometry, organic matter and chlorophyll are the minimum environmental data required. The infauna should be sampled on 0.5 mm sieve and identified to highest taxonomic level possible. Data can be processed using appropriate univariate, bivariate and multivariate statistics.

Depending on resource value surveys of fish and shellfish can be made around the structures using appropriate methods (nets, traps, visual transects). Such survey should be made at least four times per year to allow for seasonal variation.

The ecology of the hard substrates can be monitored using broad-scale rapid assessment methods (biotope mapping) compiled with more detailed stratified random non-destructive sampling of major species and categories (percentage cover of canopy forming algae, ephemeral algae, algal turfs, barnacles, mussels, number of grazers and predators (especially winkles, limpets and whelks). In addition where mussels occur biomass can be evaluated. If there are exploitable resources, then yields should be estimated by recording fishing activities. Structures should be censused 1, 3, 6, 12, 18, 24 months after construction bioannually for at least 5 years. Each survey is estimated to take 2 people times 2 days for a single structure.

8.8 MAINTENANCE PLAN

(Lamberti, Zanuttigh & Martinelli, UB; Burcharth, AAU)

Structures built for local shore protection and the accompanying beach fill must be maintained to preserve the project functionality. The maintenance plan should be part of the design procedure and should include periodic scheduled interventions (ordinary maintenance) as well as sporadic interventions after exceptional storms (extraordinary maintenance).

It is necessary to identify:

- possible «failure modes» of the intervention;
- state indicators to monitor the first signs of these «failure modes»;
- threshold values of these state indicators to trigger maintenance actions;
- the type of maintenance to be performed.

The plan is site specific and based on the information obtained from preliminary surveys of the site (see Section 8.7):

- historical records of natural shoreline evolution (regression) and of shore response to similar defense schemes;
- general environmental conditions of the littoral (tide, wind, waves, ecology);
- records of subsidence of the coastal zone including the submerged beach;
- sediment characterization and sediment budget of the protected cell;
- coast vulnerability to sea ingression.

The use of morphological/morphodynamic simulations allows:

– to quantify the frequency and the sand volume for re-nourishment;

– to anticipate local erosions close to the structures that may require reinforcement of toe protection.

The necessity of structure/beach maintenance is made evident by comparison of the state indicators with the threshold values.

For instance a failure mode may be beach erosion beyond a limit that cause damage to landward structures (dunes, seawall, buildings,...). Beach width or beach volume are appropriate indicators; they can be evaluated from surveys of the shoreline position or from bathymetric and topographic surveys of the submerged and emerged beach; the volume might be preferred because it is insensitive to temporary displacement of sand from the emerged beach to submerged bars and therefore less noisy than the beach width. A target and a threshold value of the beach width can be defined; if erosion continues so that the beach width falls below the threshold value a nourishment has to be carried out and the necessary sand volume can be estimated from the difference between the target and actual beach width (or from the loss of beach volume).

If scour holes of the order of twice the stone diameter are shown by bathymetric surveys, toe berm stability may be compromised and toe protection should be reinforced and widened.

In the Mediterranean Sea, cross-shore profiles of the structures frequently documented structure settlement. Field observations in Ostia, Pellestrina and Lido di Dante (see the description of the sites in Chapter 11) show a barrier settlement variable in the range 3 to 15 cm/year, with the greatest values occurring immediately after the works on fine sandy bottoms. Since LCS effectiveness is very sensitive to submergence, settlement can easily bring the structure out of the acceptable functioning domain and rock recharge has thus to be planned.

In case of flooding, dune maintenance (planting and fertilizing dune stabilizing vegetation and/or installing proper sand fences) should be performed.

If beach recreational value is affected by organic deposits on the beach (for instance, algae grown on the structure and drifted during storms), periodic removal of these deposits has to be done, even daily in the holiday season.

Attention has to be paid to the fact that maintenance of water and sediment quality is extremely difficult and costly compared to a design that avoids this negative effects of the intervention.

Maintenance works produce disturbance to the surrounding ecosystem; it is therefore suggested to moderate the maintenance frequency. Re-nourishment should hence be planned with a frequency not greater than once every 3rd year and the maintenance of a rocky structure is suggested to be even more rare, i.e. once every 10-20 years.

CHAPTER 9

Materials for LCSs

Materials used to construct coastal engineering projects are critically important for the success and longevity of the project.

The selection of materials for LCSs comes from knowledge of the following characteristics:

- specific gravity (self-weight of the structure to resist applied loads) and strength (determines the size, shape, and stability of component structural members);
- durability (ability to resist abrasion, chemical attack and corrosion, marine biodegradation, wet/dry cycles, freeze/thaw cycles, and temperature extremes);
- costs and availability (eg. related to quantities of material needed, construction and transportation costs);
- handling requirements;
- maintenance requirements;
- environmental impacts.

For LCS construction the following materials are generally used:

- natural rocks;
- concrete blocks;
- geotextiles (plastic filaments or fibres woven or needlepunched).

Material selection is mainly dictated by availability and cost, and execution methods.

9.1. NATURAL ROCK

(Prinos, AUTH; Franco, MOD; Moschella & Hawkins, MBA; Burcharth, AAU)

The vast majority of LCSs is built as rubble mounds armoured with quarried natural rock, since this material is generally available from nearby quarries and it is suitable for structures subjected to waves.

Rock quality is another important consideration, especially for the primary armour layers since they are subjected to severe wave action, thus requiring high strength and durability characteristics. According to current practice, when selecting suitable rock material properties such as density, water absorption, porosity, shape, discontinuities, weathering grade and intact strength should be carefully examined.

Wherever possible the common rock type in the coastal cell should be used and calcareous rocks have advantages over granitic rocks in terms of habitat provision (see Section 9.4).

9.2. CONCRETE

(Prinos, AUTH; Franco, MOD; Moschella, MBA; Burcharth, AAU)

In areas with excessive wave action, calling for big armour units (usually over 10-12 t), or where the size of the required rocks is difficult to be found or uneconomical to be transported on site or when the rock quality is poor, concrete blocks (typically cubic and parallelepipedic shapes) and special concrete armour units such as tetrapodes, accropodes, dolos, cubes, etc. can be the appropriate choice. The disadvantage of this material apart from the aesthetics (which is appearance not a problem for submerged LCSs), is that concrete may be less acceptable in the coastal environment than natural rocks for environmental reasons. In case this solution is adopted, a construction yard and a concrete plant are required on the coast or the units can be constructed close to a nearby port and transported to the site by sea. In the case of use of patented armour units royalties must be paid. Concrete used in the coastal environment must be of high strength and good quality to resist abrasion imposed by gravel moved by wave action. Great resistance to sea environment can be achieved by using sulphate resistant cement. The use of steel reinforcement of the armour units should be avoided. If absolutely necessary the steel should be protected by thick cover layer.

If necessary the concrete blocks can be given an appropriate shape and holes to provide both wave attenuation and artificial structures for fish habitat enhancement.

9.3. GEOTEXTILES

(Prinos, AUTH; Franco, MOD; Moschella, MBA; Burcharth, AAU)

Geotextiles are typically used as filter to prevent migration of finer materials into coarser materials, i.e. between a sandy sea bed and the rubble mound bedding layer. Geotextiles should always be protected by a layer of smaller stones in order to avoid damage from larger rocks or concrete blocks.

Economical considerations have recently promoted the application of bags or tubes made of geotextiles and filled with sand or gravel. The so-called Longard tubes has been used along northern Adriatic beaches, due to the lack of local rock quarries. This type of structure is relatively cheap, easy to place, flexible to allow for settlements and with little harm to swimmers. However it is relatively impermeable and reflective (inducing toe scour) and easily vulnerable to vandalism and cutting for mussel collection with knives. Experience shows that their service life is rather limited. However, they might be used as core material for rubble mounds.

Strength, elasticity, strain, creep, durability, mass density and cost, are the important parameters for the selection of type and material of the geotextile. Basic materials are polyester, polyamide, polypropylene and polyethylene. The textile can be divided in woven, non-woven and knitted types. The different types of basic material and type of textile provide different performances, Pilarczyk (2000).

9.4. ENVIRONMENTAL CONSIDERATIONS

(Moschella, Thompson, Hawkins, MBA)

It is known that the type of substratum plays an important role in the colonisation and development of benthic organisms (Richmond and Seed, 1991; Callow and Fletcher, 1994). The main feature of the substratum affecting the composition, abundance and spatial distribution of epibiota is the topographic complexity (Crisp, 1974; Holmes *et al.*, 1997; Johnson *et al.*, 2003). A rough surface with crevices and small pits provides marine organisms a better protection from wave action, desiccation and insolation stresses and refuges from predators and grazers. As a result, a higher number of species can settle and survive. In general, the rougher is the surface the greater is diversity and abundance of epibiotic species.

Natural rocks generally are characterised by these complex features, especially those that are more easily weathered, such as carbonate rock (e.g. limestone). These are subject to bioerosion, if boring species are present in the area (e.g. the date mussel *Lithophaga lithophaga* in the Mediterranean), thus further increasing complexity. The rock material used for construction should be (where possible) the same of similar to the coastal geology of the area.

Colonisation of epibiota on concrete can be very different depending on the surface roughness. Very smooth concrete blocks are poorly colonised and very few species settle on them. Results from DELOS showed that when the concrete is rough there are no differences in the epibiota between this material and the natural rock. If concrete is used, a rougher surface texture should be preferred. Cast concrete can also integrate features such as small rock pools or holes that can promote colonisation by epibiotic species, crustaceans and fishes.

Geotextiles do not offer a suitable substratum for colonisation by marine life unless they are very textured. Results from DELOS showed that organisms such as barnacles and mussels are not able to colonise smooth surfaces, and ephemeral green algae are generally the only species present. This can have an important impact also on the recreational value of LCSs such as shellfish harvesting, sport fishing and observation of marine life.

CHAPTER 10

Construction of LCSs

10.1. CONSTRUCTION METHODS

(Prinos, AUTH; Franco, MOD; Burcharth, AAU)

LCSs can be constructed with either floating or land-based equipment. The selection of the construction method depends on constraints related to transport and storage of materials and environmental conditions like water depth, tidal range and wave climate. Besides this also rapidity, safety, and accuracy plays a role.

Land based equipment (dumpers, front loaders, dozers, cranes including backhoes) is used if materials are transported by road to the site and the structures are either placed in very small water depth close to the shore (see for example Figure 7.3) or constructed on coasts with a tidal range large enough to make the site dry out in each cycle. Floating equipment (barges and cranes on barges) is preferably used in calm water more than 3-4 m deep, and when the materials are transported to the site on barges. However, depending on the local conditions many combinations of land based and floating equipment are used.

Figure 10.1. Construction activities with land-based equipment of a LCS at Casalbordino, Italy.

Many beaches are generally highly exposed to wave activity and therefore a crane on a barge cannot operate safely and accurately for long periods. This is the case on many coasts in Italy, for which reason LCSs are generally executed with land based equipment by dumping rock material from lorries and placing armour with cranes. If the structure is to be submerged the emergent crest of the mound is lowered at the end of the works when the crane is retreating and dumped at the sides of the mound. Access for the equipment to the LCSs are provided by interim access causeways, which are removed at the end of the works (see Figure 10.1).

In Italy, the water depth is generally low (2-4 m MSL) and tides are negligible. Water turbidity due to provisional causeway construction and demolition can recover quickly. Only if stringent environmental constraints exist, floating equipment may be recommended. In any case severe regulations are enforced to preserve the environment (e.g., by regulations limiting dust emissions in air and sea, and recovering of any material from demolitions and dredging operations).

In Greece, despite similar microtidal regime as in Italy, LCSs are constructed from offshore using floating cranes and barges for placement of materials, as wave activity is often moderate. Thus direct dumping from barges with the assistance of floating cranes for rock placement is the most common construction method. The material supply is from land by barges. The crane barge for placement of individual units and the material haul are usually separate, allowing the crane barge to remain on station while a material shuttle operates.

Several types of self-unloading barges can be used, differing only by the method of unloading, i.e. split barges, bottom-door barges, tilting barges and side-unloading barges. Commonly available self-unloading types have load capacities of the order of 500-800 t. The first three types do not allow great precision in placing materials but are generally adequate for core construction. For bedding layers, scour protections and berms, flat-deck barges with a bulldozer for discharge can also be used. Capacities of such barges can be much higher, typically reaching 5000 t. For all types of barges, strengthening of the surfaces in contact with rock material is normally required.

The maximum construction elevation for barge-dumped core material is governed by the maximum draught of the barges plus a safety clearance for heave (vertical motion) of the barge. In exposed sites it is important to plan the construction procedure in such a way that finer materials are not left unprotected in longer periods with high risk of erosion in stormy seas.

Conveyor systems, trucks or cranes can load the barges. It is preferable to have a stockyard at the loading area in order to make the barge transport less dependent on the supply from the quarry.

For quantification of the material placed by barge, weight measurement after loading is preferred to volume measurement, because soundings cannot account for bed settlements, scour or filling of scour holes at the placement area.

For placement of filter layers only side-unloading or flat-deck barges can offer good precision. In general if the barges do not operate with a high precision positioning system it is not possible to place thin layers (0.50 m) on the seabed or on the core. Thin layers can be laid by multiple passages of the dumping barge. Alternatively the material can be placed by a clamshell or front-end loader working from a barge.

Placing of gravel-size materials can be carried out using modern trailing suction hopper dredgers. Such hoppers are equipped with a system for pumping the material through the suction pipe with the drag head suspended only a few meters above the seabed.

Stone blankets can also be placed by crane. It is normally both convenient and economic to use containers (rock trays or skips) in order to reduce crane time.

For construction of an armour layer of relatively small rock, a side-unloading barge may still be used, but often specifications do not allow dumping because of the required accuracy of placing. The alternative method for rocks or concrete units is the use of derrick barges or pontoon-mounted cranes. The armour units have to be placed piece by piece in order to form a proper two-layer cover. For controlling the placement a positioning system has to be installed in the crane. Critical fall velocities for both rock and especially concrete armour units should be considered. For most applications where cranes are necessary, rock is mainly handled with grabs, and concrete armour units by wire slings. The latter has the advantage of adding little to the crane payload, while the former has a self-weight of about half that of the rock lifted.

The construction tolerances are related to the functional requirements of the structure and the working method. The stricter the requirements, the more sophisticated the working method. The accuracy of LCSs built by floating equipment is generally less than if built by land-based equipment, and the risk of damage to concrete armour units during placement is greater when floating equipment is used.

Generally in sheltered water (no severe currents and waves) a horizontal accuracy of 1 m can be achieved. In exposed conditions this accuracy will be less and the accuracy will also decrease with increasing water depth.

For operations the following site conditions will have to be considered: current, wind and wave, available water depth and manoeuvring space, seasonal effects, tidal variation and visibility.

Currents, waves and wind conditions obviously control any working conditions. Positioning of floating equipment is achieved by a roundabout anchoring system (usually 6

Figure 10.2. Construction activities with floating equipment of a LCS at Alaminos, Cyprus.

anchors). Dynamic positioning systems using computerized thruster propulsion is generally not used for LCS construction.

Down-time caused by waves and wind is often determined by the influence on the positioning accuracy of the stone-dumping vessel and the accuracy of the armour placement rather than on operational limitations of the equipment.

Seasonal effects are essential. Construction may not be allowed during the winter season working when severe wave conditions prevail. In case construction time has to be split across several seasons, temporary protection layers may to be applied to prevent erosion of exposed materials.

Locally generated waves having a short period (2 to 6 s) and subsequent small wave length, have less impact on the floating equipment stone dumping process from than swell conditions, having longer periods. Generally wind waves should not exceed 1 to 1.5 m, whereas swell conditions beyond 0.5 m can already impose restrictions on the dumping.

The critical limits are even lower for cranes, when barge mounted, as the maximum wave height is limited by the effect on the ringer mechanisms and the derricks. Cranes are normally not designed to take any lateral forces caused by swinging loads due to barge motions. For this reason maximum allowable tilts should not exceed a few degrees.

10.2. ENVIRONMENTAL IMPACTS DURING CONSTRUCTION OPERATIONS

(Moschella & Frost, MBA; Gacia & Martin, CSIC; Thompson & Hawkins, MBA)

During construction there will be considerable environmental impacts due to plant, machinery and the deployment of materials. These will have direct effects on the sediment structure and the associated biota. Indirect effects will occur due to suspended material. The construction impact should be significantly mitigated if the works are carried out from the sea instead of a land-based construction. This (frequent and cheap) procedure results in a severe threat to the fringe communities that are crucial to the stability of the whole coastal cell. Underwater, this construction procedure results in great disturbance infaunal assemblages and seagrass meadows due to suspended materials and accumulation of fine sediments on the seabed. After construction phase, maintenance of LCSs should be kept to minimum, to facilitate recolonisation and development of infaunal assemblages.

Part II

Appendix

CHAPTER 11

Case Studies

11.1. ELMER

(Moschella, MBA; Paphitis, Plomaritis & Collins, UoS; Aberg, Granhag & Jonsson, UGOT; UoS; Frost, Thompson & Hawkins, MBA)

11.1.1. Introduction

The Elmer study site (West Sussex, south coast of U.K.), lies on an approximately straight stretch of coastline, between Bognor Regis and Littlehampton (Figure 11.1). Elmer bulges slightly, beyond the average coastal alignment; within this context, it has been referred to as a small headland (Green, 1992). The breakwater scheme extends along 1.75 km of coastline. The first 1.25 km from west are under the responsibility of the Environment Agency (EA, formerly National Rivers Authority) and 500 m under the responsibility of Arun District Council.

11.1.1.1. Selection of Elmer defence scheme as case study for the DELOS project

Case studies for DELOS were selected to represent different coastal systems across European countries and Elmer represented the case study for macrotidal shores. Although detached breakwaters have been used as a form of coastal protection for more than four decades (King *et al.*, 2000) their use was restricted to micro- and meso-tidal. In macro-tidal areas (tidal range > 4 m), such as the UK, their use is still uncommon. The study of interaction between tidal currents and waves in the vicinity of low crested structures is important for identification of processes driving the sediment transport. Such conditions (high tidal range and wave energy) are exemplified in the scheme at Elmer, which was investigated in terms of: a field measurement programme of sediment, waves and currents (at high frequency); and the development and use of a 2-D numerical modelling approaches. Furthermore, specific engineering choices (i.e. the unusually high permeability) make Elmer an interesting study site.

Technically, the location of the scheme in the intertidal zone also allowed easy access to the structures as they are completely uncovered at low tide. Ecological investigations and experimental studies could therefore be carried out by accessing the structures on foot. The relative proximity of Elmer (South of England) to University of Southampton and Plymouth also allowed frequent field visits to the breakwaters. Furthermore, the system consists of 8 similar islands that represent ideal replicate sites for statistical comparisons.

11.1.1.2. Problems that led to decision of building a sea defence

Historically, the Elmer sea frontage suffered from fairly rapid coastal erosion (Roger

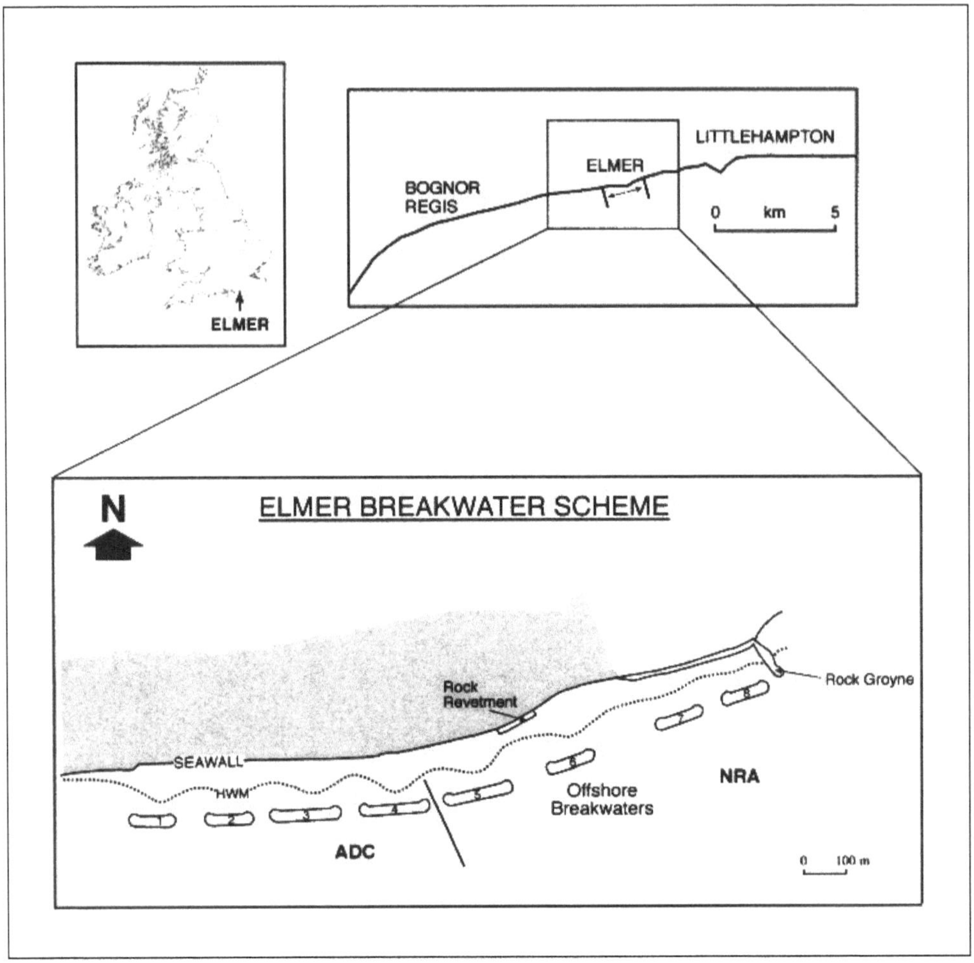

Figure 11.1. Location map of the study area, showing its regional setting, together with a sector of the coastline and the breakwater scheme.

Spencer, Borough Engineer, Arun District Council, personal communication). The area experiences substantial wave focusing and this, along with other environmental factors, produces a regime of increased wave height and potential for flooding (Green, 1992). Thus, Elmer has long been affected by wave overtopping and consequent flooding of the low-lying hinterland; most recently, in the winter of 1989/90, severe flooding occurred on two separate occasions, causing large-scale damage to the existing defences. The starvation of this part of the coastline, from littoral material, was considered to be one of the main reasons for the continued coastal problems. Following the later flooding events, a plan was conceived as a form of emergency works, to overcome the immediate problems of the area and provide coastal protection over the impending winters. These emergency works included the construction of two rock breakwaters 90 m long, with a gap of 80 m between them at approximately 120 m from the coast (to reduce incoming wave energy) and of a rock revetment on the National Rivers Authority (now EA) frontage (to provide storm protection).

11.1.1.3. Selection of shore parallel low crested structure

The defence scheme at Elmer was selected after a variety of alternative options were considered and evaluated from both engineering and socio-economic perspectives.

Erosion problems in that area were well known since 1986 and protection options were already considered at that time. The first solution of building a secondary sea wall proposed by Posford consultants was rejected by the local community, as the wall would have required the destruction of private seaward gardens. Although a timber groyne field pattern, consisting of long and short groynes for the retention of sand and shingle respectively, was historically adopted over the Elmer frontage, under the new circumstances this type of sea defences was not considered, as it was unlikely to be successful in retaining shingle. Timber groynes had periodically required a modest amount of replenished material that was deposited on the foreshore, to provide additional protection.

Alternatively, a scheme was designed, consisting of four elements: new timber groynes, restoration of seawalls, (where necessary) a rock revetment parallel to the shore and a pump to return overtopping water back to the sea, for additional protection against erosion and flooding. However, HR Wallingford modelled the revetment and contrasting outcomes in the performance were obtained. Whilst the performance at low energies was good, in extreme conditions it was actually worse – probably due to wave grouping – the first wave filled the gap between the revetment and the seawall and the second rolled over the top of the first, the beach not having time to drain. The distance from the shore was therefore set at 130 m.

Due to the pressing need to build a coastal and sea defence before the winter storms, it was decided to build a wider frontage. The first two islands were planned to be built with rock by sea delivery, but due to risks related to sea delivery companies refused to carry out the construction work in winter, thus land delivery was adopted to build the defence structures using a simple mound approach. The same approach was used for the remaining 6 rock islands.

As a result, a system of eight shore-parallel offshore breakwaters was constructed, and the area between these and the coast nourished with sediment. This scheme was considered as being the most suitable, in both environmental and engineering terms in comparison to the other scheme options: (i) Minor improvements to the existing groyne field; (ii) Minor improvements to the emergency works; (iii) Construction of fishtailed breakwaters (Robert West & Partners, 1991).

11.1.2. The defence scheme

A system of eight (incorporating the two emergency breakwaters, with only a small relocation and expansion of their initial size) shore-parallel offshore breakwaters was constructed, and the area between these and the coast nourished with sediment (Holland and Coughlan, 1994). The construction of the scheme (budgeted at £ 6.5 million) commenced in 1991 and was phased over the next few years, reaching completion in August 1993.

The eight breakwaters at Elmer come under the joint jurisdiction of Arun District Council and the Environment Agency, being responsible for breakwaters 1-4 (including the beach to the left of the structures) and 5-8 (including the beach to right of the structures) respectively (King *et al.*, 2000). Arun District Council erected two emergency offshore breakwaters (3 & 4) close to low water mark and at the same time the Environment Agency constructed a rock revetment to the east in order to prevent an earth bank from being breached (King *et al.*, 2000). The emergency breakwaters were constructed from 6-8 tonnes limestone blocks transported by road from the Mendips, West England (Pope, 2001). During the following summer 11 000 m^3 of natural sand and shingle built up in the lee of the breakwaters. In the final scheme, completed in 1993, the initial emergency breakwaters were extended and a further 6 rock islands added, as well as a terminal rock groyne at the down-drift end, (King *et al.*, 2000). For this purpose a 600 mm layer of 350-650 mm graded bedstone was placed on exposed bedrock, to provide the foundation of the breakwaters' main rock armouring (Cooper *et al.*, 1996; Pope, 2001) and 33 000 tonnes of Norwegian syenite (an igneous rock) in form of blocks of 6-10 ton each were used to build the main breakwater body construction (~ 95%), although some French quartzite was also used as a bedstone.

The eight breakwaters vary in size (Table 11.1, see also Figure 11.2), depending upon their location, and extended, overall, along 2 km of the coastline. Towards the east, the gaps are larger and the length of the breakwaters shorter; this reduction in protection was intentional, in order to produce a smoother transition between the scheme and the open beach downdrift (King *et al.*, 2000). A terminal rock groyne to the east of the system (downdrift end) acts as the beach level regulator. The high tidal range over the area created difficulties in the original location of the breakwaters, with respect to the coastline, since there was a need for scheme efficiency (towards protection) during the whole of the tidal cycle. The offshore structures are exposed completely at low tide and during high water they do not become completely submerged.

Table 11.1. Breakwater dimensions and design parameters of the Elmer «final» scheme 1.

Breakwater	Crest Elevation (m) AOD[2]	Breakwater Length (m)	Gap length (m)	Distance Offshore (m)
1	4.5	90	80	85
2	4.5	90	60	79
3	4.5	140	60	75
4	4.5	140	44	77
5	4.5	140	100[3]	88
6	4.5	80	140	54
7	3.0	80	80	68
8	3.0	80		38

[1] For locations, see Figure 11.1.
[2] Above Ordnance Datum.
[3] Opposite this particular gap is the area of the revetment.

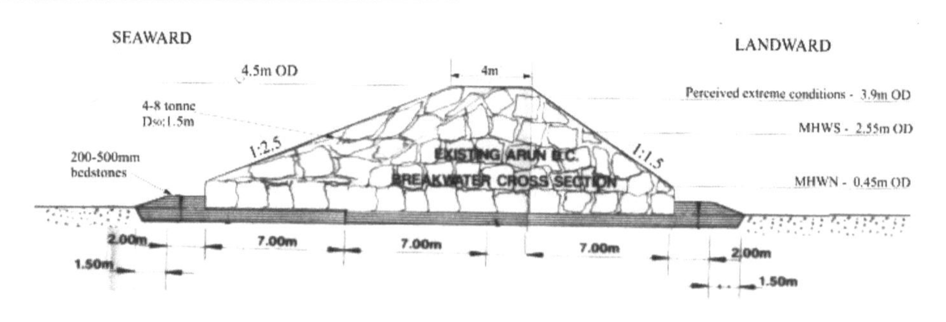

Figure 11.2. The positioning and size of the 4.5 m breakwater at Elmer with respect to different water levels.

The breakwaters are round-headed with a slope of 1:2.5 at the head, each breakwater is approximately 6 m high with a slope of 1:1.5 on the landward side and 1:2 on the seaward side with a 4 m wide crest (see Table 11.1).

11.1.3. Environmental setting

A physical and ecological description of the area where the LCS were built is provided below.

11.1.3.1. Hydrodynamics and sediment regime

Waves

The dominant wave direction is the Southwest; with 65% of the waves approaching from within the segment 180° to 220°, but with some 15% of the waves approach from the 100° to 160° (Southeast). Waves come from the sector of 180° to 200°, with a significant wave height of up to 5.5 m and a wave period of about 7.5 sec (Hydraulic Research, 1994). The sheltering effect of the Isle of Wight limits waves arriving from 220° to 260°. In response to the gently sloping bathymetry at Elmer, the waves reach the coastline with very small angles of approach; this is especially characteristic of waves arriving from the southeast direction, which are more normally aligned to the shore.

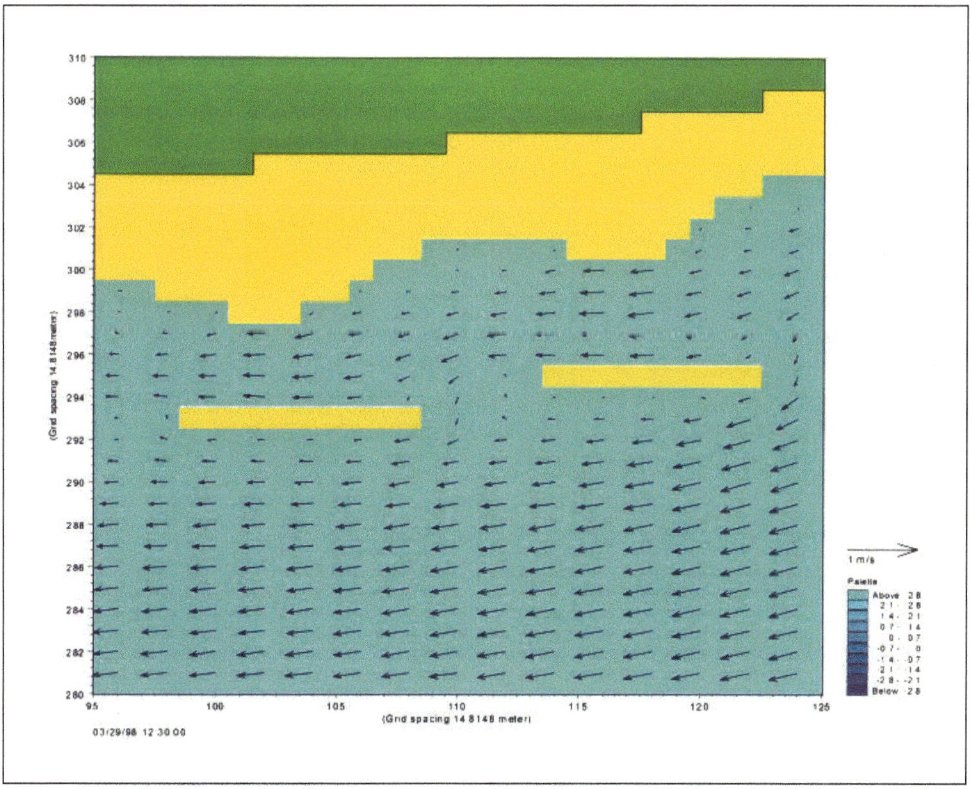

Figure 11.3. Typical high water spring tidal currents in the upper intertidal zone of Elmer.

Tides

Elmer is located within a macrotidal environment, with a semi-diurnal tide. The mean spring tidal range is approximately 5.3 m, whereas the mean neap tidal range does not exceed 2.9 m maximum. Spring tidal ranges can reach up to 6 m. Near bottom (approximately 30 cm above the bed) tidal currents over the area do not exceed 1 m/s (on spring tides); they run in a general east-west direction in the offshore areas. Tidal currents in the intertidal zone almost always flow in a westerly direction in this coastal cell (Figure 11.3).

Superficial Sediment

The coastal plain generally comprises a poorly-consolidated layer of sand, exposed during low tides, with a 115 μm median grain size. Shingle occurs on the upper part of the beach, on top of the thin sand veneer, median diameter of 20 mm (King *et al.*, 2000). The longshore sediment transport in the area is to the east, with possible temporal reversal during long periods of Southeast winds and associated waves (Bray *et al.*, 1995).

11.1.3.2. Ecology of the surrounding area

The area around the LCS at Elmer can be divided in three zones: the vegetated shingle beach, the intertidal zone and the subtidal zone. The vegetated shingle is located at the top of the shore and is characterised by a wide variety of wild plants, some of them being artificially seeded as mitigation measure soon after the construction of the rock islands. The plants living on the shingle ridge are generally typical of this habitat and include babington's orache *Atriplex glabriuscula*, sea kale *Crambe maritima*, yellow horned poppy *Glaucium flavum* and tree mallow *Lavatera arborea* and other common coastal species. Apparently the vegetated shingle backing the structures is the only site in West Sussex where little robin *Geranium purpureum*, a rare plant, can be found. These plants attract invertebrates of particular scientific or conservation interest such as the toadflax brocade moth *Calophasia lunula*, which is included in the Biodiversity Action Plan and is also a Data Book species. This zone is also used as a nesting site by birds such as the ringed plover . The intertidal zone is typical of moderately exposed sandy shores. Polychaeta and amphipods dominate the infaunal assemblages. In particular, the most common species are the lugworm *Arenicola marina* and the amphipod *Bathyporeia* spp. In the lower intertidal natural boulder fields and rocky outcrops are colonised by ephemeral algae (*Ulva lactuca*, *Enteromorpha* spp.), gastropods (slipper limpets, *Gibbula cineraria*), crustaceans such as amphipods, shrimps and crabs, and benthic fish (gobids). The subtidal is a mixture of sand, shingle and rocky areas, probably hosting a variety of organisms.

11.1.4. Environmental effects of Elmer defence scheme

11.1.4.1. Effects on hydrodynamics/sediment transport

Numerous studies, using a range of techniques, have been undertaken in the area mainly after the construction of the offshore breakwater scheme. The main focus of the studies was the investigation of hydrodynamic processes introduced by the scheme and the associate sediment dynamics.

The general wave-induced circulation pattern observed inshore of the breakwaters is characterised by a clockwise pattern, with its core inshore of the gap (Sterlini, 1997). The magnitude of the wave-induced currents depends upon the direction of wave approach and

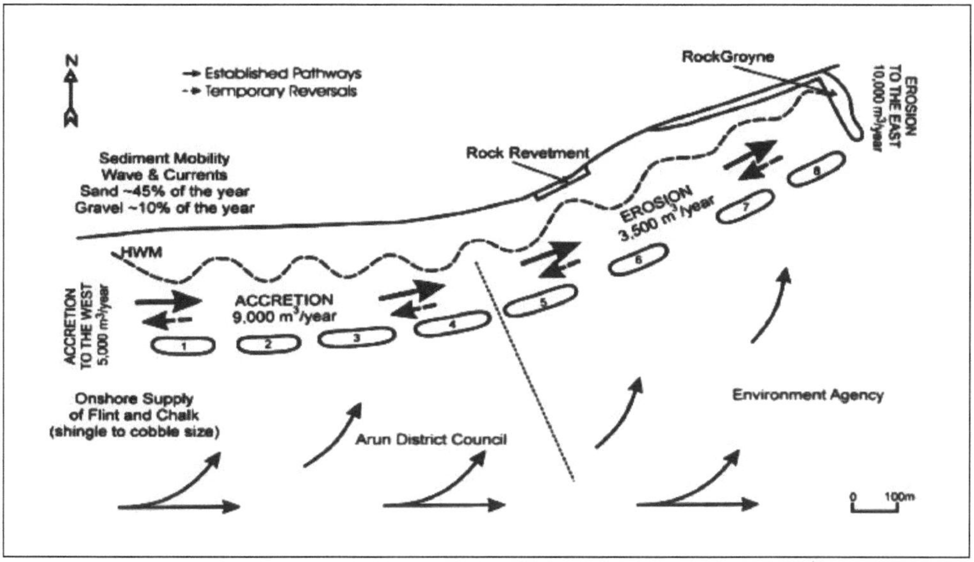

Figure 11.4. Localised sediment transport in the vicinity of the Elmer (offshore breakwater) scheme (adopted from various sources).

their characteristics. As mentioned earlier tidal currents in the upper part of the intertidal zone i.e. in the area of the scheme are flowing mainly in a westerly direction (Figure 11.3). The magnitude of spring current in the area is low at the beginning of the tidal cycle, increasing before high water and degreasing slowly again, during the ebb phase of the tide. However, the flow appears to reverse under high-energy wave conditions (Pope, 1997), this flow reversal is an important factor controlling mainly the net sediment transport close to the breakwaters. The tidal currents accelerate in the lee of the breakwater as they flow over the salient feature enhancing the sediment mobility (Figure 11.3); this mechanism probably is controlling the salient growth behind the structures.

Fluorescent pebble tracer studies have revealed that sediment in the immediate lee of the breakwaters remained immobile during storm conditions, highlighting the degree of protection afforded by the structures; likewise, their ability to maintain the beach. In addition, these experiments revealed that, under calm conditions, movement from the west into the scheme was negligible; however, movement out of the scheme at the eastern end did occur (King, 1996a; Cooper *et al.*, 1996). Notwithstanding these observations, the terminal rocky groyne at the eastern end (Figure 11.4) is proving to be somewhat successful in retaining the sediment along the defended frontage. Beach profile analysis, undertaken after completion of the scheme (King, 1996b), for the evaluation of longshore sediment transport, has revealed accretion to the west of the scheme (an increase in beach volume of around 5000 m^3/year), and in the area controlled by Arun District Cancel (approximately 9000 m^3/year). Throughout the remaining of the scheme (area controlled by the Environment Agency), the beach volume was reduced by 3500 m^3/year. Down-drift of the scheme, after the terminal groyne, a reduction in the beach volume of 10000 m^3/year has been estimated.

Aluminium tracer experiments revealed that with predominant waves from the southwest, net transport directions recorded were from west to east, with recorded rates of up to 2 m^3/day, under the most typical wave conditions. The maximum rate of transport recorded in the lee of the breakwater was 57 m^3/tide (for shingle), during a storm (King *et al.*, 2000).

However, this rate of transport, as opposed to that on natural beaches under the same conditions, is an order of magnitude lower; this demonstrates the efficiency of breakwaters in reducing the wave energy that reaches the beach. Under mild wave conditions the net sediment transport pathways, in the close vicinity of the structures, were inferred using grain size trend methods. Offshore of the structures the pathways had clearly onshore direction. Between the structures and the coastline the direction of transport was diverted East and West feeding the salient features.

In the offshore areas of the breakwaters, over the inner continental shelf, sand was found to be mobile for approximately 40-50% of the time over a typical year (Velegrakis, personal communication). The mobility of gravel for the same area is around 10% of the time, over a year.

All experimental, literature and morphological evidence on the sediment transport in the area of Elmer is suggesting littoral drift from West to East; which is consistence with the general trend observed in this coastal cell. However, tidal currents in the upper part of the intertidal zone (i.e. in the area of the scheme) are flowing mainly in a westerly direction (Figure 11.3). That difference in the direction of the peak tidal currents and the net long-shore transport is due to the effect of the incoming waves (dominant direction Southeast-South-southeast) creating, as mention earlier, a flow reversal that is driving the sediment transport to the East.

11.1.4.2. Effects on the ecology

Introduction

The construction of the defence scheme at Elmer has produced a series of changes to the surrounding environment. Environmental impacts include aesthetic effects on the landscape, recreational value, ecological effects on soft- and rocky bottoms, fish assemblages and other mobile fauna and birds. Many of these were investigated and assessed over the 3 years activities of the DELOS project. The structures were built with the sole purpose of protecting that part of the shore from beach erosion and flooding of the residential area located behind the beach. There were no primary ecological objectives set up for the construction of the LCS, therefore the ecological effects observed must be considered only as a bi-product of the construction of these structures. Some effects, although negative from the ecological viewpoint, can have positive consequences from a socio-economic perspective.

Effects on sediment infauna

The effect on sediment-dwelling biota surrounding the LCSs at Elmer was investigated during two studies, in summer 2001 and 2002. The first study was restricted to the effects of LCS on infauna and sediment characteristics, whilst the second investigated the extent of these effects along the shore and the effect of tidal level.

Results from the two studies were consistent. All the areas investigated were characterised by a high degree of spatial variability that affected both sediment descriptors and biotic features. This variability made it difficult to detect small changes in the sediment descriptors (chlorophyll, organic matter, granulometry, anoxic layer), and may explain why no significant differences were detected. However, some changes in the sediment features could be observed on the landward side. Chlorophyll in sediment was generally less on the landward side than in the surrounding area. Organic matter was evenly distributed in the locations investigated, except for the landward where a slightly higher value was recorded. On this side of the structures sediment was also finer, including a small amount of silt/clay.

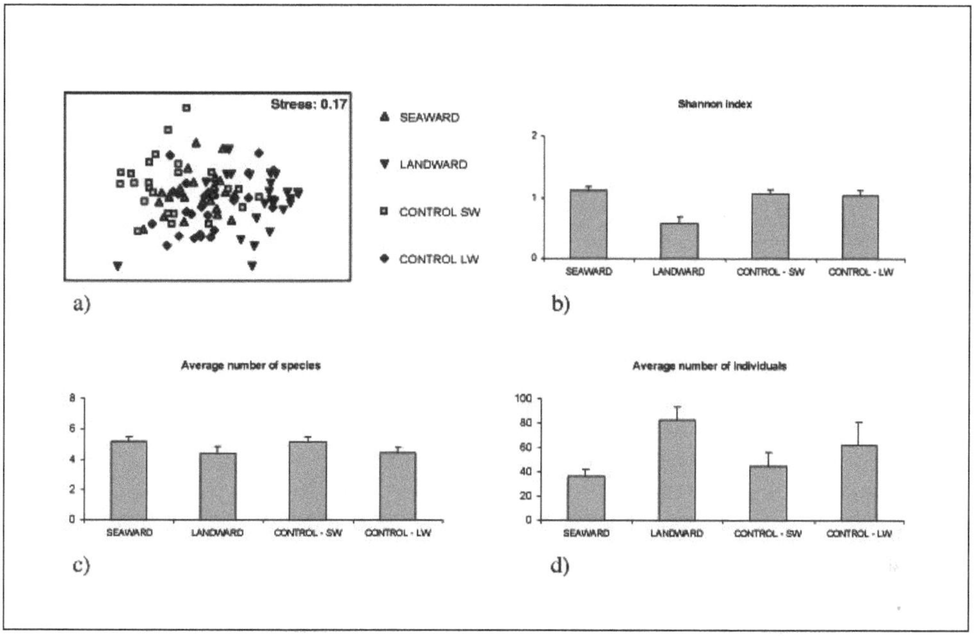

Figure 11.5. a: nMDS plots of infaunal communities at Elmer showing differences between the landward, seaward and control areas. b, c and d: comparison of diversity, expressed as Shannon index (b), mean number of species (c) and total abundance, expressed as number of individuals (b) and (c) on the seaward, landward and control areas sampled.

The absence of clear patterns in the sediments surrounding the breakwaters and in control areas along the coastline can be attributed to two factors: the characteristics of the beach where the structures are located and the porosity of the structures. The beach at Elmer is a typical sandy shore with moderate exposure to waves and moderately reflective. This is a much more dynamic system than more dissipative, sheltered beaches. For example, the investigations conducted on LCS located on a sandflat in the Wirral, showed less variability and stronger effects on the landward side. The peculiar design of the LCS at Elmer, lacking a central core and having a high porosity allows greater water flow from the seaward to the landward side of the structures. Therefore the hydrodynamics is not so strongly reduced, thus also the effect on sediments is not severe. The multiple structure scheme also probably contributes to create zones of turbulence and local currents on the landward side.

As a result, sediment characteristics on the landward and on the seaward side and control areas are relatively similar providing therefore similar habitats. More clear patterns were shown in the infaunal assemblages present and the LCSs had apparent effects on the composition and abundance of infaunal communities. There were significant differences in infaunal assemblages between the landward and the seaward side and control (Figure 11.5.a). Crustaceans dominated the infaunal communities at all locations considered. On the landward side of the structure, the average abundance of amphipods was approximately ten times higher than that of polychaetes. The main dissimilarity between landward and seaward and control areas was attributed to the amphipod *Bathyporeia* spp., which was 5 times more abundant on the landward side than the other locations. Although not statistically significant, diversity (indicated by Shannon's index and total number of species) tended to be lower whilst abundance was higher than in other locations (Figure 11.5.b,d).

The effect of LCSs on the soft-bottom community appeared to be evident only on the landward side, as the seaward side and the other control areas along the coast were very similar in diversity and abundance of organisms and sediment characteristics. Also, the effect appeared localised within 100 m or so around the structures as no effect was detected at increasing distances. On the landward side of the structures the formation of tombolos and salients, which can alter considerably the tidal level between the two sides of the structures, appeared to have only a minor effect on the soft bottom communities as minimal differences where observed in control areas at similar tidal elevations.

These studies showed that the environmental setting is extremely important in determining the magnitude of impacts on the soft-bottom habitat and communities. On relatively reflective and exposed beaches such as Elmer, LCSs seem to have a minor but significant impact on sediments and infaunal communities. On dissipative shores, such as on the Wirral (West England), the impact of LCS on surrounding soft-bottoms was more apparent, and the effects on sediment characteristics and infaunal communities were similar to those observed at Elmer but markedly amplified. Also changes in sediment characteristics and infaunal assemblages still occur at Elmer but are probably less evident and often obscured by the natural variation. Design features of LCSs, however, can partially reduce the effects, for example through increased porosity.

Provision of rocky habitats

A major effect of LCSs at Elmer is the creation of artificial habitats for species naturally living on rocky shores. Elmer is located on a stretch of coastline which lacks of natural rocky shores With only patchy boulder fields and small rocky outcrops. The area is also characterised by low recruitment of common rocky shore species such as mussels. The epibiota colonising the blocks of the structures is relatively poor in terms of diversity (21 species). The most common organisms observed are fucoids, ephemeral algae, limpets, littorinids snails and barnacles. Distinct differences between landward and seaward side were observed on all the structures. On the seaward side limpets and barnacles were dominant whilst on the landward side permanent patches of fucoid and ephemeral algae were present (Figure 11.6.a and b). The absence of algae on the seaward side was probably the combined result of physical factors (strong exposure to waves, higher dislodgement forces) and biological interactions (higher grazing pressure). Rock pools were also present at the base of the structures on the seaward side. These had extremely high diversity (72 species), with numerous species typically found on the lower intertidal/subtidal zone. One of the reasons for the significantly lower diversity on the structures than in the rock pools is probably the low complexity of the blocks and their freely draining nature, which does not provide enough micro-habitat diversity as on a natural rocky shores. Experiments that was carried out on the structures showed that more complex surfaces with holes and pits significantly increased species diversity, particularly for species that are more sensitive to desiccation and insolation stresses occurring at low tide. A more complex topography also promotes settlement of juvenile marine invertebrates and provides algae and refuges for mobile fauna. Several south-western species that reach their limits in the English Channel have been able to colonise further east by using the breakwaters at Elmer. These include the snakelock anemone *Anemonia viridis*, the periwinkle *Melaraphe neritoides* and the purple top shell *Gibbula umbilicalis*.

The conservation value of the Elmer site has been recognised by the proposed designation as a Site of Special Scientific Interest (SISI). This is largely because of the vegetated shingle

Figure 11.6. Epibiota colonising the rocks on the seaward (a) and landward side (b) of one LCS at Elmer. The close-up pictures showed limpets and barnacles on the seaward side and fucoid (brown) and ephemeral (green) algae.

Figure 11.7. Stratum of pebbles and gravel on the landward side of LCS at Elmer. From the close-up pictures it is possible to observe colonisation of fucoids and ephemeral green algae, indicating the relative stability of the sediment.

but also because of the animals and plants colonising the breakwaters.

Another special feature of the LCS at Elmer is the formation of a relatively stable layer of pebbles which extends for a few square meters from the base of the structures on the landward side (Figure 11.7). These pebbles consist of chalk and flint probably transported during storms from offshore through the gaps to the landward area of the beach. They then got trapped behind the structures, probably because hydrodynamic forces on the landward side were not sufficient to transport the rocks back to the sea. These small rocks provide a new rocky habitat for colonisation mainly by ephemeral algae, fucoids and sometimes also littorinids snails.

The structures are of considerable recreational value for the area. Local users and seasonal tourists enjoy observation of marine life on the rocks and in the pools. Thus, in the case of the Elmer defence scheme, mitigation measures to enhance diversity on the structures would be beneficial and appreciated by those using the breakwaters for informal recreational activity. Epibiota contributes not only to the amenity value of the structures but it provides natural resources for fish and mobile fauna.

Effects on fish and mobile fauna, including birds

The LCS appeared to have some effects on fish and mobile fauna. In a similar way to the results obtained for the soft-bottoms, effects were more evident on the landward side of the structures. Surveys of fish and mobile fauna were carried out over the three years of the DELOS project. The composition of fish and mobile fauna around the LCS consisted of species typical of both rocky and soft-bottoms. This suggests that LCS, especially when built in coastal areas dominated by soft-bottoms, can have a strong influence on the structure of fish communities, attracting species typical of rocky shores therefore increasing local diversity. Several of these species are of commercial importance such as sea bass (*Dicentrarchus labrax*), mullet (*Chelon labrosus, Liza ramada*), sole (*Solea solea*), plaice (*Pleuronectes platessa*) and other flat fish.

More importantly LCS provide a nursery ground for fish, particularly for commercially

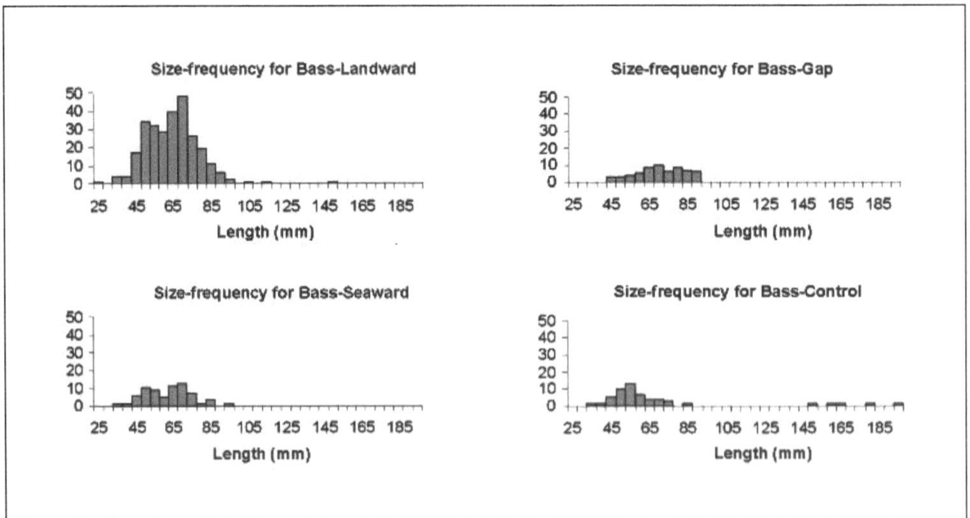

Figure 11.8. Size-frequency plots of sea bass caught around the LCS at Elmer (from fish survey 2002).

and recreationally important species, the sea bass *Dicentrarchus labrax* and several flat fish (e.g. *Solea solea*, *Pleuronectes platessa*). So, potentially LCS could have an enhancement effect on local fisheries. The landward side of the structure appears to provide a better habitat for juvenile fish (Figure 11.8). This could be partially a consequence of the more sheltered conditions occurring on the landward side. Also, on this side, the accumulation of drift algae appears to provide a suitable habitat for juvenile species. Crustaceans such as shrimps and crabs are particularly abundant in the structures and represent further food resource for fish and birds.

On the basis of the investigations carried out, it was not possible to formally assess the effect of LCS on birds. However there is evidence that the rock islands attract birds that generally are found on rocky shores, such as cormorants and oystercatchers; these use the structures as resting sites and for feeding resources (e.g. limpets). In contrast, the LCS could negatively affect other species of birds by modifying the species composition of infaunal assemblages which these birds feed on. For example, on the landward side of LCS at Elmer, amphipods become considerably more abundant than polychaetes such as lugworms (*Arenicola marina*).

Effect on accumulation of seaweed detritus on the beach

The stretch of coast where the Elmer defence scheme is located is periodically affected by large amounts of seaweeds that are detached from the offshore reefs and washed onto the shore after stormy weather. This phenomenon, however, seems to be particularly evident around the LCS, as more seaweed detritus accumulates on the landward side of the structures than in the adjacent areas of unprotected beach. The algae are probably pushed inshore by waves and inshore winds, but they eventually get trapped by the LCS. The accumulation of seaweed causes recreational and ecological problems. Strong unpleasant smells develop as a consequence of the seaweed decaying and the underlying sediment becoming highly anoxic (Figure 11.9). In

Figure 11.9. Accumulation of seaweed detritus on the landward side of LCS at Elmer and consequent sediment anoxia.

addition, during summer flies are also abundant on the rotten seaweed. This is detrimental for beach users and several complaints have been made by the local community. Accumulation of seaweed detritus also has ecological consequences. The sediment covered by the seaweed detritus becomes anoxic as a consequence of changes in the redox potential. This is likely to have an impact on the infaunal assemblages, especially for the more sensitive species. At high tide, however, some of the algae float and seem to provide an attractive habitat for juvenile fish, thus they may enhance the local fish assemblages.

11.1.4.3. Socio-economic perspective

Introduction

Since the late 1950s extensive residential development has taken place in the low-lying Elmer foreshore area. In common with other coastal areas of SE England this development has been in the form of private estates providing predominantly retirement homes. Coastal protection measures, to limit erosion and to control flooding, were first instigated in 1932 and in the late 1950s came under the control of Arun District Council (ADC). This coastal defence, which protected an increasing number of residential properties against tidal inundation, was largely achieved by the means of groynes, together with various constructions at the back of the shingle beach, the majority of it constructed before the advent of planning control. However, by the late 1980s some of the existing defences were coming to the end of their useful life, and erosion of shingle from in front of the sea walls and breastworks highlighted the very real risk of a breach of the defences. During the winter of 1989/90 severe storms caused a significant further deterioration in the shingle beach, overtopping of the sea defences and flooding to properties on two occasions. Responsibility for protection of the low-lying residential development and agricultural land along the 1750 m Elmer frontage is split between the ADC and the National Rivers Authority (NRA), now the Environment Agency (EA), in line with their statutory responsibilities for coast protection and sea defence. The ADC frontage extends some 500 m westwards from the house called «Opal Tide», the NRA frontage extends eastwards to the Poole Place groyne. ADC, NRA and Robert West & Partners (RWP) jointly developed the solution to these problems as a three-stage scheme.

– Stage 1: Emergency Works in the winter of 1990/91 consisting of the construction of a rock revetment (NRA), two shore parallel offshore breakwaters (ADC) and a limited amount of beach nourishment. A coastal defence study was also initiated to determine the design of a permanent scheme to guarantee the integrity of defences for the next 50 years.

– Stage 2: The reconstruction of the Poole Place groyne, which is the terminal groyne supporting the eastern and downdrift end of the Elmer shingle bank.

– Stage 3: Implementation of a permanent scheme resulting from the coastal defence study, which considered the benefits, costs and preliminary environmental impacts of four possible scheme options. The preferred project option was the extension of the two existing shore parallel offshore breakwaters, the construction of a further six similar structures (four NRA, two ADC) and a large beach nourishment with shingle.

The total costs of the scheme were approximately £ 8.5 m, grant aided by the Ministry for Agricultural Fisheries and Food MAFF, now Defra (Department for Environment, food and rural affairs).

The stated purpose of the works was to reduce coastal erosion, prevent overtopping of seawalls by storm driven high tides and to reduce the risk of a breach of the coastal defences

Table 11.2.a. Perceived environmental impacts of the Elmer scheme.

Coastal Asset[1]	*Benefit/Cost*[1]	*Positive Impact*[2]	*Negative Impact*[2]	*No Impact*[2]	*N/A*[3]
Land	Loss of land	Prevention of erosion			
Buildings (non-heritage)	Inundation (complete loss)	***Prevention of erosion and flooding***			
	Flooding Risks	***Significant reduction in flood risk***			
	Other amenities				X
Heritage Buildings	Preservation				X
Temporary Structures (mobile home, caravan etc.)	Complete or partial loss				X
Utilities Infrastructure (road, rail, cable, sewer etc.)	Complete or partial loss				X
Sea Defence infrastructure	No Threat	Restoration of a protective beach in front of seawall			
Navigation Channels	Sedimentation			Littlehampton harbour entrance	
Human Health	Storm control (physical casualties) Flooding discomfort	Reduction of hazards and improved safety on frontage	Potential hazard to persons using the beach, esp. LW		X

Table 11.2.b. Perceived environmental impacts of the Elmer scheme.

Coastal Asset[1]	*Benefit/Cost*[1]	*Positive Impact*[2]	*Negative Impact*[2]	*No Impact*[2]	*N/A*[3]
Ecosystem and natural heritage (wetlands, dunes, reefs, beach, water quality etc.)	Preservation	Support of environmental conservation aims	Rock islands impact sea views from the beach and seaside houses	Supply of littoral material to down drift beaches [M1]	
		Rock islands are a significant landscape feature Creation of small inter-tidal pools Seaweed deposition reduced [M4] Improvement of soil conditions above MHWS Creation of new and valuable inter-tidal habitat Increase in bird numbers (addn. food sources)	Deposition of fine sand on landward side of rock islands – soft and relatively unstable	Climping S.S.S.I. Sub-tidal benthos Water Quality Beach amenity value Beach Access Archaeology Air Quality Climate Noise & Vibration	
Tourism	Loss or gain of activity			Coastal leisure activities	
Agricultural activity	Yield change (storm, erosion, salinity)	Protection of land from saline intrusion		Land drainage system	
Commercial Fisheries	Yield Change	Colonisation of intertidal zone by marine flora and fauna-fish/shellfish feeding opportunity		Damage to sea bed Hazard to commercial fishermen	
Recreational Fishing Scuba Diving Shellfish collection	Preservation of site Preservation of site Preservation of site			Amateur Angling	 X X

Table 11.2.c. Perceived environmental impacts of the Elmer scheme.

Coastal Asset[1]	*Benefit/Cost*[1]	*Positive Impact*[2]	*Negative Impact*[2]	*No Impact*[2]	*N/A*[3]
Wildlife watching	Preservation of site				X
Sailing	Preservation of site	Sheltered water for sailing training at HW			
Boating	Preservation of site		Launching/landing small boats [M2]		
			Rock islands are a potential hazard to small craft		
Water-skiing	Preservation of site				X
Windsurfing	Preservation of site		Windsurfing opportunities limited by rock islands		
Waterfowl hunting	Preservation of site				X
Beach visitation	Preservation of site				X
Heritage Buildings (recreational use)	Inundation (complete loss)				X
	Flooding risks				X

[1] Reference DELOS Work Package 4.1, www.delos.unibo.it
[2] Reference Elmer scheme ES.
[3] Either not applicable to the Elmer scheme, or not considered by the ES.

on the Elmer frontage. The works would also protect adjacent properties and highways from damage by flooding.

Cost Benefit Analysis

The following documents form the basis of this review of the BCA for the Elmer scheme:

– DELOS Work Package 4.1 «Extracting a Benefit Transfer Function from CV studies». www.delos.unibo.it;

– the Environmental Statement (ES), prepared for the NRA Sea Defences at Elmer, West Sussex, by Environmental Assessment Services Ltd in April 1992. This covered the stage 3 works along the NRA frontage;

– the «Joint Engineer's Report for the Elmer Coastal Defences - Stage 3», prepared by RWP in April 1992.

Coastal Defence Impacts

A coastal defence scheme has many kinds of consequences on the seafront and on its residents. For example, on top of changing erosion patterns and flood risk, a breakwater will change the appearance of the landscape, offer some recreational opportunities and modify the local biodiversity. Therefore the value of the coastal defence scheme is composed of the sum of the values for each of these changes.

DELOS Work Package 4.1 identified a comprehensive list of coastal assets and their benefit/cost values. These relate to mitigation, enhancement, preservation and other indirect benefits or costs that may, or may not, have value at a particular location. The perceived impacts of the Elmer scheme are considered within this framework (Table 11.2).

Perceived Impacts of the Elmer Scheme

The ES for the EA frontage perceived a number of long-term impacts of the stage 3 permanent scheme (Table 11.2) and a number of short-term impacts relating to the construction of the scheme, which are not considered in this report. Planning procedure covered the ADC works; an ES was not required at the time as no SSSI was impacted. The

Table 11.3. Estimated benefits of the Elmer scheme.

Impact	*Benefit*
Flooding: Value of properties[1] permanently lost to habitation[2]	£ 14.9 m
Flooding: Damage to other properties[3]	£ 0.041 m
Erosion: NPV of properties lost[4]	£ 2.17 m
Total Benefit	**£ 17.15 m**

[1] Property was valued based on average property values in Elmer in 1992. These ranged from £ 50 000 (Flat) to £ 115 000 (Detached House).
[2] An assessment of the value of property that would be flooded more often than once in two years and therefore assumed to be rendered uninhabitable. Comparison made between the 1 in 2 year flood level (3 375 m OD) and the upper ground floor level of each property.
[3] An assessment of the flood damage to properties that would be flooded less often than once in two years.
[4] An assessment of the value of property that would be lost to erosion over the 50 year design life of the scheme. The assumption was made that two key segments of the existing coastal defences were at the end of their useful life and would provide no further defence against erosion and that erosion would radiate from these locations at the historical rate of retreat of 2.6 m per year.

ES concluded that the permanent works would have very few adverse environmental effects. The most severe impact would be that of the rock islands on the sea views. However, when balanced against the impacts of flooding, it was concluded that overall the proposals would be of benefit to Elmer and its locality.

Economic Benefit Analysis

The Joint Engineers Report, 1992 estimated that the total discounted benefits of the scheme (Table 11.3) ranged from £ 17.15 m, assuming an immediate breach of the NRA frontage, reducing to £ 10.5 m in the unlikely event of the breach not occurring for 10 years. Details of this benefit calculation are shown in Chapter 15 of the tools. Compared to the total costs of the scheme of £ 8.5 m, these benefits indicated a benefit cost ratio for the overall scheme of between 2.1 and 1.3.

Benefits Assessed

Benefits of the scheme were calculated based (only) on the positive impacts shown in ***bold italics*** (Table 11.3). The methodology adopted was that presented in Middlesex Polytechnic Flood Hazard Research Centre (1987) and supported by detailed methodologies presented in Penning-Rowsell *et al.* (1987) and Parker *et al.* (1987).

Limitations of the Elmer BCA

The economic value of a significant number of the impacts identified in the ES (Table 11.2) was not assessed. These included:

– the agricultural benefits of preventing flooding to land adjoining the eastern part of the frontage; this was considered insignificant in comparison with the residential flooding benefits;

– the indirect benefit of removing property development constraints. There is evidence of recent development of high value property at Elmer;

– ecological benefits, such as the creation of a new inter tidal habitat;

– tourism and leisure related benefits. In the case of Elmer this is probably justified, as there is little visible attempt to encourage visitors to the area. Public access is limited; there is a lack of parking and nowhere for visitors to spend money.

Monitoring

The Elmer ES called for monitoring of the following potential impacts (Table 11.2) of the scheme:

– [M1] the supply of littoral material to Climping beach, which is downdrift of the scheme;
– [M2] the impact of beach nourishment on the launching and landing of small boats;
– [M3] all aspects of the environment and coastal engineering issues;
– [M4] predicted patterns of seaweed transport and deposition.

Key elements of the monitoring programme implemented and managed by ADC, as part of a District monitoring scheme, included:

– monitoring of the Elmer frontage and 1 km updrift (west) and 2 km downdrift (east), based on 69 shore normal profiles and 4 shore parallel profiles;
– profiling updated monthly for the first 15 months, then every 3 months since 1994/5;
– profiles derived from 1:3000 scale stereoscopic aerial photography.

This monitoring programme focussed on the physical performance of the scheme. ADC considers that it addresses the first three of the ES monitoring requirements. No environmental monitoring or specific monitoring of seaweed transportation/deposition has been carried out, but the ADC view is that construction of the breakwaters has made the problem slightly worse. There has been no monitoring or substantiation of the perceived impacts on recreational activity.

ADC is no longer directly responsible for the monitoring programme, as the scheme is now covered by the SE England Regional Monitoring programme, operated by the Channel Coastal Observatory, based in Southampton University. However this scheme of monitoring is solely physical parameters and does not cover other areas of benefit.

11.1.5. Conclusions

11.1.5.1. Hydrodynamics/sediment transport

In terms of hydrodynamic regime, two significantly different hydrodynamic conditions were revealed in response to differences in the incident wave energy. Under low wave energy the tidal currents are dominant; however, flow reversal appears under higher energy conditions (Pope, 1997).

The wave-induced circulation pattern observed inshore of the breakwaters (Sterlini, 1997) are characteristic of surface piercing breakwaters with a clockwise pattern, with its core inshore of the gap (Pechon *et al.*, 1997).

The sediment mobility behind the structures was found to be reduced in comparison with natural unprotected beaches.

The scheme appears to be successful in protecting the low-lying areas from flooding. However the increasing gap dimensions and the decreasing length of the breakwaters to the east led to the need for further scour protection at the revetment. Further, the east part of the scheme undergoes a net loss of material of 3500 m^3/year.

Downdrift erosion is estimated to be 10 000 m^3/year significantly different from the updrift accession rate of 5000 m^3/year.

11.1.5.2. Environmental considerations

The LCS showed several effects on the surrounding environment, including changes in the composition and abundance sediment infaunal assemblages, increased diversity of epibiotic species and enhancement of juvenile fish. These effects, however, were localised on the landward side of the structures and seem of reduced magnitude in comparison with other case studies such as Lido di Dante (Italy). This might be due to the high permeability of the structures and the increase in gap length, which allows a higher level of water movement and sediment transport on the landward side. The geographical location and the type of shore, however, are likely to influence the magnitude of these impacts.

The main effect of the Elmer defence scheme is probably represented by the accumulation of seaweed detritus on the landward side. This has a negative effect on the recreational value of the area but could also severely impact the sediment characteristics and the associated infaunal assemblages.

The Elmer defence scheme is apparently a success in terms of protecting from flooding and coastal erosion. From a socio-economic perspective, the impacts of the Elmer scheme seem to be compensated by the high amenity value of its structures.

11.1.5.3. Elmer scheme benefits

The Elmer scheme has been successful, in that there have been no breaches of the sea defences, or flooding of the residential areas, since its inception in 1992. It can therefore be considered to have met its socio-economic objectives. In addition maintenance costs of the scheme have been lower than anticipated. The monitoring programme has identified that further beach nourishment is now required, whereas It was originally thought that this would be necessary after 5 years, thereby delaying expenditure over 6 years.

11.2. ALTAFULLA

(Sierra, UPC; Martin, Satta Gacia, Mc Pherson, CSIC)

11.2.1. Introduction and background

Altafulla is a typical Mediterranean beach located on the tourist coast of Tarragona (Spanish Mediterranean), 70 km south of Barcelona. The beach of Altafulla is facing South and surrounded by two rocky salients enclosing the considered morphodynamic system. The length of the beach is about 2300 m and it has an average slope of 1.6%. In 1965 a defence concrete seawall with a length of 250 m was constructed, being extended to 450 m in 1972. In 1983 the seawall suffered from increasing scour and failed. The failure area was then protected with a conventional rubble mound.

However, and due to the high tourism value of the place, in 1991 a LCS was built, complemented with a 160000 m^3 sand nourishment to increase the width of the emerged beach. The LCS was placed in the middle of the coastal cell, in front of the «Roca de Gaià» which splits the beach (Figure 11.10) in two parts. The structure was located between –4 and –5 m water depth and it is 110 m long, 5 m wide and the stillwater freeboard is less than 1 m. The nourishment took place at the East of the coastal cell (right part of Figure 11.10) where there was a lower amount of sand due to the E-W (right to left in Figure 11.10) net sediment transport pattern. Due to the lack of precise knowledge on the actual hydrodynamic conditions, the nourishment did not behave as expected and two years later, in 1994, another recharge was required to maintain the sub-aerial beach surface. This time 250000 m^3 of sand were fed at the East side of the beach. The cost of the structure and the first nourishment was of 1002934 €. The second beach fill cost 1681649 €.

11.2.2. Description of the defence scheme

The LCS in Altafulla is a simple, single structure, unclosed by lateral groins and built to protect a single large beach, which joined the two previously separated northern and

Figure 11.10. Aerial view of the Altafulla beach in 1983 (above) and 2001 (below).

southern ones as a result of beach nourishment after LCS building.

This detached breakwater is parallel to the coast and it has a length of 116 m at the base and 100 m at the crest and a width of 21 m at the base and 5 m at the crest. The initial distance to the shoreline was 180 m and it was located at a water depth between 3.5 m and 4.5 m. The freeboard is of + 0.50 m. Figure 11.11 shows a ground plan and longitudinal sections of the

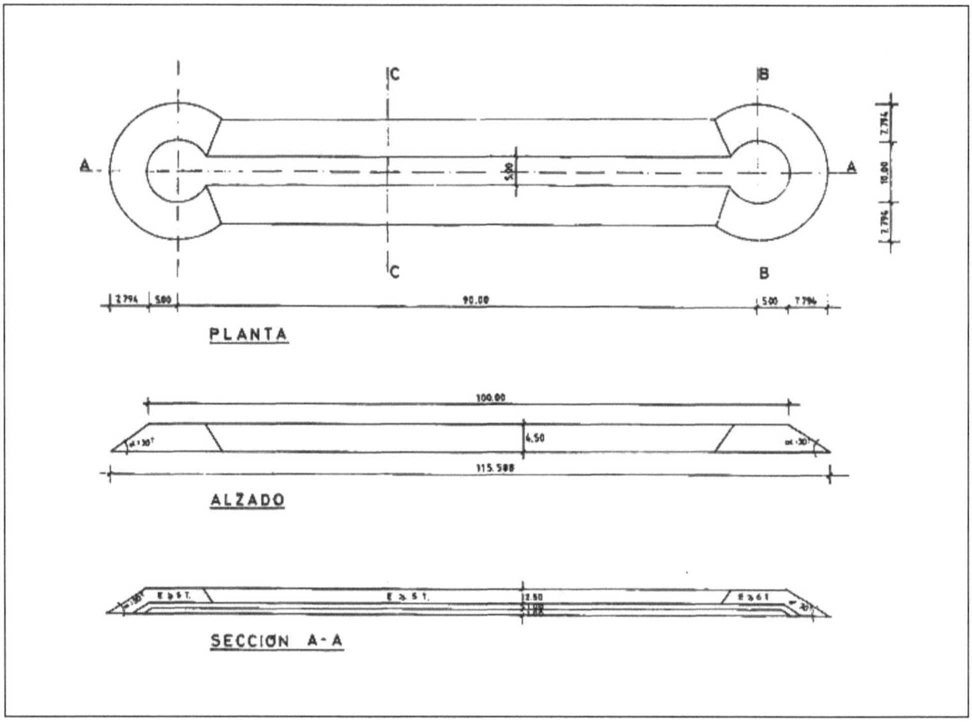

Figure 11.11. Ground plan and longitudinal section of the Altafulla structure.

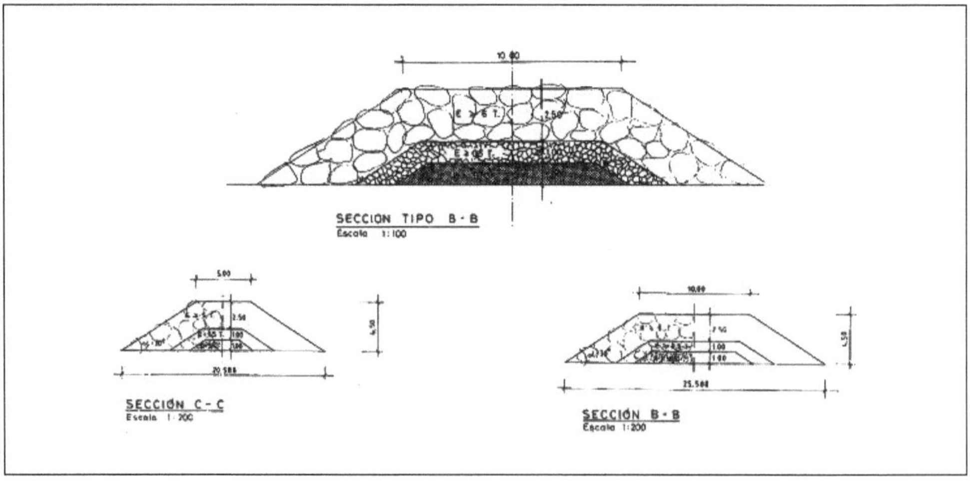

Figure 11.12. Cross sections of the Altafulla structure.

structure, while in Figure 11.12, different cross-sections are plotted.

The breakwater is round-headed with a slope of 1:1.6 along all the structure. The cross section is constituted by a core of quarry run, a filter layer with stones of 0.5 *T* (nominal diameter of 0.57 m) and a width of 1 m and, finally, by a armour layer with stones of 6 *T* (nominal diameter of 1.31 m) and a width of 2.5 m. The employed material is granite with a density of 2.65 T/m^3.

The stones were placed with barges and, since its construction, no structural problems were observed. Other problems such as structure settlement or scour were not observed either.

11.2.3. Hydrodynamics and sediment regime

No wind or currents have been measured, although wind data are available from a meteorological station located at the Tarragona harbour (about 15 km to the SW). Data recorded from January 1st 2002 to March 6th 2003 are summarized in Figure 11.13.

Wind climate shows that the most frequent winds are those from the N (18%) followed by WNW (8.4%), NW (8.1%) y SW (7.5%). The strongest winds are those from the W-NW, with speeds greater than 11.6 m/s. Winds from the N rarely exceed 5.4 m/s. This pattern is the usual one during all the year, although in summer winds from the NW increase and those from the SW decrease. In autumn, winds between NE and S are almost nonexistent. In winter, winds from the E are the most frequent and strong, exceeding many times velocities of 11.6 m/s. In spring, winds from the W-NW decrease their frequency while those from the SW and E increase, with few episodes of strong winds.

The local wave climate has been studied from forecasted data (1996 to 2003) supplied by the Spanish Ministry of Public Works («Puertos del Estado»), obtaining the distribution of significant wave heights and directions.

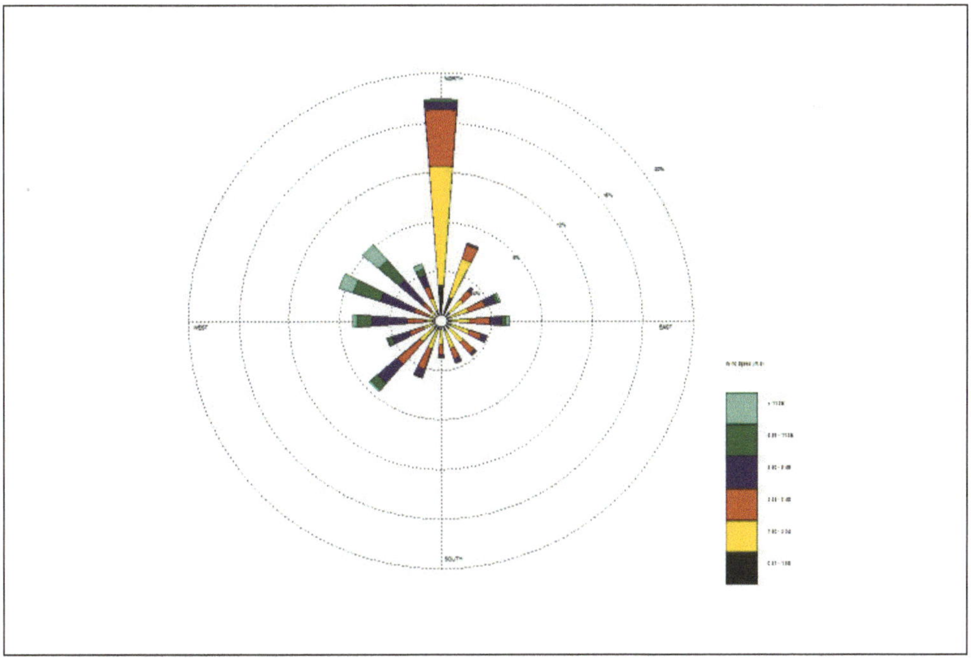

Figure 11.13. Wind distribution in Tarragona harbour.

Table 11.4. Local wave heights (m) and directions. Percentages of occurrence.

	Wave height (m)					
Direction	*0-1*	*1-2*	*2-3*	*3-4*	*4-5*	*% Total*
N	5.70	0.37	0.01	-	-	6.08
NE	4.93	0.51	0.03	-	-	5.47
E	16.87	3.04	0.44	0.03	0.02	20.41
SE	16.27	0.86	0.05	-	-	17.18
S	22.87	1.74	0.17	0.02	-	24.79
SW	5.89	0.96	0.02	-	-	6.87
W	3.29	0.48	0.01	-	-	3.78
NW	4.62	0.37	-	-	-	4.99
Total	80.44	8.32	0.74	0.05	0.02	89.57
Calms						10-43

Table 11.5. Local wave peak periods (s). Percentages of presentation.

T_p *(s)*	*%*
< 2	0.90
2-3	14.74
3-4	29.72
4-5	18.01
5-6	13.52
6-7	11.55
7-8	4.03
8-9	3.03
6-10	2.02
> 10	2.49

The analysis of these wave data shows a typical Mediterranean wave climate, with mild conditions most of the time. The significant wave height is lower than 1 m for about 91% of the time and more than 99% of the time it is lower than 2 m (including the calm periods). The prevailing wave conditions are those between E and S (more than 62% of the time). Wave periods also show typical Mediterranean values, with peak periods ranging between 3 and 7 s about 73% of the time. Tables 11.4 and 11.5 summarize this wave information.

Concerning the tides, Altafulla is located in a microtidal environment, with a semi-diurnal tide. The spring tidal range is smaller than 0.3 m. Due to these limited tides and the location of the LCS in a relatively open coast area, tidal currents are negligible in this area.

The general circulation in the Catalan coast goes from NE to SW. As a consequence and due to the local orientation of this coast sector, the general circulation, as well as the predominant littoral transport, goes from E to W.

The coastal plain is constituted by a layer of fine sand, with a medium grain size of 0.12-0.2 mm, although during the beach nourishments, finer fractions of sand were employed.

11.2.4. Effects on hydrodynamics/sediment transport of the Altafulla LCS

There are no hydrodynamic field campaigns carried out in this area, so the effect of the LCS on the hydrodynamic pattern can only be inferred from numerical studies. With this goal, different numerical simulations have been performed. Figure 11.14 shows the wave field computed for a wave train with $H_s = 1$ m and $T_p = 4$ s, travelling with normal incidence towards the LCS on the 1992 bathymetry.

From this figure a clear wave diffraction pattern can be observed, giving rise to a significant reduction of wave heights as well as the apparition of wave height gradients, in the leeside of the structure.

Figures 11.15 and 11.16 show the circulation obtained for 1992 and 1999 bathymetries. The main changes between both bathymetries are the narrowing of the sheltered area and the depth decrease also in the area where the salient is appearing. These changes modify the

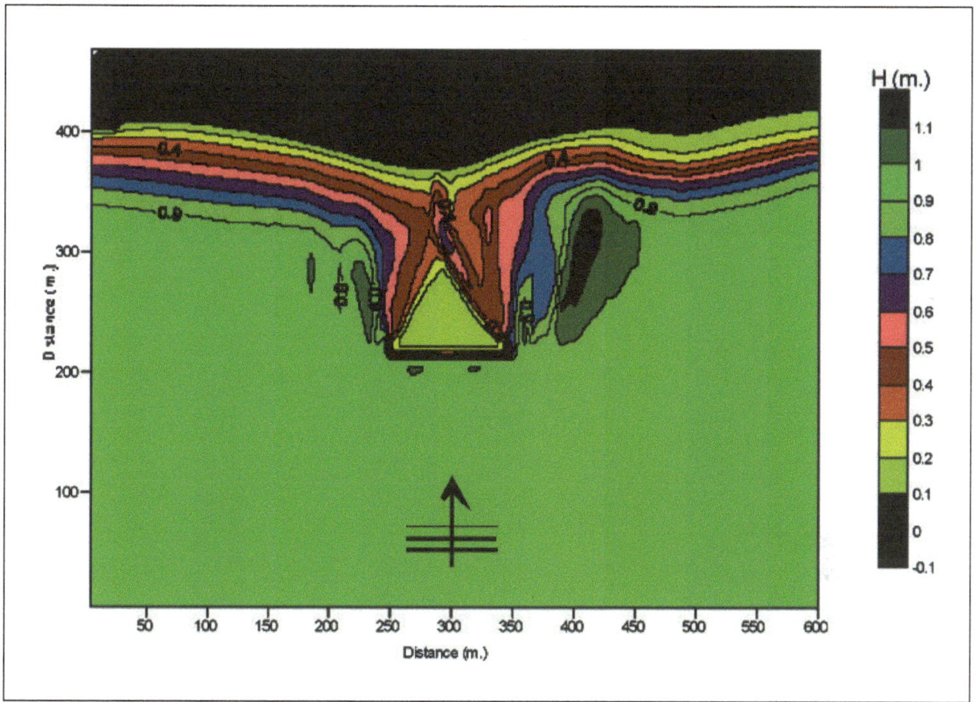

Figure 11.14. Wave field for the 1992 bathymetry, with normal wave incidence and $H_s = 1$ m.

wave field and the induced circulation under the same wave conditions.

Starting at the 1992 bathymetry, two vortices appear at both sides of the LCS generating a convergence of water fluxes into the sheltered area close to the shore (Figure 11.15). There, a component towards the structure appears in the central section together with an offshore returning water flux close to the structure closing the eddy circulation as previously indicated.

Analyzing the obtained results for the circulation induced with the 1999 bathymetry, employing the same wave conditions as for the 1992 case, it should be stressed that at first sight there are some changes in the eddy intensities and the displacement of the upper eddy to the sheltered area. Since all conditions are the same, the circulation variation can only be induced by bathymetric changes (the increase of the salient dimensions and the decrease of the depth in the sheltered area).

On the other hand, bathymetric surveys were carried out in the area in July 1991, February 1992, July 1992, November 1992, June 1993, December 1993, July 1994, May 1995, March 1996, October 1997, February 1998, June 1998, November 1998 and February 1999.

In 1989, before the LCS construction, there was a rectilinear beach with isobaths reasonably parallel to the shoreline. The rocky outcrop «Roca de Gaià» placed near the middle of the beach interrupted this shoreline. The LCS was constructed (1991) at 180 m from the head of the «Roca de Gaià», and the distance from the LCS to the initial shoreline was about 230 m.

In July 1991 (3 months after the first nourishment) significant bathymetric changes and a fast redistribution of sediment near the structure were observed. The LCS modified the sheltered shoreline (and corresponding bathymetry), decreasing water depths and acting as a sediment trap. The distance between the LCS and the shoreline reached a mean value of

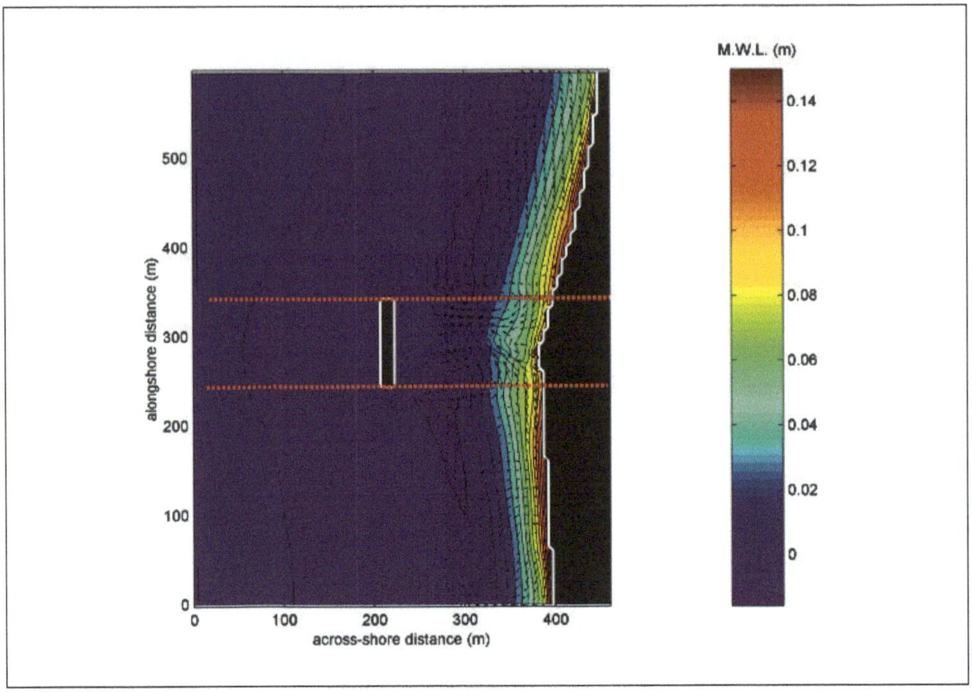

Figure 11.15. Circulation pattern induced, for the 1992 bathymetry, by normal wave incidence and $H_s = 1$ m.

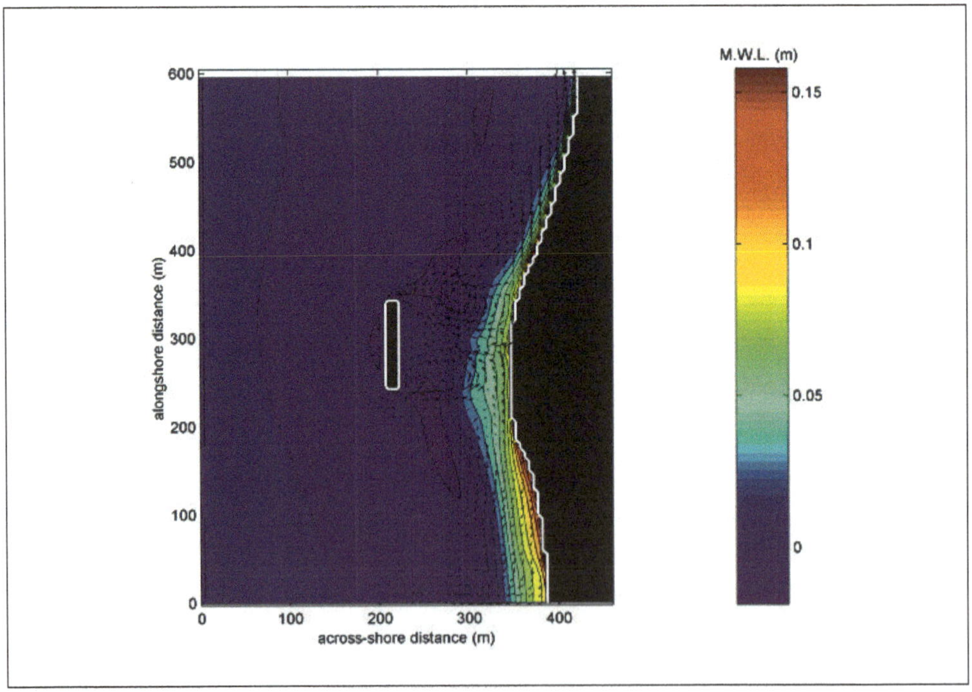

Figure 11.16. Circulation pattern induced, for 1999 bathymetry, by normal wave incidence and $H_s = 1$m.

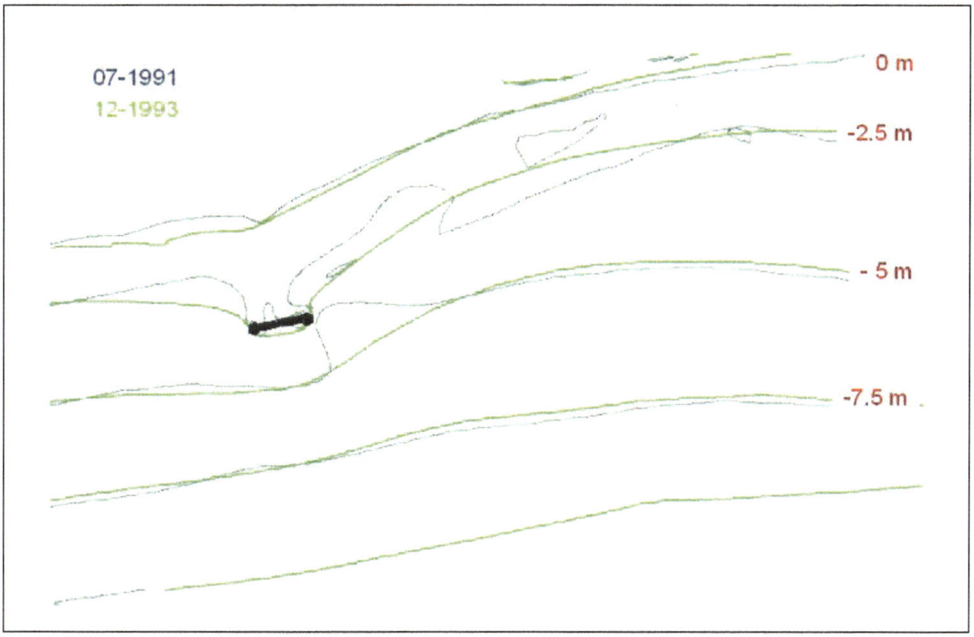

Figure 11.17. Comparison of the bathymetric surveys in July 1991 (blue line) and December 1993 (green line).

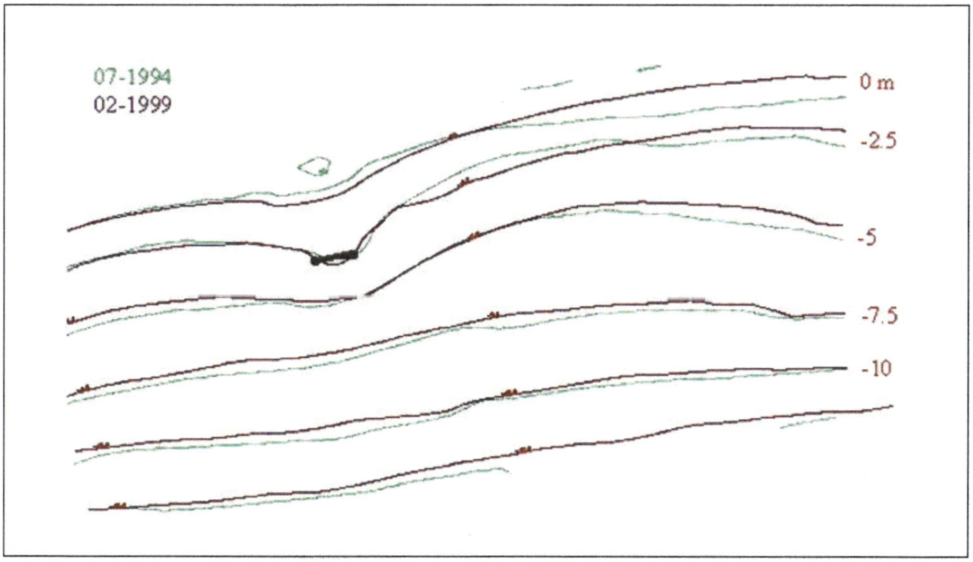

Figure 11.18. Comparison of the bathymetric surveys in July 1994 (blue line) and February 1999 (red line).

162 m. The outcrop had been by then completely buried.

In December 1993, before the second nourishment, the fast movement of sand (placed in the first nourishment) observed in the first bathymetry after the LCS construction was evolving more slowly. In Figure11.17, it can be observed how the isobath of – 5 m close to the place where the LCS was constructed in 1991 had moved 88 m seaward while the one

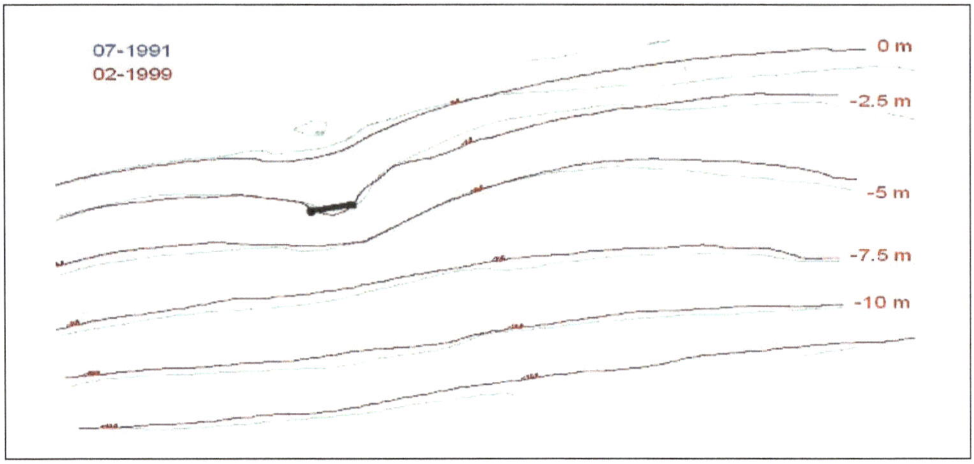

Figure 11.19. Comparison of the bathymetric surveys in July 1991 (blue line) and February 1999 (red line).

of – 2.5 m has been smoothed by the better distribution of the sediment coming from the nourishment in the elapsed years, completing the bottom reshaping process. Then the isobaths located on the seaward side, in front of the LCS, remained almost rectilinear and parallel to the coast, while those located close to the structure, in the sheltered area, showed a clear offshoreward advance.

The second recharge (1994) introduced an important reserve of sand in the East part of the beach. The 250 000 m^3 of sand incorporated to the system, helped the beach in the last years to avoid scour near the water front, while the sediment movement continued from the East to the West as can be seen in Figure 11.18.

In February 1999 the shoreline was located at 130 m from the LCS, while the beach and bathymetric changes were smoothly shaped behind the structure. The depth at the leeside of the LCS had been dramatically reduced from – 3 m in 1991 to less than – 1 m in 1999.

In Figure 11.19, the first and last available bathymetries have been plotted. As it can be observed there, the greater changes occurred in the leeside of the structure. The irregularities observed on the right side of the 1991 bathymetry are attributed to the nourishment done 3 months before the measurements and the subsequent redistribution of the spilt sediments.

11.2.5. Effects of the Altafulla LCS on the existing populations, colonisation and biodiversity

11.2.5.1. Soft-bottoms

In Altafulla, the landward side of the structuretends to be deeper than the seaward side; the sediments are slightly but consistently coarser on the seaward than on the landward side and finer in controls and when deeper and far from the LCS. This last trend coincides with an increase of microphytobenthos. The hemitombolo is narrow near the LCS, this giving rise to a sharp decrease in depth from the centre towards the laterals.

The infaunal assemblages were typical of the fine sand with *Spisula subtruncata* assemblage. These assemblages consist mainly of polychaetes and amphipods, contributing to the abundance of individuals, and bivalves, contributing considerably to total biomass. There was a characteristically patchy spatial distribution, however, significant differences

Table 11.6. Percentage change in biological descriptors of the soft-bottom assemblages and the abundance of species indicator of organic enrichment such as *Capitella capitata* around the Altafulla LCS relative to Assemblages at control sites. sp: number of species; abu: abundance; bi: biomass; div: diversity.

ALTAFULLA					
	sp	*abu*	*bio*	*div*	*Capitella*
Landward	– 33	– 46	– 81	– 28	6800
Seaward	– 57	– 82	– 91	– 49	0

were apparent between seaward and landward sides of the LCS and between controls and the landward side. Most infaunal biological descriptors tended to increase with depth and to decrease with the increasing grain size. The presence of the LCS induces a disruption in the normal progression of biotic and abiotic variables from the shoreline to deeper waters in three ways: 1) a markedly higher spatial variability at landward; 2) lower values in seaward sites facing the LCS than in the corresponding sites along control transects; and 3) trends not-strictly perpendicular to the coastline (southern areas differing from northern). The decrease of all biological descriptors relative to the controls (Table 11.6) is particularly evident for biomass and is especially drastic at seaward (less than 50% compared to controls).

Taking into account the whole region, however, the presence of the LCS only results in a slight increase of biodiversity (3.4% of the species present around the LCS were absent at control sites) In particular, there were 7 and 21 species present at seaward and landward respectively, which were absent in the controls. At seaward, however, most of these species were present with very few individuals. At landward, some of them (e.g., *Spisula subtruncata*) are indicators of more calm waters.

The response of some species to environmental changes can help assessing the impact of LCS. For example, the polychaete *Capitella capitata*, which is a typical indicator of organic enrichment was very abundant on the landward side (with a proportional increase compared to controls), reaching about 200 worms per m^2 in deeper and more protected zones (either by the LCS itself or by the hemitombolo). This may indicate that landward conditions were more delicate and may easily be perturbed by changes in the sediment characteristics in parallel with a reduction of water circulation. The more protected they are, the more fragile is the equilibrium.

Changes in sediment characteristics and infauna seem a predictable consequence of the presence of LCS, which tend to induce changes in the level of hydrodynamics. In principle, some effects seem not necessarily negative, such as the overall increase in species diversity. In Altafulla, however, this is mainly caused by the presence of species accidentally found in the sediment but belonging to the newly added hard bottoms or from species often associated to increasing disturbance conditions, so that the increase biodiversity is virtually not-relevant from an ecological point of view, and may even be considered as a negative transformation of the environment.

11.2.5.2. Hard-bottoms

Natural rocky shore assemblages differed from those in the artificial substrate, which, in turn, significantly differed depending on the orientation (i.e. between blocks, seaward and landward). The number of species tended to be higher in the reference sites than at landward, particularly in late spring. However, this pattern was not significant overall. Moreover, no consistent significant differences in species diversity are found between the artificial structures and the natural rocky shores, in contrast to results of similar studies in Australia (Glasby and Connell, 1999).

Species diversity describes quantitatively the nature of an assemblage but it does not necessarily give an indication of the functioning of the system. In Altafulla, some of the key species in the natural substrate (i.e. *Cystoseira mediterranea*, *C. compressa*) do not grow on the LCS, that is occupied by opportunistic fast growing species such as *Ulva rigida*, *Cladophora coelothrix*, and very abundant *Ceramium* spp. dominating the artificial substrate. The former are typical of more stable conditions while the later may reflect a more disturbed environment.

Different factors may contribute to disturbance of the epibiotic assemblages on LCS. On the exposed, seaward side, the lack of complexity of the substrate does not help dissipate strong wave energy or to create diverse habitats for long living species to grow. On the landward side, beach nourishment, confinement and strong human impact from collecting bivalves and gastropods prevented the community from developing to more stable stages of succession. Finally, between the building blocks there is very strong water flow that restricts the settlement and growth of many taxa. However, other factors such as consequences of confinement (e.g. slightly higher water temperature or nutrients) may enhance the development of fast growing epiphytes keeping diversity values relatively high on the LCS.

The absence of *Cystoseira* species on the LCS may be related to their low reproductive output and success. Hence they would seldom be able to recruit to LCS that are isolated by long sandy beaches. Conversely, rocky shores that have continuity of hard substrate may be able to retain populations of this key species.

In summary, the diversity of the community growing on the natural and artificial hard-bottoms is not informative on the impact of the LCSs on the constructed coast. To approach how the introduction of the new substrate may change the epibiont communities in the area, there is a need of background studies on the composition of the hard-bottom assemblages in the area. These should help to identify key species from opportunistic ones and, thus, predict the evolution of the new potential communities on the substrates based on the results shown here. As a general pattern, the proximity of natural-rocky shores would enhance the development of epibiont communities on the LCS more similar to natural substrates. By contrary, in coasts dominated by sandy beaches, the presence of opportunistic-fast growing species and easily dispersed would be enhanced.

11.2.5.3. Mobile fauna

The number of fish species recorded in the Altafulla LCS was clearly smaller (19) than other LCS systems in Spain (> 30). However, there were no significant differences in the number of fish species recorded among the LCS systems and natural sites. The low species diversity is probably attributable to environmental conditions at Altafulla, where the LCS is located in an open area surrounded by sandy beaches and with wave abrasion. Abrasion influences the abundance of branched algae, which is an important habitat for many small fish species and which is used by adults in reproductive (nesting) activities. As a consequence, numerous species cannot settle or reproduce on the LCS. Significant differences were also found between landward and seaward of LCS. The protected zones at landward provides the ideal habitat for settlement of some common species of fish, such as *Diplodus sargus* (in summer) and *D. vulgaris* (in winter). These settlers are absent from shorelines that lack of protection from the dominant winds. Other common species settling on the seaward side (e.g. *Oblada melanura*, *Thalassoma pavo*, *Chromis chromis*) do not show this pattern.

The LCS does not provide habitats that maintain structured fish populations, because of the small size of the structure but, also, because of the intense sport fishing activities around the LCS. The populations of the different species mainly consist of juveniles no older than two years. The presence of the LCS in Altafulla does not increase the biodiversity of the area, allowing only the development of local assemblages that remain at early stages of succession. None of the species occurring on the LCS are different to those of the local fish fauna. In this particular area of the Mediterranean, other factors such as eutrophication or proximity to major boat traffic are more relevant in terms of a potential enhancement of introduced species than the creation of artificial habitats in areas near to natural rocky shores.

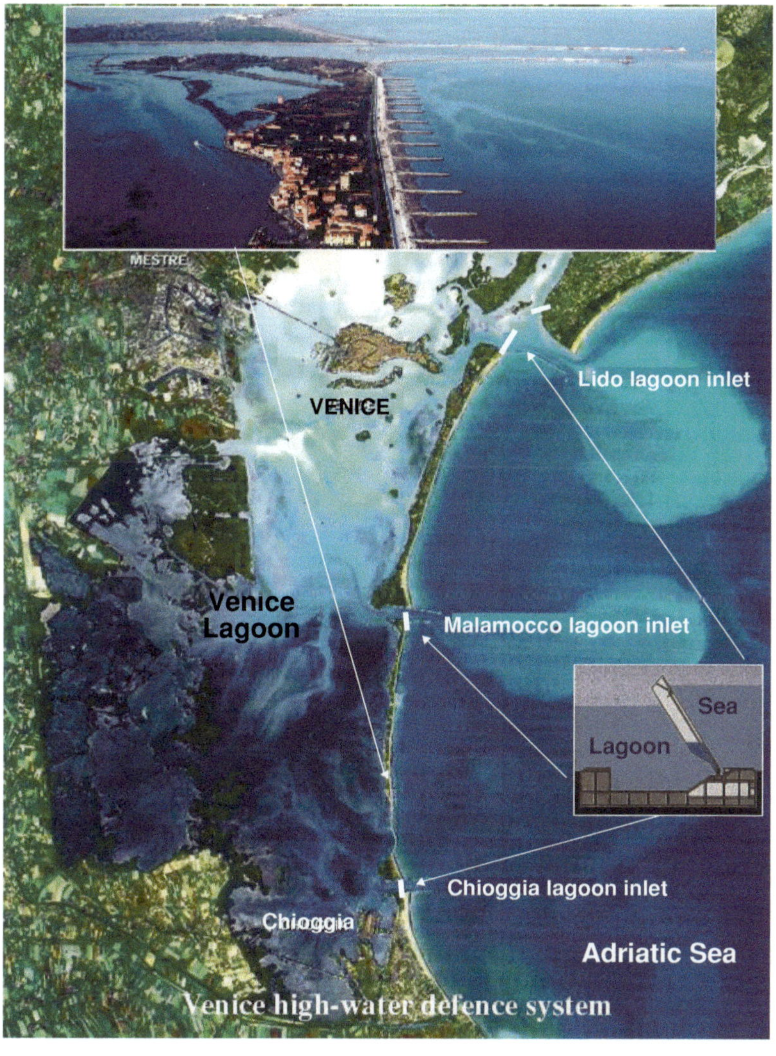

Figure 11.20. Aerial image of Venice Lagoon with position and view of Pellestrina Island.

11.3. PELLESTRINA

(Lamberti, Zanuttigh, Archetti, Marzetti, UB)

11.3.1. The site

The island of Pellestrina is the southernmost barrier dividing Venice Lagoon from the Adriatic sea; it is separated from the mainland by Chioggia lagoon inlet southwards and from Lido Island by Malamocco inlet northwards (Figure 11.20).

Pellestrina is about 13 800 m long in N-S direction and has a minimum width of 25 m and a maximum one of 210 m.

11.3.2. Environmental conditions

11.3.2.1. Bathymetry

Pellestrina littoral is characterised by a closure depth of 5 m. The average steepness of the beach is about 1:60 and becomes milder (1:90) southwards due to sedimentation caused by the maritime dike of Chioggia.

Natural grain size dimension between the shoreline and – 3 m depth is $D_{n50} = 0.175$ mm, with greater values northwards and finer southwards.

11.3.2.2. Winds

The major winds blowing in front of Pellestrina are: *Bora* (NE), which is the strongest in frequency and velocity during autumn and winter rising up to 70 knots; *Scirocco* (SE) that dominates during spring and summer with maximum intensity of 55 knots. Figure 11.21 shows the wind-rose for data acquired in about 15 years of measurements at the CNR tower.

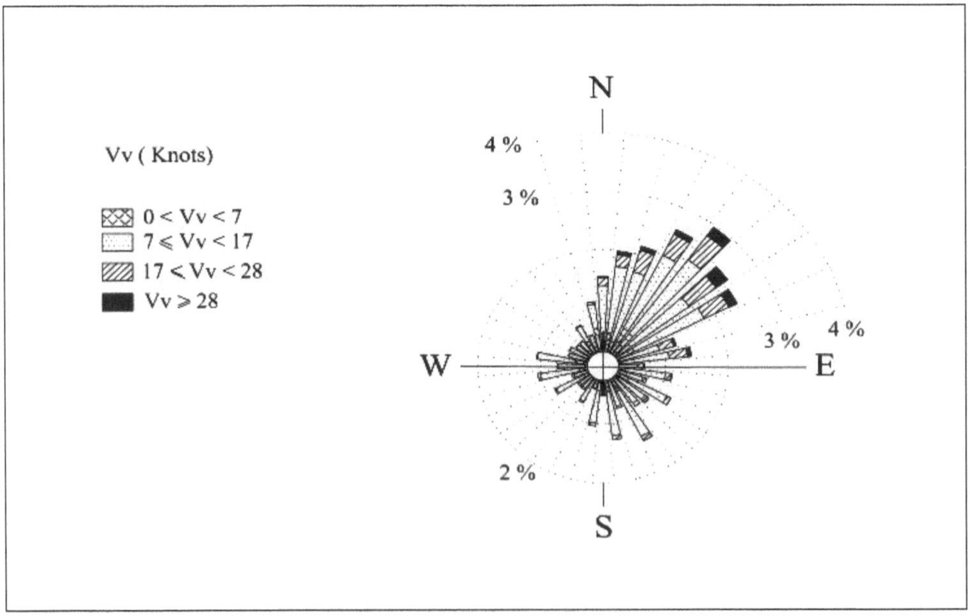

Figure 11.21. Wind rose at CNR tower (Venice). Period: October 1987- December 2002.

11.3.2.3. Waves

The most frequent waves are induced by *Scirocco* winds, come from 130°-140° N and reach in average 1 m; higher waves (up to 3 m high) come from 110°-120° N. The highest waves are due to *Bora*, come from 80°-90° N and rise up to 3.5 m.

The typical annual wave climate, expressed by significant wave heights and frequencies, is summarised in Table 11.7. Figure 11.22 shows the wave-rose for data acquired in about 15 years of measurements at the CNR tower.

11.3.2.4. Water level

Venice Lagoon is frequently flooded, particularly during winter, due to the phenomenon of *acqua alta*, which occurs whenever sea level exceeds 0.8 m above datum (– 0.23 m a.s.l).

Spring tidal range is about 1 m, however the highest water levels are due to storm surges caused by *Scirocco*.

The closed and narrow shape of the lagoon allows the rising of seiches that are usually characterised by an oscillation period slightly shorter than tide (11 and 22 hours).

11.3.2.5. Current system

The littoral current system is mainly driven by wind and waves, so wind coming from NE (*Bora*) leads to a southwards directed current whereas wind coming from SE (*Scirocco*) leads to a northwards directed current.

Table 11.7. Annual wave climate at CNR tower. Wave frequencies with varying wave height and direction.

	Significant Wave heights Hs [m]								
	0.125	0.375	0.75	1.25	1.75	2.25	2.75	3.25	0.25/4.0
Direction from North									
50°		0.45							0.45%
60°		0.54	0.30	0.30	0.10				1.24%
70°		0.64	2.10	1.30	0.60	0.20	0.10		4.94%
80°		0.64	3.50	1.40	0.80	0.40	0.10	0.30	7.14%
90°		1.74	1.60	0.70	0.30	0.20	0.10	0.10	4.74%
100°		0.85	1.80	0.70	0.30	0.10			3.75%
110°		0.97	1.80	0.60	0.20	0.10	0.05		3.72%
120°		1.44	2.00	0.80	0.30	0.10	0.05		4.69%
130°		5.05	3.50	0.90	0.30	0.10			8.85%
140°		5.31	3.40	0.90	0.30				9.91%
150°		0.84	0.30	0.10					1.24%
160°		0.44	0.10						0.54%
TOTAL	48.2%	18.9%	20.4%	7.70%	3.20%	1.20%	0.40%	0.40%	100%

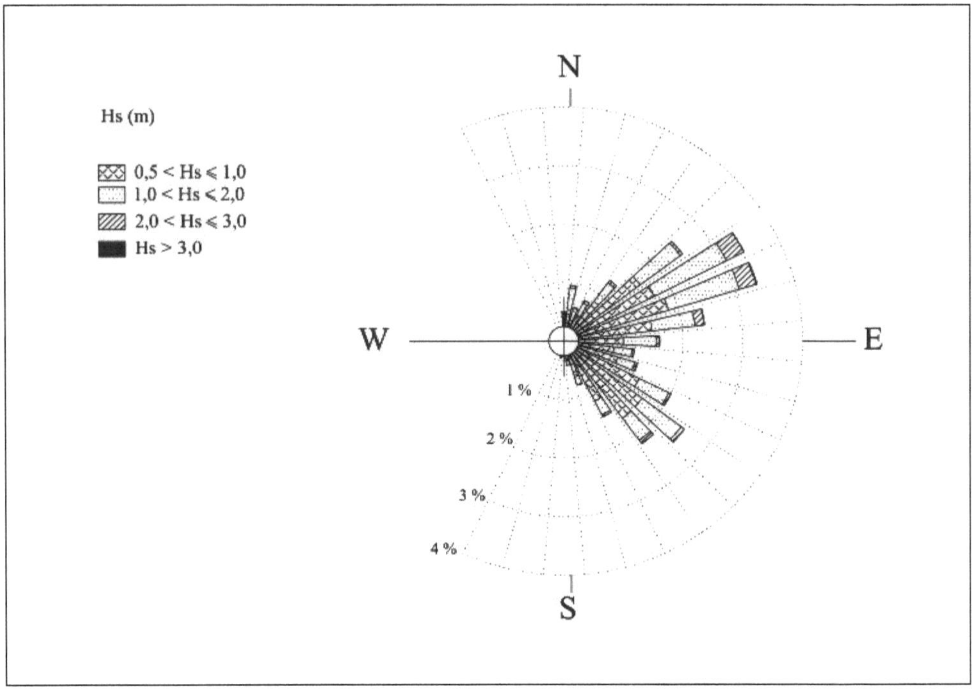

Figure 11.22. Wave rose at CNR tower (Venice). Period: October 1987-December 2002.

11.3.2.6. Sediment transport

Before the 1997 works, the littoral of Pellestrina was typified by a strong long-shore sediment transport of about 13 000-15 0000 m^3/year, directed from North to South, and by a significant cross-shore transport due to reflection caused by stone walls (Murazzi).

11.3.3. The defence scheme

Pellestrina is one of the most evocative examples of the combined effects of erosive wave forces and subsidence in absence of sediment feeding.

About 6 000 years ago, long submerged bars were formed and developed into bar islands separating Venice Lagoon from the Adriatic Sea.

After the Middle Age, several anthropogenic interventions (the diversion of rivers Brenta, Sile and Piave outside the lagoon and the protection of Chioggia, Malamocco and Lido inlets) caused an important loss of sediments.

The coastline was so seriously exposed to the risk of being submerged that the Venetian Water Authority decided in 1751 to construct the «Murazzi» system, which are 5 m high massive Istrian stone sea walls. This kind of defence reduced sea ingressions but did not stop erosion of the submerged beach (Figure 11.23).

«Murazzi» became inadequate around 1900, when long jetties, reaching depth around – 8 m, were built to defend the Malamocco and Chioggia inlets. These dikes interrupted the natural long-shore sediment transport provoking a small recession of the northern coastline and a strong accretion southwards.

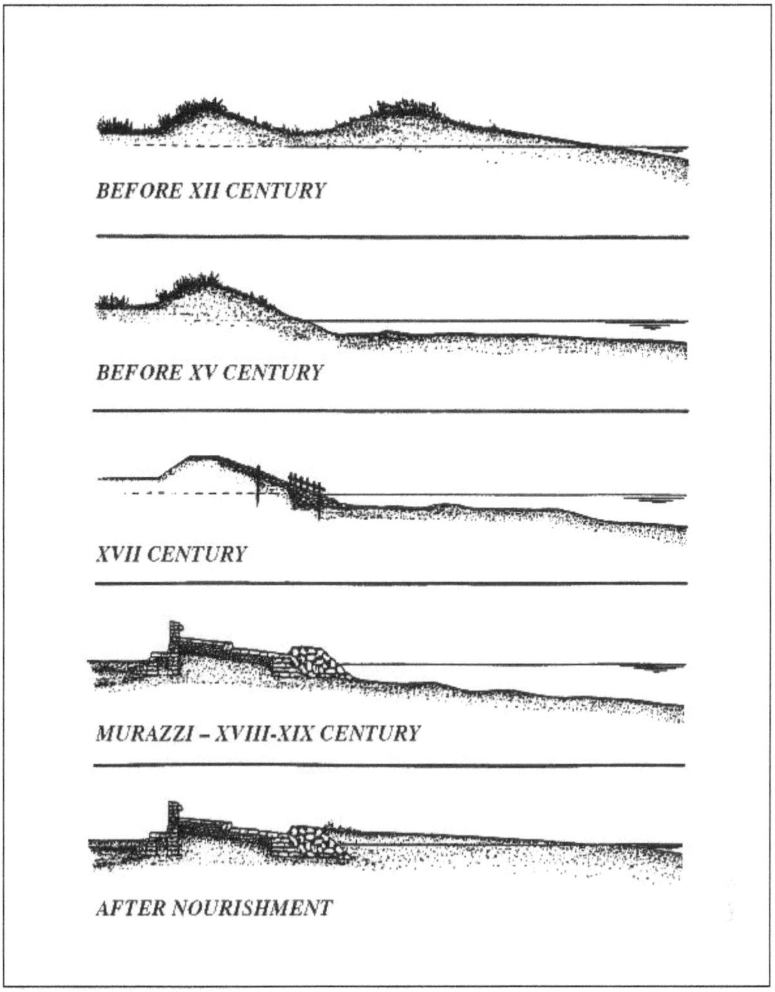

Figure 11.23. Historical evolution of Pellestrina cross-shore profile.

An exceptional storm surge in 1966 provided clear evidence of the fragility of Pellestrina defence system: severe overtopping occurred and the sea walls were damaged in several points. After this storm, the toe defence was significantly reinforced and a new protection system was finally designed.

The works done in Pellestrina in 1996-1998 (Brotto and La Terza 1996) were aimed at protecting the island from coastal erosion and, at the same time, at creating a sheltered wide beach.

The composite intervention covered 9 km and consisted of:

- a submerged barrier, parallel to the coastline, placed 290 m far from the shore on a – 4 m depth, with crest level – 1.5 m a.s.l; 50-500 kg stones compose the leeward side of the barrier and bigger 500-2 000 kg stones the seaward side, lying on a geotextile;
- 18 emerged groynes, forming 17 cells, each of which is 500 m long;
- 18 submerged groynes, 150-210 m long, that connect the barrier to the emerged groynes; the groynes are made of 50-500 kg stones lying on a geotextile;

- a nourishment performed with 4 600 000 m^3 of sand that is characterised by a $D_{n50} = 0.2$ mm and was dredged 20 km far from the littoral.

The plan view of the intervention is sketched in Figure 11.24; cross-sections of the barrier and of the groyne are shown in Figure 11.25.a and b respectively; view of Pellestrina before and after the composite intervention is presented in Figure 11.26.a and b respectively.

It can be noticed (Figure 11.26.b) the dark sand colour that produced immediately after the intervention a negative reaction of the residents; during the years, the sand colour has progressively become lighter because of sun exposure and residents have appreciated the presence of the beach to which they were not familiar at the beginning.

The construction of the beach created some problems to residents, due to sand transported by the wind inside houses and, more dangerous aspect, in the streets, requiring sometimes direct interventions to remove sand deposits. Tamarisks planted after the works between the beach and «murazzi» were obviously too small to actively retain sand; a successful solution was then found by placing fences on the beach (Figure 11.27) that will be removed when the plants have sufficiently grown.

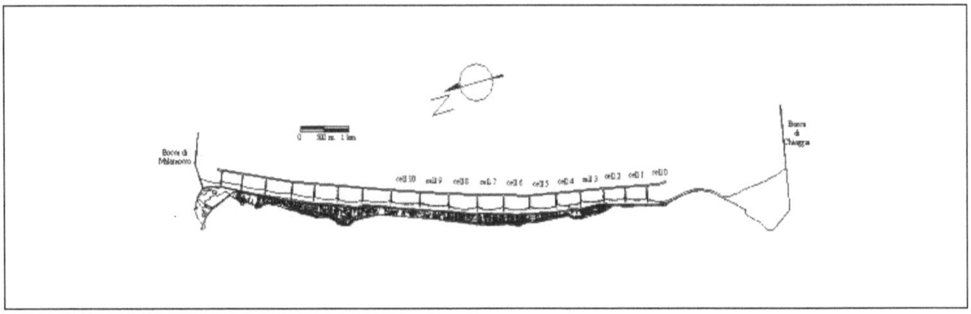

Figure 11.24. Plan view of Pellestrina defence scheme built in 1997-1998.

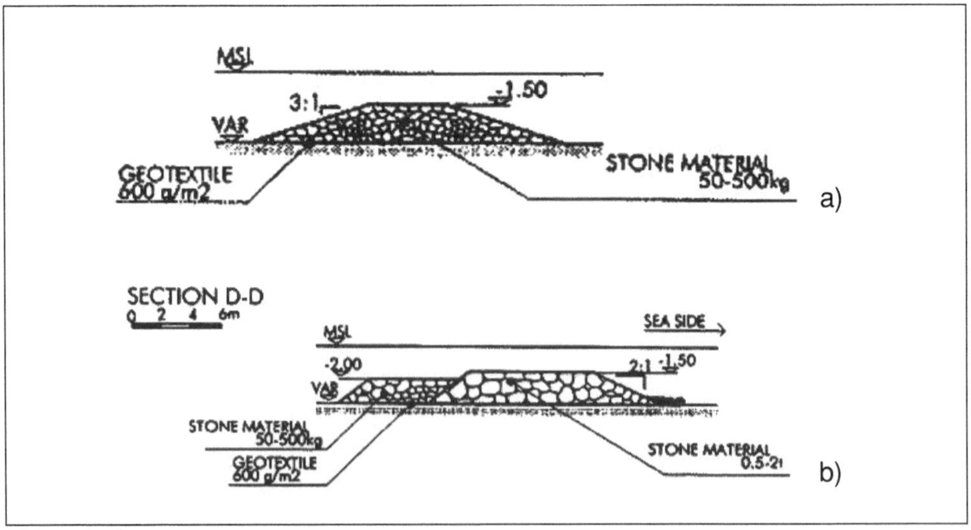

Figure 11.25. Cross-sections of the submerged connectors (a) and of the submerged barrier (b).

Figure 11.26. The littoral of Pellestrina at 1994 (a) and at 1999 after composite intervention (b).

Figure 11.27. View of Pellestrina beach, showing tamarisks and fences, from one of the emerged groynes.

11.3.4. Currents induced by the composite intervention

Interaction of the main current system with the composite intervention leads to formation of eddies and rip-currents at the roundheads of the emerged groynes, intense currents along the submerged barrier and adequate water mixing inside the protected area.

Numerical simulations with MIKE21 (Zyserman *et al.*, 2005) representing wave and current fields due to two typical wave attacks coming from NE and SE are shown in the Figures 11.28 and 11.29, at the right and left hand-side respectively. When looking at the results, an important aspect to account for is that waves generated by *Scirocco* is coupled with high tide (0.8 m a.s.l), whereas no tide is present under *Bora* conditions.

Both in presence of *Bora* and *Scirocco* (and of null and high tide) the submerged barrier works properly in reducing wave energy (Figure 11.28). In particular, for the *Scirocco* condition, waves are all breaking at the beach, whereas for the no-tide *Bora* condition waves are all breaking over the structure, producing a more variable wave agitation inside the protected area and close to the shore.

Inside the protected area a calm region develops with marked long-shore current along the shoreline. The maximum current intensities, both for *Bora* and *Scirocco* conditions, are reached along the submerged barrier and at the round-heads of the emerged-groynes (Figure 11.29), where long-shore currents, interacting almost perpendicularly with obstacles, generate rip-currents and/or vortexes.

It can be noticed that the long-shore current along the barrier is well-defined and parallel to the shore under the *Scirocco* attack, whereas, under *Bora* conditions, currents have more or less a sinusoidal shape, which induces a similar sinusoidal distribution of finer-coarser sediments in the protected cell as it was observed during field campaigns after *Bora* storms. Wave set-up increases with incident wave height, reaching 0.3-0.4 m a.s.l. for *Bora* attack (Figure 11.29).

11.3.5. Beach evolution after the composite intervention

The Consorzio Venezia Nuova (CVN) performed regular surveys twice a year for monitoring the bottom and the shoreline profiles.

Figure 11.30 presents the results of these surveys for the 9th cell, which was selected as representative of the defence system. A significant regression occurred immediately after the protected nourishment, in the years 1997-2000; the last surveys performed in the period 2000-2003 show a stable shoreline position.

Comparing the barrier profiles in 1997, immediately after the construction, and in 2000, it can be seen that the barrier crest level changed from – 1.5 m a.s.l to between – 1.8 and – 2.0 m, due to stone sinking and settlement (Figure 11.31).

Within DELOS, two detailed bathymetries with multi-beam system were performed in October 2002 in a representative cell (the 9th) and at the southern roundhead (Fig. 11.32). Figure 11.32.b shows that erosion occurred in the 9th cell landward of the barrier and close to the submerged connectors; a significant scour hole can be seen in the leeward of the roundhead (Fig. 11.32.a) and can be explained by the action of plunging breakers (Sumer *et al.*, 2004).

A field campaign carried out within DELOS, again during October 2002, showed that after nourishment the sediment grain size did not change in the sheltered area. This fact proves that the submerged barrier works properly in reducing the wave energy incident on the beach. The Skewness distribution in the 9th cell gives a concentration of negative values

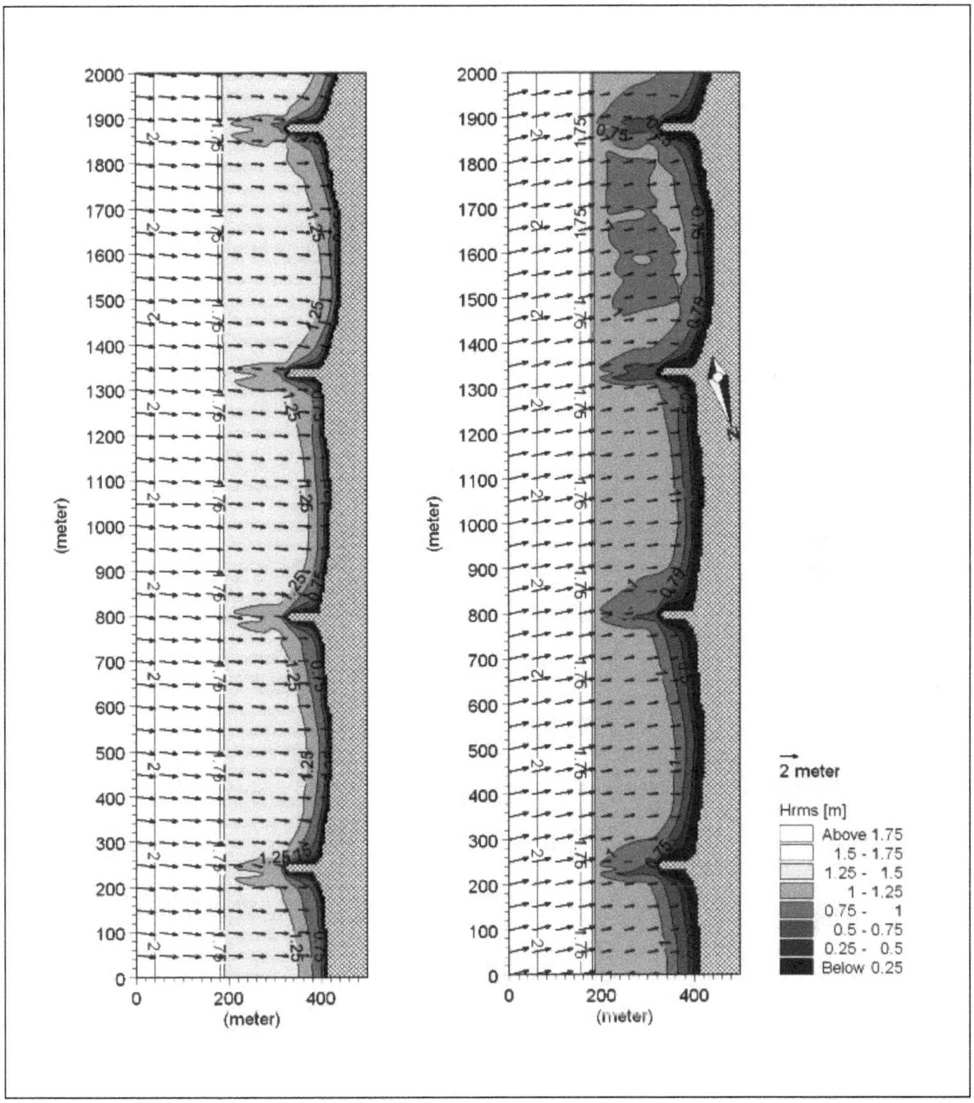

Figure 11.28. Wave intensity from MIKE21 PMS module: at the left hand-side a *Scirocco* wave attack (112°N, wave height 2.0 m, wave period 6.0 s, water level 0.8 m a.s.l), at the right hand-side a *Bora* attack (91°N, wave height 2.2 m, wave period 8.1 s, water level 0.0 m a.s.l). From Zyserman *et al.*, (2005).

only in correspondence of a central eroded zone leeward of the barrier, showing a general morphological equilibrium in the protected cell. The sediment transport analysis indicates that a very limited amount of nourished sediment is lost, which can be estimated less than 3% per year.

11.3.6. Ecological effects induced by the composite interventions

No ecological survey was performed during the project. The area inshore the barrier is presently under analysis, looking in particular at environmental restoration through sea-grass transplanting and fish breeding enhancement.

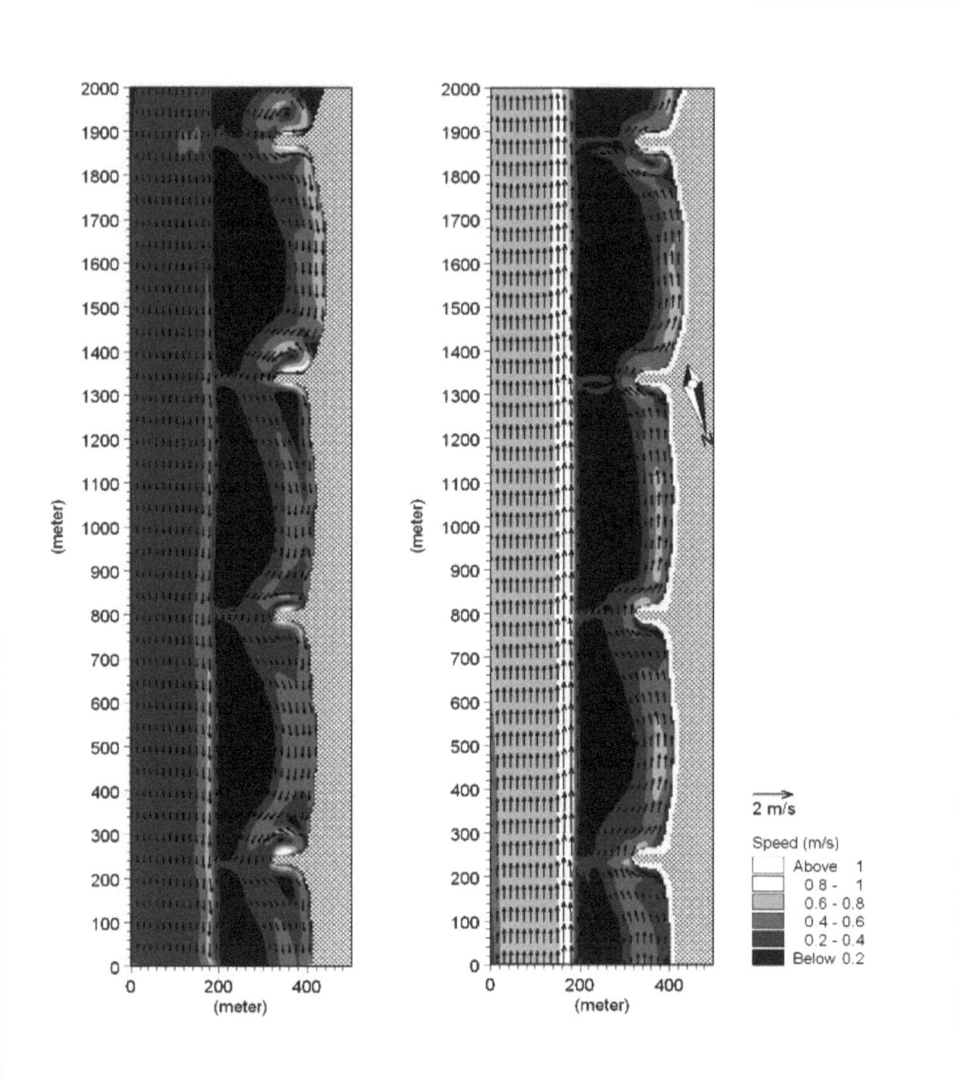

Figure 11.29. Current field obtained from MIKE21 HD module: scale colour represents surface elevation, vectors denote current speed; at the left hand-side a *Scirocco* wave attack (112°N, wave height 2.0 m, wave period 6.0 s, water level 0.8 m a.s.l), at the right hand-side a *Bora* attack (91°N, wave height 2.2 m, wave period 8.1 s, water level 0.0 m a.s.l). From Zyserman *et al.*, (2005).

11.3.7. Economic relevance of beach defence

Pellestrina's artificial beach is used for informal recreational activities such as sunbathing, walking, relaxing, swimming and so on. It is an undeveloped beach mainly used by residents and day-visitors. In summer 2002 an experimental Contingent Valuation Method (CVM) survey of 80 residents and 75 day-visitors was carried out with the purpose of evaluating non-marketable recreational benefits of the artificial beach in its present state (Marzetti, 2003a; Marzetti and Lamberti, 2003; Polomè *et al.*, 2004). The Value of Enjoyment (VOE) questionnaires of the *Yellow Manual* (Penning-Rowsell *et al.*, 1992) were adapted to the

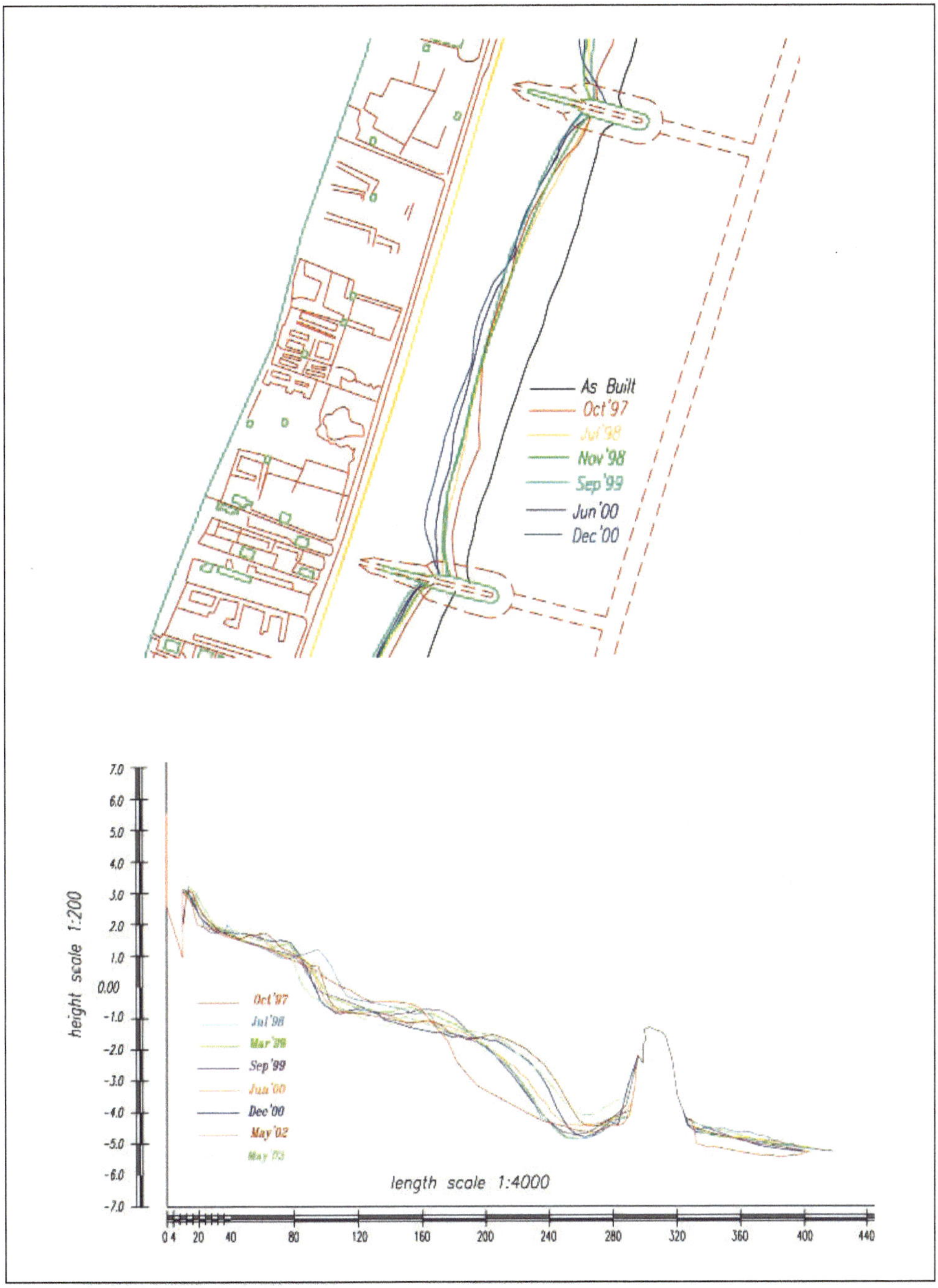

Figure 11.30. Shoreline and profile evolution in the 9th cell in the period 1997-2003.

characteristics of this site for estimating the recreational value of Pellestrina beach, and they were further developmend by asking the recreational use not only in spring/summer but also in autumn/winter.

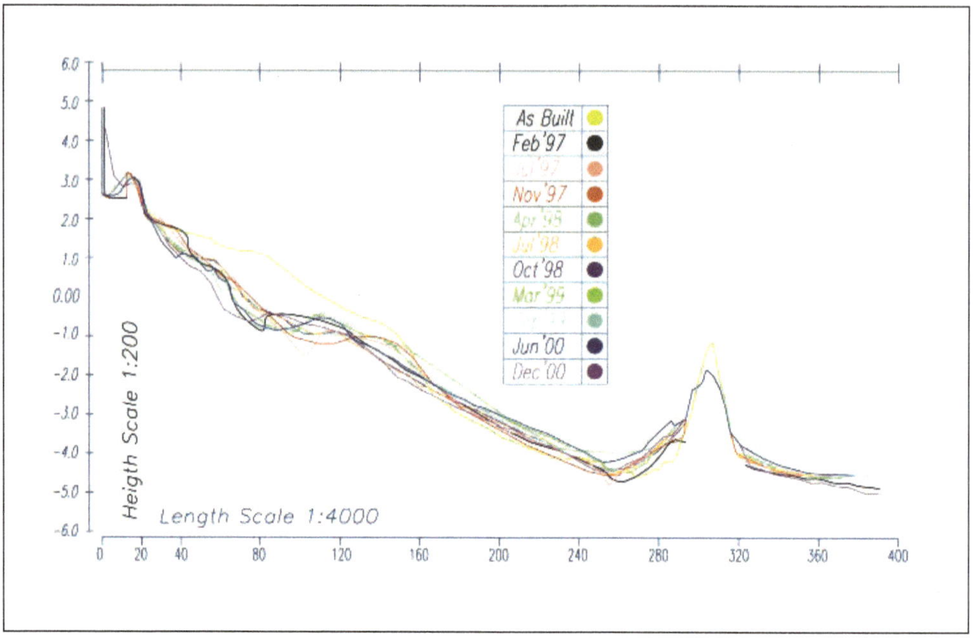

Figure 11.31. Barrier profile evolution for the 1st cell.

The Pellestrina survey results show that the great majority of respondents are in favour of the defence of the beach. Pellestrina beach is evaluated higher by residents than by day-visitors, both in spring/summer and in autumn/winter. In addition the beach use value is considerably lower in autumn/winter than in spring/summer. Therefore, it is generally appropriate to distinguish the recreational value according to the different seasons.

In order to obtain useful information for project researchers, questions about the preferences on the design of different defence structures and beach materials were added to the CVM questionnaire of day-visitors, as shown in Section 12.4.8.1, (Marzetti *et al.*, 2003; Polomè *et al.*, 2005; Marzetti and Lamberti, 2003). Of four different defence structures, the composite intervention (nourishment, groynes and submerged breakwaters), such as the defence works on Pellestrina Island is the most preferred. This preference was mainly justified by suitability for recreational activities and aesthetic reasons. In addition a medium-high level of preference was assigned to the fine sandy beach and groynes.

11.3.8. Conclusions

The composite intervention performed in 1997-1998 significantly reduces wave energy and thus currents induced in the protected area. The more intense currents occur along the submerged barrier and at the roundheads of the groynes, where, in presence of the highest waves, rip current or vortexes may form.

The defence system appears able to solve the erosional problems providing the formation of a stable beach. Sediment transport is strictly correlated to the hydrodynamic conditions and results partially blocked cross-shore by the submerged connectors and long-shore by the submerged barrier. Based on field surveys, the sedimentary budget presented an equilibrium trend, with an average erosion of about 3% per year.

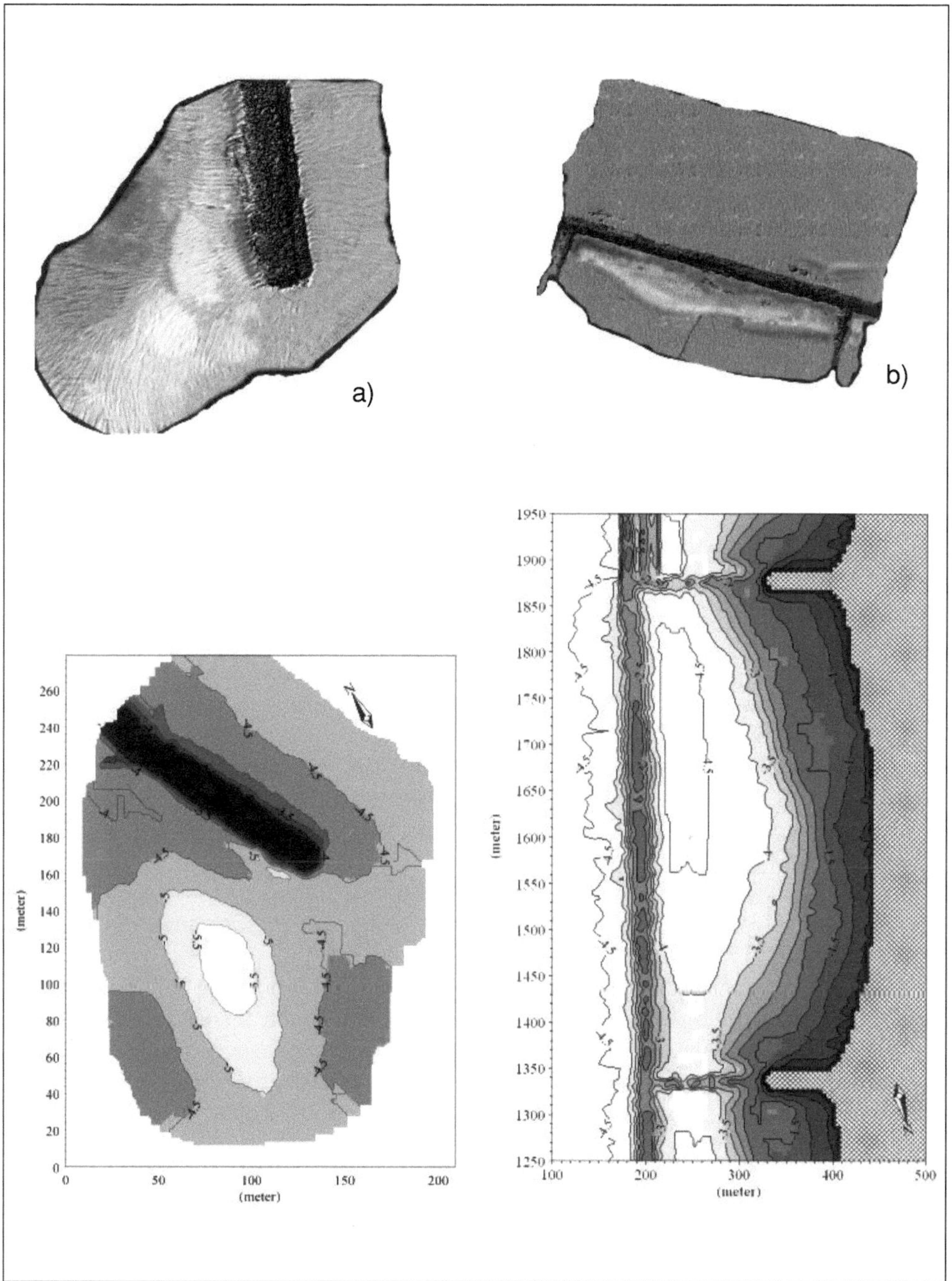

Figure 11.32. Detailed bathymetry of the roundhead (a) and of the 9th cell (b); multi-beam surveys performed within DELOS in October 2002.

Figure 11.33. Location of Lido di Dante.

The nourished beach plays an important role in defending Venice Lagoon from hight waters known as *acqua alta*.

All human activities in the zone benefit from the beach safeguard and the beach itself may promote tourist development.

11.4. LIDO DI DANTE

(Lamberti, Archetti, Zanuttigh UB; Airoldi, Bertasi, FF; Marzetti, UB)

11.4.1. The site

Lido di Dante (Lamberti and Zanuttigh, 2005) is a seaside resort in the Emilia-Romagna coast, 7 km far from the city of Ravenna. It is located in the area between Fiumi Uniti to the North Bevano River to the South (Fig. 11.33).

The beach in front of Lido di Dante is about 1 300 meters long and has a surface of about 70 000 square meters. It is classified as a dissipative beach characterised by a sandy, flat and wide surf zone; it presents a concave shape of the cross-shore profile with orientation NW-SE. It is still possible to find some dunes in the back of the beach. Nowadays this system is pretty narrow due to the development of tourist facilities and erosion problems.

11.4.2. Environmental conditions

Lido di Dante is part of a wide coastal area undergoing erosion problems whose causes started around the 1950s.

Erosion has both natural and anthropogenic origins. Land subsidence is one of the main causes: the youth (geologically speaking) of the sediments which characterize the Pianura Padana together with underground water and gas extraction enhanced this process. Low rates of sediment transport associated with the location of Lido di Dante, near to a closed estuarine river mouth, do not allow the natural support of sand to preserve a constant beach width. Furthermore, the tourist development has modified the natural dynamics of the beach.

The area can be divided into two parts: the norhern beach (almost 600 m long) was subjected to great erosion and therefore it has been protected by groyne, nourishment and a semi-submerged breakwater; instead the southern beach is undergoing only slight erosion and is in a very natural state.

11.4.2.1. Bathymetry

The seabed has a quite gentle slope reaching about 6 m/km, whereas the slope decreases offshore, where is of about 0.96 m/km. The mean sediment diameter varies from 0.20 mm near the shoreline to 0.08 mm at a depth of 6 m.

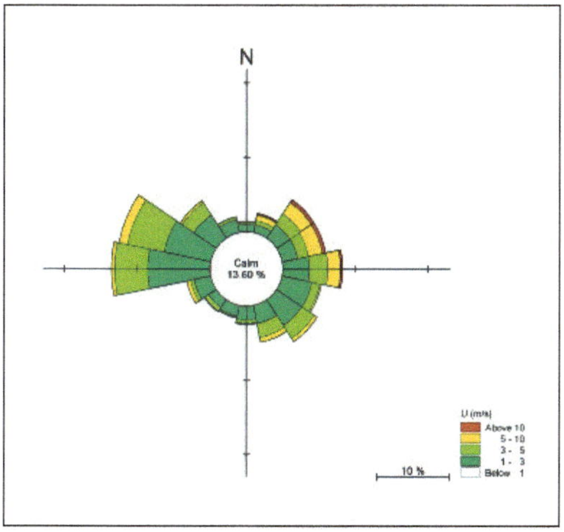

Figure 11.34. Wind rose in Porto Corsini. Period: June 2002-December 2003.

11.4.2.2. Winds

The strongest winds occur during winter (more than 24 knots) from NW-N-NE; summer, on the other hand, is characterized by a high frequency of southern winds. The different distribution and intensity of the winds are due to the different dimension of the fetch area characterising the two main wind directions. In this area another important wind, coming from land (S-W), *Libeccio*, creates some effects, but it is not relevant on the littoral area. A wind rose representative of more than one year measurements is given in Figure 11.34.

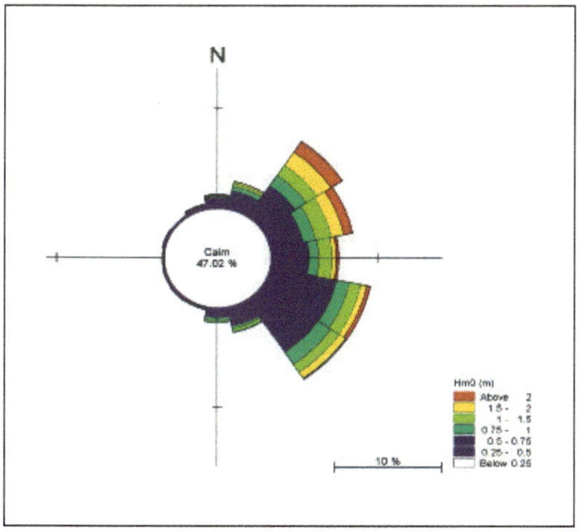

Figure 11.35. Wave climate offshore Lido di Dante. Period: June 2002-December 2003.

11.4.2.3. Waves

A set of wave data, from wave gauges installed on the offshore structures of the gas supply company AGIP is available and provides an important source of data collection. Data cover the period 1 January 1992-31 December 2000.

The most frequent storms come from *Scirocco* (S-E) but the strongest ones come from *Bora* (N-E). The analysis of measurements carried out in the period 1996-2002 shows that waves reach 3.5 m average height every year and around 6 m every 100 years.

A wave rose representative of more than one year measurements is given in Figure 11.35.

11.4.2.4. Water level

Two principal effects cause variation in water level: astronomical tide, reaching 80 cm range at spring tide and 30 cm at neap tide, and storm surge that is more relevant in North Adriatic. Currents generated by these processes are estimated to be ~ 0.05 m/s, one order lower than wave-generated currents.

High water level in the North Adriatic is due to the effects of storm surge contemporary to a high astronomical tide. Winds blowing from South, South-East (*Scirocco*) are responsible of the exceptional high water level, known as *Acqua Alta*. Statistics of the extreme high water level and extreme low water level based on a

Table 11.8. Statistics of annual extreme water level (Idroser, 1996).

	HW (cm)	LW (cm)
Mean annual extreme	84.2	– 75.6
Standard Deviation	9.8	7.0

time series of 54 years (1934-1989) are given in Table 11.8.

11.4.2.5. Current system

The littoral current system is mainly driven by wind and waves, so wind coming from NE (*Bora*) leads to a South directed current whereas wind coming from SE (*Scirocco*) leads to a North-directed current.

Prevalent offshore currents due to tidal residuals and wind are directed Southwards, transporting fine sediments from Po and Reno rivers that have built the low and silt-bed that can be found 1 km from the shoreline.

11.4.2.6. Sediment transport

The study area is characterized by sand transport diverging from the Fiumi Uniti outlet at the scale of littoral morphology, whereas northward directed sediment transport prevails near the shoreline.

11.4.3. The defence scheme

The submerged breakwater is part of a more complex project realized in 1995 mainly to solve the problem of the extensive erosion.

The defence of the littoral is composed of:

- three groins, the first was built at the northern site in 1978 and two were constructed 300 m and 600 m south from the first in 1983;
- a parallel submerged breakwater 770 m long placed at 180 m from the coast on a 3.5 m depth, interrupted by a surface opening 30 m wide and 1 m deep from the LCS crest level (– 0.5 m);
- 2 submerged groynes linking the emerged groynes head to the barrier (1995);
- beach nourishment using sand with D_{n50} = 0.23 mm; 60 000 m^3 in 1993 and 74 400 m^3 in 1996.

The plan view of the shore protection system at Lido di Dante in 1995 is sketched in Figure 11.36 and the typical cross section of LCS is presented in Figure 11.37.

In 2001 the following works were performed, in order to increase the efficiency of the shore protection system:

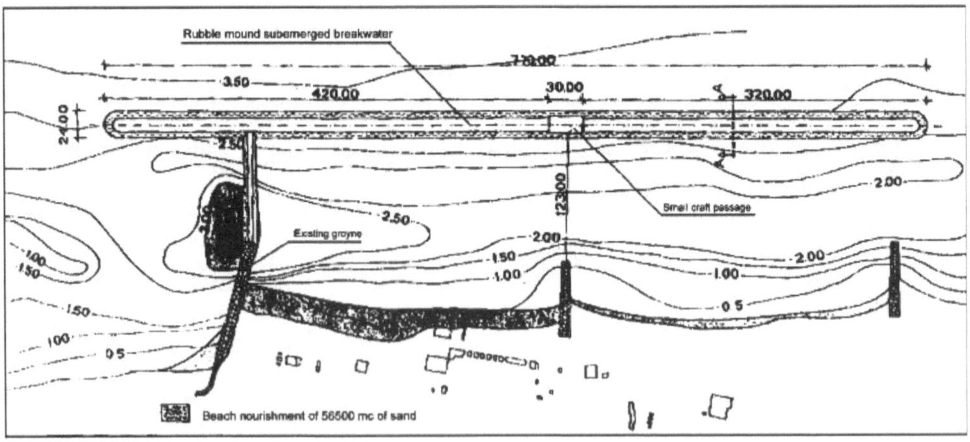

Figure 11.36. Sketch of the littoral defence constructed in 1995.

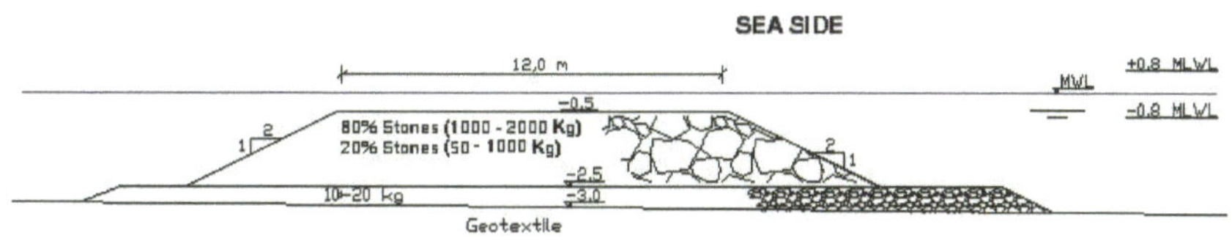

Figure 11.37. LCS cross section at Lido di Dante.

- increase of the crest height of the barrier by one stone layer, approximately 0.80 m;
- construction of a submerged groin 120 m long connecting the southern groyne to the barrier;
- scour protection of LCS roundheads;
- protection of the central gap.

Figure 11.38 shows the aerial view of Lido di Dante, with indication of all the works carried out from 1978 till 2001.

In June 2003, maintenance works supported by the local council (Comune di Ravenna) were performed. The works are:

- placement of stones on the crest of the barrier, to contrast structure settlement;
- increase of the submerged transects crest to the SWL;
- protection of the central gap.

In the actual lay out, the freeboard of the LCS and of the two boundary groynes is emerged approximately 20 cm a. s. l.

Figure 11.38. Aerial view of Lido di Dante after the 2001 works.

11.4.4. Currents induced by LCS

Interaction of the main current system with the LCS and groyne system of the area leads to formation of eddy circulation at both heads of the LCS, and rip-current towards the gap in the middle of the LCS (Zyserman *et al.*, 2005). Due to this current pattern, several changes in the bottom morphology occurred since the LCS was built, above all erosion at both heads of the structure caused by the eddy circulation.

Numerical simulations carried out with MIKE21 represent in the Figures 11.39 and 11.40 the typical wave and current system during a wave attack from SE in presence of different works.

Comparison of the flow fields in Figure 11.40 to the surveys (to be discussed in the next section) of June and October 2001 (see Figure 11.41) indicates that the crescent-shaped erosion holes observed around the southern roundhead can be linked to blocking of the northward longshore current by the southern connector under *Scirocco* wave events. The presence of the barrier further concentrates the deflected flow at its roundhead, which results in locally increased transport capacity and consequent erosion. Eddy formation and flow concentration behind the northern roundhead of the barrier under *Scirocco* waves results in far-field erosion shaped as shown in Figure 4 for the surveys of June and October 2001. The analysis and these surveys are consistent with the fact that *Scirocco* is predominant during spring and summer.

In a similar way, erosion patterns like the ones shown in Figure 11.41 for the June 2002 and January 2004 surveys can be linked to *Bora* events, which predominate during autumn and winter.

The generalised erosion observed behind the submerged barrier in the June 2001 bathymetry can be linked to the significant flow that existed behind the structure before construction of the southern connector. This flow accelerated towards the northernmost groin, resulting in increasing transport capacity and erosion leeward of the barrier. Figure 11.40 shows that the current behind the barrier was largely eliminated following construction of the southernmost connector, which in turn reduced the erosional trend along the protected beach, as shown in Figure 11.41. Following recharge of the barrier, return flow is concentrated at the gap, which explains the erosion shown in Figure 11.41 for the January 2004 survey.

Currents have been monitored in Lido di Dante (Drei *et al.*, 2001; Archetti *et al.*, 2003) using an ADCP, which provides the punctual (Eulerian) measurements of waves and currents, and dropping floating drifters (Langragian) at the edge of the study area and following their patterns with several techniques. Both methodologies appeared to be essential to obtain a reliable representation of velocity fields and for the calibration of the numerical models.

11.4.5. Beach evolution

From a very detailed bathymetry carried out with the multi-beam system (June 2002), a deeply eroded area at about 70 m from the two roundheads is recognisable. This is due to the strong vortexes that are induced at the roundheads during strong storms from *Scirocco* (at the southern roundhead) and *Bora* (at the northern roundhead).

In Figure 11.41 (the bathymetry in 2004), the erosion at the heads and at the gap is more evident. It is interesting also the accumulation on the seaward side of the LCS.

The topography and bathymetry of the site have been monitored before and after the construction of the structures, in order to study the changes and the evolution of the beach. Figure 11.41 compares four bathymetries carried out in the years 2001-2004.

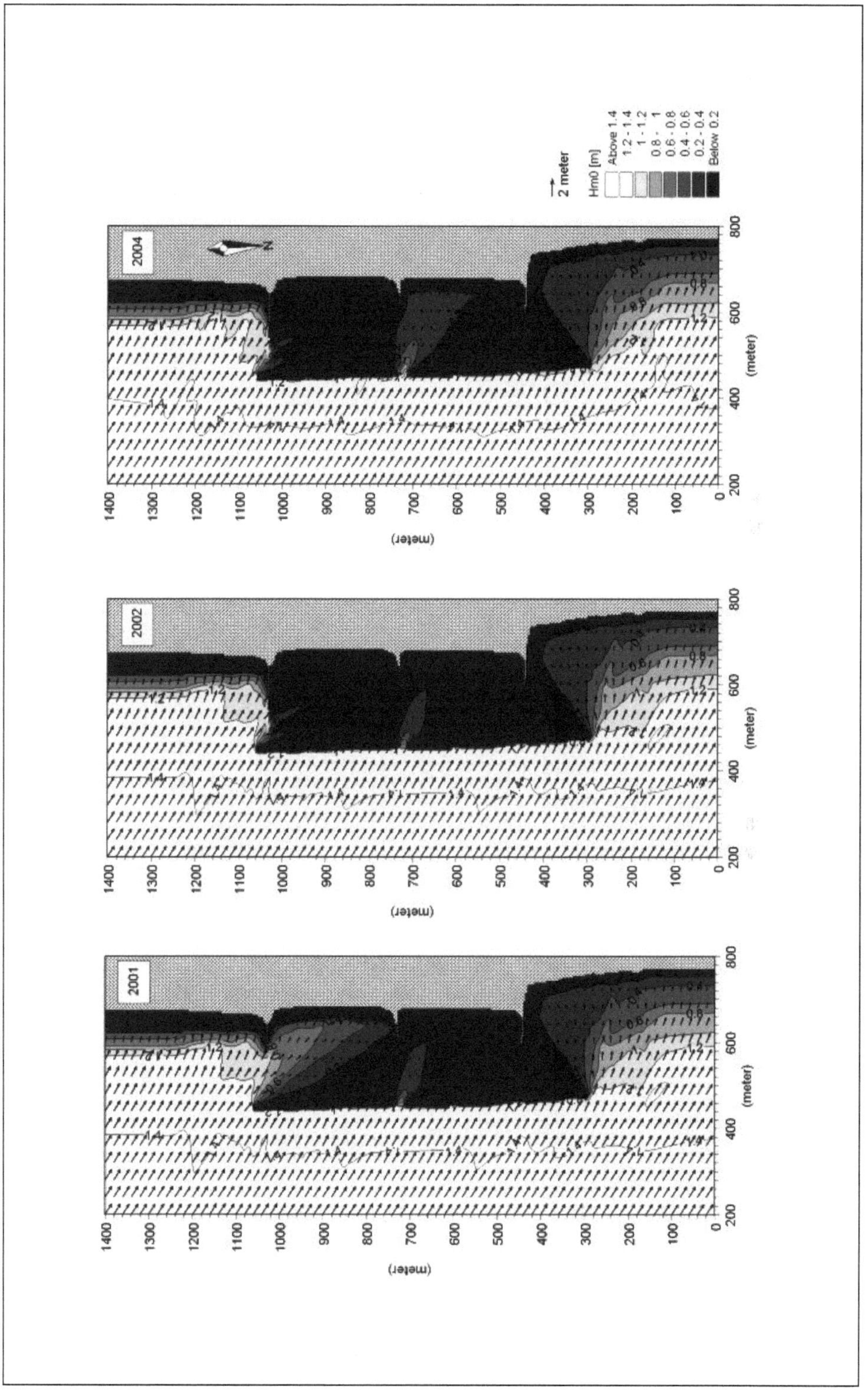

Figure 11.39. Waves due to a 1.5 m high wave coming from SE (135°N), with period 8.05. Results of simulations carried out with MIKE21 on different bathymetry. From Zyserman *et al.* (2005).

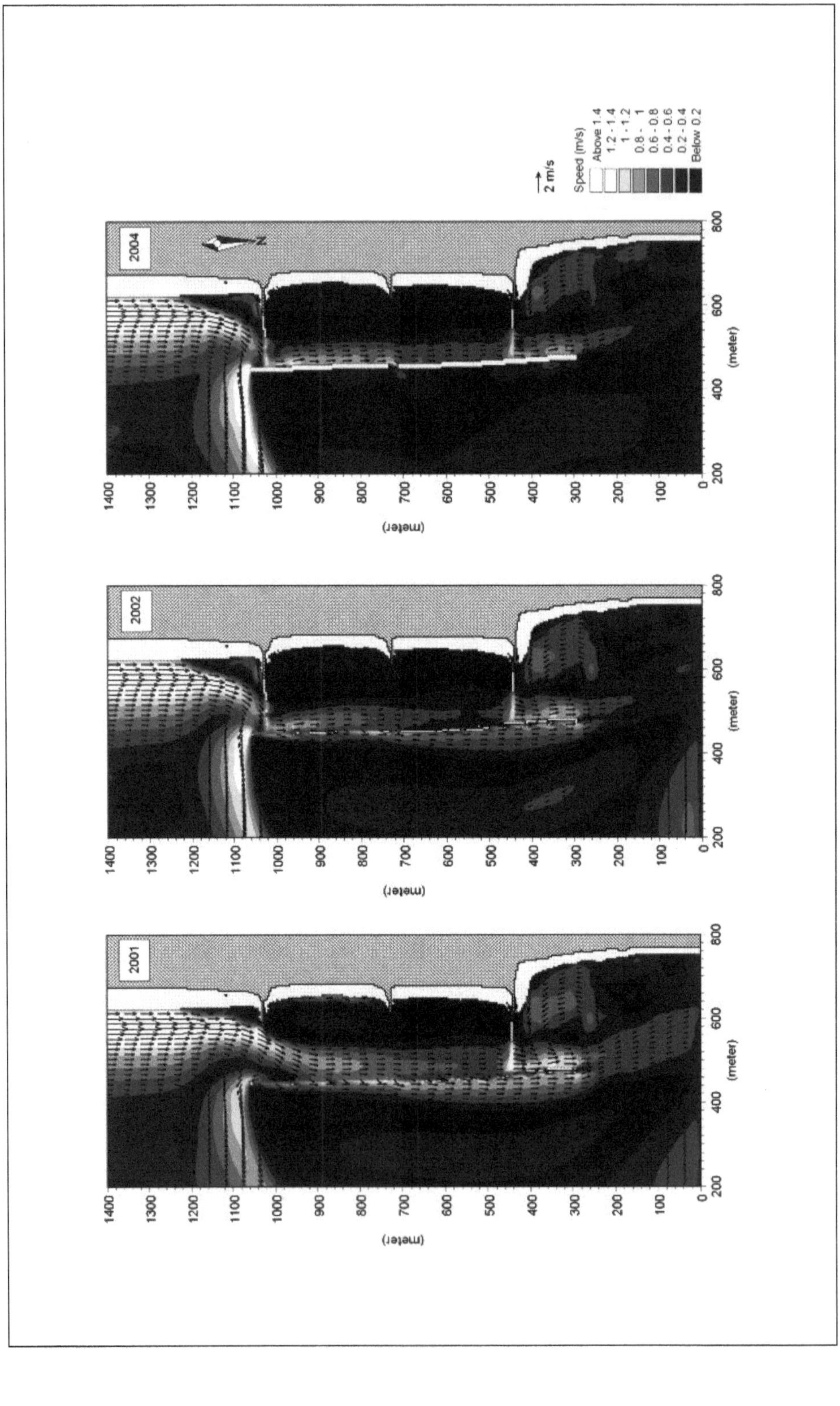

Figure 11.40. Currents induced by a 1.5 m high wave MIKE21 coming from SE (135°N) with period 8.03, as above. From Zyserman *et al.* (2005).

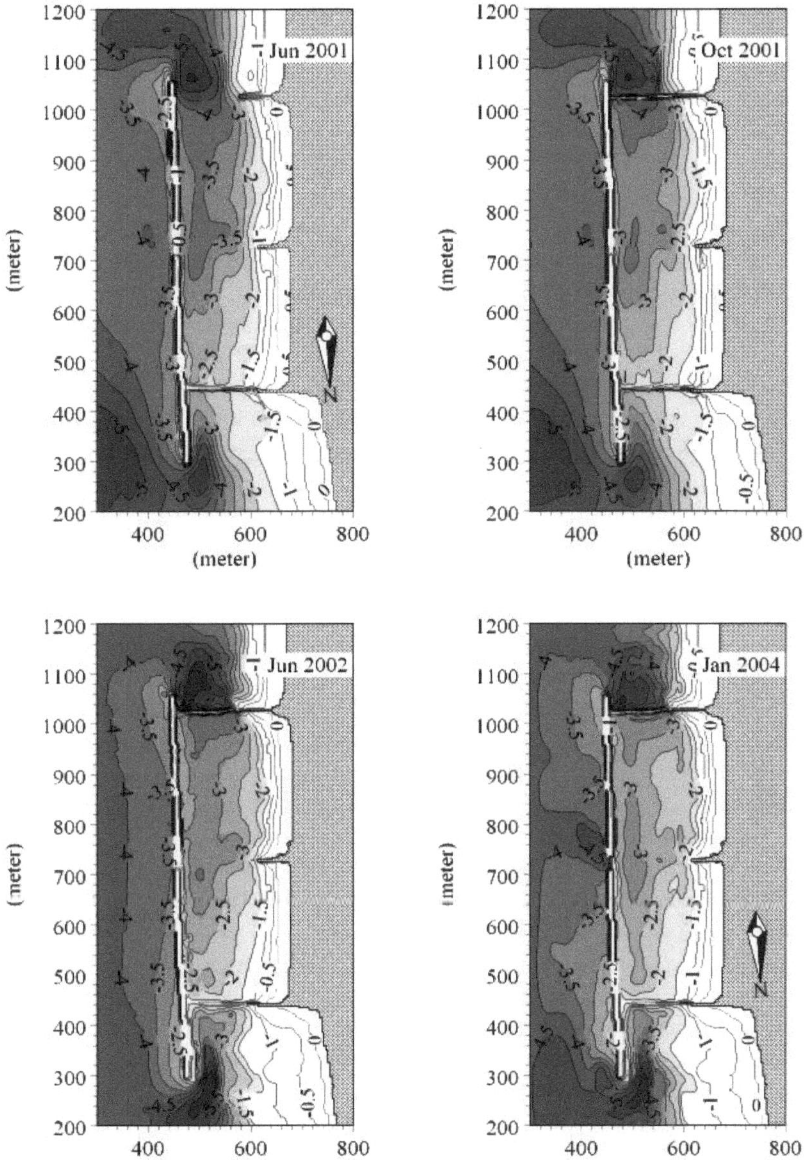

Figure 11.41. Bathymetry maps derived from surveys carried out, from top to bottom and from left to right: in June 2001, October 2001, June 2002 (multi-beam) and January 2004 (multi-beam). From Zyserman *et al.* (2005).

Two bathymetric surveys performed in 2001 are presented, one month before (Figure 11.41, top, left-handside) and three months after (Figure 11.41, top, right-hand-side) the works carried out in July 2001. The bathymetry of June 2001 shows an intense erosive process at the North of the protected area, with a shoreline retreat between 12 and 17 m; an erosive trend is clear also at the Southern beach. Inside the protected area, the behaviour of the northern and the Southern cell is different. The Northern cell seems to be in equilibrium, erosion inside the barrier stopped and minor sedimentation took place, whereas the southern cell is still under erosion due to the high currents flowing between the barrier and the shore. From the second survey of October 2001, it appears that the southern submerged groyne works properly in the reduction of the erosion trend in the southern cell; this is confirmed by also by the following surveys in 2002 and 2004 (bottom of Figure 11.41).

From the first multi-beam bathymetry carried out in June 2002 (Figure 11.41, bottom, left-handside) a deep erosion at the two barrier roundheads is recognisable, due to the vortexes induced during strong storms from *Scirocco* at the southern and *Bora* at the northern roundhead.

From the bathymetry carried out on January 2004 (Figure 11.41, bottom, right-hand-side), a reduction of erosive trend is visible in the northern hole and in the southern cell, while erosion is increasing seaward the central gap that is now the only way for water inside the cell to flow offshore. Inside the protected cells, the sedimentation process is appreciable.

11.4.6. Ecological effects induced by LCSs

From an ecological viewpoint, the Italian coast of the north Adriatic Sea represents a particularly interesting case study, both for the environmental peculiarities of the area (a sandy flat coastal system almost uninterrupted except for one isolated rocky promontory, the mouths of rivers, channels and lagoon systems and for human-made structures) and for the dramatic proliferation of LCSs and groynes that has affected the whole coastline. The ecological consequences of these constructions can be seen on a local scale at each single site, but have also propagated up to affecting coastal assemblages at a regional scale (Bacchiocchi & Airoldi, 2003). For this reason, the analysis of the ecological implications of LCSs cannot be restricted to the site of the case study alone, but needs to be expanded to cover the whole geographical area (i.e. about 400 km of coast from Trieste south to Ancona).

The main ecological consequences of the construction of LCSs along the Italian coast of the north Adriatic sea can be summarized as follows.

1) The loss of natural soft bottom habitats and associated assemblages of animals and plants as a direct consequence of the construction of LCSs. Although the surface covered by any individual structure or schemes of structures is limited, in some areas, such as the coasts of Emilia Romagna, construction of LCSs has affected over 60% of the natural landscape in intertidal and shallow subtidal habitats. Thus the losses sum up to a significant surface-area.

2) Changes in the surrounding soft bottom habitats and associated assemblages as a consequence of the reduction of wave energy on the landward side of LCSs and in some cases of the enhanced sediment loads due to beach nourishment. Specifically for Lido di Dante, such alterations have directly influenced the characteristics of the sedimentary habitats (i.e. grain size, percentage of silt/clay, content of organic matter). This has resulted in changes in the composition and/or abundance of animal assemblages living in the sediments. These changes reflected both the peculiarity of the benthic biocenoses typical of the North Adriatic Sea and the design of the defence scheme at Lido di Dante, where the presence of groynes in addition to the LCS creates a water enclosure that approximates to a lagoon system (this

is also indicated by the presence of species typical of lagoon habitats (e.g. *Musculista senhousia*, *Hediste diversicolor*), coupled with large numbers of opportunistic worms (e.g. *Capitella capitata*, *Spio decoratus*) and species from deeper waters (e.g. *Corbula gibba*, *Owenia fusiformis*) as a consequence of an increased abundance of fine sediments and reduced water flow on the landward side of the LCS). Overall, natural communities inhabiting the surf zone of the Adriatic coast were relatively species poor. Conversely, a more structured community, characterized by a higher richness and diversity of species than the natural assemblages, was present on the landward side of the LCS up to the shoreline (Figure 11.42).

3) The extensive introduction of LCSs, providing hard substrata as well as sheltered habitats, has considerably changed the identity, abundance and distribution of hard bottom species within the region. LCSs have become colonised by animals and plants that are typical of natural rocky coasts. The composition and distribution of these assemblages is largely influenced by location of the LCSs (with a trend of increasing species richness from North to South) and by the orientation within structures. Overall, assemblages on LCSs, and particularly along the coasts of Emilia Romagna, were structurally simple, dominated by only few species and with a large amount of unoccupied space (Figure 11.43). Most LCSs in this region were colonised by extensive beds of the mussel *Mytilus galloprovincialis*. This species is also abundant in coastal lagoons of the region as well as on other types of artificial coastal structures (Ceccherelli & Rossi, 1984; Bombace *et al.*, 1995; Relini *et al.*, 1998), and is a species intensively harvested. Green ephemeral algae (i.e. *Ulva* spp.), that are also common in coastal lagoons, and filamentous algae were the only other abundant species, and their growth is a major problem for local tourism, as these fragile algae are torn off of the LCSs and washed up along the shoreline, where they accumulate and begin to decompose. The accompanying smell reduces the amenity value of the beaches, hence they need to be periodically removed. Although LCSs were colonised by rocky bottom species, the assemblages differed from those on nearby natural reefs (Figure 11.44), and their composition was not related to the age of the LCSs. These differences are probably related to the frequent disturbances of LCSs by maintenance works. Maintenance of structures by adding new blocks to the crest has dramatic effects on epibiota, effectively reducing biodiversity to an early stage of succession, with few species compared to that on structures which have not been maintained, and favouring the development of green ephemeral algae.

4) Considering large scale effects, the proliferation of LCSs, by providing extensive new habitats for colonisation of rocky-shore species, has allowed the dispersal of hard bottom species using LCS as stepping stones to areas where they would not previously have occurred because of the unavailability of suitable natural habitats and failure to disperse. One of the consequences of these corridor effects is that in this area LCSs and other man-made structures have acted as a vector for the spread of exotic species (e.g. *Codium fragile* ssp. *Tomentosoides*). Spatially explicit population dynamic models have been developed that predict the rates and pathways of dispersal and persistence of hard-bottom species that can result as a consequence of the proliferation of LCSs over large stretches of coasts. The model developed for the sedentary gastropod *Patella caerulea* showed approximately 60% occupation of the available habitat for this particular species in this region. Changing the spatial distribution of habitat patches by either adding or removing breakwaters is bound to change the dynamics. Adding virtual breakwaters to the area between Cesenatico and Lido di Savio has in principle little effect more than increasing the proportion of occupied habitat. Removing breakwaters show non-linear results depending on the specific location of each breakwater.

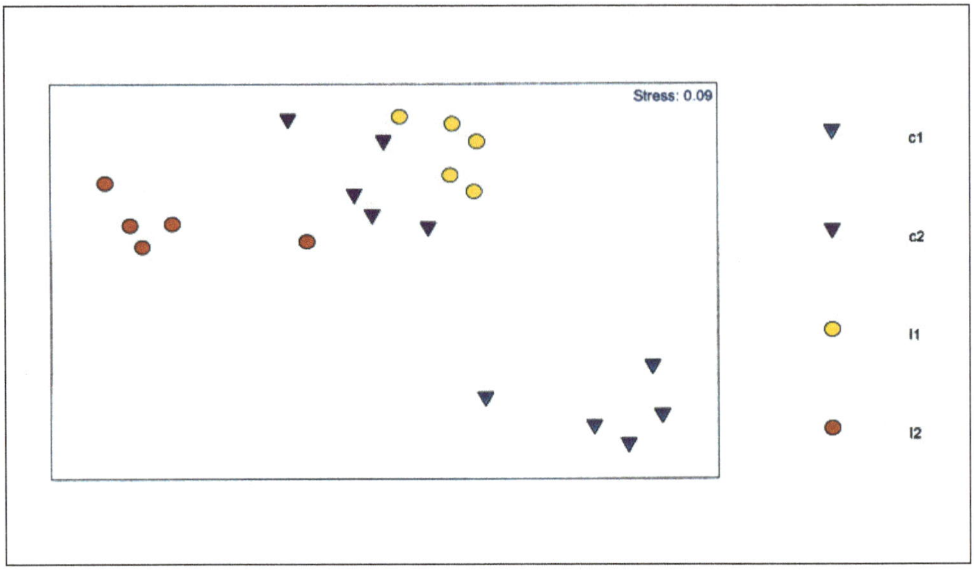

Figure 11.42. nMDS plot of macrobenthic communities based on fourth-root transformed abundance data C = control site; L = landward site; S = seaward site; 1 = 1.0 m depth; 2.5 = 2.5 m depth.

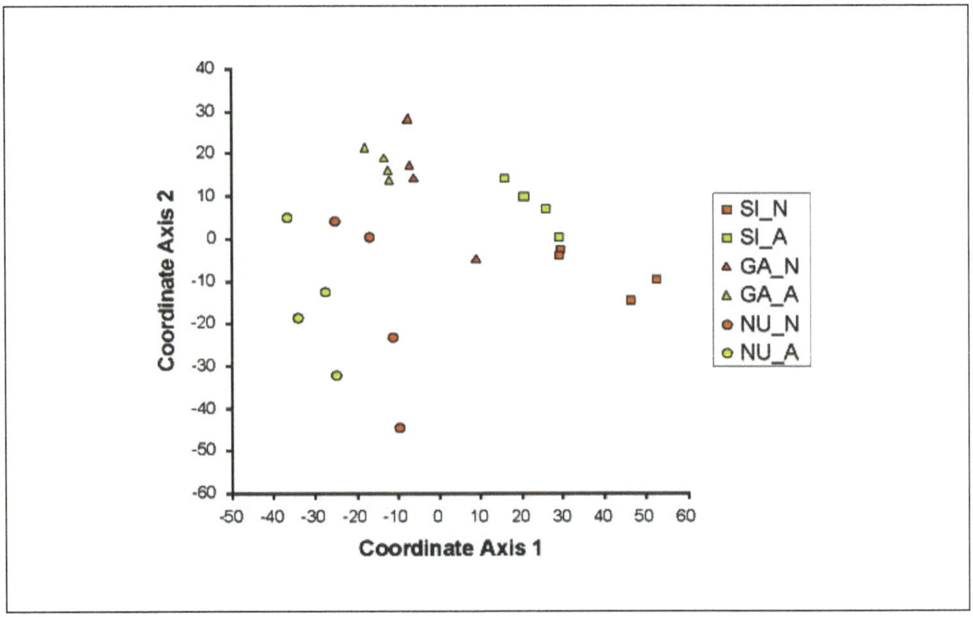

Figure 11.43. Plot of the analysis of the principal coordinates (PCCOORD, or metric MDS) showing centroids of areas sampled on coastal defence structures (labelled A) and on natural rocky reefs (labelled N) at 3 different locations along the Italian shores of the North Adriatic sea (SI = Sistiana, GA = Gabicce and NU = Numana). Results show how assemblages on natural reefs and defence works were notably different at each of the 3 locations.

Overall, two main considerations emerge from the evaluation of the ecological impacts of LCSs along the shores of the North Adriatic sea:

Figure 11.44. Benthic assemblages growing on the LCS at Lido di Dante.

1) For any new LCS introduced into the marine environment it will take time for the biological assemblage to reach a stable climax community that is most likely to resemble that of a natural shores. For mature biological communities to develop, LCSs need to be stable and built in such a way that maintenance will be minimal. Unless LCSs meet these criteria, there is little point in introducing additional features to enhance diversity, as attempts to repair the structure will result in considerable degradation of developing communities.

2) The Italian coast of the North Adriatic Sea represents an example of poor management particularly at a regional scale. By piecemeal local defence interventions, planned without an overall consideration of the regional environmental conditions, erosion problems have been extended to other parts of coast, and in some cases have magnified the original problem which defence works set out to resolve. The proliferation of defence structures has substantially changed the identity and nature of the coastal landscape of this region. Only by taking an holistic approach, and treating the whole coast as a natural unit, can successful management ever occur.

11.4.7. Economic relevance of beach defence

The Lido di Dante beach is visited by local residents, day-visitors and tourists mainly for informal recreational activities. Tourism is well developed and foreign tourists are numerous, mainly attracted by the natural state of the Southern beach. Within the Cost-Benefit Analysis framework, a contingent valuation method (CVM) survey of 600 interviews was carried out in Summer 2002 (Marzetti *et al.*, 2003a; Marzetti and Zanuttigh, 2003; Polomè *et al.*, 2005) which main aims were i) to estimate the Value of Enjoyment (VOE) of a daily visit to the beach in the status quo, after a hypothetical erosion of the beach, and after a hypothetical protection of the beach; ii) to find out whether in these two hypothetical situations of the beach respondents would change their number of visits and would go to another beach.

The basic structure of the VOE questionnaire used for the Lido di Dante case-study is the standard site user questionnaire published in the *Yellow Manual* (Penning-Rowsell *et al.*,1992). It was adapted to the specific characteristics of this site, and innovated by including specific questions about the VOE in autumn/winter.

The results show that, compared with the mean economic value of the present beach state, in spring/summer the change in the mean value of enjoyment due to erosion is considerable (from 27.67 to 13.26 €), while there is little change as regards the situation of protection (from 27.67 to 28.37 €). In particular, as regards the different areas of the Lido di Dante beach, the undeveloped or natural area is also evaluated highly in the hypothetical situation of erosion. Foreigners were also interviewed, and the majority of them elicit higher values than Italian visitors. In addition, the daily use value of the Lido di Dante beach in the low season is considerably lower than in the high season, justifying in this way the seasonal distinction of the beach use value for this beach. These results, contingent to the specific scenarios described in the Lido di Dante survey, confirm the conviction that beach visitors are very sensitive to the defence of beaches from erosion: not only is the daily reduction of enjoyment for the hypothetical situation of erosion fairly high, but also the percentage of visitors who would reduce the number of visits because of erosion is high; while, in condition of beach protection very few respondents would go to another beach (Marzetti, 2003a).

Finally, in order to design defence projects which also satisfy beach visitors' preferences, some specific questions about respondents' opinion on four different defence projects were included in the survey questionnaire. Among the different defence techniques, respondents prefer the composite intervention, consisting of nourishment, groynes and submerged breakwaters; aesthetic reasons mainly justify their preference (Marzetti *et al.*, 2003; Marzetti and Zanuttigh, 2003).

11.4.8. Conclusions

The protection built in Lido di Dante in 1995 seems to have produced benefit almost only to the northern cell, while the southern cell and littoral remained exposed to the erosive power of currents induced by overtopping and flowing out the southern gap. Current circulation around the structures was active and complicated, causing a strong mixing of water, erosion near the roundheads and apparently a positive effect on water quality.

The rocky barrier induced a change of assemblages in the area, increasing biodiversity of the littoral zone. Wave energy reduction in the protected area and the higher sediment loads due to periodic beach nourishment have directly influenced the characteristics (composition and/or abundance of assemblages) of the sedimentary habitats.

The construction of the southern submerged connector in 2001 has produced the stabilisation of beach bottom inside the northern cell and a progressive sedimentation in the southern one. Water mixing appears sufficient to guarantee an adequate water quality.

A contingent valuation method survey on beach visitors, carried out during summer 2002, showed that the users did appreciate the beach defence system in use at the time.

The latest evolution of the beach defence, mainly due to the strong pressure exerted by the owners of the bathing facilities, has produced an almost complete closure of the system in 2003, which has perhaps reached an excessive defence level. The water enclosure, approximating to a lagoon system, presently affects both water quality and habitat characteristics.

11.5. OSTIA

(Franco, UR3; Marzetti, UB)

11.5.1. Introduction

The shallow (1% slope) sandy beaches of Lido di Ostia stretch along the southern delta cusp of the river Tiber, some 25 km from Rome on the Tyrrhenian Sea, and represent a very popular holiday resort for the Roman community for a long time. It is exposed to waves from West to South (Figure 11.45). The tidal range is very small (+/– 0.2 m) with setup up to 0.5 m. The depth of closure is 7.0 m MSL.

The cuspated delta was formed by alluvial sediments carried by the river, producing a progressive coastline advance of more than 4 km from the Roman age until the last century. Then, particularly in the last 35 years, a severe erosional process has been taking place reverting the evolution trend to a recession rate of 1.7 m/year. The main cause has been the strong reduction of river sediment supply (due to upstream dams and extraction of building material from the river bed) with a consequent deficit in the coastal budget and a trend towards the cusp straightening and smoothing out, due to the gradient of alongshore sediment transport to the southeast. Coastal protection works, such as the system of detached breakwaters constructed near the river mouth, have shifted erosion downdrift, mainly affecting the southern beach between the Vittoria Pier and the Pescatori Canal, causing damage to the beach clubs and to the littoral road during storm events.

11.5.2. The perched beach project

An innovative beach nourishment project was then designed in 1988 by the competent Authority, the Office of Civil Engineers for Maritime Works of Rome (Ministry of Public Works) with the support of 2D model stability tests and one-line shoreline evolution modelling, both performed at DH (Ferrante *et al.* 1993). The aim of the project was to recreate a wide protective beach with an efficient technical defence solution complying with the economical, managing, political and environmental requirements. In fact the local community rejected any traditional emerging coastal structure to favour tourism, aesthetics and ecology. Indeed the project represented a new approach of the administration toward a global view in coastal defence, also taking account the environmental aspects. Given the existing high deficit of the littoral sand budget, the proposed beach nourishment needed to be protected by some coastal structure able to dissipate part of the wave energy and reduce the littoral transport, and to retain the new fill material. The most suitable solution then included an offshore underwater rock barrier fixing the natural dynamics sandy bar, as a *perched beach* scheme. The submerged bar should hold the artificial beach at a shallower slope, reducing both offshore sand losses and longshore transport, enhancing the development of marine fauna, without endangering bathing and leisure navigation. Important constraints were also resulting from the scarcity of marine sand for nourishment. The dark native beach sediments have a fine grain size with D_{50} = 0.15-0.3 mm. Fill material needed to be quarried inland on the alluvial Tiber delta at 20 km distance from the beach: the available material is a poorly sorted mix of well rounded sands and gravels. The works were carried out in 1990 by means of land-based equipment.

The protection scheme covers a beach length of 2.8 km and basically consists of:

- a **sill** made with a submerged rubble mound parallel to the shoreline at a distance of some 150 m, with toe level at about MSL – 4.0 m, a 15 m wide crest berm at – 1.5 m, seaward slope of 1:5, a multilayer rock mound (maximum stone weight of 1 t) placed above a geotextile and a 5 m wide rock toe protection in a 1 m deep trench. The material required was about 300 000 m^3 of rock (basalt and limestone from different quarries). The barrier crest was actually built at – 1.8 m MSL and settled rapidly at – 2.0 m MSL and later at – 2.3 m MSL.
- A **fill** with a double layer of quarry material: a lower layer of mixed sand with grading of 0.08-120 mm, and a 1 m thick upper layer of yellowish sand with grading 0.3-1.3 mm; the underlayer also acts as a 5 m thick filter between the sand and the rock bar; the beach equilibrium slope is 2.5% and the berm crest located at MSL + 1.0 m. The average design shoreline advance is about 60 m. The material quantities were about 1 360 000 m^3 of sand

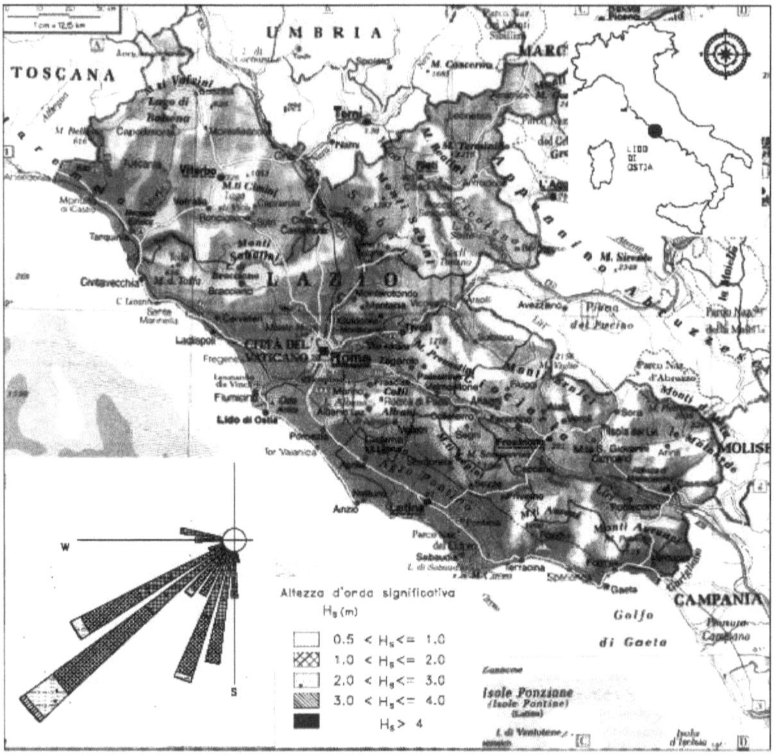

Figure 11.45. Lido di Ostia location and wave climate (from Ferrante *et al.* 1993).

and selected mixed sandy-gravel.

Later on additional works were performed as described in Figure 11.46. In 1998 a 235 000 m^3 beach nourishment (D_{50} = 0.2 mm supplied from land quarries) was placed from Repubbliche Marinare Way to Lido (1 220 m), in 2000 a new 70 000 m^3 sand backpassing fill (dredged from Pescatori Canal inlet) was added onto the beach from Magellano Square to Belsito (680 m), in 2003 further 366 000 m^3 beach nourishment (grey fine sands from offshore quarries) were delivered from Vittoria Pier to Belsito. Also maintenance works have been made by 1-3 t rock recharging over the barrier along partial stretches (2001 and 2003/4) raising the crest up to – 1.0 and – 0.5 m MSL.

11.5.3. Monitoring programme

Given the innovation of this technical solution and the unusual length of nourished beach without groynes, the Supreme Council of the Ministry of Public Works attributed an experimental character to the works and imposed the setup of a monitoring programme since the construction start in 1990. More recently the monitoring surveys are carried out by the Centro di Monitoraggio of the Osservatorio dei Litorali of Regione Lazio now in charge of the coastal defences. The periodic acquisition of field data includes: aerial photographs, beach profile surveys, sediment sample analysis and, just for first 3 years, directional wave recordings (see Table 11.9).

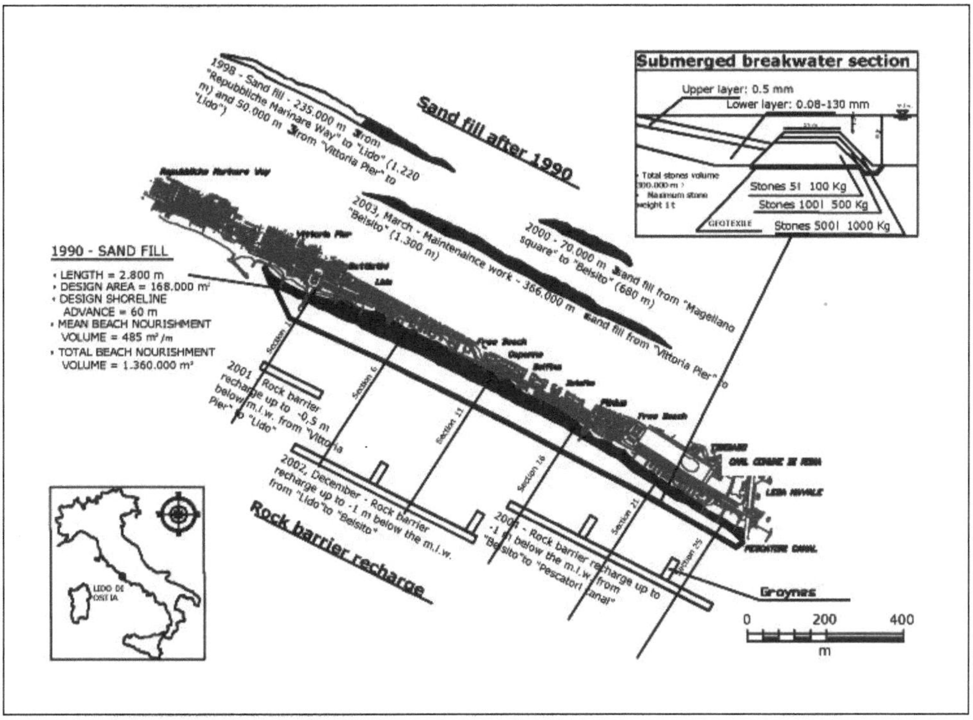

Figure 11.46. Planimetric view and submerged breakwater section scheme.

11.5.4. Analysis and observations on beach morphology and rock mound

Figure 11.47 shows the aerial photo of 1944 with superimposed retreated shorelines of 1955 and 1967 and a double bar system under the transparent water (at 70 and 300 m distance from shore).

Historically reconstructed shorelines have been diachronically analysed to derive the aerial variations of the emerged beach compared to the 1944 reference situation (Figure 11.48). Before the 1990 works the 2.8 km long dry beach had lost nearly 60 000 m^2 as compared to the 1944 condition. After the works of 1990 an erosion rate of some 16 m^2/m was observed in the next 8 years.

The analysis of the topographical beach surveys has shown a marked rotation of the shoreline with shoreline advance (at southern end) and retreat (at northern end), due to the southbound littoral drift. In 2003, after the last fill, the emerged beach area is almost equal to that of 1944.

Historical beach profiles were compared for 6 representative sections at 500 m spacing (Figure 11.49), where the rock barrier position is also indicated. The disappearance of the offshore bar is noted.

Volumetric computations carried out with Beach Morphology Analysis Package (BMAP) by Coastal Engineering Research Center (CERC) show the beach reduction in the first period 1992-96 with an erosion peak of 234 m^3/m at p11 (Figure 11.50), while accretion obviously occurred after additional recent fills, particularly at the downdrift sections (due to the expected deposit against the Canal groyne) and at the most updrift section (due to the LCS raising at – 0.5 m MSL).

Table 11.9. Summary of work and monitoring activities.

Year	*Works*	*Beach Profile survey*	*Shoreline survey*	*Grain size data*
1944			RAF photo (may)	
1955			IGM (photo)	
1967			SARA photo (april)	
1990	Construction of the submerged breakwater up to –1.8 m below m.l.w. and 1 300 000 m^3 beach nourishment from «Vittoria Pier» to «Pescatori Canal» (2 700 m)		CTR	Design data
1992		May to –4 m	RILTER photo	
1994		July to –4 m	VOLO ITALIA	
1995		September to –7 m	Foto RILTER	28 sections (each 100 m). Samples at elev. +1; 0; –1; both barrier toes
1996		February to –8 m	AIMA	
1997				
1998	235 000 m^3 beach nourishment from «Repubbliche Marinare Way» to «Lido» (1 220 m)		CGR	
1999			SIDRA photo	
2000	70 000 m^3 beach nourishment form Magellano square to Belsito (680 m)	October to –10	Aeroplane Photo (May)	
2001	Submerged breakwater rock recharge up to –0.5 m below m.l.w. from «Vittoria Pier» to «Lido» (340 m)		CM (July) local survey (October)	7 sect. samples at el. 0; 2.5; 5;7.5 m
2002	Submerged breakwater rock recharge up to –1 below m.l.w. from «Lido» to «Belsito» (1 000 m)	May to –10 m December to –10 m	AGEA photo	7 sect. samples at el. 0; 2.5; 5;7.5m
2003	366 000 m^3 beach nourishment from «Vittoria Pier» to «Belsito» (1 300 m)	February. to –10 m May to –10 m	Satellite photo	
2004	Submerged breakwater rock recharge up to –1 m.l.w. from Belsito to Pescatori Canal (1 150 m)			

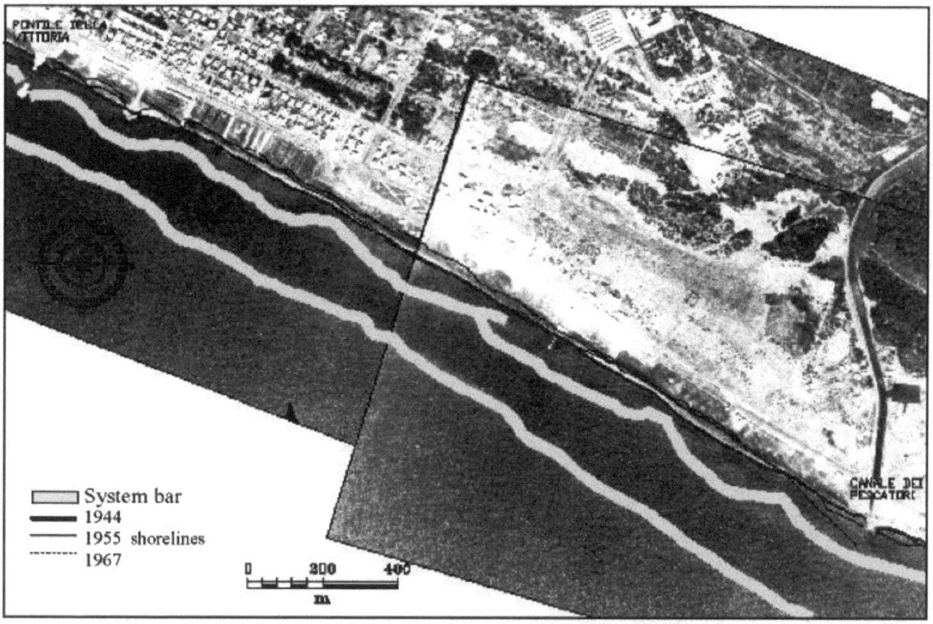

Figure 11.47. 1944 photograph with bar system and 1944, 1955, 1967 shorelines.

Grain size analysis confirms the migration of sands both offshore and downdrift, only reduced after the rock recharges of the submerged barrier. The LCS has been reshaped in time by both settlements and wave action, with an average crest lowering of 0.5 m in a decade. A computation of the actual damage was made by comparing negative differences (eroded areas)

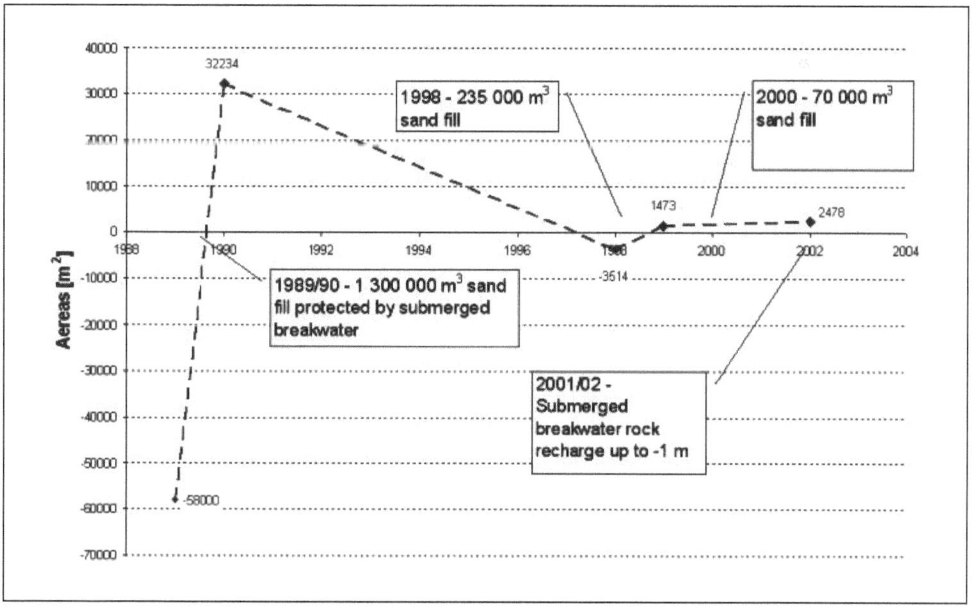

Figure 11.48. Temporal evolution of the beach area with respect to 1944 situation.

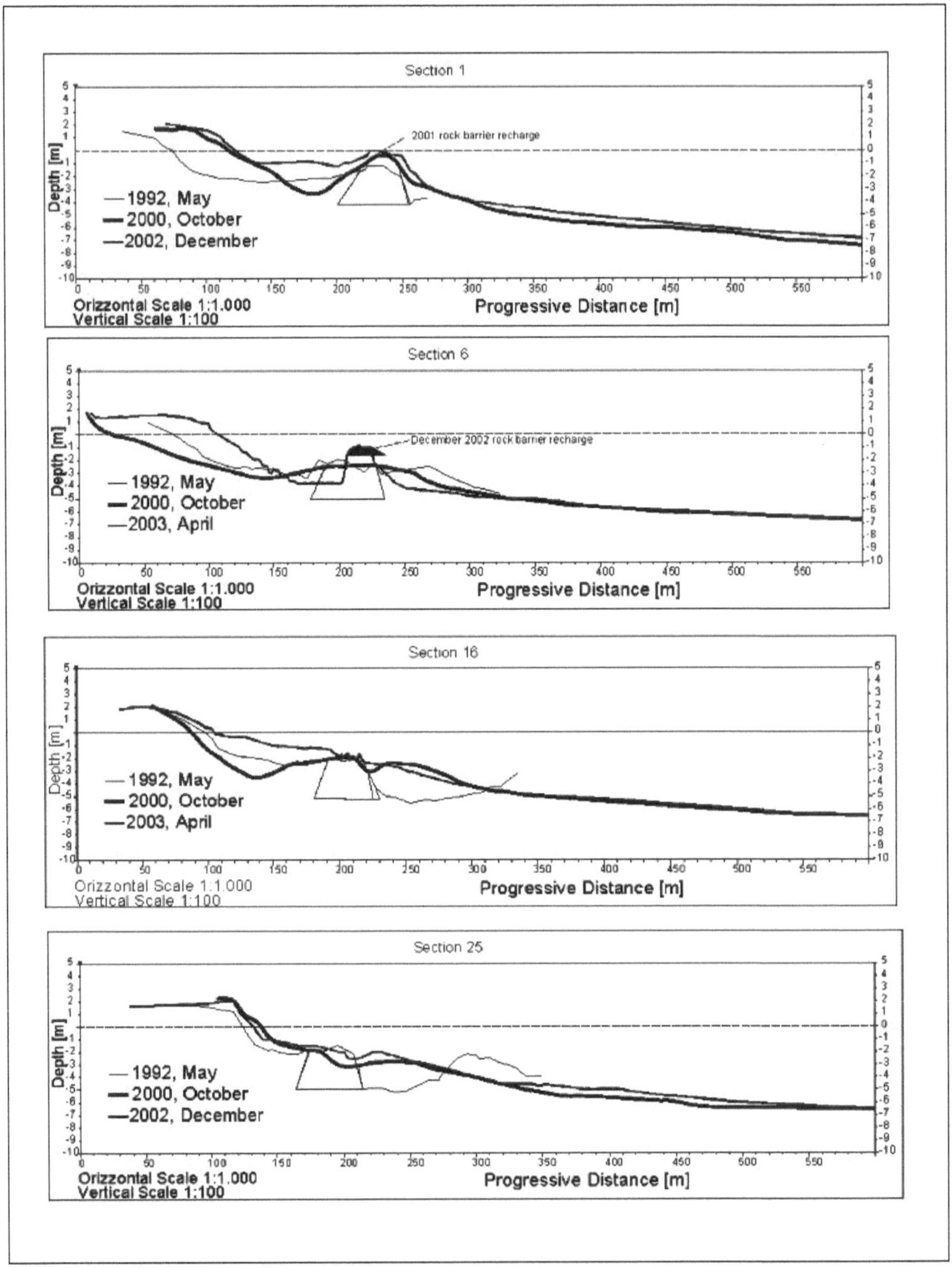

Figure 11.49. Beach profile surveys comparison.

of barrier cross sections with the «as built» geometry of 1992 survey. The average damage over the 6 representative sections is plotted in Figure 11.51. There is an obvious tendency to equilibrium with a maximum mean damage of 12.5%. The most damaged section is p1 with 25%, while p11 and p16 only show a 4% damage. This damage is well predicted by Van der Meer formulae, assuming $D_{50} = 0.5$ m ($W_{50} = 0.35$ t) and depth-limited breaking waves. The progressive barrier siltation from both shoreward and offshore transport reduces the rock barrier porosity and efficiency, and increases its reflectivity.

In conclusion the original rock LCS has a weak protection effect due to its low crest elevation, (average of – 2.3 m MSL) after settlement (despite geotextile) and erosion due to direct wave action and scour; the size of the rock also appears to be underdesigned. The old barrier only provides a transmission coefficient of about 0.6 under typical storm conditions.

The strong wave obliquity still produces a significant drift, which is now being slowed by few semi-submerged groynes.

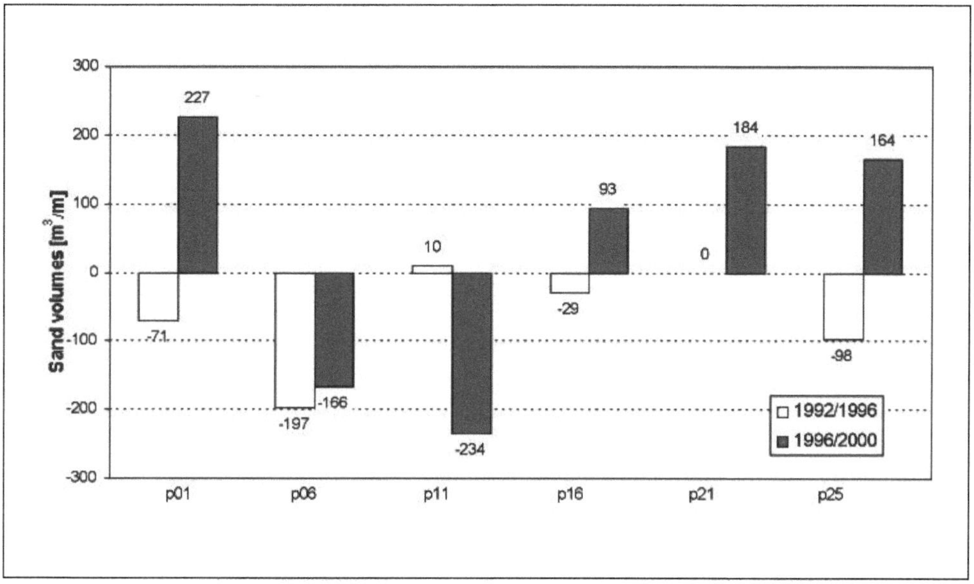

Figure 11.50. Sand unitary volumetric variations at six transversal sections.

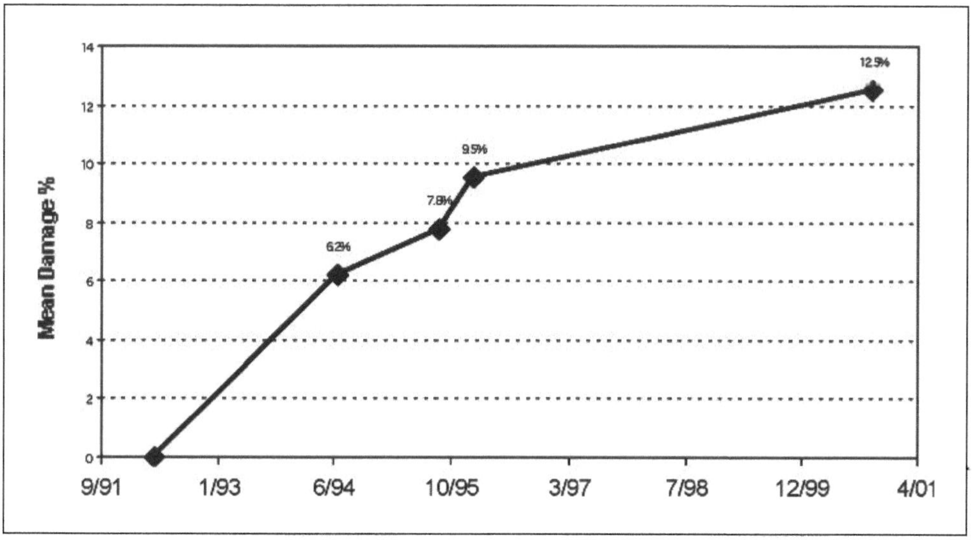

Figure 11.51. Submerged breakwater mean damage.

11.5.5. Socio-economic investigations

In order to perform a Contingent Valuation Method (CVM) survey about the use value of Ostia beach, similarly to Lido di Dante and Pellestrina (Marzetti *et al.*, 2003a; Marzetti and Franco 2003; Polomé *et al.*, 2005) a questionnaire (with 40 questions including photos and figures) was created. A few technical questions related to the preference about coastal protection works and sediment type have been added with the aim to find out users' preferences, Marzetti *et al.*, 2003.

Some 100 interviews were made at Ostia Beach in summer 2002 with good responses: 50% of the approached people accepted the interview and generally showed interest and good understanding (especially the more sensitive residents). Ostia is a popular beach town just 25 km from Rome (3 million people) from where most beach visitors come (67% of interviewed people). In general the residents showed more concern for the overall sea defence issue, while the summer visitors from Rome paid more attention to visual impacts and water quality. With regards to the preferred type of beach protection scheme nearly 50% favoured the inclusion of some kind of rigid structure (14% emerged detached breakwaters, 22% submerged barriers, 6% groynes, 5% a mixed box-type system) since they believe they last longer and are more effective for the beach defence. However the remaining 50% prefers a pure soft option as sand nourishment, especially for aesthetical reasons, but also to favour recreation activities.

With regards to the preference about sediment characteristics it is noted that nearly 80% of users prefer fine light-coloured sands and just 14% like the dark sand which was the original one at Ostia beaches. Some 10% prefers coarse sand and no one likes a gravel beach. This quite obvious response can be useful for nourishment projects. With regards to the fundamental question about the amount of money users would spend for one day at the replenished beach the average value was around 23 euros in developed areas and just over 6 euros within free undeveloped areas, but this drops to only 1-2 euros in case of severely eroded beach. This analysis can quantify the loss of enjoyment due to erosion problems and thus can be used to quantify the benefits of coastal protection works.

Finally, as regards a hypothetical beach change because of erosion, 39% of respondents would reduce the number of visits and 36% would never visit the beach.

11.5.6. Ecological aspects

Specific biologic studies were started only recently by the Central Institute for Marine Research (ICRAM). Various diving inspections and two video films (Nov 2003 and Feb 2004) were carried out. Observations show that the rock barrier has lost its porosity and is mostly filled with sand (also coming from the new artificial fill) and well naturalized with the seabed, resembling a natural reef with active marine life (fishes, octopuses, vegetation, mussels, etc.). However the existing fine sand beach did not experience hard bottom structures before. In general the water quality at Ostia Beach is improved in the last years and the general attitude towards the rock barrier is positive.

CHAPTER 12

An example of environmental design of coastal defence

Zanuttigh, Martinelli, Lamberti, Marzetti, UB; Moschella, Hawkins, MBA

12.1. PREFACE

The aim of this Chapter is to apply the knowledge achieved within DELOS to an existing prototype case, in order to provide an example on how the guidelines can be used.

In order to assure consistent boundary conditions, a real well-documented case which suffers from erosion was selected. This case is Lido di Dante, Ravenna, Italy, already presented in the previous Section 11.4. The guidelines will be applied to the site as it was in 1994, subjected to great erosion and protected only by small groynes (see Figure 12.1), in order to allow the investigation of many realistic design alternatives.

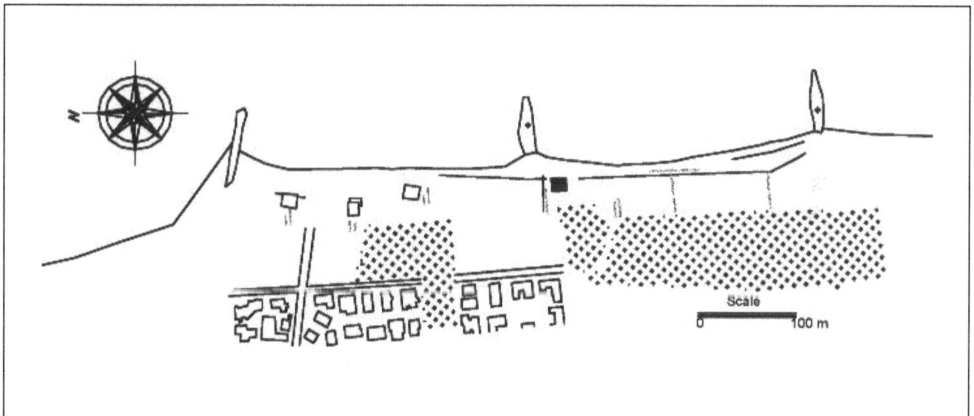

Figure 12.1. Plan view of Lido di Dante, 1994.

12.2. INITIAL CONSIDERATIONS

12.2.1. Relevant policy and legislation

The EU directives have been adopted in Italy and form the standards at national and local (Regione Emilia Romagna) scale. The relevant policies and legislation are given in Table 12.1.

The current Italian technical recommendation for maritime works are:

Istruzioni tecniche per la progettazione delle dighe marittime/Technical instructions for breakwater design, Consiglio Superiore del Ministero dei lavori Pubblici & Consiglio Nazionale delle Ricerche, CNDCI, 1996, Roma (in Italian and in English).

Table 12.1. Relevant legislation.

Code of directive	*Directive/Convention*	*National and/or regional legislation (modifications are not quoted)*	*Main subject of Italian legislation (in italian)*
85/337/EEC and 97/11/EC	EIA (Environmental Impact Assessment	D.P.R. 12.04.96 (technical standards); D.Lgs. 31.05.1998, n. 112, L. 31.10.2003 n. 306 (application of most recent directives) L.R. 18.05.1999, n. 9, L.R. 16.11.2000, n. 35 (for regional implications)	Procedura di valutazione dell'impatto ambientale
2001/42/EC	SEA (coastal works against erosion and works that alter the coastline)	PROGETTO DI LEGGE REGIONALE (under discussion)	Disciplina della programmazione energetica territoriale ed altre disposizioni in materia di energia
2000/60/EC	Water framework	D.Lgs. 11.05.1999, n. 152; D.Lgs. 18.08.2000, n. 258	Tutela delle acque dall'inquinamento
76/160/EEC and 91/692/EEC	Bathing water	D.P.R. 26.07.1082, n. 470; L. 29.12.2000, n. 422	Qualità delle acque di balneazione
79/409/CEE 92/43/EEC	Conservation of wild birds; Habitat;	L.R. 15.02.1994, n. 8; L.R. 21.04.1999, n. 3; L.R. 16.02.2000, n. 6	Protezione della fauna selvatica e per l'esercizio dell'attività venatoria; Riforma del Sistema Regionale e Locale;
91/271/EEC and 91/676/EEC	Waste water treatment; Pollution by nitrates	REGOLAMENTO REGIONALE 20.11.2001, n. 41; D.Lgs 11.04.1999, n. 152	Regolamento per la disciplina del procedimento di concessione di acqua pubblica
90/313/EEC	Access to environmental information	D.Lgs. 24.02.1997, n. 39	Libertà di accesso alle informazioni in materia di ambiente
79/923/EEC	Shellfish water directive	D.Lgs. 27.01.1992, n. 131	Requisiti di qualità delle acque destinate alla molluschicoltura
Barcelona Convention (1976, revised in 1995)	Protection of the marine environment and the coastal region of the Mediterranean	L. 25.01.1979, n. 30 L. 29.05.99, n. 175	Prevenzione ed eliminazione dell'inquinamento del mar Mediterraneo
RAMSAR Convention (1972)	Wetlands of international importance	D.P.R. 13.03.1976, n. 448	Zone umide di importanza internazionale, in particolare come habitat di uccelli acquatici

Regional coastal plans are available (IDROSER, 1996 and ARPA, 2001) with a description of the coast at regional scale, individuation of critical points and suggestion of preliminary designs.

12.2.2. EIA Constraints

Social preferences led to motivate the choice of fine yellow sand (based on the results of the CVM survey carried out in Lido di Dante during Summer 2002, see Marzetti *et al.*, 2003; Marzetti and Zanuttigh, 2003; Polomè *et al.*, 2005).

In the surrounding area, natural rock is extensively used, whereas no artificial blocks are present and this constitutes a technological constraint.

12.2.3. Definition of technical, environmental, and socio-economic objectives

The main objective of the design is the maintenance of an adequate beach for recreational activities; desired features for the resort include:

- sufficient length of the beach (50 m is generally required in the region);
- use of material which is typical of the surrounding areas (yellow sand of medium grain size, approx. 0.2 mm, and natural rock);
- appropriate swimming conditions (reduce risk to swimmers of possible injuries or drowning);
- small visual impact (structure should not be such as to obscure the horizon);
- good water quality (avoid colonisation of the sheltered habitats by organisms such as ephemeral green algae, which also cause a drift algae on the beach).

The achievement of this objective also provides a proper protection of land and infrastructures. It is indeed necessary to avoid possible floodings, to protect the residential properties and streets; the northern part in correspondence of the urban area is more critical than the southern, where the dune system is more consistent.

It is also desired that the intervention:

- minimise impact on cultural heritage;
- minimise impact on ecosystem, habitat and species;
- if possible, enhance natural living resources for food and recreation.

12.2.4. Project service lifetime and safety classification

Although the functional lifetime may be considered to be 30-60 years, the expected economic lifetime may be assumed to be much smaller, since a proper maintenance programme is foreseen and scheduled. A lifetime L of 15 years is more appropriate.

Possible damages to the structure are not likely to cause human injury or immediate large economic losses, and therefore a structural failure probability P_f of 25% or more may be tolerated.

The return period of the design wave load becomes:

$$T_{rp} = \frac{L}{-\ln(1-P_f)} = 52 \approx 50 \text{ years}$$

The main design load is then characterised by a 50 years return period. The actual load on the structure is due to a combination of wave height, wave period, wave direction, water level and tidal currents. The probability of occurrence of the combination of all these factors together is of course higher than the occurrence probability of each single load, and the joint statistics should be referred to. It is seen in the next sections, however, that knowledge of the joint statistics of waves and water level is absent.

When both such rare loads contrast the stability, two cases are analysed, for simplicity: 100% probability of the first rare load (waves or tide) plus 70% of the second rare loads (tide or waves), plus 100% of all other permanent or very frequent loads.

In some cases it is not known *a priori* the effects of water level on submerged structure stability. In this case the rare load effect should be investigated in more detail considering all possible effects of water level ranging from a 70% of minimum to 70% of maximum.

An initial phase of 1-2 years will also be considered, relative to a particular configuration in which the structure has not yet reached a final settlement; with similar structural failure probability, a return period of 5 years should be assumed.

12.2.5. Consideration of environmental context

Lido di Dante is a small seaside resort in the Northern Adriatic Sea, 7 km far from the town of Ravenna, between the mouth of the rivers Fiumi Uniti Northwards and Bevano Southwards. The two rivers drain basins of very different size and characteristics: Fiumi Uniti basin is much wider and contains an important mountainous part contributing to a significant sediment load in the past; Bevano river is essentially a natural drainage channel of the plain with little sediment transport.

The Adriatic Sea in this area is characterised by a maximum depth around 50 meters and normally eutrophic conditions caused by waters drained by the Po river from the highly inhabited and cultivated Po plain.

The sandy beach of Lido di Dante has a concave shape and is more than 2500 m long. It can be divided into two parts: the Northern beach (almost 600 m long) was subjected to much erosion and therefore it has been protected by groynes, nourishment and semi-submerged breakwater; the Southern beach instead has undergone slight erosion and is in a very natural state.

Shore protection in Lido di Dante was the result of several successive interventions aiming to stop littoral regression starting around 1960. The first work was carried out in 1978, when a single Northern groyne was constructed to retain sediment transport due to littoral drift. In 1983, other two groynes were constructed South of the previous one, forming two cells; a beach nourishment protected by a submerged barrier made of sand bags completed the intervention (many bags were destroyed and found on the beach during the following years). Erosion however continued: the greatest erosion occurred North of the defence system (90 m), but it was significant also in the Northern cell (40 m) and in the Southern one (30 m), requiring a further intervention in 1994 (the start year of this exercise) before the nourishment protected by a semi-submerged barrier.

Present shoreline retreat is mainly caused by the low sediment transport rates of the rivers in the last decades and by the anthropogenic and natural subsidence, which justifies recent beach recession rate of 3 m/year. Erosion has disrupted beach equilibrium, with major damage when storm surges are coupled with high tides. Littoral recession, such as erosion of dunes and land subsidence, together with building of tourism facilities, has altered and partially destroyed the maritime pinewoods behind the dunes.

12.2.6. Status, vulnerability, sensitivity and resilience of coastal ecosystem

The results of the Coast Project on indicators/indices for monitoring and assessment of European coastline/marine eutrophication showed that the North Adriatic, in particular the Emilia-Romagna region, is characterised by the highest sensitivity to eutrophication.

Surveys carried out within DELOS on the Emilia Romagna coasts demonstrate that in general chlorophyll *a* content landward of rocky structures was higher than in outside/ seward the protected areas, indicating increased eutrophic conditions in presence of such defence structures.

Based on surveys carried out in 2002, the Emilia Romagna coast is typified by artificial rocky bottoms provide additional habitat for species.

In Lido di Dante, marked differences occurred in the structure of macrofaunal communities. A significant increase in the number of species occurred landward the barrier at 1.0 m depth. Moreover, a gradual change of the community structure was observed following a progressive decrease of the hydrodynamic stress on the sediments.

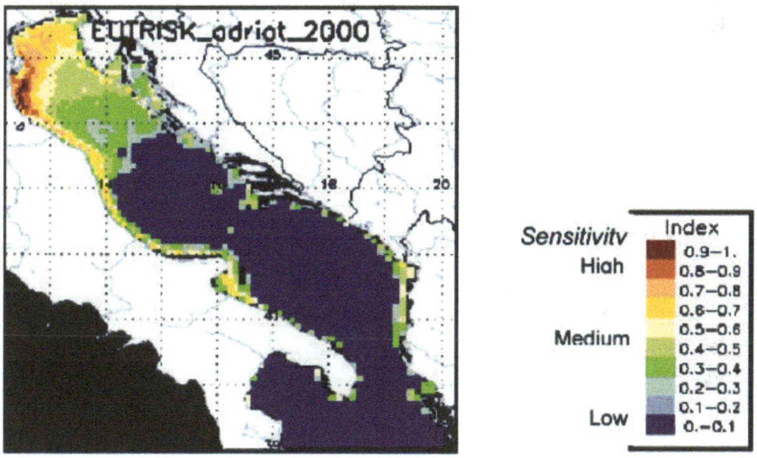

Figure 12.2. Eutrophication in the Adriatic Sea.

From a qualitative standpoint, the increased biodiversity on the landward side was due to colonisation by species commonly living in lagoons or saltmarsh habitats (e.g. *Musculista senhousia*, *Neanthes succinea*, *Cirriformia tentaculata*). Moreover, the polychaete *Capitella capitata* typically associated with organically enriched environments, where low hydrodynamics tend to lead to the accumulation of muddy sediments, showed significantly higher abundance landward than in the control, where it was only occasionally detected.

These results cannot be considered as an improvement of the benthic environment, but rather as a substantial modification of the natural characteristics of the biotope. The presence of species typical of the lagoon fauna, coupled with large numbers of opportunistic worms and specimens coming from deeper environments indicates a substantial transformation of the benthic communities in the protected site. Most of these species are known to be indicative of increasing disturbance (e.g., organic enrichment, presence of stagnant or brackish waters).

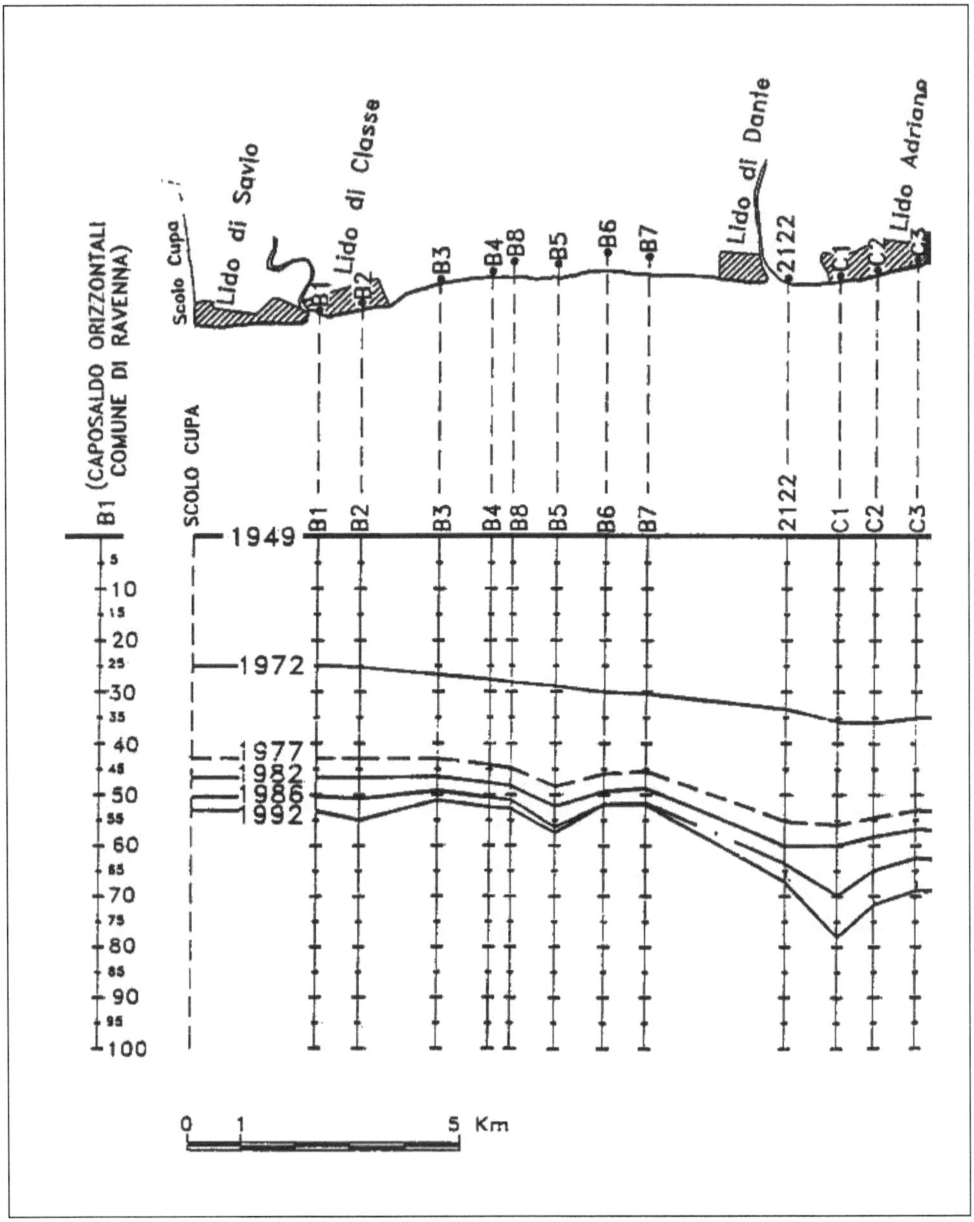

Figure 12.3. Subsidence data collected in the period 1949-1993.

12.3. ENVIRONMENTAL CONDITIONS

12.3.1. Bathymetry, topology and geology

Several bottom surveys, described in Section 11.4, are available.

From a geological viewpoint, compaction of deeper layers due to liquid extraction is an important issue that must be considered with special attention. The subsidence in the site is the combination of a small natural contribution, of the order of 3 mm/year, and an additional term mainly due to extractions of liquids from the subsoil. The regimentation of water extraction in Ravenna area, started in the 1980s, succeeded in reducing the anthropogenic components in many areas along the cost, but not in Lido di Dante. The subsidence is therefore assumed to be 20 mm/year, which determines, assuming a 1:100 mean slope (from shoreline to depth of closure), a mean shoreline retreat of 4 m/year. In order to compensate subsidence, the necessary volume is approximately 20 000 m^3/year, e.g. a 20 mm/year multiplied by the active profile length, 1.1 km long, and by the width the beach requiring more protection (0.9 km). The nourishment compensating the apparent erosion due to subsidence can be reduced by limiting the causes which determine the subsidence and by reducing the active profile length, like for instance by use of coarser sand or by defending the beach with parallel barriers.

12.3.2. Wind and Wave climate

The climate data are derived from information and measurements taken after 1983; by assuming the statistic to be stationary, these data can still apply to the beach.

The meteorological climate of Lido di Dante (Ravenna) is characterised by hot summers with occasional heavy storms, persistent high pressure and thermal inversion, cold winters with possibly some snow, rainy springs and even more rainy autumns characterised by low pressure (cyclonic circulations).

Metereological and wave observations were carried out on the numerous gas platforms just in front of Lido di Dante beach. The analysis of measurements from years 1996-2002 shows that most intense events come from Bora and Scirocco with similar intensity; waves reach 3.5 m average height every year and around 6 m every 100 years. Wind intensity is stronger from the shorter fetch sector of Bora (NE) where it reaches frequently 35 knots intensity, whereas from the long fetch sector of Scirocco it seldom exceeds 30 knots. The waves resulting from Bora winds are steeper and break far offshore than waves from Scirocco winds.

Frequency of occurrence of Bora and Scirocco winds range from 20 to 30%. Thermal gradient winds characterise the summer.

The representative wind and wave climate is given in Table 12.2; Table 12.3 gives the extreme wave conditions.

The design wave height (50 years return period) is given for different sectors, in order to define the critical conditions for the structure stability (refraction reduces waves approaching obliquely).

Table 12.2. Representative climate.

Condition *n°*	*Wave direction* *[°]*	H_s *[m]*	T_m *[s]*	*Wind velocity* *[m/s]*	*Frequency* *[%]*
1	45°	1.5	5.0	12	4.74
2	45°	4.0	8.0	20	0.53
3	90°	1.5	5.0	12	5.86
4	90°	3.5	8.0	18	0.81
5	135°	1.5	5.0	12	4.80
6	135°	3.5	8.0	18	0.47
7	120°	0.3	3.0	5	40.00

Table 12.3. Extreme wave values.

TR	*1 y*		*2 y*		*5 y*		*10 y*		*25 y*		*50 y*		*100 y*	
Dir	H_s	T_s	H_s	T_s	H_s	T_s	H_s	T_s	H_s	T_s	H_s	T_s	H_s	T_s
60°	3.6	7.4	4.0	7.8	4.5	8.1	4.9	8.3	5.4	8.7	5.8	9.0	6.2	9.2
90°	3.5	8.4	3.9	8.7	4.4	9.1	4.9	9.6	5.4	10.1	5.8	10.5	6.2	10.7
120°	2.8	7.8	3.3	8.4	3.8	8.9	4.2	9.4	4.7	9.9	5.1	10.3	5.5	10.8

The coast is approximately aligned in the North-South direction, facing East, and the structure is exposed to waves coming from direction 90°.

The critical off-shore conditions, relative to 50 years return wave load, marginal over direction, is only a little higher than $H_s = 5.8$ m, of the order of $H_s = 6.0$. Location of the measurement is at depth of 30 m, which are indeed deep water conditions ($h/L_0 = 0.17$).

In practice such wave in the beach is depth limited, of the order of 50%-70% of depth, with lower values associated to flat foreshores; since the foreshore slope is mild, the mean value, 60%, may be considered. For simple considerations the details of wave climate may be abandoned and 60% of depth may be assumed as the highest wave condition.

12.3.3. Currents

Currents generated by tide are estimated to be small in comparison to the site dynamics.

12.3.4. Water level

The area under study is subject to small astronomical tide. Water level statistics is given in Table 12.4. Depths are usually described with reference to the mean level of low water at spring tide.

Table 12.4. Water Level Variations.

Parameter	*Description*	*Level [m]*
	Extreme high level (50 years return period) Extreme high level (10 years return period) Expected maximum annual level	1.09 0.97 0.80
MHWS	Mean high water springs	0.40
MWL	Mean water level	0
MLWS	Mean low water springs (most frequently used chart datum) Expected minimum annual level Extreme low level (10 years return period) Extreme low level (50 years return period)	– 0.40 – 0.72 – 0.84 – 0.93

12.3.5. Sediment transport by winds and waves

The study area is characterised by sand transport diverging from the Fiumi Uniti outlet at the scale of littoral morphology, whereas northwards directed sand transport prevail near the shoreline, specifically in the first 1-200 m from the coast, where breaking of the long and frequent waves due to *Scirocco* winds takes place.

In the more off-shore region, up to a depth of 6 m, the neat sediment transport is south-directed, due to a combination of the currents driven by the more intense and steep *Bora* wind waves. In total, the sediment transport in the area is still south directed, of the order of 100 000 m^3/year (assessment based on wave climate and valid for a free beach configuration, IDROSER, 1996).

From comparison of cross profiles 7 years distant, cross-shore sediment transport appears limited to the depth of – 8 m, which is placed 1.1 km far from the shore.

12.3.6. Water quality

Periodic surveys in the area are carried out by Agenzia Regionale per l'Ambiente (ARPA) Ravenna, by monitoring different indicators of organic (Coliform, Streptococci) and factory pollution (pH, phenol, mineral oils), oxygen, colour and transparency that can be related to eutrophication phenomena. Based on the data collected in the last ten years, it can be deduced that the values of dissolved oxygen few times per year are lower than the limits fixed by DPR 470/82; moreover, few cases of too high microbiological parameters are usually identified

Table 12.5. Information obtained on the basis of 25 surveys between 2002 and 2004 (ARPA ER).

Investigated property		*150 m South of Fiumi Uniti mouth*	*2,15 km South of Fiumi Uniti mouth*
Total coliforms. n°/100 ml (max 2000/100 ml)	Minimum value Median Maximum value	0 0 500	0 0 250
Faecal coliforms n°/100 ml (max 100/100 ml)	Minimum value Median Maximum value	0 0 95	0 0 80
Streptococci faecali UFC/100 ml (max 100/100 ml)	Minimum value Median Maximum value	0 0 18	0 0 10
Dissolved oxygen [%]	Minimum value Median Maximum value	77.9 106 129	38 107 141
pH	Minimum value Median Maximum value	7.8 8.1 8.7	7.8 8.1 8.7
Colour [Pt/Co scale]	Same for all samples	0	0
Turbidity by Secchi depth [m]	Same for all samples	1	1
Mineral oils [mg/l]	Same for all samples	0	0
Surface actives agents	Same for all samples	Absent	Absent
Phenols [mg/l]	Same for all samples	0	0

during the bathing season, but insufficient for bathing prohibition. In both cases water hyper-oxygenation is usually found out together with algae hyper-trophication.

Result of 25 surveys between April 2002 and April 2004, in the beach of Lido di Dante are presented in Table 12.5. The limits associated with the organic indicators are fixed by DPR 470/82.

The presence of the Po river to the North, with its load of nutrients, determines a North-South gradient of most water quality parameters along the Coast of Emilia Romagna. There is a general tendency to eutrophication, extended to 10 km from shore, in winter conditions. The winter euthrophic state is usually suddenly removed by the water recirculation induced by storms. During summer the euthrophic conditions are confined closer to shore and from the Po River to Ravenna. The discharge of the Savio River is concentrated over only a few days and has some influence on Lido di Dante. The chlorophyll «*a*» and the algal biomass is found in average below 10 μg/l (data from 1992 to 2002 from Ravenna to Cesenatico).

12.3.7. Ecosystems, habitat and species

Data on ecosystems, habitat and species are derived from the field monitoring carried out by FF during DELOS project. Data to be used for the design (1994) are assumed to be the same collected in the period 2001-2003 in the Lido di Dante control site, which is located outside the boundaries of the protected area (data from Bacchiocchi *et al.*, 1999; Bacchiocchi and Airoldi, 2003).

A total of 106 species were identified and were grouped into 17 major taxa (Table 12.6).

Control site is almost completely dominated by *Lentidium mediterraneum* (96% and

Table 12.6. Total contribution to the abundance, biomass and number of species of the major taxonomic taxa in each treatment.

TAXON	*Abundance [ind/m²]*	*Biomass [mg/m²]*	*N. of species*
	C	*C*	*C*
Anthozoa	0	0	0
Turbellaria	0	0	0
Nemertea	26	32	1
Sipunculida	3	8	1
Gastropoda	52	1 807	4
Bivalvia	49 508	4 995	14
Polychaeta	598	945	24
Clitellata	0	0	0
Amphipoda	65	11	10
Anisopoda	0	0	0
Isopoda	7	1	1
Cumacea	95	15	4
Mysidacea	0	0	0
Thoracica	0	0	0
Decapoda	13	623	3
Insecta	0	0	0
Echinodermata	0	0	0
TOTAL	50 367	8 436	62

86%, respectively), a species known to be well adapted to energetically dynamic habitats. This suggests that the environment is mainly structured by physical factors and, therefore, characterized by simplified macrobenthic assemblages.

12.4. CONCEPTUAL PRE-DESIGN ALTERNATIVES

12.4.1. Definition of local conditions and constraints

A plan view of the site is given in Figure 11.1.

Main physical constraints are the Northern and Southern river, the urbanised area and a pinewood in the rear. The dune system is generally poor, almost absent in the north.

The constraints are detailed in the following list.

- A urban area in further expansion is located behind the northern part of the beach. Some bathing establishment are placed very close to the shore and their change of position is not practical.
- A pine forest is present in the southern part of the area, just behind the dunes; it has some natural heritage interest (the pine is the symbol of Ravenna) and has a well developed undergrowth.
- Fiumi Uniti River in the north discharges mainly during spring, with a significant amount of sediment transport (fine sand).
- Bevano River, in the south, is on the contrary very short, the outlet branch migrating toward North, thus eroding the natural dune, not having sufficient energy to clear the natural sand bar at the mouth.

Biological and socio-economic constraints are typical of the region and given in the previous chapters.

12.4.2. Identification of alternatives

The following alternatives for beach defence can be considered:

- nourishment (no intervention);
- nourishment with gravel or pebbles;
- revetment;
- submerged structure;
- submerged multi-structure;
- emerged structure;
- emerged multi-structure;
- groynes.

It is immediately seen that the use of pebbles or gravel contrasts with one of the requirements, which is the use of sand of small grain size. Similarly, the revetment does not provide a beach for recreational use. Finally, a single or multiple high crested structures will be not accepted by the local community for aesthetic and ecological reasons.

Based on these simple observations, five design alternatives can be selected from the list above:

- sand nourishment (Alternative 0);
- submerged single structure (Alternative 1);
- emerged multi-structure (Alternative 2);

- prolongation of existent groynes (Alternative 3);
- composite intervention, with submerged barrier and connectors to existent groynes (Alternative 4).

All the Alternatives suggesting the construction of structures also include a beach nourishment with sand.

12.4.3. Preliminary investigation of design alternatives

The basic design and the morphological response of the five alternatives selected in the previous section is outlined below:

0) no intervention solution (see Figure 11.1);

1) submerged continuous barrier, 670 m long; depth at barrier (axis) is 3.5 m, mean distance from shore is 185 m; the single structure is meant to uniformly reduce wave action; the typology is suited in case currents in the protected area remain small;

2) emerged barriers parallel to the coast, made of 4 sections 150 m long and separated by small gaps. The barrier is continuous at level – 2.0, providing a protection to the toe and to the gaps. Depth at barrier (axis) is 3.0 m, mean distance from shore is 125 m; the type is suited in case of strong waves, associated to high tide;

3) northern and southern groyne extension (80 and 40 m, respectively); this option is suitable where there is large long-shore sediment transport and where the reduction of transport toward adjacent beaches is not critical;

4) submerged barrier 530 m long, connected to the beach by submerged groynes; depth at barrier (axis) is 3.5 m, mean distance from shore is 185 m; the configuration is similar to n. 1, except land connections to the longitudinal LCS are planned; this option is appropriate where strong long-shore currents are induced by overtopping and aims at reducing the loss of material from the protected area.

12.4.3.1. Preliminary investigation on sediment transport

The following simple considerations are used to preliminarily investigate the sediment transport in the area.

As an example the simple CERC formula is applied to the series of waves representative of the wave climate defined in Table 12.2:

$$I_1 = c_f K/16\ (\rho_w\ g^{1.5}/\gamma_b^{\ 0.5})\ H_{b,rms}^{\ \ 2.5} \sin(2\alpha_b)$$

$$Q_1 = I_1/((\rho_s - \rho_w)\ g(1-n))$$

In practice the formula does not account for the complexity of the phenomenon, and the uncertainty of the result is so high that it may be used only as a very preliminary investigation.

The immersed weight transport rate I_1 and volume transport rate Q_1, given in Table 12.7, are obtained with the following parameters:

ρ_s = mass density of the quartz sand (2 650 kg/m^3);
ρ_w = mass density of water (1 030 kg/m^3);
n = in-place sediment porosity (0.4);
γ_b = breaking condition for H_{rms} = 0.78;
c_f = conversion factor for use of H_s instead of H_{rms} = $32^{0.25}$;
K = coefficient based on utilizing the rms breaking wave height ($H_{b,\,rms}$) = 0.92.

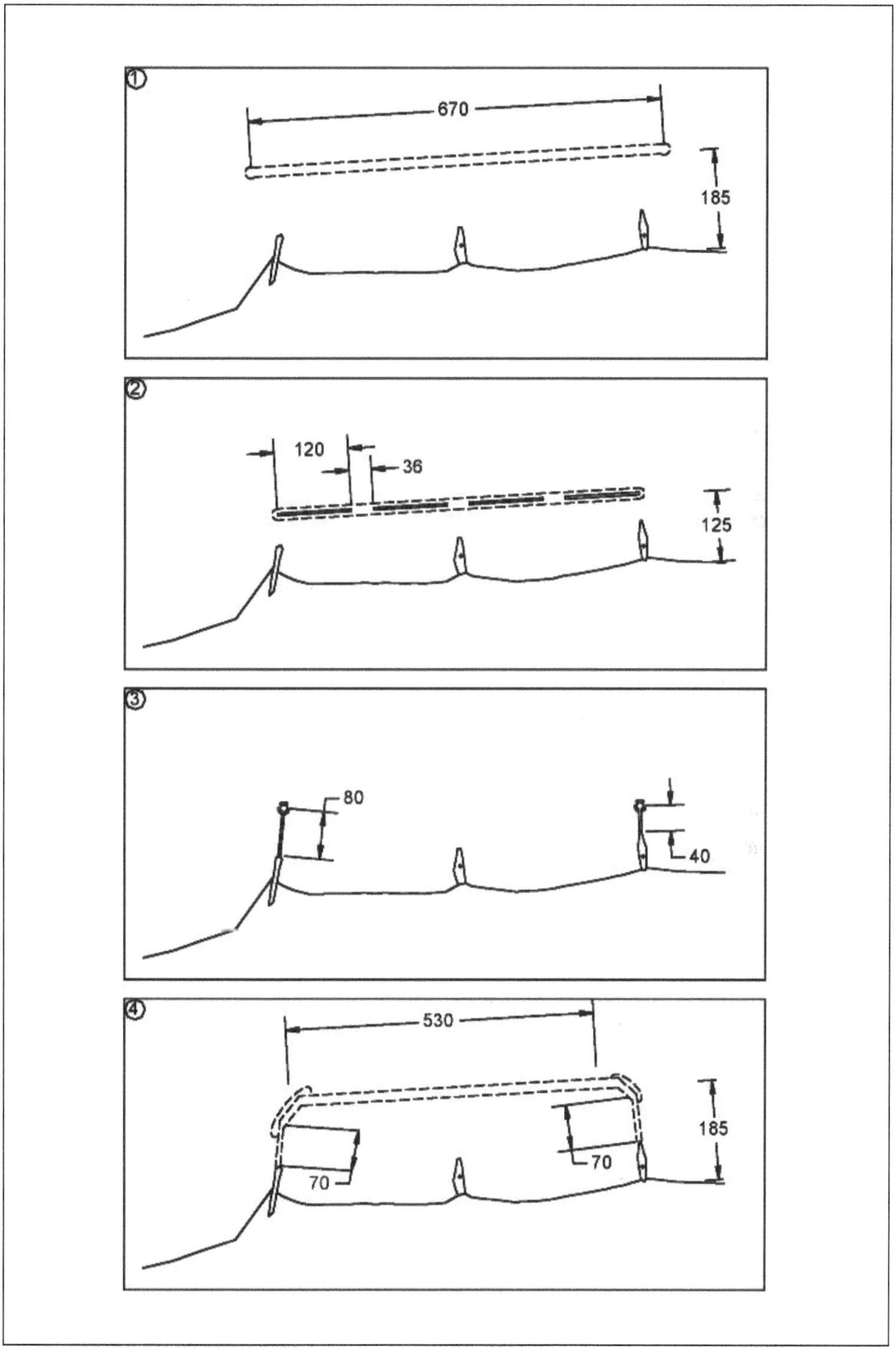

Figure 12.4. Plan view of four alternatives (dashed line = submerged).

Table 12.7. Potential Sediment transport evaluated with CERC formula.

H_s	a	I_l	Q_l	*Frequency*	*Transport*
[m]	*[deg normal to the beach]*	*[kgf/s]*	*[m³/s]*	*[%]*	*north directed [m³/year]*
1.5	– 41°	– 2 352	– 0.2459	4.74%	– 367 628
4	– 41°	– 27 316	– 2.8559	0.53%	– 477 340
1.5	4°	330	0.0346	5.86%	63 849
3.5	4°	2 748	0.2873	0.81%	73 398
1.5	49°	2 353	0.2460	4.80%	372 365
3.5	49°	19 568	2.0458	0.47%	303 226
0.3	34°	39	0.0041	40.00%	51 954
					19 824

The choice of $K = 0.92$ is, according to Del Valle *et al.* (1993), the best value for sand of diameter of 0.2 mm.

12.4.3.2. Submerged single structure

Submerged structures are in general less efficacious than emerged structures, and their wide adoption is justified by the water quality constraint, which requires that some fraction (30%-40%) of the incoming wave energy enters the protected area.

A submerged single structure, parallel to shore, is designed as first alternative. The main design variables are the distance from shore (i.e. the depth at the structure), the crest freeboard and the crest width.

The cross section is designed in resemblance of the design of Pellestrina (described in Chapter 11.3), subjected to wave conditions and constraints similar to Lido di Dante:

- depth at the structure = 3.5 m, which determines a distance from shore of 185 m;
- crest freeboard = – 1.5 m;
- berm width = 16 m.

The assumed cross section is presented in Figure 12.7, including the stone dimensions.

In this preliminary phase we will assume for simplicity that extreme waves are depth limited, with $H_{si} = 0.6\, h = 2.1$ m, and absence of tide. The investigated phenomena are: set-up (or piling-up), overtopping and transmission.

Experimental studies in wave flumes give some indications of overtopping (although not for submerged structures) in absence of piling-up, and piling up in completely confined conditions (absence of return flow). The actual piling up and overtopping depends on the degree of confinement of the structure (gap to barrier length ratio and friction), see Section 13.5.

Lamberti *et al.* (2003) showed that Van der Meer and Janssen (1995) formula, designed for high crested structures, may be extrapolated up to null freeboard. In conditions of null piling up and therefore in absence of a return flow over the structure, discharge for negative freeboards, at least until waves break on the barrier, is assumed similar to discharge in case of null freeboard, and the overtopping is assessed by using the available formula (Van der Meer & Janssen, 1995). The following input values are used:

- $R_c = 0$ (although actual crest freeboard is $R_c = -1.5$ m);
- $\xi_{op} \approx 0.5/\sqrt{0.04} = 2.5$;

- $\gamma_b = 1.0$ (influence of the berm is small for low berms);
- $\gamma_f = 0.6$ (reduction factor for rough slope);
- $\gamma_b = 1.0$ (reduction factor for oblique wave attack);
- $\gamma_v = 1.0$ (reduction factor for presence of vertical wall on the slope).

The assessed overtopping is $Q_{Max} = 2.0$ m³/m/s, associated to a null set up (frictionless return flow). For a 670 m long barrier, total discharge is approximately 1 340 m³/s, that in stationary conditions must return off-shore. Gaps are absent, the barrier is distant 185 m from shore, and the only return paths are lateral, on a mean water depth of 1.2 m, for a total section of 450 m². The rip current velocity is therefore of the order of 3.0 m/s.

Next step is the evaluation of set-up induced in absence of recirculation. Such value depends from the permeability of the structure, and therefore details of the structure cross section are needed (see Figure 12.5). From experimental data (Debski and Loveless, 1997)

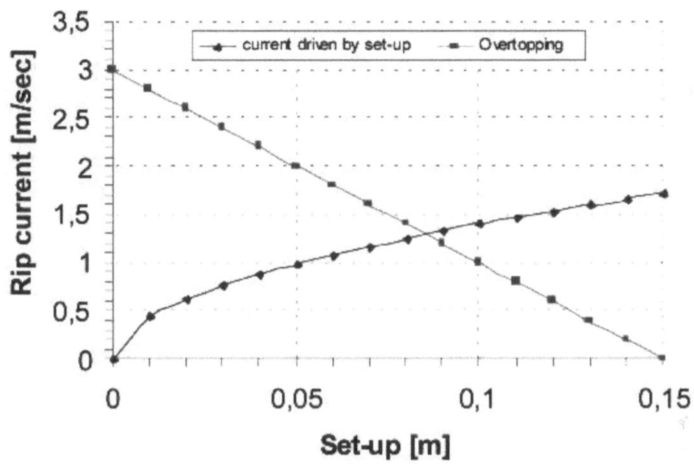

Figure 12.5. Total overtopping and rip current as function of set-up.

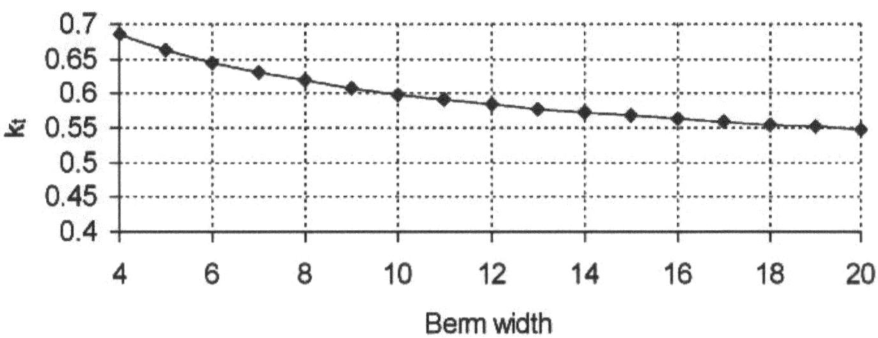

Figure 12.6. Effect of berm width on transmission (geometry relative to submerged structure, Fig. 12.7).

a set-up (or piling-up) of 15 cm is associated to a structure in similar conditions (very submerged). According to Bellotti (2004) formula, piling-up results 32 cm (a slight overestimation is a consequence of the postulation of impermeability).

Total overtopping and return flow, which are strongly dependent on piling-up, must be equal in stationary conditions. The actual piling up is indeed found imposing the mass balance. Figure 12.5 tentatively describes the two functions with a simple approach: 1) the rip current velocity is driven by set-up as through a weir; 2) the equivalent velocity due to overtopping is the difference between a constant shoreward component, determined above, and filtration return flow, proportional to piling up, with zero discharge associated to a piling up of 15 cm.

The complex effect of lateral confinement is not accurate and should be considered, accounting for the appropriate head loss. In Figure 12.5 the overtopping discharge per meter of barrier is converted into rip current velocity using as conversion factor the ratio between barrier length and (contracted) gap section area. From Figure 12.5 the resulting actual set-up in the area is 9 cm, with rip currents of 1.2 m/s.

Transmission is presented in Figure 12.6, where the effect of the berm width is pointed out. In order to allow only 30% of incident wave height, the transmission coefficient $k_t \approx 0.55$ based on Eq. (13.50) and (13.51) in Section 13.3.

A submerged single structure, parallel to shore, is designed.

The main design variables are the distance from shore (i.e. the depth at the structure), the crest freeboard and the crest width. The optimal parameters allow for the desired amount of wave transmission, overtopping, set-up and currents.

Cross section (depth at the structure, crest freeboard) is similar to the design of Pellestrina, a resort in Venice subjected to wave conditions and constraints similar to Lido di Dante, see Section 11.3

The optimal design should avoid big currents and reduce high waves. High mean currents are induced by high overtopping rates and very strong currents may be expected in case of high piling-up. It is therefore desired to reduce both these effects together with incident wave energy.

The design is carried out in order to have currents of 0.5 m/sec, piling-up of 10 cm, transmission of 0.63 (allow 40% of energy in the protected area):

Extreme conditions are depth limited, e.g. $H_s = 0.6 \cdot 3.5$ m = 2.1 m. No-tide conditions are assumed for simplicity.

U = mean long-shore current;
A = lateral area where the current exits the protected zone;
L_s = length of the barrier;
Q = overtopping discharge;
ξ_{op} = breaker parameter = $\tan(\alpha)/\sqrt{s_{op}} = 0.5/0.03 = 2.9$.

According to Van der Meer formula (1988):
$\gamma_b = 0.95$ (influence of the berm is small for low berms);
$\gamma_f = 0.5$ (reduction factor for rough slope, presence of 2 rubble mound layers);
$\gamma_b = 1.0$ (reduction factor for oblique wave attack);
$\gamma_v = 1.0$ (vertical wall on the slope).

12.4.3.3. Emerged multi-structure

Emerged structures are typical along the nearby coast.

The main design variables are the distance from shore (i.e. the depth at the structure), the crest freeboard and width, the gap extension and the number of gaps.

The distance from shore should be as small as possible, in order to minimise impact to the adjacent beach. On the other hand, depth should be sufficient to allow normal bathing activity and extend to the sediment active region. A depth of 3.0 m is therefore assumed.

The crest freeboard is designed in order to be always emergent even in high tide, $R_c = 1.5$ m.

Small gaps are desired, in order to reduce the part of the shore directly exposed to the waves and thus possibly subjected to erosion. The gap length L_g should on the other hand allow for passage of boats. A value of $L_g = 36$ m agrees with the guidelines indications, according to which the gap width is generally in the range $L - 0.8\ L_s$, where:

$$L = T \cdot (\gamma \cdot d_s)^{0.5} = 37 - 43 \text{ m } \{T = 5 - 8 \text{ s}, d_s = 3.0 \text{ m}\};$$

$$0.8\ L_s = 96 \text{ m } \{L_s = \text{barrier length} = 120 \text{ m}\}.$$

Supposing the overtopping has little relevance, (K_t for emerged LCS is null for small waves and tend to 0.2 for high waves), the total energy enters only from the gaps and is totally dissipated at the beach. The amount energy in the protected area is therefore given by the length to gap ratio. The design ratio is (4 barriers of length $L_s = 120$ m, 3 gaps 36 m long) equal to 18%.

The amount of energy allowed in the protected area should be sufficient to keep in suspension the fine material in the deeper parts behind the barriers, thus avoiding deposition of the silty fraction. In the following, the minimum necessary wave energy that avoids such deposition is assessed.

The condition that should be fulfilled is that the friction velocity due to waves at the bottom U_{mo}^* exceeds the falling velocity of small material w:

$$U_{mo}^*(H) > w$$

From Table 12.2 it can be observed that wave height of 0.3 m is exceeded 57% of the time. We require that for such wave the silty fraction should be re-suspended.

Input (in brackets) and results are:
$w \approx 0.005$ m/s for $\{D_{50} = 0.0625$ mm, silt$\}$
$U_{mo} = (\pi H)/(T \sinh(k\ d)) = 0.12$ m/s $\{H = 0.3$ m, $T = 3$ s, $d = 3.5$ m, $L = 13$ m$\}$
$a = U_{mo} T/(2p) = 0.058$ m
$f_w = 0.04\ (a/k_n)^{-1/4}\ 0.026\ \{k_n = 0.01$ m$\}$
$U_{mo}^* = U_{mo} \sqrt{(f_w/2)} = 0.014$ m/s
and therefore $U_{mo}\ \mu\ H^{8/7}$.

The bottom friction velocity condition results:

$$U_{mo}^*(H) \approx 0.014(H/0.3)^{8/7} > w$$

which requires H to be higher than 0.10 m.

In conclusion, where the wave exceeds 0.1 m, the silty fraction remain in suspension. It is therefore enough that 9 – 10% of the incident energy (with $H > 0.3$ m) is allowed in the sheltered area in order to avoid deposition for most of the time (note that energy is

Table 12.8. Conditions for formation of tombolos (c1 > c2) and salients (c1 < c3 or c4 > c5).

Ref.	*Parameter*	*Expression*	*Value*
c1	Parameter characterising single structure	L_s/X	0,96
c2	Condition for tombolos	$(1\div1,5)/(1-K_t)$	1.25÷1.875
c3	Condition for salients	$1/(1-K_t)$	1.25
c4	Parameter characterising multi-structures	$G\,X/L_s^2$	0.3125
c5	Condition for salients	$0.5(1-K_t)$	0.625

proportional to the square of wave height).

In practice, the energy is not constant in the sheltered area, and although some reflection of the beach may contribute in increasing the waves, some stagnation points (and formation of salients) are expected. Salients of some relevance are indeed expected to develop according to the guidelines present some expressions which can be used to predict the formation of salients and tombolos in case of small transmission (Table 12.8). Tombolos are expected if c1 > c2 (see the tag in column 1 of Table 12.8), whereas for smaller values of c1, the expected coastline projection has dimension that increases with the ratio c1/c3, so that when c1 ≈ c3 salients may look almost like tombolos, and when c1/c3 is smaller than 0.1-0.3 no shoreline response is expected.

12.4.3.4. Groynes

The groynes are intended to trap a significant percentage of the long-shore sediment transport, to reduce long-shore currents and to stabilise the nourished beach.

As indicated in Sub-section 12.3.5, the transport closer to the beach is north directed, whereas in a fore-shore region the transport is south-directed. This depends on the fact that waves coming from south are more frequent, longer and generally less intense than waves coming from north; the breaking process is then concentrated closer to shore.

The length of the groyne is designed in order to trap a fraction of the transport. The northern groyne, 40 m long, is therefore extended of 80 m. Also the Southern groyne is extended, just 40 m, with the aim of stabilising the coast orientation.

Table 12.9. Potential sediment transport trapped by a 120 groyne.

Hs	*T*	*Sediment transport derived in Table 12.7*	*Off-shore limit of transport[1] (depth)*	*Off-shore limit of transport[2] (depth)*	*Assumed off-shore limit of transport (distance from shore)*	*Trapped transport*
[m]	*[s]*	*[m³/year]*	*[m]*	*[m]*	*[m]*	*[m³/year]*
1.5	12	– 367 628	3.3	2.4	250	– 264 704
4	20	– 477 340	8.8	6.3	1 000	– 190692
1.5	12	63 849	3.3	2.4	250	45 973
3.5	18	73 398	7.7	5.5	800	29 102
1.5	12	372 365	3.3	2.4	250	268 114
3.5	18	303 226	7.7	5.5	800	123 716
0.3	5	51 954	0.7	0.5	50	51 954

[1] Value assessed applying Hallermeier (1978, 1981)
[2] Value assessed applying Birkemeier (1985)

The groynes should reflect as little as possible, and have an appropriate roundhead to prevent scour. A 1:3 slope is designed in order to reduce reflection, with the same crest freeboard of Alternative 2 ($R_c = 1.5$ m).

The preliminary design may benefit from a simplified representation of the sediment transport distribution. In first approximation we imagine that, during a single storm, the transport takes place between the shore and the breaking point, or in a region slightly wider. The breaking point can be assessed using a ratio between depth and significant incident wave height of the order of 1.8 - 2. A confirm that this is the area where the transport takes placed is found observing that similar coefficients relate the depth of closure to the significant wave height of a characteristic storm, according to Hallermeier (1978, 1981) or Birkemeier (1985).

For each wave condition presented in Table 12.9, the transport is assumed to be parabolically and symmetrically distributed; the groyne is supposed to trap all the sediment occurring between shore and the roundhead, 120 m off-shore.

12.4.3.5. Submerged cell

In this case, the cross section of Alternative 1 is completed by two submerged groynes connecting the structure to shore. This should increase piling up and reduce the rip currents.

12.4.4. Structural design

Only rock and stone material is considered for design as it is available, widely used in the area and environmentally acceptable.

For the actual conditions of the site the simple rule of thumb for armour layer design ($D_{n50} = 0.3\ H_c$ see Subsection 13.11.1) is applicable and has been used, cf. Table 12.10.

In practice structures receive much damage, due to toe collapse, even for stability number $N_s = H_s/(\Delta\ D_{n50}) < 1$, which, in shallow water (typical of LCS) corresponds to big stones $D_{n50} > 0.37\ d$; note that where the toe is not firm, the bigger the armour stones the quicker they sink in the sand.

Design of alternative cross sections are given in Figures 12.7, 12.8, 12.9 and 12.10. For the groyne with 1:3 slope (Alternative 3), the designed size of armour stone is slightly smaller, than for the groyne with 1:2 slope (Alternative 2).

Table 12.10. List of relevant designed parameters.

		Alternatives			
Parameter		*1*	*2*	*3*	*4*
Distance of structure from shore	X [m]	185	125		185
Length of the barrier	L_s [m]	670	120		530
Length of the groyne	L_s [m]			80/40	
Length of the gaps	G [m]		36		
Depth at the structure	d_s	3.5	3.0	2.5-3	3.5
Freeboard	R_c [m]	– 1.5	1.5	1.5	– 1.5
Structure height	H_c [m]	2.0	4.5	4.0-4.5	2.0
Armour (30% H_c)	D_{n50}[m]	0.60	1.35	1.35	0.60
Transmission	K_t	0.55	0.18	1	0.55

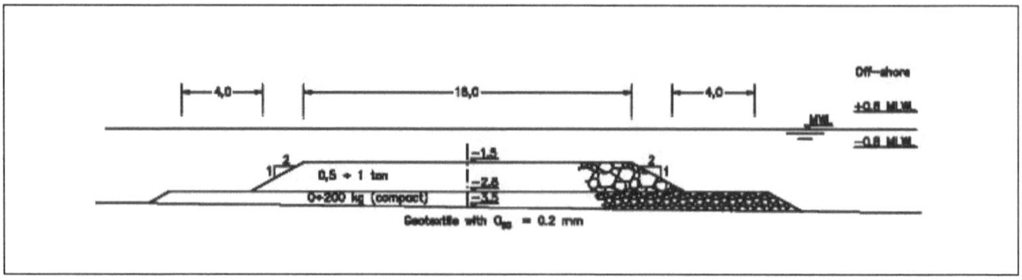

Figure 12.7. Cross section of submerged barrier, Alternatives 1 and 4.

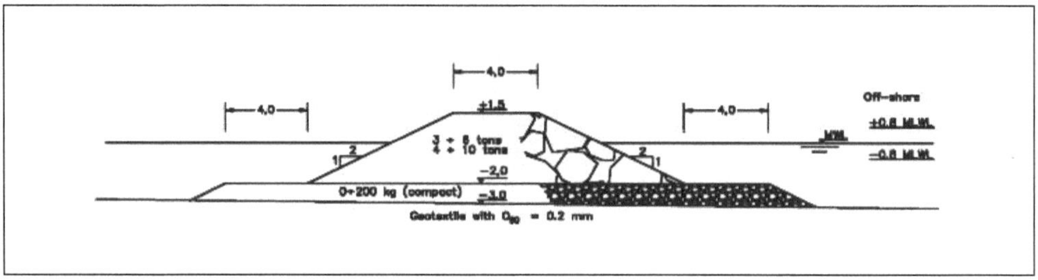

Figure 12.8. Cross section of emerged barrier, armour slope 1:2, Alternative 2.

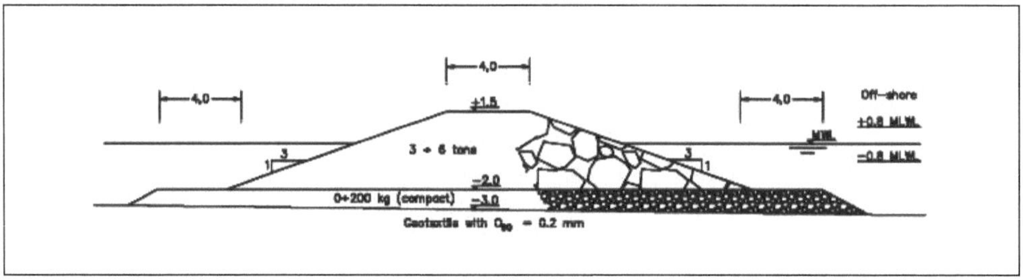

Figure 12.9. Cross section of emerged groyne, armour slope 1:3, Alternative 3.

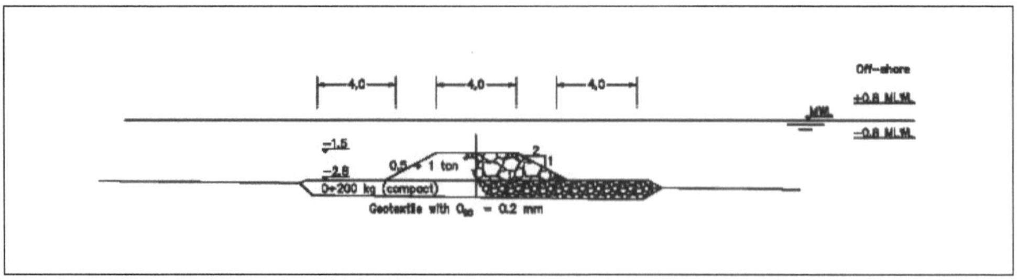

Figure 12.10. Cross section of submerged transverse connectors, Alternative 4.

12.4.5. Analysis of waves, currents and sediment transport induced by each design alternative by means of 2DH numerical simulations

12.4.5.1. Numerical model: settings and results

Numerical simulations presented here were performed with MIKE 21, a 2DH numerical modelling suite developed by DHI Water & Environment. In particular, the Near-shore Spectral Waves (NSW), the Parabolic Mild Slope (PMS), the Hydrodynamic (HD) and the Quasi-3D Sediment Transport (ST-Q3) modules of MIKE 21 were applied.

The NSW model is a wind-wave model, which describes the growth, decay and transformation of wind-generated waves and swell in near-shore areas. The model is a stationary, directionally decoupled parametric model and takes into account the effects of refraction and shoaling, local wind generation, energy dissipation due to bottom friction and wave breaking, wave-current interaction. The basic equations in the model are derived from the conservation equation for the spectral wave action density and are solved using an Eulerian finite difference technique. The PMS module is based on the parabolic approximation to the mild-slope equation of Kirby (1986) which assumes a predominant wave direction and neglects wave diffraction and back-scattering in the direction of wave propagation. The HD module solves the full time-dependent non-linear equations of mass and momentum balance. The solution is obtained using an implicit ADI finite-difference second-order accurate scheme, see e.g. Abbott *et al.* (1973) for details.

The ST-Q3 module calculates the rates of non-cohesive sediment sand transport for both pure current and combined waves and current situations, on the basis of the hydrodynamic conditions that correspond to a given bathymetry. No feedback is given of the bed level change rates on the waves and the hydrodynamics, as in the case for a full morphological model. Hence, the results provided by ST-Q3 can be used to identify potential areas of erosion or deposition and to get an indication of the initial rate at which bed level changes will take place, but not to determine an updated bathymetry at the end of the simulation period.

Offshore wave conditions in Table 12.2 were tested for each design alternative. In particular, waves from 1 to 6 reconstruct the typical wave attacks during a year, whereas Wave 7 is representative more or less of calm periods, with low waves coming from Scirocco that have been documented to induce sediment transport close to the shore-line from South to North. Wave 7 was also chosen to look in details at stagnant zone formation for ecological purposes.

Simulations account both for a sinusoidal tide variation in the range ±0.5 m and for wind as it is reported in Table 12.2.

Bottom bathymetry was reconstructed following field observations and detailed multi-beam surveys performed during DELOS (see Fig. 11.41). Based on sediment samples collected within Lido di Dante monitoring, bottom D_{50} was assumed to be equal to 0.28 mm inshore the structures and 0.22 mm offshore; structure D_{50} was fixed as 0.8 m.

NSW and PMS boundaries were assumed to be «symmetrical» (i.e., uniform conditions), whereas at HD boundaries fluxes and levels derived from radiation stresses were imposed.

Wave breaking was evaluated both in NSW and PMS modules according to Battjes & Janssen (1976) model, with default suggested values: $\gamma_1 = 1.0$ (controls steepness breaking), $\gamma_2 = 0.8$ (controls depth limited breaking) and $a = 1.0$ (controls breaking dissipation rate).

In the HD module, eddy viscosity was imposed to be constant with dissipation coefficient equal to 0.8.

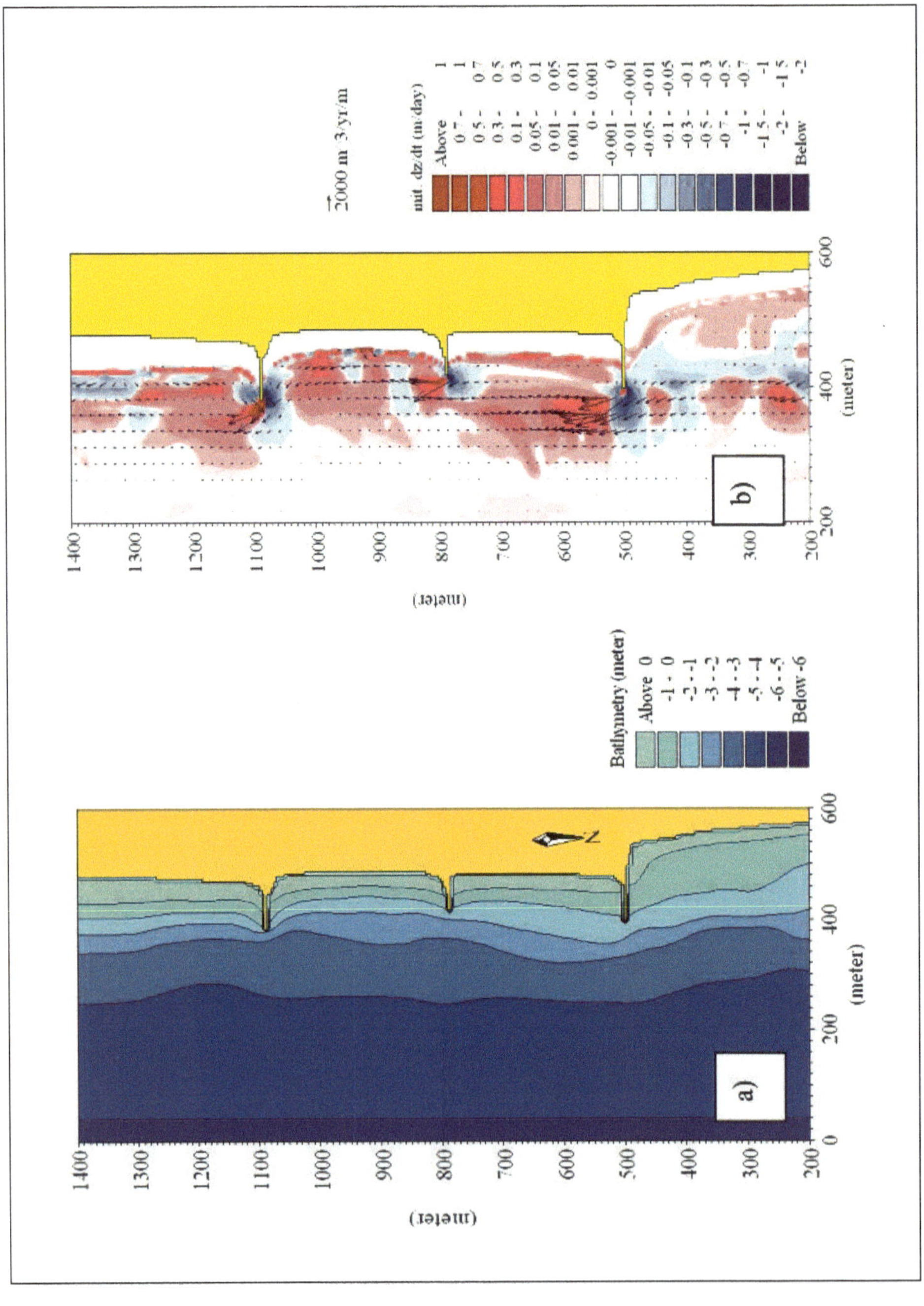

Figure 12.11. Alternative 0, no intervention: a) bathymetry; b) average bottom level variation per day.

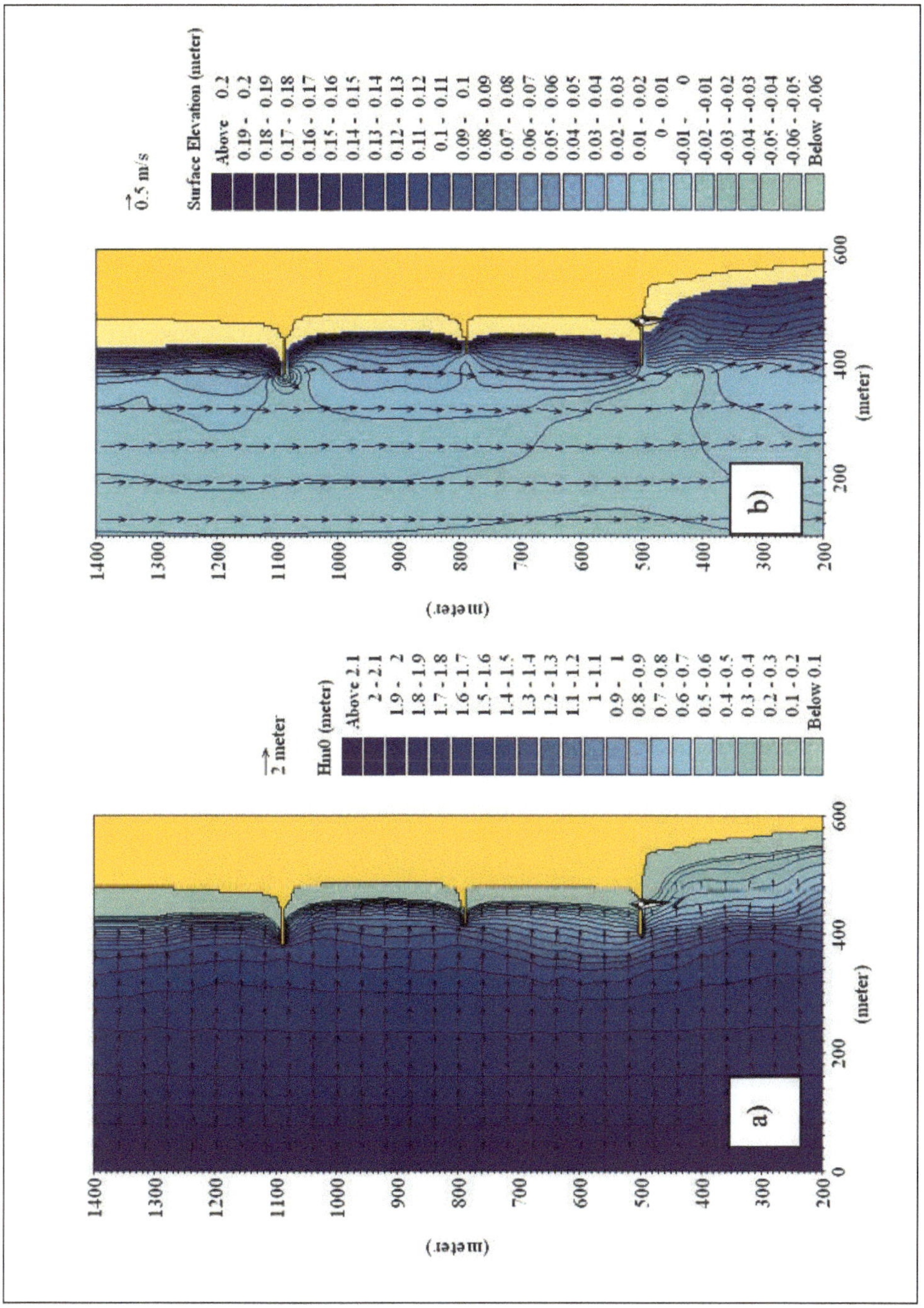

Figure 12.12. Alternative 0, no intervention, wave 6: a) wave field; b) surface elevation.

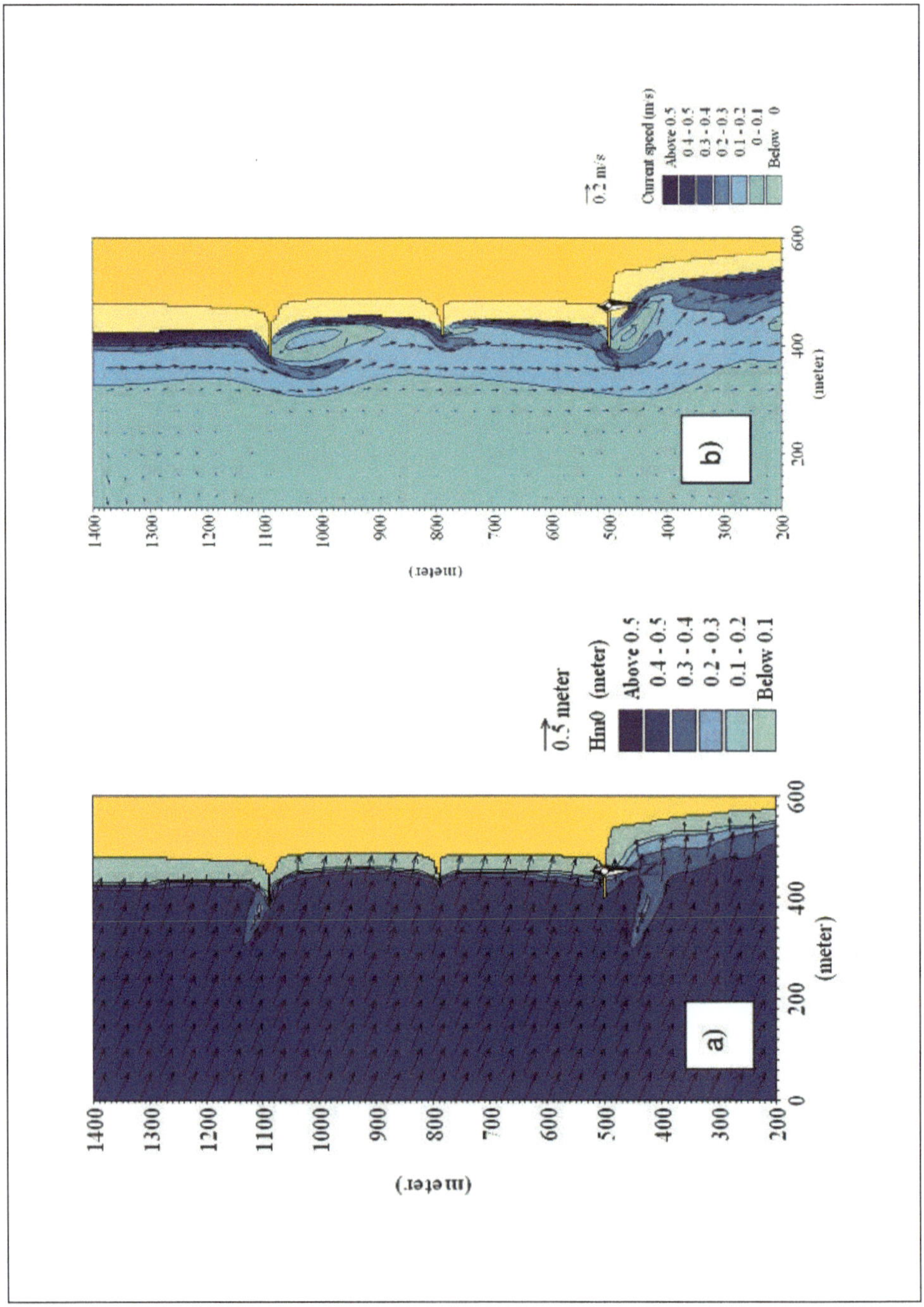

Figure 12.13. Alternative 0, no intervention, wave 7: a) wave field; b) current speed.

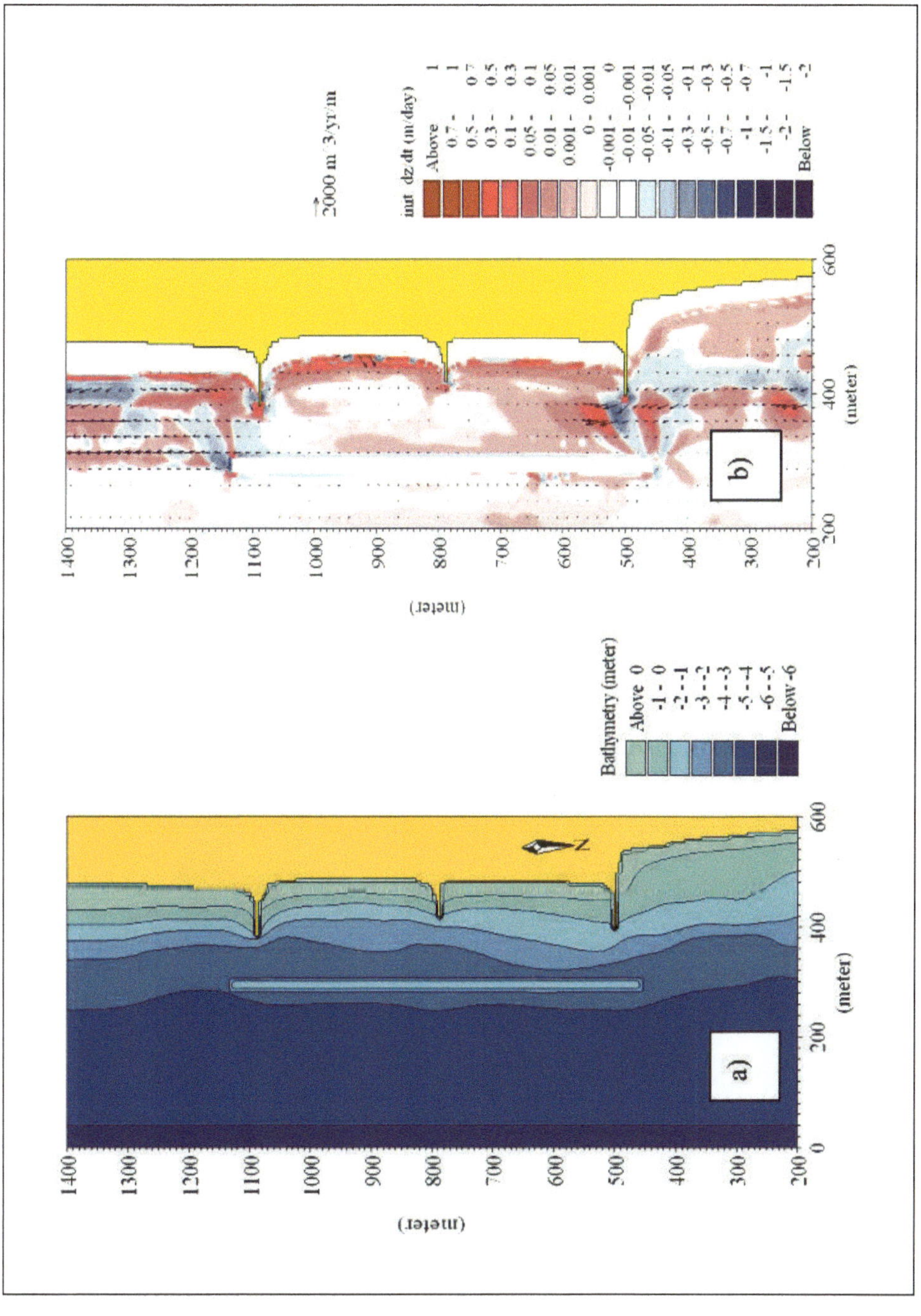

Figure 12.14. Alternative 1, submerged barrier: a) bathymetry; b) average bottom level variation per day.

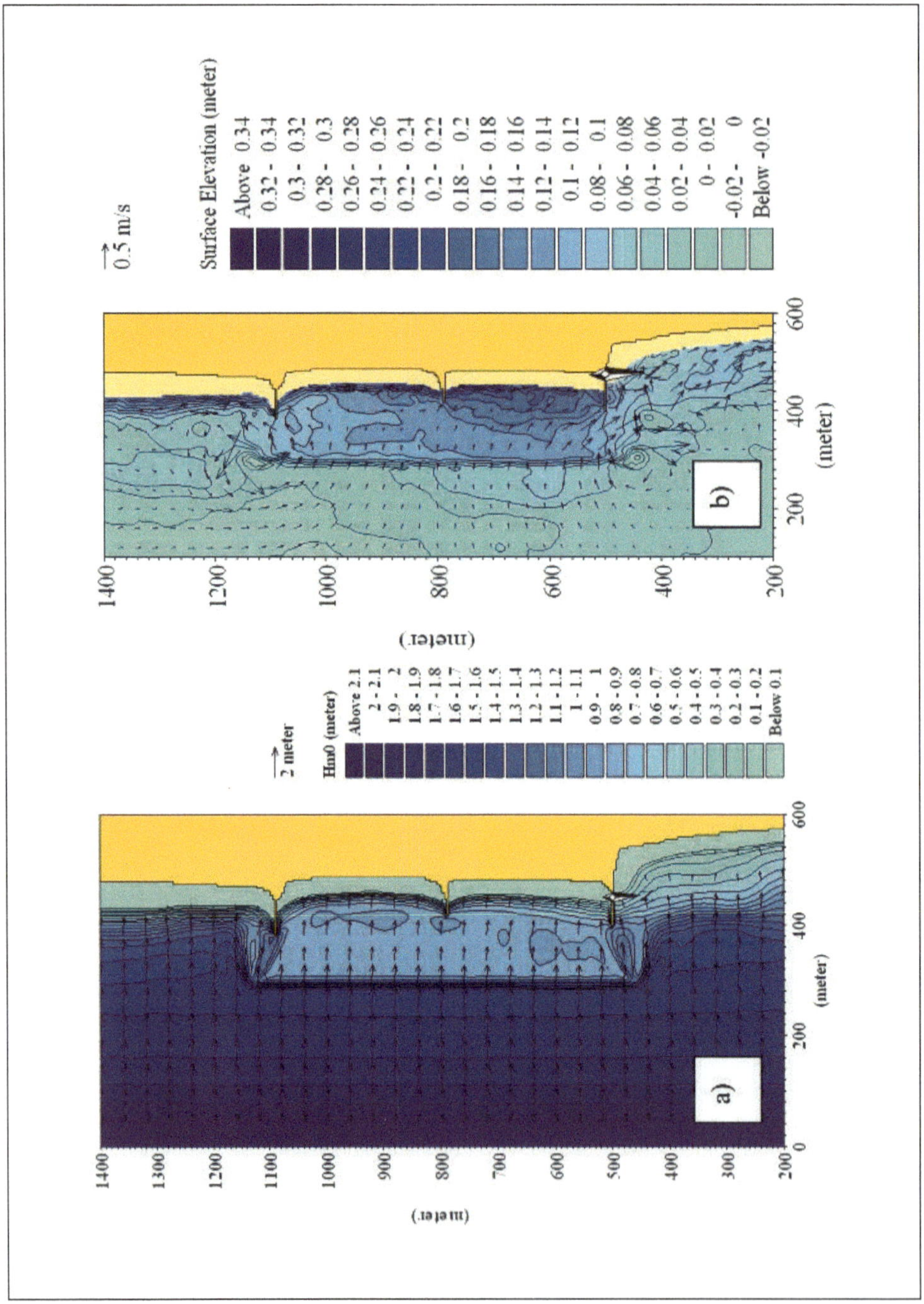

Figure 12.15. Alternative 1, submerged barrier, wave 6: a) wave field; b) surface elevation.

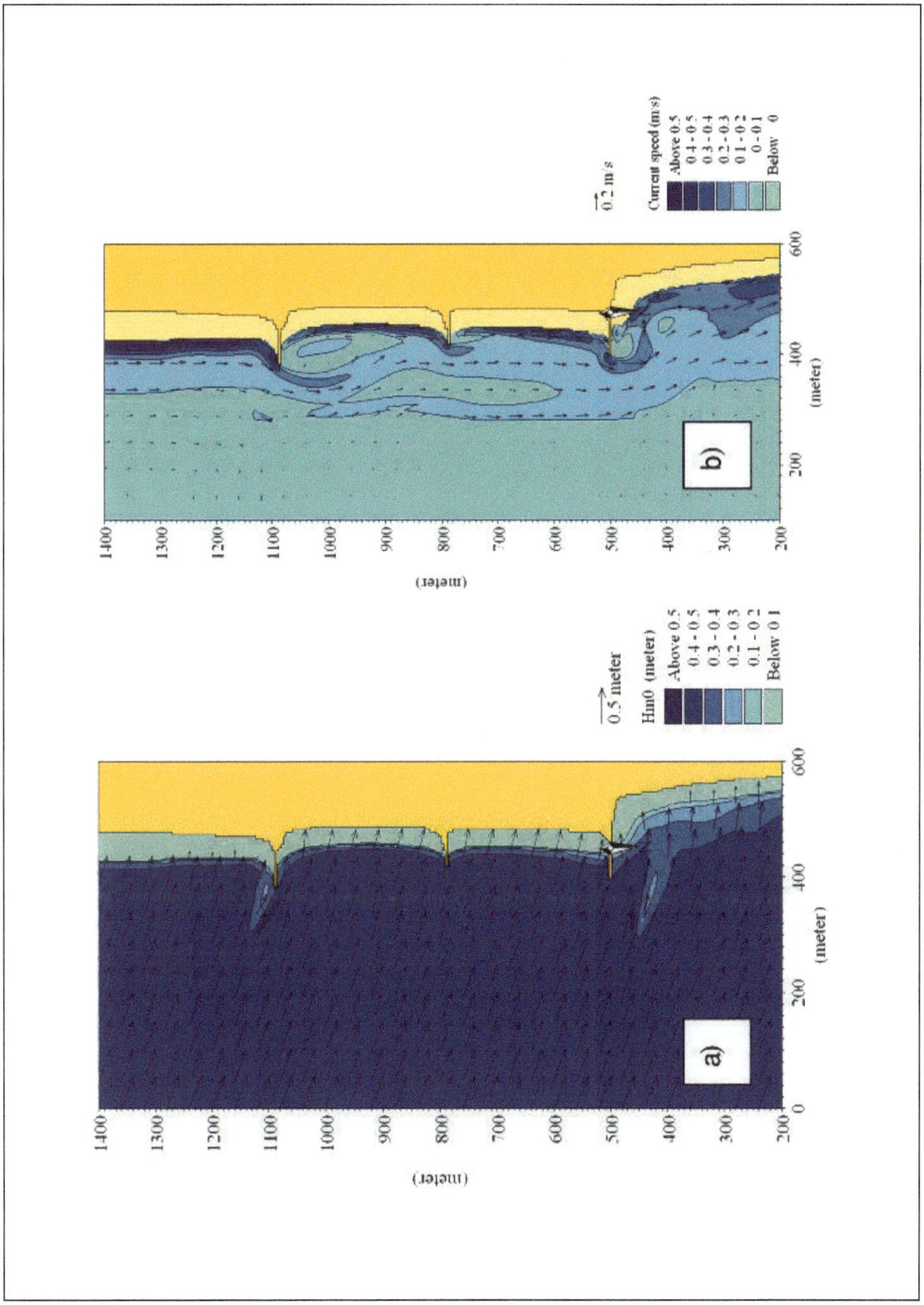

Figure 12.16. Alternative 1, submerged barrier, wave 7: a) wave field; b) current speed.

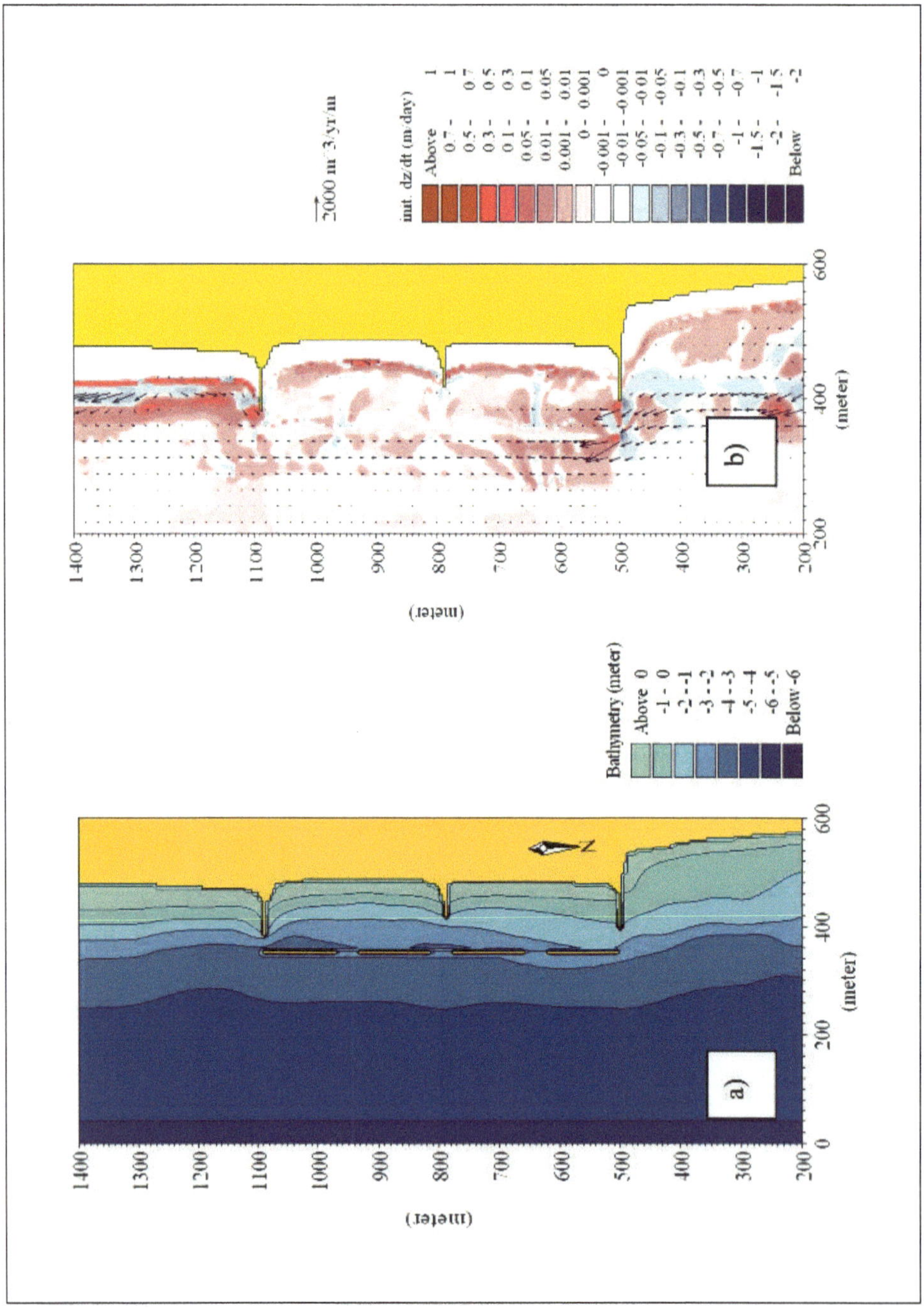

Figure 12.17. Alternative 2, emergent barriers: a) bathymetry; b) average bottom level variation per day.

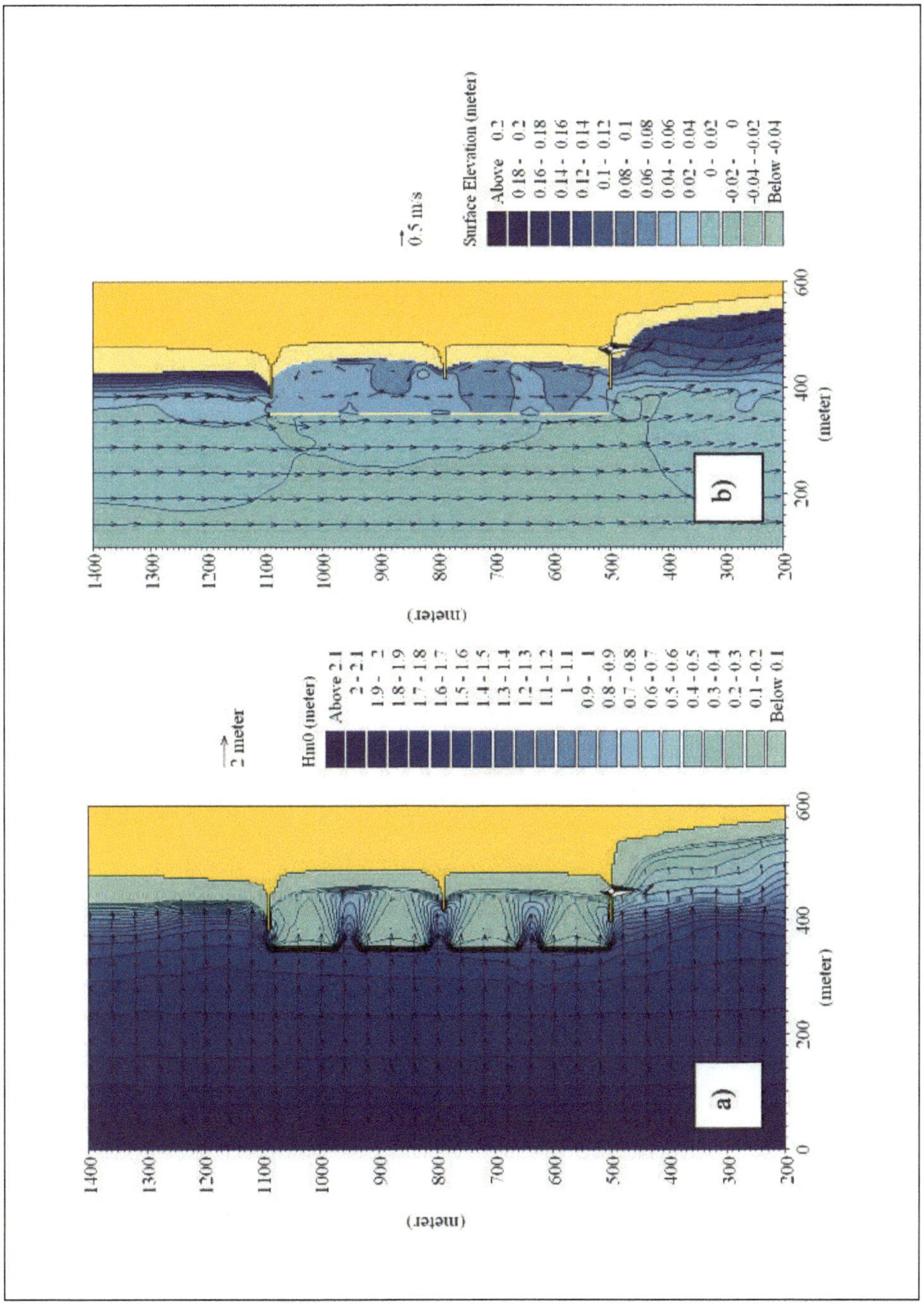

Figure 12.18. Alternative 2, emergent barriers, wave 6: a) wave field; b) surface elevation.

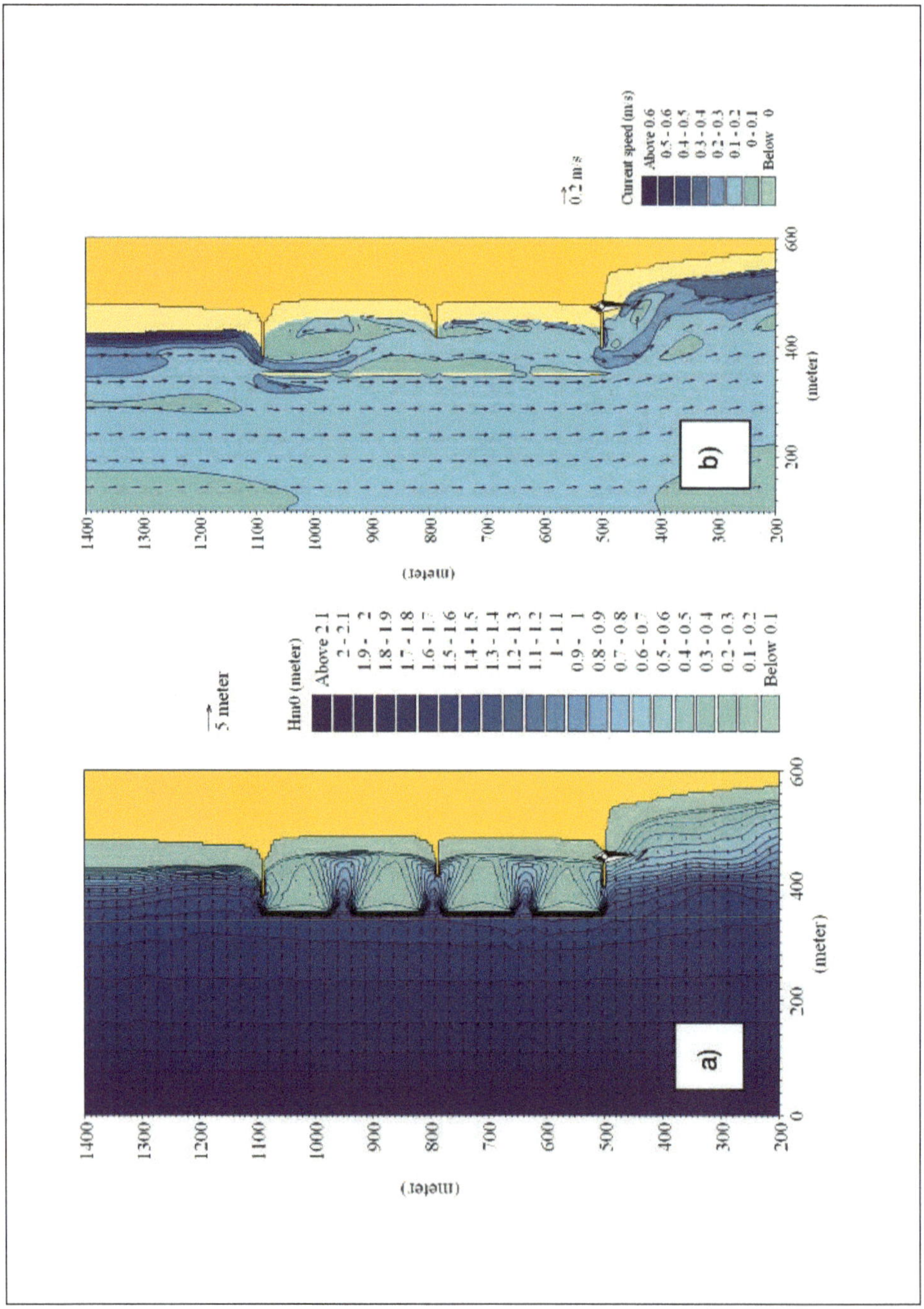

Figure 12.19. Alternative 2, emergent barriers, wave 7: a) wave field; b) current speed.

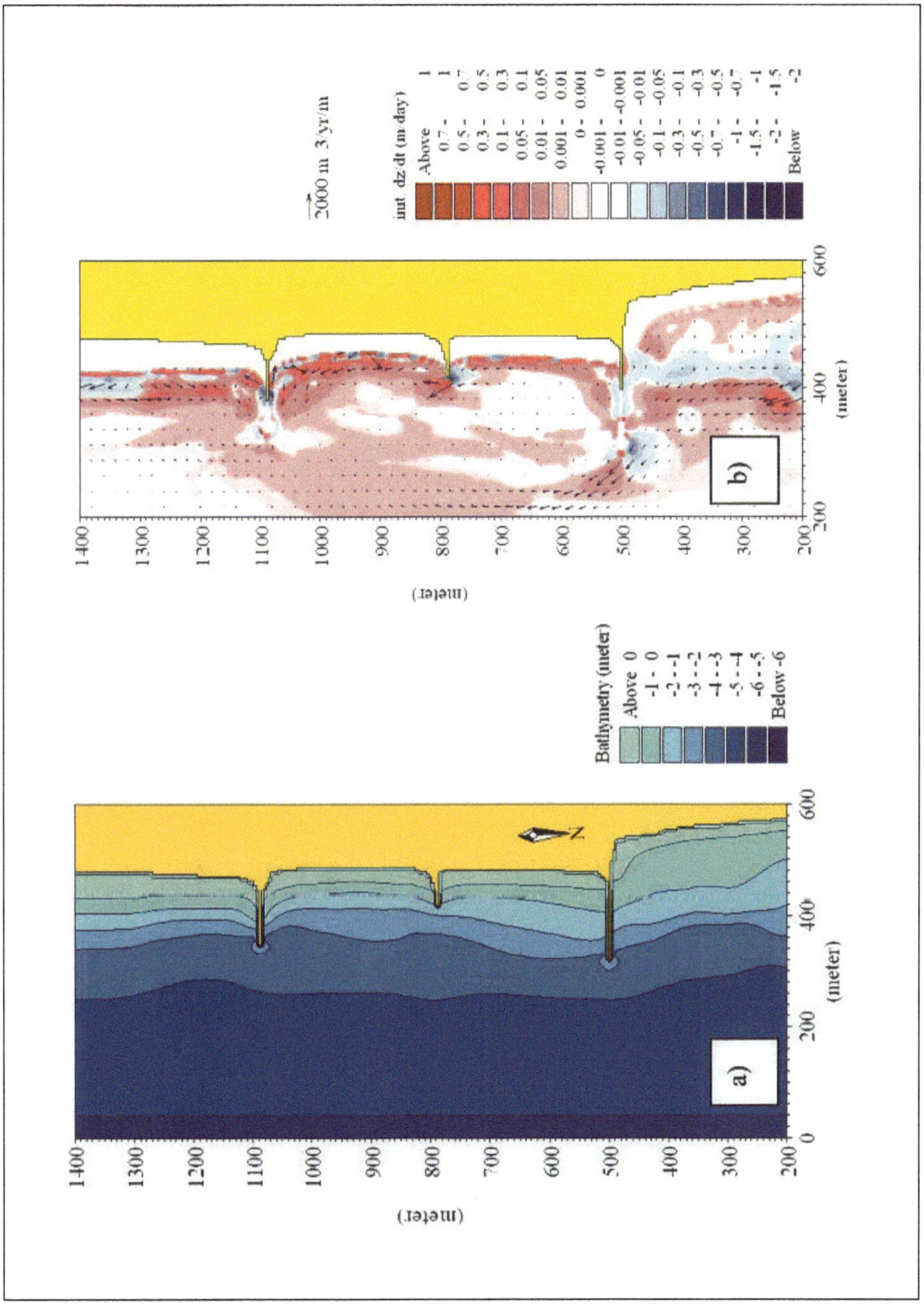

Figure 12.20. Alternative 3, groynes: a) bathymetry; b) average bottom level variation per day.

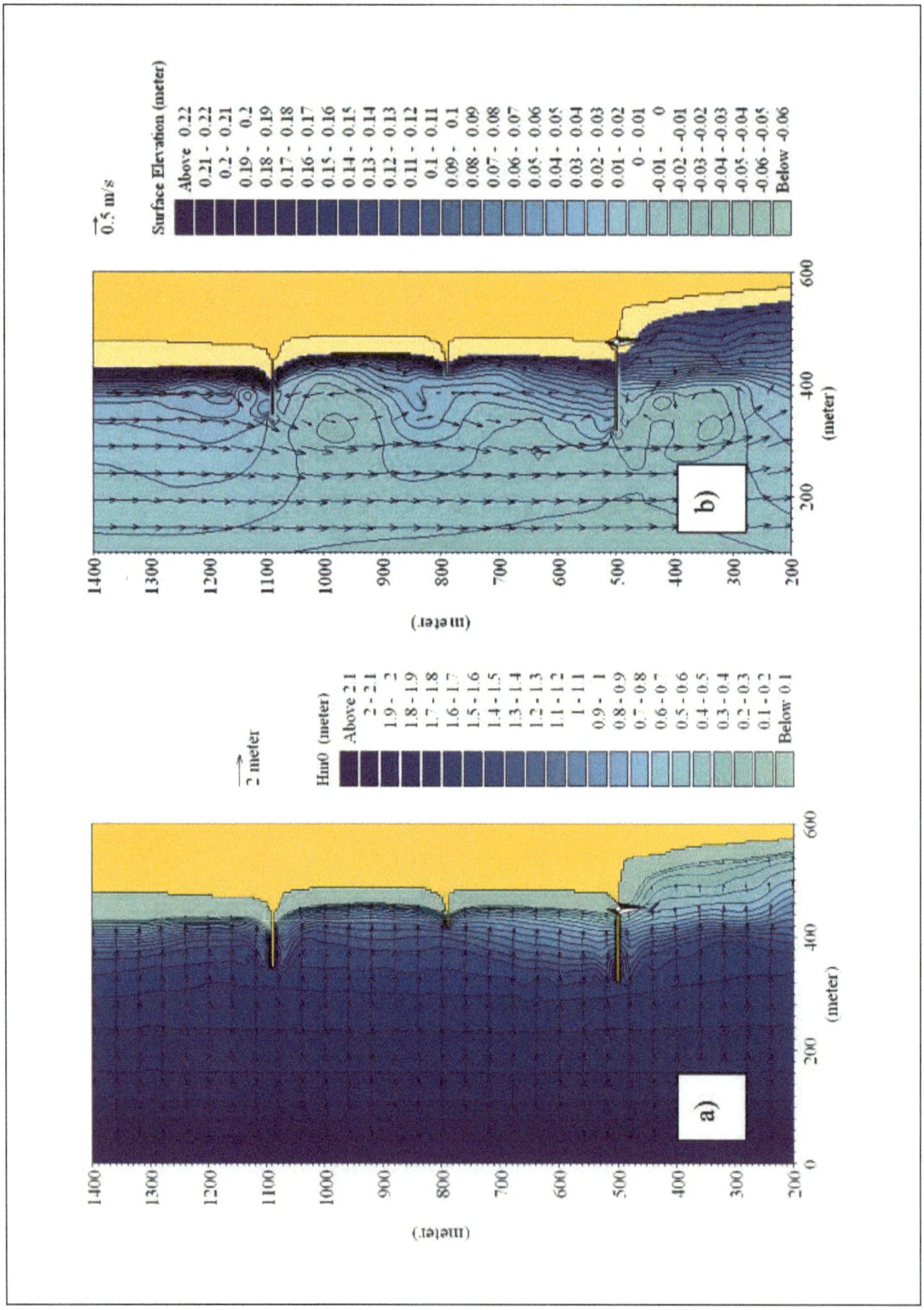

Figure 12.21. Alternative 3, groynes, wave 6: a) wave field; b) surface elevation.

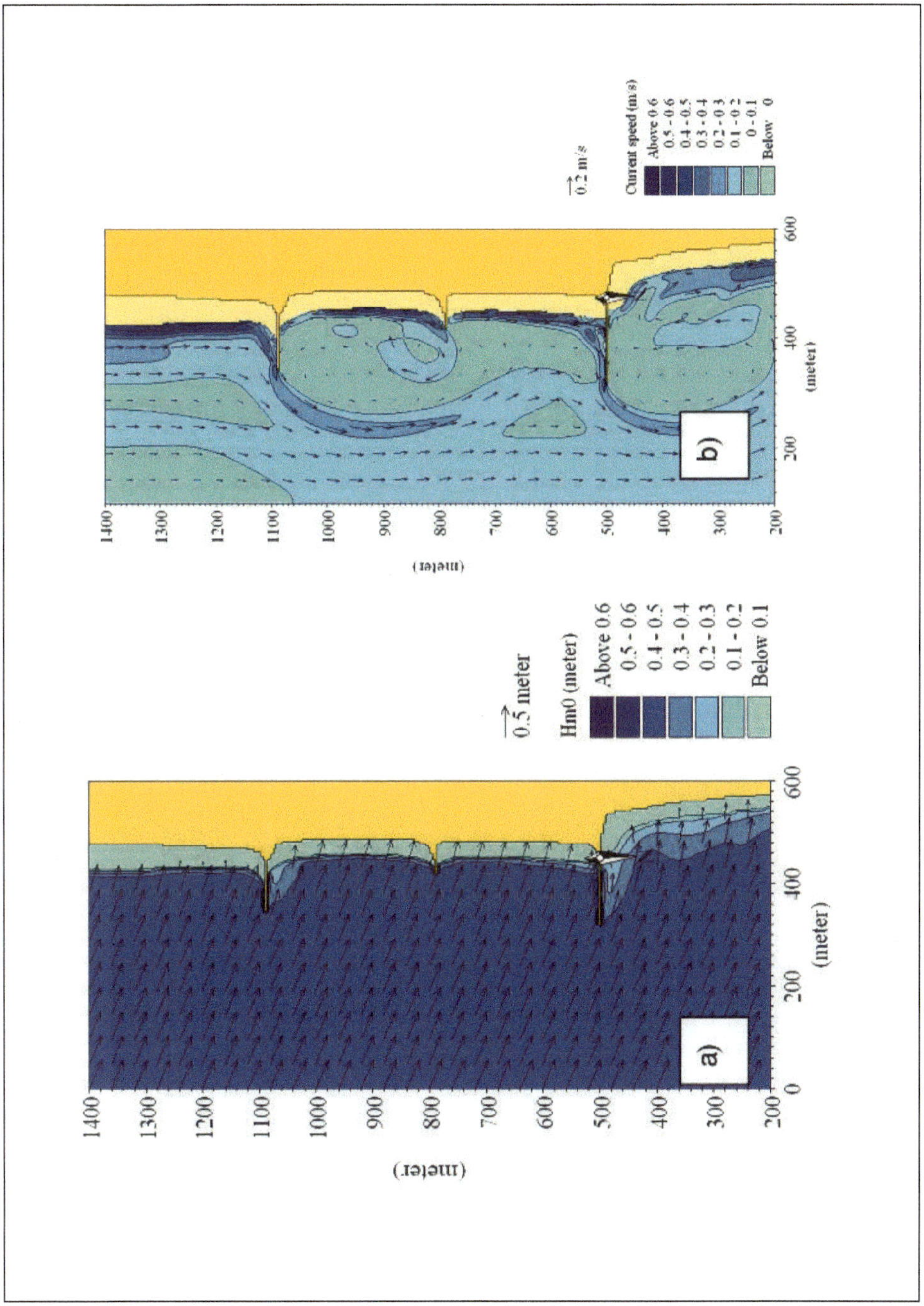

Figure 12.22. Alternative 3, groynes, wave 7: a) wave field; b) current speed.

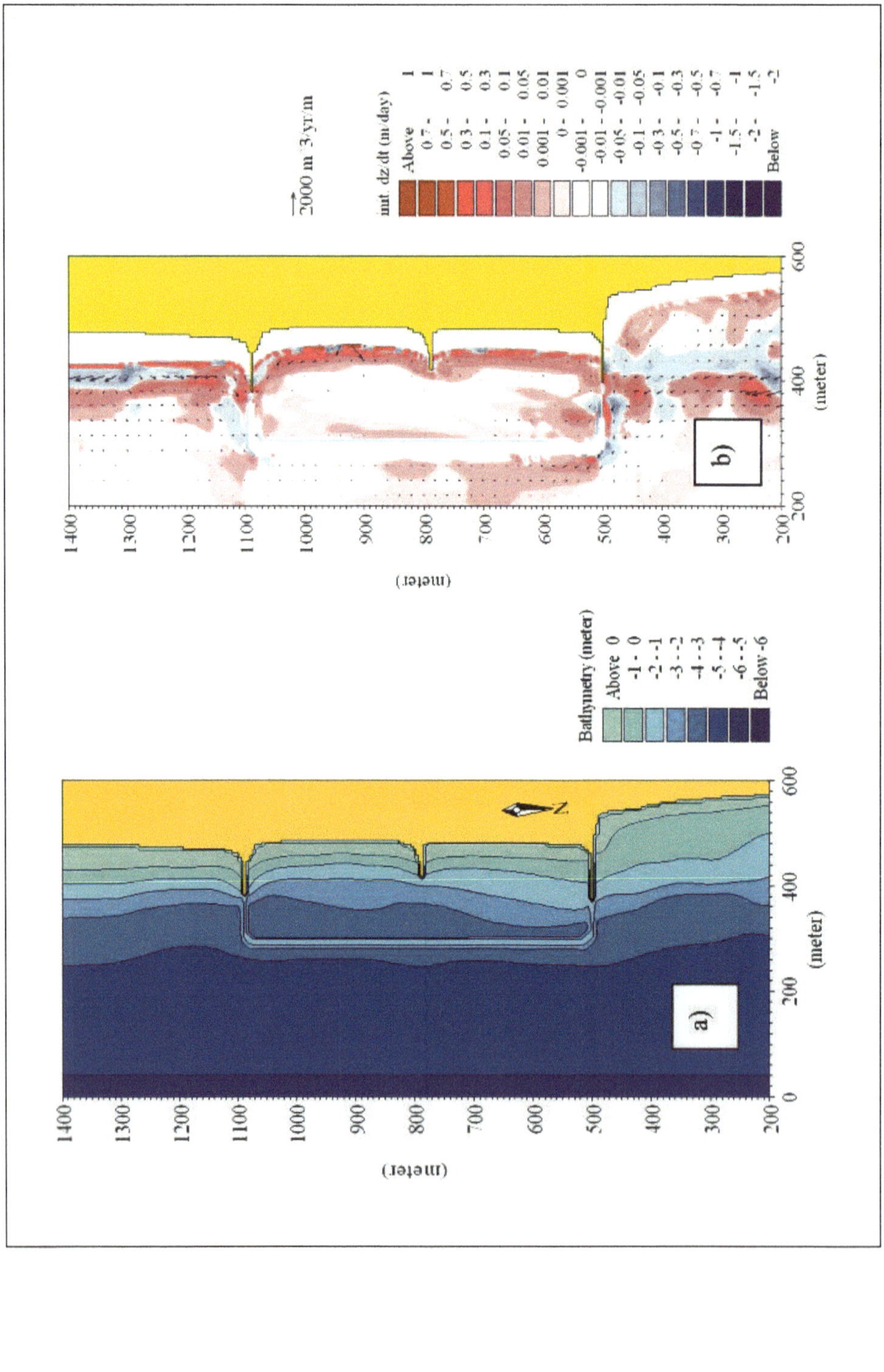

Figure 12.23. Alternative 4, submerged barrier and connectors: a) bathymetry; b) average bottom level variation per day.

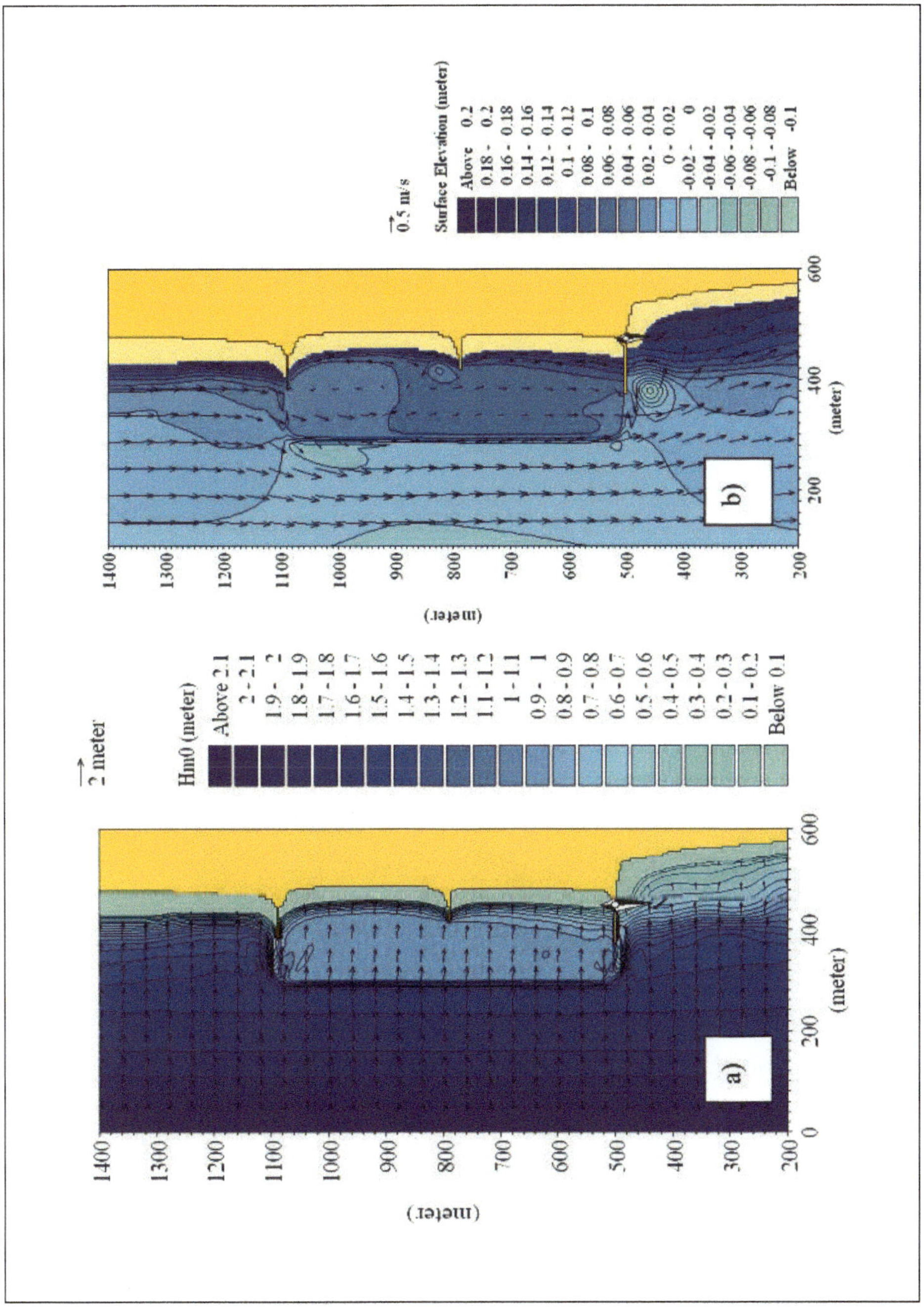

Figure 12.24. Alternative 4, submerged barrier and connectors, wave 6: a) wave field; b) surface elevation.

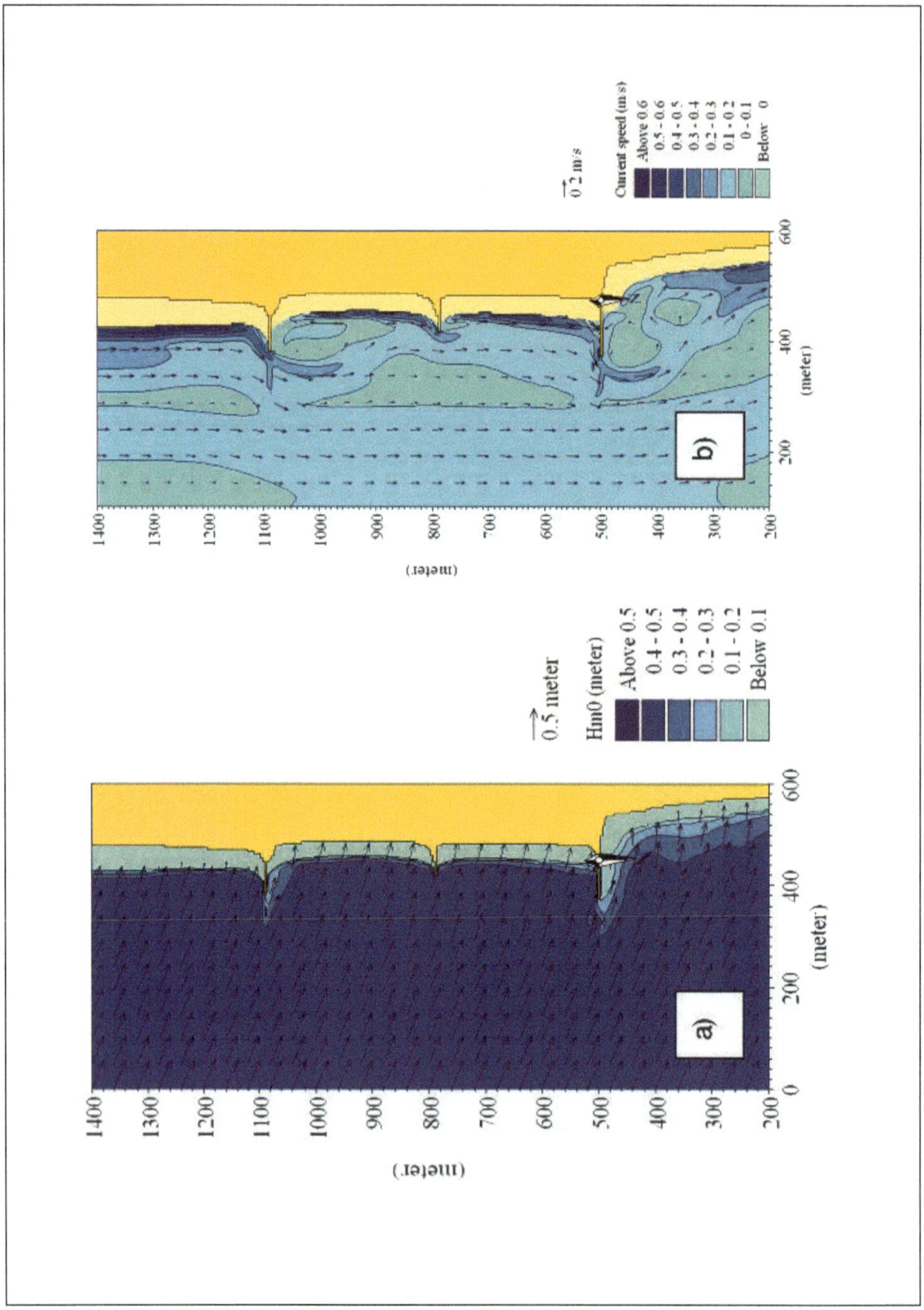

Figure 12.25. Alternative 4, submerged barrier and connectors, wave 7: a) wave field; b) current speed.

Figures 12.11 to 12.25 present, for each design alternative, the following plots in the order:

- Bathymetry of the intervention, see Figures 12.11.a, 12.14.a, 12.17.a, 12.20.a, 12.23.a;
- Average bottom level variation per day (erosion/deposition intensity in blue/red scale and sediment fluxes denoted by vectors). The deposition/erosion trend is obtained by a weighted integration (weights in Table 12.2) of all tested conditions, see Figures 12.11.b, 12.14.b, 12.17.b, 12.20.b, 12.23.b;
- Wave field (wave height intensity in both colour scale and vectors) for the most severe condition identified by Wave 6 (waves breaking at the submerged barrier, highest wave height around 1.55 m in front of the structure itself), see Figures 12.12.a, 12.15.a, 12.18.a, 12.21.a, 12.24.a, 12.12;
- Current field (set-up in colour scale; current speed intensity and direction as vectors) again for Wave 6, see Figures 12.12.b, 12.15.b, 12.18.b, 12.21.b, 12.24.b;
- Wave field (wave height intensity in both colour scale and vectors) for the lowest wave, Wave 7, to show the residual water agitation level inshore the structures in the worst conditions, see Figures 12.13.a, 12.16.a, 12.19.a, 12.22.a,12.25.a;
- Current field (speed intensity in both colour scale and vectors) for the lowest wave, Wave 7, to identify areas interested by worst circulation conditions, see Figures 12.13.b, 12.16.b, 12.19.b, 12.22.b, 12.25.b.

A summary of numerical results useful for ecological purposes is reported in Table 12.11. which presents extreme values of wave agitation and water residence time inside the protected area. These values are obtained as average values of wave height and hydrodynamic flux to water volume ratio over the protected area in correspondence of Waves 6 and 7. These values can be regarded as indicators of the intensity of residual agitation in the protected area and water exchanges with the adjacent areas, factors that can strongly affect the existing habitat.

Effects of the design alternatives on sediment fluxes are summarised in the Table 12.12, which contains long-shore and cross-shore average fluxes in correspondence of the boundaries of the protected areas and in the neighbour beaches, North and South of the two extreme groynes. Cross-shore fluxes are positive if directed inshore and long shore fluxes are positive if directed Southwards.

12.4.5.2. Comments on numerical results

Wave agitation. Both in Alternative 0 and 3 waves propagate inshore undisturbed. In the protected cell, wave energy is reduced more or less of 50% both by Alternative 1 and 4. In Alternative 2, wave agitation is almost null behind the barriers, whereas is still of importance at gaps (separated values in Table 12.3). Reduction of incident wave height on the shore is responsible of two opposite effects: one, positive, the reduction of offshore sand transport from the emergent beach; another, negative, the landward reduction of wave agitation, inhibiting deposition of fine sediments.

Currents. Current intensities induced by the Alternatives is similar, except for Alternative 2 were they are lower. Current speeds landward the structures are in the range 0.1-0.3 m/s with peaks of 0.5 m/s at the shoreline for all the Alternatives except for Alternative 2 where the maximum is 0.3 m/s. Currents in correspondence of the groyne roundheads are in the range 0.4-0.5 m/s for all alternatives except for Alternative 3, for which are in the range 0.3-0.4 m/s. These currents are directed offshore in Alternative 0 and this effect is moved more offshore in Alternative 3 by the groyne prolongation; in Alternatives 1, 4 and in a more marked way in Alternative 2 they appear to be redirected towards the beach. In Alternative 1, vortexes are induced at the submerged barrier roundheads.

Set-up. Set-up at the beach, compared to the no-structure case (Alternative 0) increases with increasing the beach protection level, in ascendant order, from Alternative 3 to 4 and 1. The only case for which set-up decreases is in presence of emerged barriers (Alternative 2).

Water mixing. Considering the values of the residence time t_r in Table 12.11, all the interventions with hard-structures imply the growth of t_r with respect to the existing situation. Alternatives 1 and 4 are the only designs that allow to maintain the range of t_r very close to the one computed for Alternative 0: t_r for lower waves (Wave 7) is nearly not affected at all, whereas for higher waves (Wave 6) is about 1.5 times the t_r for Alternative 0. In Alternative 3, the prolongation of the groynes break currents northwards directed and induced a very calm area; Alternative 2 is likely to produce the strongest effects on water circulation due to the very close environment produced by the emerged barriers.

Sediment transport. The erosion inside the protected cell, which is very high for the no-structure case (Alternative 0), is strongly reduced by the introduction of hard structures.

Alternative 1 shows a deposition tendency landward the submerged barrier, with still some shoreline erosion; seaward the barrier there is in average a deposition process whereas at the roundheads erosion takes place.

In Alternative 2, deposition occurs in average along te coastline, although erosion takes place inside gaps. The mixture of erosion and deposition patterns that seems to characterise the protected cell has to be interpreted on the basis of the more or less calm conditions produced by Wave 7 that lasts the 40% of the year (Figure 12.19.a): the global tendency is an accumulation process that can be responsible of salients/tombolos as in other places defended by breakwaters in Emilia Romagna coast, like Igea Marina, or in Marche coast, like Gabicce. The salient formation is also confirmed by applying to this design alternative the formula by Herbich (2000).

Both in Alternative 3 and 4 the deposition process is more marked near the shoreline and in the Southern part than in the Northern part of the protected area. In Alternative 4, deposition takes place both landward and seaward the submerged barrier, whereas erosion occurs in vicinity of the roundheads and of the submerged connectors.

Erosion at the groyne roundheads is present in all the alternatives.

Considering the effects on the adjacent beaches, all the alternatives induce erosion, in particular at the Northern beach.

Alternative 0 produces the highest erosion; by introducing hard structures, the erosion process is strongly reduced especially near the shore close to the Southern groyne, where some deposition takes place for Alternatives 2, 3 and 4. In Alternative 3, the sediment flux from the Northern beach is deviated far off-shore by the groyne prolongation.

Quantitative comments can be derived from Table 12.12. Alternative 2 guarantees the highest entrapment of sediments inside the protected area, followed in descendent order by Alternative 1, 4 and 3.

Table 12.11. Extreme value of wave agitation H_s and residence time t_r inside the protected cell; values are obtained as average over the cell in correspondence of Wave 6 and 7 respectively.

Alternative	**Wave agitation H_s**		**Residence time t_r**	
	Wave 6 [m]	Wave 7 [m]	Wave 6 [s]	Wave 7 [s]
0	0.92	0.44	1043	5760
1	0.84	0.40	1438	5833
2 (gaps)	0.31 (1.30)	0.05 (0.40)	2667	9600
3	0.92	0.44	2143	9130
4	0.78	0.35	1667	5676

Table 12.12. Sediment transport for each design alternative.

Alternative	*Protected Area*		
	Long-shore flux (m^3/y)	*Cross-shore flux* (m^3/y)	*Inside the cell* (m^3/y)
0	+ 51856	– 82320	– 30464
1	+ 26896	+ 3284	+ 30180
2	+ 33527	+ 4960	+ 38487
3	+ 7283	+ 3985	+ 11268
4	+ 5285	+ 9180	+ 14465

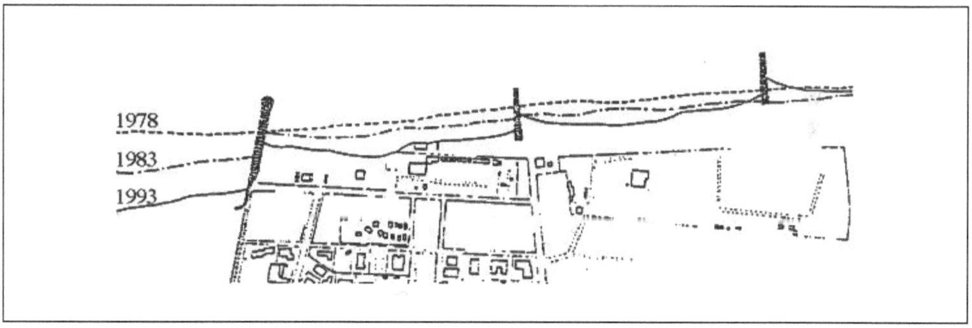

Figure 12.26. Evolution of shoreline in Lido di Dante, 1978-1993.

Alternative 0 is the only one that produces a sand loss, as expected on the basis of historical data. This sand loss for the examined cell (600 m long × 5m deep) is equivalent to 10 m/year. Data on shoreline retreat collected from 1978 (construction of the first groyne) to 1993 show an average recession of about 35 m in the protected area. Moreover, the nourishment performed in 1983 (after the shoreline survey presented in Figure 12.26) should have produced a shoreline advancement of 25 m. Surveyed shorelines in Figure 12.26 shows that shoreline retreat in the protected area is about 12 m in the period 1978-1983 and 23 m in the period 1983-1993, to which the 25 m of beach advancement have to be added. This proves that immediately after the nourishment the erosion rate is higher and the shoreline recession can be estimated as 5 m/year, corresponding to an offshore flux of 15 000 m^3/year. The overestimation of about twice in numerical simulations can be explained – even if not completely justified – by two considerations: first, simulations are carried out on a nourished and advanced profile, which was derived from a detailed 2001 bathymetry of the area; then, other nourishment of smaller entities, a part from the intervention in 1983, were perhaps performed but not recorded. In conclusion, an overestimation of about 50% shall be considered when interpreting values in Table 12.12.

12.4.6. Construction costs

12.4.6.1. Initial costs

The building costs are evaluated in a simple way, considering a tentative unit cost for the supply (from Croatia) and the placing (with a floating equipment) of each part of the structure (armour 17-21 €/m^3, dense filter 17 €/m^3, geotextile 12 €/m^2) multiplied by the actual volumes. A detailed analysis is indeed behind the scope of the example. An initial nourishment of 100 m^3 per metre of beach (40-50 m of beach advancement), giving a total of 110 000 m^3 equal for all alternatives, is also foreseen. The cost for the initial nourishment, assuming 12 €/m^3 is 1 320000 € and exceeds the building costs for all the alternatives.

Results of the calculations are reported in Table 12.13.

Table 12.13. Construction costs.

Alternative 1	*Quantity*	*Unit Cost*	*Total*
Structure (cross section Figure 12.7)	641 m	1 231.20 €/m	789 199.20 €
Roundhead with radius increased of 4 m	n. 2 (r = 14.5 m)	21 850.00 €/n	43 700.00 €
Total cost			**832 899.20 €**
Alternative 2	*Quantity*	*Unit Cost*	*Total*
Structure (cross section Figure 12.9)	376 m	1 644.50 €/m	618 332.00 €
Gaps (no armour)	108 m	836.00 €/m	90 288.00 €
External roundhead (radius increased of 4 m)	n. 2 (r = 13.0 m)	33 177.00 €/n	66 354.00 €
Roundhead at gaps (radius increased of 4 m)	n. 6 (r = 13.0 m)	16 989.00 €/n	101 934.00 €
Total cost			**876 908.00 €**
Alternative 3	*Quantity*	*Unit Cost*	*Total*
Structure (cross section Figure 12.10)	87 m	2 054.00 €/m	178 698.00 €
Additional toe protection	400 m^3	17 €/m^3	6 800.00 €
Roundhead (radius increased of 4 m)	n. 2 (r = 16.5 m)	56 222.00 €/n	112 444.00 €
Total cost			**297 942.00 €**
Alternative 4	*Quantity*	*Unit Cost*	*Total*
Structure (cross section Figure 12.7)	600 m	1 231.20 €/m	738 720.00 €
Submerged groynes (cross section Figure 12.8)	140 m	823.20 €/m	115 248.00 €
Additional toe protection	400 m^3	17 €/m^3	6 800.00 €
Total cost			**860.768.00 €**

12.4.6.2. Total Costs (including maintenance)

Maintenance for a reasonable period should also be considered for a proper analysis. The historical information suggests that the site is subjected to constant shoreline regression: the part of the beach included within the existing groynes requires a nourishment of 15 000 m^3/year in order to maintain a stable shoreline, whereas the adjacent beaches to the North and South require approximately 9000 and 1 000 m^3/year, respectively. This fixes the maintenance plan for Alternative 0.

It is suggested to moderate the frequency of maintenance, which negatively affects the development of the ecosystem, reducing the development of mussels and enhancing the ephemeral green algae. The nourishment is therefore planned every 3 years. For Alternative 0 a nourishment of 45 000 m^3/3 years for the protected areas and 30 000 m^3/3 years for the South and North beaches are planned.

On the basis of comparisons between the numerical simulations and on the basis of experience on similar sites a specific nourishment plan is formulated for all alternatives.

Maintenance is distributed in time in order to obtain an equivalent initial cost, after applying a proper interest rate. The applied interest rate (free from inflation) is 4%. Lower

Table 12.14. Initial and maintaining costs.

30 years lifetime (4% interest)	*Alternative 0*	*Alternative 1 (submerged)*	*Alternative 2 (emerged)*	*Alternative 3 (groynes)*	*Alternative 4 (multistructure)*
Building costs [€]	–	832 899.00 €	911 756.00 €	296 898.00 €	860 768.00 €
Initial nourishment [m^3]	110 000 m^3	110 000 m^3	110 000 m^3	110 000 m^3	110 000 m^3
Costs of initial nourishment (12 €/m^3)	1 320 000.00 €	1 320 000.00 €	1 320 000.00 €	1 320 000.00 €	1 320 000.00 €
Initial cost [€]	**1 320 000.00 €**	**2 153 000.00 €**	**2 197 000.00 €**	**1 618 000.00 €**	**2 181 000.00 €**
Periodic nourishment (beach between groynes)	40 000 m^3/3years	20 000 m^3/3years	10 000 m^3/3years	30 000 m^3/3years	15 000 m^3/3years
Periodic nourishment (South and North beaches)	30 000 m^3/3years	25 000 m^3/3years	35 000 m^3/3years	40 000 m^3/3years	25 000 m^3/3years
Structure maintenance		6 700 m^3/9years	5 880 m^3/9years	1 200 m^3/9years	7 400 m^3/9years
Maintenance costs (anticipated) [€]	**4 394 000.00 €**	**2 883 000.00 €**	**2 876 000.00 €**	**4 405 000.00 €**	**2 575 000.00 €**
Total costs [€]	**5 714 000.00 €**	**5 036 000.00 €**	**5 073 000.00 €**	**6 023 000.00 €**	**4 756 000.00 €**

values may also be reasonable, leading to higher equivalent initial costs.

The period considered is 30 years, which can appear a long time if compared to the usual political horizon, but is actually very short if compared to the existing structures in Emilia Romagna Region, some of them built more than 90 years ago and still under periodic maintenance.

The maintenance of the rocky structure is supposed to be rare (once every 10 years, i.e. 3 times in the considered period) and quantified in a tentative value of 10 m^3 per metre of structure (for a cost of 20 €/m^3).

It is assumed that the value of the structure at the end of the 30 years is zero. Indeed the building cost is small compared to the total and it is difficult to know whether at the end of the period the structures are still efficient or whether it will be necessary to remove them, causing additional costs.

The periodic nourishment (planned every 3 years, i.e. 9 times in the considered period) results the main cost entry in terms of equivalent initial costs. Cost for damage to adjacent beaches is not included and is similar for the different alternatives. Note that the beaches immediately adjacent to the protected area are included in the simulation and their maintenance is considered. The cost for maintenance dominates for Alternatives 0 and 3, which would appear cheaper judging on the basis of the initial costs.

Results are in Table 12.14.

12.4.7. Ecological comments to design alternatives

12.4.7.1. Preliminary considerations

Every type of LCS that is built on the coast will change the surrounding environment. Results from DELOS have shown that the severity and extent of the impacts on the habitats and associated biota depend on the physical and biological features of the coastal environment as well as the design of the LCS scheme (Martin *et al.*, 2005; Moschella *et al.*, 2005).

In Lido di Dante, the relatively shallow seabed, the eutrophic state of water and the considerable input of organic material and sediments from the nearby rivers make the area more sensitive to changes in the environmental conditions (Correggiari *et al.*, 1992). For example, under such conditions, a reduction in water circulation could indirectly facilitate the formation of toxic algal blooms and anoxic bottom sediments via nutrient retention on the lee of the structure.

The proposed design alternatives will all produce some modifications in the physical environment. These will in turn change the type of habitats present in the area, with likely consequences on species and ecosystem function. Biological responses to physical changes in the coastal environment are not linear, but can vary in time and space. Predicting ecological impacts of design alternatives with high level of confidence is therefore difficult. It is possible to forecast, however, in qualitative terms, the relative magnitude of impacts caused by each type of LCS scheme on the various components of the ecosystem (epibiota, sediment infauna, fish and shellfish) and water quality. These can be assessed on the basis of the degree of changes in the physical conditions predicted by the model, results from DELOS and the background knowledge on the ecology of sandy and rocky shores.

12.4.7.2. Forecast environmental impacts of structures

Scores indicating the magnitude of changes (from 1 being no changes to 4 being marked changes) in water movement (waves, residence time), currents and sediment transport are assigned to each design alternative (Table 12.15). Changes are assessed using the Alternative 0 as reference situation, where no intervention to hydrodynamic conditions was made.

The ecological considerations of each design alternative described below are only indicative and should be verified by studies and monitoring of real design applications. It seems clear however, that at local scale design options can induce very different ecological effects.

12.4.7.3. Alternative 2 – Emerged barriers with gaps

This design option is likely to cause the strongest changes in the surrounding environment, particularly on the landward side. The reduction in hydrodynamics on this side of the structures will markedly affect the sediments and water quality, which will in turn influence the abundance and diversity of the sediment infauna.

Water movement is considerably reduced during most of the year, leading to periods of stagnant water in summer. This will also result in deposition of very fine sediments (silt/clay) with likely increase in organic matter and decrease in oxygen. These features are not characteristic of an open beach but reflect typical lagoonal conditions, thus the species assemblages will change accordingly. In contrast, water circulation in the gaps between the structures is not affected, independently of wave conditions (summer or winter situation). The landward side is therefore characterised by areas of fine, muddy sediments with areas of coarser sand, particularly in proximity of the roundheads. The habitat patchiness is likely to increase species diversity, although this effect will depend also on the temporal stability and disturbance of these areas. For example, erosion is higher in the gaps than in normal open beach conditions, resulting in higher disturbance for infaunal species.

The presence of emerged portions of the barriers increases the diversity of rocky habitats. In respect of Alternatives 1 and 4, where only subtidal habitats are created, this design option include the intertidal zone, thus a higher number of species can colonise the barriers, including mussels and oysters. Also, different types of epibiotic assemblages will colonise the different areas of the barriers, ranging from species typical of exposed shore (seaward side, ends) to species of more sheltered habitats (landward side). On a microtidal system such as the Adriatic coast, however, the intertidal zone is very narrow, thus the increase in species diversity will be minimal. The increase in habitat diversity will also raise the risk for invasion of non-native species, which can permanently change the identity of the native species assemblages.

The lack of water mixing will also affect water quality, as turbidity will increase as consequence of sediment suspension and trapping of organic material. More importantly, the limited water circulation will facilitate formation of algal blooms, particularly during summer, when water temperature and nutrient concentration increase considerably. This will in turn cause anoxia and light depletion in the water columns with detrimental consequences for the soft-bottom benthic fauna and flora.

Potential mitigation effects of this design option might include the increase of habitat and species diversity (for appreciation of marine life), promotion of natural resources such as mussels and oysters and mobile fauna (for leisure food harvesting and fishing), and easy accessibility to the structures by beach users.

12.4.7.4. Alternative 4 – Submerged barriers with connectors

The reduction in wave transmission of almost 50% produced by this LCS design and the cell system created by the connectors and the shore-parallel barrier will create a fairly stable and homogenous sedimentary habitat on the landward side, despite the structures being submerged. Sediments on the landward side will have similar characteristic to those already observed in

Alternative 2, with fine, muddy sediment accumulating the behind the barrier. Under these conditions, diversity is likely to increase in comparison with adjacent more exposed sandy beaches, but species more sensitive to environmental changes will disappear. Siltation will also increase and hence disturbance to epibiotic species on the building blocks located in proximity of the seabed.

The submerged barriers will provide new rocky habitats for colonisation by epibiotic species, and in particular shellfish, for example mussels. The barriers will also attract fish and crustaceans by providing food resources and refuges in the cavities and gaps between the rocks. The semi-enclosed system created on the landward side can, however, prevent fish moving into this area, taking also in consideration the reduction in water depth on this side of the barriers.

Turbidity of waters will probably increase, as a consequence of sediment resuspension and siltation. Water quality can be negatively affected as nutrients, pathogens and pollutants are likely to be retained and hence accumulate on the landward side due to lack of water mixing.

The likely increase in fish and mobile fauna can be seen as a positive effect for leisure fishing and food harvesting. However, as the structures are only subtidal, appreciation of marine life will be possible only by divers or snorkellers. Furthermore, the increased siltation on the landward side can significantly reduce visibility and thus make it more difficult visiting the structures.

12.4.7.5. Alternative 3 – Extended groynes

Sediment processes appear markedly affected near the northern groyne and the southern groyne. Similarly to the landward areas of shore-parallel barriers, the habitat behind the northern groyne will be characterised by accumulation of fine grained and organic-rich sediments. In the southern groyne, erosion of sediment creates a more disturbed environment for the infaunal assemblages. The central sedimentary area between the two main groynes

Table 12.15. Magnitude of environmental changes from the reference situation (Alternative 0) induced by each design option. Both Wave 6 (winter conditions) and Wave 7 (summer conditions) simulations were considered when scoring wave agitation, residence time and currents. Scores represent degree of effects: 1 = minor, 2 = medium, 3 = marked and 4 = very marked.

	Alternative			
	1	*2*	*3*	*4*
Physical changes				
Waves	2	4	1	2
Residence time	2	4	4	2
Currents	3	2	2	3
Sediment processes	3	4	3	4
Environmental effects				
Sediment infauna	2	4	3	4
Epibiota	2	4	1	2
Shellfish & mobile fauna	3	4	2	2
Water quality	2	4	3	2

seems less affected, as frontal waves are not stopped by offshore barriers and wave energy is still high. Similarly, water quality will be less affected than in option 2 and 4, as water movement is mainly reduced in the sheltered areas behind the groynes.

The impacts of this design option appear to be more localised than with respect of the design Alternatives 2 and 4. In contrast, erosion of the adjacent beaches outside the protected coastal cell is considerably high. This defence scheme seems to produce more important large-scale effects than the other design alternatives.

The extended groynes also provide additional rocky habitats that can be colonised by both subtidal and intertidal epibiotic species, crustaceans, fish and birds. The habitat and species diversity and the easier access to the structures by beach users and in particular by children increases the recreational value of this defence scheme.

12.4.7.6. Alternative 1 – Submerged barrier

This design option seems to cause the least impacts on the surrounding environment. The ecological effects, although very similar to those of Alternative 4, are much reduced in magnitude. The absence of shore connectors makes the landward area a less enclosed environment, thus reducing problems of water quality and sedimentation. As a result, differences in the infaunal assemblages between the landward area and the adjacent beaches should be relatively smaller.

Similarly to Alternative 4, mitigation effects are limited, as the structure cannot be easily accessed by people. However, the structures still provide new habitats for fish and mobile fauna, thus promoting natural resources.

12.4.7.7. Concluding comments

The first, *a priori* environmental consideration would be to avoid any change from the original, natural conditions of the site. This is, however, a rather unrealistic option, as several engineering interventions to prevent coastal erosion had already been made in Lido di Dante since 1978, before our reference situation (Alternative 0). Therefore a more appropriate approach for such modified environment should be adopted, identifying the LCS design alternative that represents the best trade-off between engineering performance, conservation of ecological conditions and socio-economic value.

The choice of an LCS scheme should include design criteria that minimise and mitigate ecological impacts. Mitigation effects (e.g. LCS design promoting shellfish resources) can be considered as byproducts of the construction of LCS and their importance in the evaluation of design alternatives will depend on the management goals. From an ecological viewpoint, however, minimisation of impacts should be given the highest priority in the final choice of LCS design (see Table12.17). Furthermore, any potential impacts and mitigation effects of design alternatives should be considered in a geographically broader context rather than the single coastal cell where the LCS is being built. This is particularly important on the Adriatic coast, where local environmental impacts are amplified at a regional scale, due to the extensive coastal defence protection (Colantoni *et al.*, 1997; Airoldi *et al.*, 2005). Also, mitigation effects become negligible in respect of the cumulative impacts caused by the proliferation of coastal defence structures, thus overengineering should always be avoided.

All the design alternatives proposed here could be improved by modifying selected design features, as shown in several ecological studies and experiments carried out during DELOS. These include:

– making the structures more stable, thus reducing disturbance by frequent maintenance works. On the Adriatic coast, this causes great disturbance to epibiotic assemblages, which are kept at an early successional stage characterised by low diversity and patchiness. Reducing maintenance works will therefore increase diversity in epibiotic assemblages.

- Creating or increasing gaps between barriers, to facilitate hydrodynamics around the structures. Increasing porosity of the barriers, perhaps by reducing or eliminating the core. This will reduce water stagnation on the landward side.
- Increasing habitat and surface complexity, for example by creating pits and small holes or creating rock pools.
- Using limestone as building material. This is more easily weathered than other types of rocks offering therefore a rougher surface that promotes settlement of epibiotic species.

12.4.8. Socio-economic comments to design alternatives

Lido di Dante beach is characterised by a significant development of tourism facilities, due to the widespread availability of rented accommodation and the existence of campsites. Data collected during the period 1978-2001 from the Tourism Office of Ravenna show that the mean annual night stays of tourists in the area is about 90000, with a minimum of about 51000 in 1989. This reduction may be related to the severe algal blooms caused by water eutrophication in that year (see Drei, 1996). For this reason, particular attention shall be paid to the impact of design alternatives on water quality and eutrophication risk.

In summer 2002 a Contingent Valuation Method (CVM) survey (600 face-to-face interviews) was carried out here (Marzetti *et al.*, 2003, Marzetti and Zanuttigh, 2003). To the specific question about the main activities on the beach (as shown in Figure 12.27), the data about the beach use value are presented in section 11.4.7; 47.5% of respondents said that they

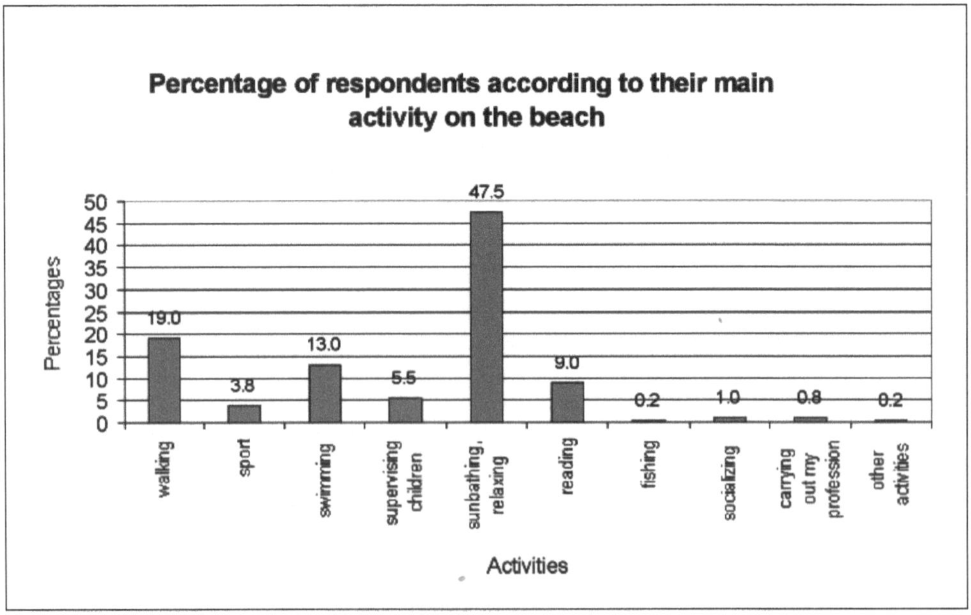

Figure 12.27. Percentage of respondents.

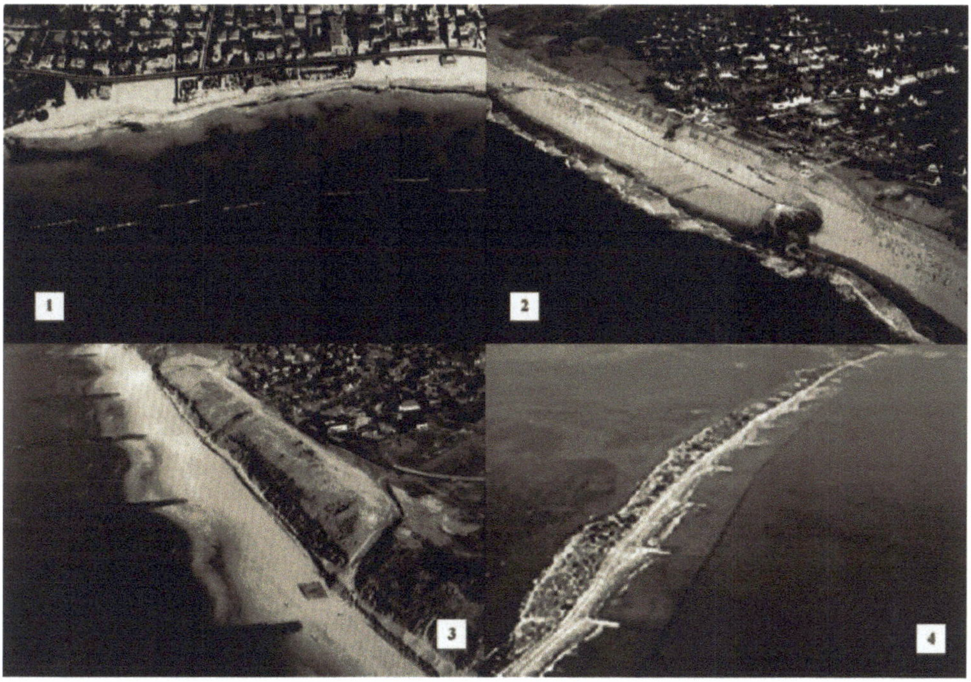

Figure 12.28. Four different kinds of defence structures. 1) Emerged parallel breakwaters; 2) nourishment; 3) groynes; 4) composite intervention.

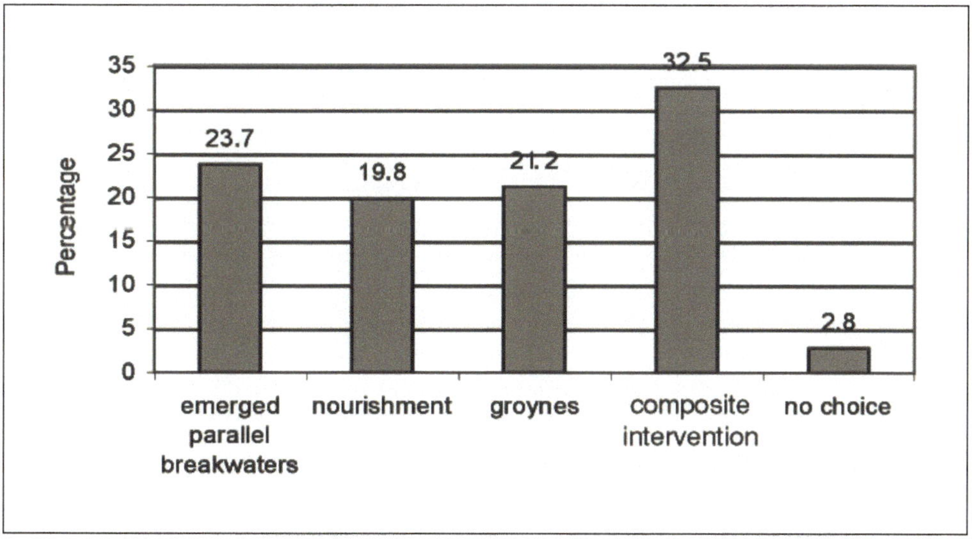

Figure 12.29. Preference about beach defence techniques: percentage of respondents.

go to the beach mainly to sunbathe and relax, 19% to walk and 13% to swim. Only 0.2% of respondents go fishing. Of those who did not choose it as their main activity, the second most preferred activity was still sunbathing and relaxing (24.2%). 32.5% of respondents practise only one activity.

12.4.8.1. Visitors' preferences regarding different kinds of defence structures and beach materials

To save time and money a CVM questionnaire is also a good opportunity to collect information other than the economic data (Marzetti *et al.*, 2003). In order to design sustainable LCS to satisfy beach visitors' preferences, some specific questions about respondents' preferences for different kinds of beach defence structures were added to the CVM questionnaire of Lido di Dante (Marzetti *et al.*, 2003):

- «The beach can be protected from erosion with different techniques. Which of these techniques do you prefer?». The photomontage presented in Figure 12.28 was shown to respondents. It shows four kinds of LCS: parallel breakwaters nourishment, groynes, and composite intervention (submerged breakwaters + groynes + nourishment).
- «Why did you choose this technique?»
- «Could you indicate a second technique together with the first one?»
- «How do you rate the presence of groynes on a beach?».

Amongst the defence techniques, as first choice, 32.5% of respondents prefer composite intervention, 23.7% emerged parallel breakwaters, 21.2% groynes (longer than those in

Table 12.16. Number of respondents according to their preferences and motives for preference.

	Aesthetic impact	*Recreational use*	*Water quality*	*Suitable for children*	*Best solution*	*Other reason*
Nourishment	82	7	23	7	7	11
Emerged breakwaters	71	12	32	13	27	2
Groynes	71	15	36	6	8	6
Composite intervention	141	2	31	2	19	5

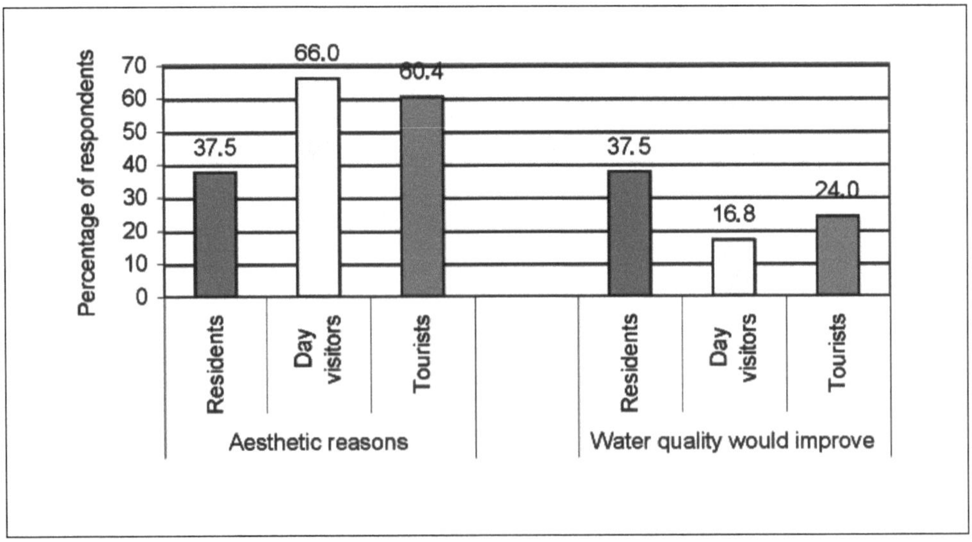

Figure 12.30. Aesthetic reasons and water quality –percentage of respondents distinguished into residents, day-visitors and tourists.

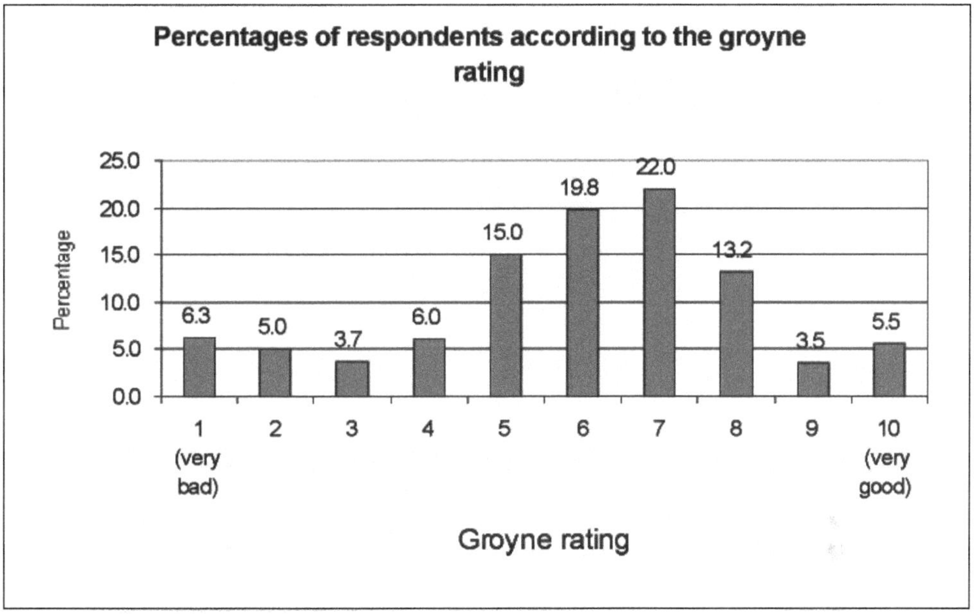

Figure 12.31. Percentage of respondents according to the groyne rating.

photo 4) and 19.8% nourishment (see Figure 12.29). Only 2.8% of respondents claim they are not able to express a preference.

As second choice, the majority (62.4%) of interviewees did not give a second preferred technique. As regards those who did, 13.4% prefer «composite intervention» and 12.9% «groynes». In addition, 25.4% of people preferring «nourishment» and 21.3% of people preferring «groynes» choose «composite intervention» as second option.

12.4.8.2. Motives for preference

As regards the main motives of preference according to the different defence structures, Table 12.16 highlights that aesthetic motives prevail for all the defence structures. The second motive of preference is «water quality» for all the different structures.

Figure 12.30 shows the different percentage of residents, tourists and day-visitors and their preferred protection technique for «aesthetic reasons» or «water quality» respectively. Residents are less interested in aesthetic characteristics than other groups of people and more interested in water quality. The majority of tourists (60.4%) and day-visitors (66.0%), instead, declared that their choice was dictated mainly by aesthetic reasons.

Finally, to the question «How do you rate the presence of groynes on a beach?» the mean rating is 5.91 on a scale from 1 to 10. More specifically, the mean rating for residents is 5.30; for day-visitors 5.62 and for tourists 6.17. Figure 12.31 shows that 64.0% of respondents expressed a rating equal to or higher than 6.

12.5. SELECTION OF THE SUSTAINABLE SCHEME

In the selection of the design alternative, each aspect presented in the previous section is accounted for and is evaluated with an appropriate weight (see Table 12.17).

Table 12.17. Evaluation rank of design alternatives.

Alternative	*Beach protection*		*Ecological effects*		*Social effects*			Total costs	Global Mark
	Shoreline maintenance	*Effects on adjacent littoral*	*Ecological impacts*	*Mitigation effects*	*Recreational use*	*Aesthetic impact*	*Swimming safety*		
0	1	3	5	1	3	4	1	2	**10.67**
1	4	5	4	2	2	5	2	4	**15.00**
2	5	2	1	3	4	2	5	3	**11.92**
3	2	1	3	3	5	3	4	1	**9.50**
4	3	4	2	2	2	5	3	5	**13.83**
Partial Weight	1/2	1/2	3/4	1/4	1/3	1/3	1/3	–	
Global Weight	**1**		**1**		**1**			**1**	

«Beach protection» weight is equal to 2 (twice the weight for ecological and social effects) as this is the main aim of the intervention. Moreover, «beach protection» is divided into two tasks. «Shoreline maintenance» refers to the results obtained with numerical simulations on sediment transport fluxes inside the protected cell. «Effects on adjacent littorals» considers the erosion/deposition effects induced in the areas close to the protected one and is based both on numerical simulations and on the experience on effects due to different defence types (as breakwaters, emergent barriers, nourishment) all along the Emilia Romagna coasts where several protection works have been built during the last 50 years. In particular, the prolongation of harbour defences like Porto Garibaldi, Rimini and Cesenatico appeared to produce strong and negative effects on the littoral zone downdrift.

«Ecological effects» have weight equal to 1 and ranking of the design options is based on the lowest ecological impact and highest mitigation effects. Ecological impacts refer to sediment infauna, epibiota and water quality; values in Table 12.17 increase with decreasing impact on present conditions. Mitigation effects refer to promotion of natural resources, habitat and species diversity with respect to the existing situation, Alternative 0. In the composite ranking, different partial weights are given to impact and mitigation effects (3 to 1 respectively).

«Social effects» are weighted as the ecological ones and again include three tasks: recreational use, aesthetic impact and swimming safety. Recreational use and aesthetic impact have been ranked in Table 12.17 on the basis of the results of the socio-economic survey. In particular, beach «recreational use» is mainly related to sunbathing and relaxing, walking and swimming (in order of importance); for this reason, this rank is strictly related both to «beach protection» and «water quality» ranks. Alternatives 1 and 4 are considered as having the same aesthetic impact and recreational use. «Swimming safety» has been evaluated looking at current intensities and directions (offshore or inshore) close to the shoreline and in some critical points as the breakwater/barriers trunks and roundheads.

Finally, «Total costs» are again weighted 1. Although not listed in the project objectives, some economic optimisation is implicit in any significant work. Indeed no particular budget restriction was indicated in the constraints and the weight of the economical aspects avoid *a priori* exclusions. Moreover, this term represents only building costs; maintenance costs are not considered as a separate item because it would have rather been a duplication of the «beach protection» term.

The sum of each weighted item in Table 12.17 indicates that the scheme to be preferred is Alternative 1.

12.6. DETAILED DESIGN

The detail design phase is applied to the preferred alternative.

The following aspects are considered:
- optimisation of functional design;
- structural design (including toe protections, bottom protections, roundhead);
- construction phases;
- maintenance plan;
- monitoring plan.

12.6.1. Optimisation of functional design

The weak points of Alternative 1 that need special care for optimisation are:
- biodiversity: the structure is characterised by a too homogenous design, with the same crest level, which does not enhance habitat and species biodiversity;
- bathing security: eddies at the barrier roundheads may be unsafe for bathers and dangerous for rescue boats;
- recreational usage: bathers can not take advantage of the structure as it is everywhere submerged without special facilities for boats;
- water quality: water circulation close to the barrier and the groynes can be improved to avoid stagnation zones;
- effects on adjacent beaches: erosion, in particular at the South of the protected cell, is enhanced by the sediment flux paths.

In order to answer to these disadvantages, the design is modified by:
- extending the barrier at the roundheads with two very low crest long aisles;
- building two small emerged islands just in front of the two external groynes roundheads;
- enlarging the width of the existing groynes to provide a walking path on them.

The following improvements are expected, with reference to the above aspects:
- both subtidal and intertidal epibiota can colonise the structure;
- the presence of the emerged islands is a clear sign of the extension of the submerged barrier and of the limits of the aisles, with increasing human safety;
- the two aisles become a secure passage for boats with clear limits and advantage to navigation;
- both islands and existing groynes can be 'colonised' by people for sunbathing and walking respectively;
- diffraction induced by the islands should generate long-shore fluxes in presence of small waves;
- negative effects on adjacent beaches can be reduced by extending the submerged defence at the sides of the protected area.

Figure 12.32 presents the final design of the structure (as built) that accounts for a foreseen 30 cm settlement. A detached barrier 800 m long is placed at 185 m from the

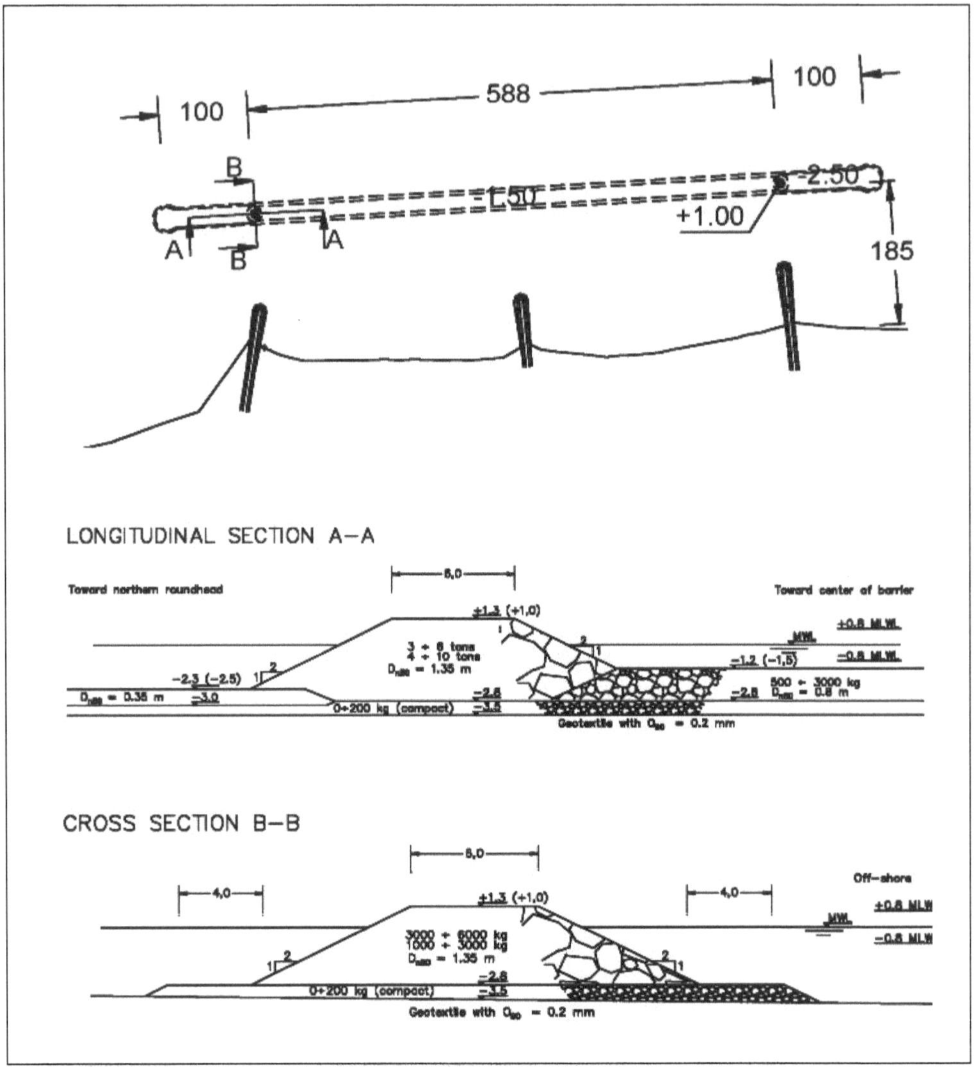

Figure 12.32. From up to down: plan view of the optimised design; longitudinal barrier section A-A; cross section of the small emerged island B-B.

shoreline on a 3.5 m depth. The structure is symmetrical and formed by three different cross sections: a central submerged part with height $H_c = 2$ m, crest level – 1.2 m, crest width $B = 6.0$ m, length $L_c = 588$ m; two emerged islands with height $H_c = 4.5$ m, crest level +1.3 m and diameter equal to 6 m; two side extensions with height $H_c = 2$ m, crest level – 2.3 m and length 100 m each; armour slope is 1:2 in all cases.

12.6.2. Structural design

The load conditions are determined by an unknown combination of water levels and waves, whose joint return period is 50 years (see Sub-section 12.2.4).

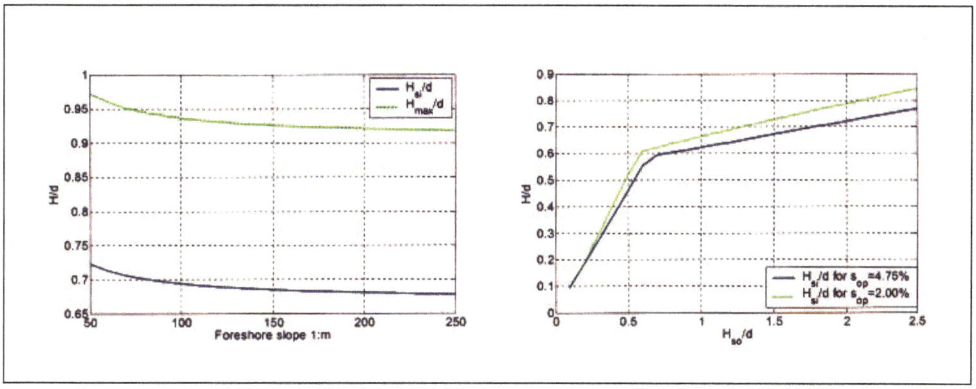

Figure 12.33. Incident wave conditions as function of a) foreshore slope $\{H_o/d = 1.6, s_{op} = 4.75\%\}$; b) wave height and wave steepness $\{\tan \alpha = 1{:}100\}$.

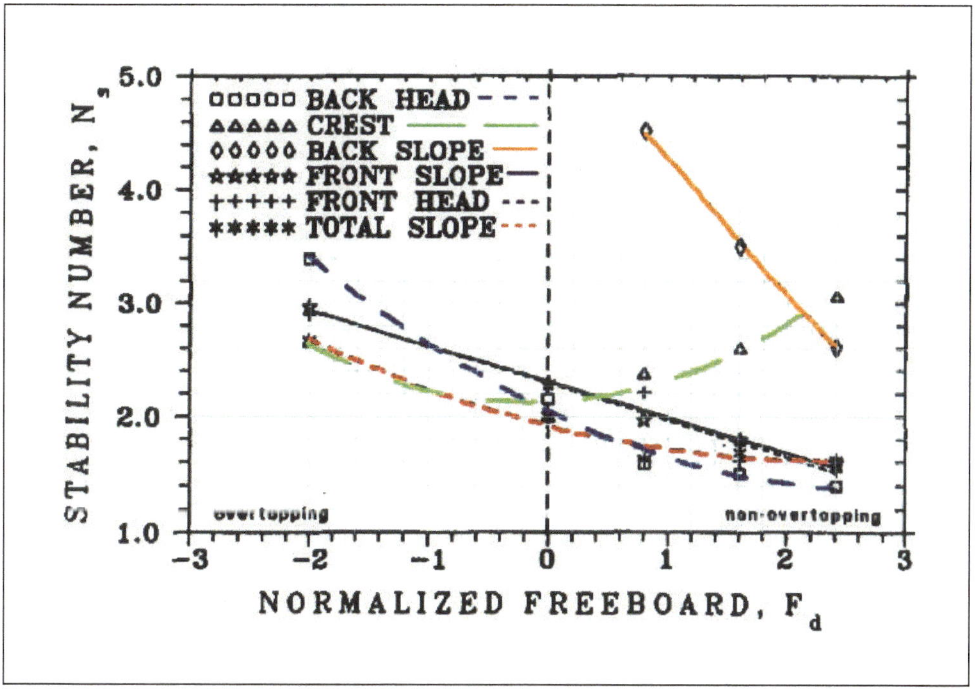

Figure 12.34. Irribarren's damage (exposure of filter layer) from Vidal *et al.* (1995), see remarks in 13.11.1.1.1.

Highest wave conditions, for 50 years return period are: $H_{o50years}(d = 30\text{ m}) = 6.0$ m; $T_s = 9$ s. The tidal extremes (including storm surge) for the design return period are: – 0.93 m a.s.l. and + 1.09 m a.s.l. A likely value of off-shore wave height, expected simultaneously to extreme water level, is $H_o = 5.0$ m, $T_s = 8.5$ s, whereas a likely value of water level, expected simultaneously to the extreme wave, lies in the range – 0.65 m + 0.78 m a.s.l. Foreshore slope is 1:100.

In order to obtain the target submergence, the structure is built assuming that 30 cm bottom settlement will take place in the first year(s). The structure stability must be verified also in this initial condition, with crest freeboard 0.3 m higher than in long term design condition. The return period for the loads in this initial phase is 5 years, with same tidal range: $H_{o5years}(d = 30\text{ m}) = 4.5$ m; $T_s = 8.5$ s.

The load is known off-shore. Wave height incident on the structure is evaluated according to Goda formula for wave transformation (see Sub-section 13.2.4). Figure 12.33

Table 12.18. Design of armour layer – initiation of damage – Structure after settlement (target R_c).

50 years return period		*Cross section*	*Island*	*Side extensions*
Geometry and D_{n50} by rule of thumb, Eq. (13.112)				
H_c	[m]	2.0	4.5	1.0
d	[m]	– 3.50	– 3.50	– 3.50
D_{n50} {Rule of thumb}	[m]	0.60	1.35	0.30
Critical combination of tide and incident wave load related to Eq. (13.111)				
H_s	[m]	2.08	2.84	2.08
$h = -3.5 + z_m$	[m]	2.85	4.28	2.85
Stable stone according to Eq. (13.111)				
$R_c = H_c - h$	[m]	– 0.85	+ 0.22	– 1.85
D_{n50}	[m]	0.79	1.36	(not applicable)
$N_s = H_s/(\Delta D_{n50})$	–	1.68	1.32	(4.40)
Stability at Iribarren damage level (trunk ÷ roundhead), see Fig. 12.34				
R_c/D_{n50}	–	– 1.1	0.2	– 6.1
N_s for Iribarren damage	–	2.2 (trunk) 2.2 (roundhead)	1.8 (trunk) 1.9 (roundhead)	(not applicable)
Design				
D_{n50}	[m]	0.8	1.35	0.35
W_{n50}	[t]	1.3	6.5	0.1
Design composition		2 layers (40% 0.5-1 ton 60% 1-3 tons)	1 layer 3-6 tons + 2 layers 4-10 tons	2 layers 50-200 kg
Obtained thickness of armour	[m]	1.6 ($\approx 2 \cdot D_{n50}$)	4.1 ($\approx 3 \cdot D_{n50}$)	0.7 ($= 2 \cdot D_{n50}$)
Obtained thickness of filter	[m]	0.7	0.7	0.5
Expected settlement	[m]	– 0.3	– 0.3	– 0.2
Obtained height of structure	[m]	2.0 + 0.3	4.5 + 0.3	1.0 + 0.2

Table 12.19. Verification of armour layer stability – initiation of damage – structure «as built» (H_c 30 cm) higher than target in view of possible settlement.

5 years return period		*Cross section*	*Island*	*Side extensions*
H_c (0.3 m)	[m]	(2.3)	4.5 (4.8)	(1.3)
$D_{n50\ \text{Rule of thumb}}$	[m]	0.69	1.44	0.39
Critical combination of tide and incident wave load related to Eq. (13.111)				
H_s	[m]	1.96	2.72	1.96
$h = -3.5 + z_m$	[m]	2.85	4.28	2.85
Stable stone according to Eq. (13.111)				
$R_c = H_c - h$	[m]	– 0.55	0.52	– 1.55
D_{n50}	[m]	0.81	1.35	0.36

shows the result of the transformation and shows the sensitivity to the foreshore slope and to the off-shore wave conditions.

12.6.2.1. Design of Armour layer

Table 12.18 gives details of the armour stone design, carried out with Eq. (13.111) given in Sub-section 13.11.1.2.2, with $\Delta = 1.57$, $\rho_s = 2.65$ t/m^3, $\rho_w = 1.03$ t/m^3. The rule of thumb (13.112) used for the preliminary design is basically confirmed. Table 12.19 verifies the stability immediately after construction, before settlement occurs.

For the permanently submerged parts of the structure, the most extreme condition occurs for low water levels, since the presence of a water cover shelters the structure from the wave impact. On the contrary, for the parts of structures always emerged, high water levels are more critical, since the most important effect of mean water level is to limit by breaking the incident waves and, in case of high water level, waves transferred form offshore to the structure are higher.

The suggested safety factor of 1.1 on the diameter (i.e. 1.3 on weight), expresses the uncertainty level for armour stability, provided that the toe is stable. In this example, like in many other cases, the crest level is a design requirement, and further security on stone geometry involve thicker armour layer, which requires a significant bottom excavation. When bottom excavation is not desired, over-design of stones is not geometrically possible, and the risk of structure damages should be accounted for in the maintenance plan.

For design optimisation, it may sometimes be convenient to differentiate the trunk section from the roundhead. Figure 12.34 shows the stability number in different parts of the structure: the trunk section damage is indicated by the «total slope curve», whereas the roundhead is the minimum between «crest», «back head» and «front head». In the present design the stability number for trunk section or roundhead is similar, so that in such conditions a differentiation of the armour along the barrier is not suggested.

The selected armour is a combination of different classes of stones, available on the market. The final grading has ratio D_{85}/D_{15} lower than 2 (as recommended in van der Meer *et al.*, 1996). The armour stone size designed for the emerged structure (4-10 tons) is not easily available in the area. In order to use stones of smaller dimensions, the emerged islands may be built with a milder slope of the armour. This shape requires bigger volumes of material and is advantageous with respect to reflection. The beneficial effects of a milder slope can be roughly assessed by computing the ratio between the stable armour layer stone (based on van der Meer, 1992) for the 1:3 and the 1:2 slopes. The reduction factor results to be 82%.

12.6.2.2. Design of toe berm

For the sake of construction simplicity, the filter layer and the toe berm are formed by the same material. The compatibility with the foundation is investigated in the following, when filter design is investigated.

The stability criterion for toe berm is given by Eq. (13.120), Sub-section 13.11.3.1.

The berm is 4.0 m wide, and therefore formed by many stones in order to tolerate some damage. A wide berm is also useful to support possible stones displaced from the armour. Should this happen, the berm will retain the removed stones, reducing the effective slope of the armour layer which then becomes more resistant.

Different tide conditions are investigated. In high tide, since waves are depth limited, the load on the structure increases. It is seen that the stability number, representing the structure resistance, also increases, but not so much. The critical conditions are indeed found in this design for high water levels.

12.6.2.3. Design of filter layer

The median stone designed in the previous paragraph can be adopted only as filter layer. According to the filter role this layer is compatible with the armour.

In the following, the toe berm/filter compatibility with the underlying sand is investigated, considering that only one layer is geometrically feasible.

For the filter-bottom interface the filter rule ($D_{15F} < 4D_{85B}$; $D_{50B} = 0.2$ mm) results in a condition which is not internally stable. Design practice suggests that internal stability condition is $D_{60F}/D_{10F} < 10$ (with no further requirements). Actually the internal stability rule can be obtained, at least conceptually, applying repeatedly the filter rule, if the amount of

Table 12.20. Design of toe berm for start of damage ($N_{od} \approx 2$).

		Submerged			*Emerged*		
H_c	[m]	2.0			4.5		
h_t	[m]	0.7			1.0		
h	[m]	2.85 (low tide)	3.50 (no tide)	4.26 (high tide)	2.35 (low tide)	3.00 (no tide)	3.76 (high tide)
H_s		2.08	2.42	2.82	1.91	2.16	2.56
D_{n50F}	[m]	0.42	0.45	0.48	0.48	0.47	0.50
W_{n50F}	[kg]	196	241	293	293	275	331
$N_s = H_s/(\Delta D_{n50})$	[m]	3.13	3.45	3.77	2.52	2.90	3.24

fine material in the bedding layer is sufficiently controlled. This is suggested for instance in Pilarczyk (2000), where, for the internal stability, it is suggested $4D_{05} > D_{10}$, $4D_{10} > D_{20}$, $4D_{20} > D_{40}$, etc., which can produce a compact material with small pore size D_P ($\approx D_{05}/5$, e.g. 1 mm) compared to the larger stones ($D_{80} \approx 250 \cdot D_{05} > 1$ m).

A small advantage in the design of the filter-foundation interface, when the bottom is made of non-cohesive fine material, relies in the application of hydraulic stability conditions. The shear stress in the fluid flowing in the filter layer is induced by hydraulic gradient and its intensity is conditioned by the pore diameter. It is desired that such shear stress is not sufficient to move the material of the foundation, possibly present in the pores (hydraulic filter condition for the bottom material). Such requirements is less strict than the geometrical filter rule.

Table 12.21 shows the characteristics of the designed filter.

Table 12.21. Design of filter layer.

Armour and foundation geometry		
D_{n50A} {Table 12.18}	[m]	0.80÷1.35
D_{50B}	[mm]	0.2
Hydraulic condition for interface with bottom		
ψ_{cr} {see for instance Pilarczyk, 2000}	[-]	0.06
H_{so}	[m]	5.0
z_m	[m]	1.09
H_{si}	[m]	2.9
k_t	[m]	≈ 0.5
B	[m]	30
j {$\approx H_{si}(1+k_t)/(2B)$}	-	0.07
Δ {$= (\rho_s-\rho_w)/\rho_s$}	-	1.57
D_P {$= 4\,\psi_{cr}\,\Delta\,D_{50B}\,/\,j$}	[mm]	1.03
Design of filter (D_{50F} is chosen in order to be stable also as toe berm)		
D_{50F} {$D_{50F} > D_{50A}/4$}	[mm]	480
D_{25F} {$\approx D_{50F}/4$}	[mm]	120
D_{10F} {$\approx D_{25F}/6.25$}	[mm]	20
D_{05F} {$\approx D_{10F}/4$}	[mm]	5
D_P {$\approx D_{05F}/5$}	[mm]	1

12.6.2.4. Design of geotextile

Placement of geotexile is planned for additional security. The geotextile is designed in HDPE (polyethylene) non woven (flexible and permeable, resistant to punctures) for $O_{90} = D_{50B} = 0.2$ mm, 600 g/m^2. It is placed by rolling it down across the section by divers, assuring a 50 cm overlapping, and anchoring it to the toe berm.

12.6.2.5. Design of roundhead

The roundhead is designed with a radius 4 m wider than the barried, in order to ensure stability and reduce the currents.

12.6.2.6. Design of details

The submerged barrier must be properly signalled to navigation. Although the structure has nominal crest level of $R_c = -1.1$ m with respect to MLWS, a controlled path of – 1.40 m is foreseen and signalled, whereas the remaining part can not be crossed. The passage is relevant with regards to bathing safety, surface.

In order to increase the recreational use of the site, the existing groynes should be maintained, providing a smooth surface.

12.6.3. Verification of expected optimisations

The expected improvements, already identified two sections above, have been verified through numerical simulations carried out with MIKE21 as already done previously for each design alternatives. By comparing the results obtained for the optimised design (Figures 12.35 to 12.37) with simulations for Alternative 1 (Figures 12.14 to 12.16), it can be seen that in the optimised design:

- sediment fluxes produce everywhere sedimentation close to the shoreline and a strong reduction in erosion induced at the Northern beach;
- erosion persists at the barrier and groyne roundheads;
- erosion is present also landward the barrier and inside the protected cell far from the shore;
- wave heights are reduced ($H_s = 0.2 - 0.8$ m);
- eddies at the barrier roundheads, in particular in presence of Wave 6, are characterized by lower intensity;
- currents inside the protected cell are characterised by lower intensities, especially close to the Southern groyne and to the shoreline. Maximum values are reached close to the groyne roundheads and rise up to 0.4 m/s.

In conclusions, numerical results confirm the desired improvements and enhance an additional improvement in deposition trends close to the shoreline.

12.6.4. Maintenance plan

Possible failure modes of the works are beach erosion and structure damage and settlement.

A suitable state indicator for beach erosion is the beach width. Accounting for tidal excursion, wave climate and beach slope, the beach shall be at least 35 m wide up to the first infrastructures or the dunes, whereas its target value is 40 m. The maintenance action is renourishment aiming to obtain the target width; since the beach is approximately 5 m high and 700 m long, the necessary sand volume is 16.500 m^3 and numerical modeling shows that intervention should be scheduled every 3 years.

The breakwater performance is strictly related to its crest height and width, whose target

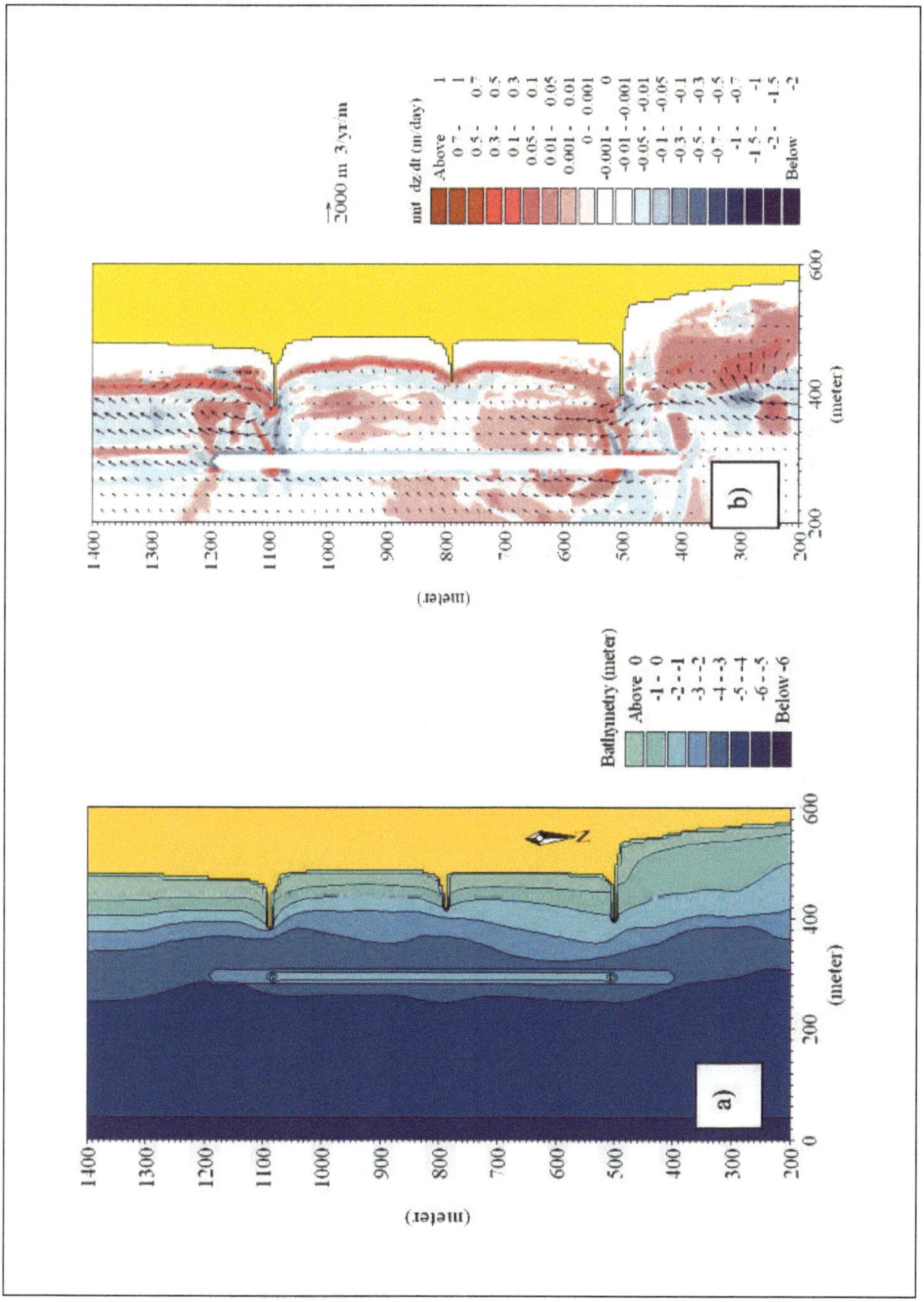

Figure 12.35. Optimised Alternative: a) bathymetry; b) average bottom level variation per day.

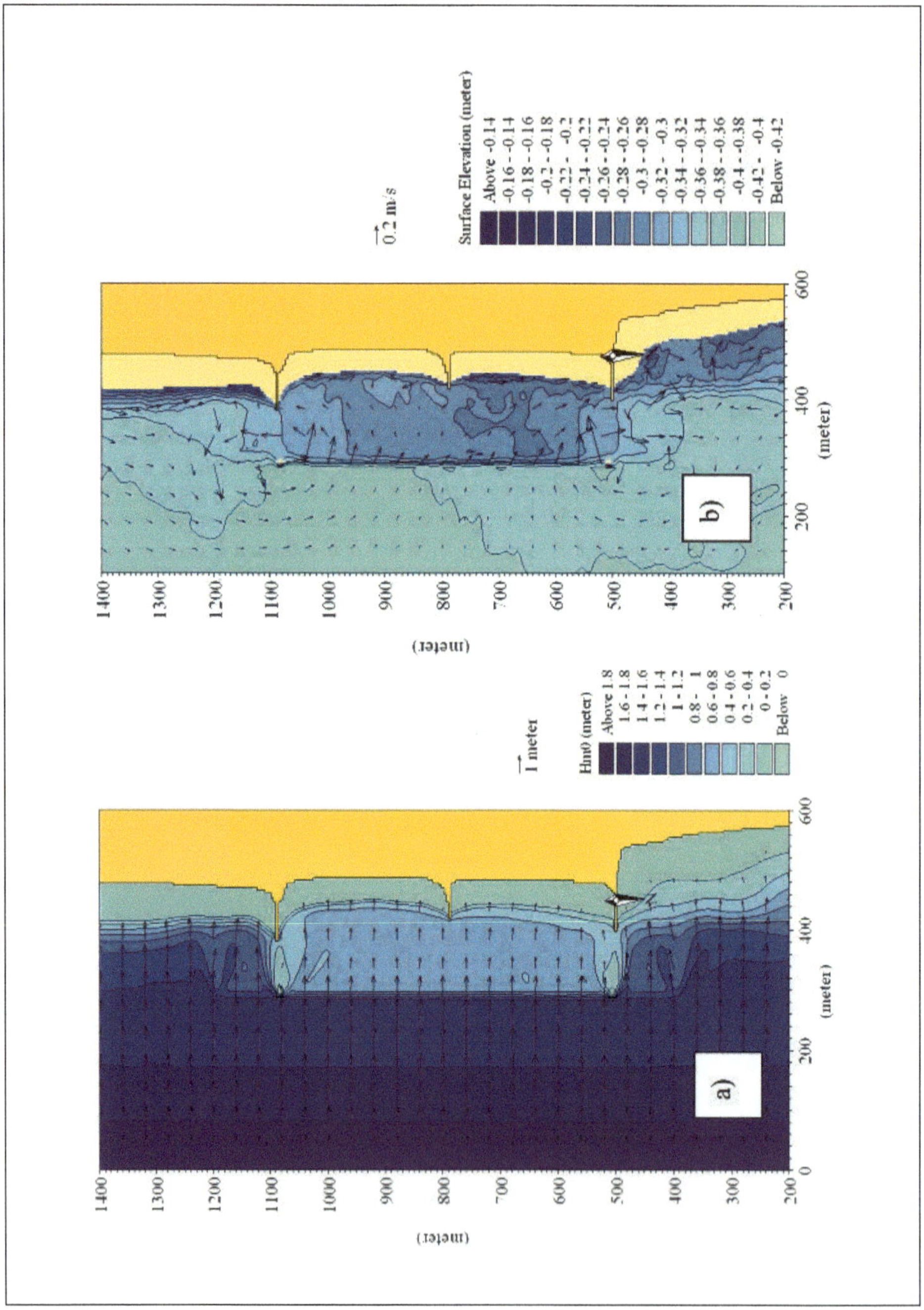

Figure 12.36. Optimised Alternative: Wave 6: a) wave field; b) surface elevation.

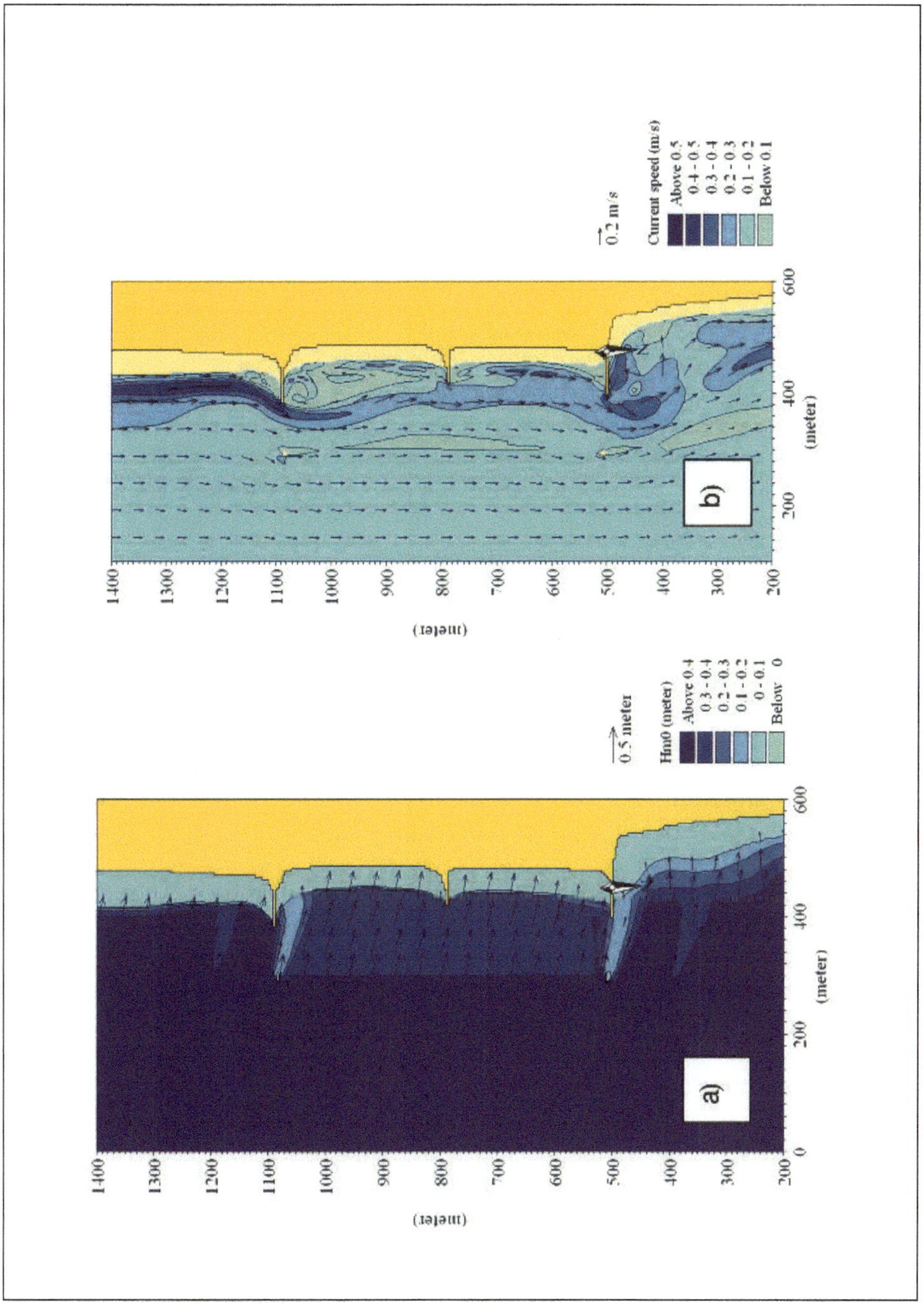

Figure 12.37. Optimised Alternative, Wave 7: a) wave field; b) current speed.

values are provided in Fig.s 12.7 and 12.32. Considering also stone size, a significant loss of functionality and possible reintegration is foreseen when any cross-section is reduced more than 6.4 m^2 (half stone size times crest width). Stones shall be placed in the most eroded part of the profile.

In order to avoid regressive erosion from the structure toe, if scour holes exceed twice the berm stone size along the trunk, i.e. 1 m, and twice as much at the roundheads, toe design profile shall be restored.

The global stone reintegration volume is estimated to be around 7.000 m^3, and the maintenance frequency is once every 10 years approximately (after 50 significant storms).

12.6.5. Monitoring plan

A monitoring plan includes:

- evaluation of transmission, piling up and rip currents during first significant storms. This can be achieved by a set of instruments measuring simultaneously waves and currents at both sides of the barrier and at the gaps;
- continuous monitoring of direction and intensity of waves. Available ondametric buoys in the North Adriatic do not cover the Emilia Romagna region. The set up of an off-shore buoy is to be considered;
- shoreline evolution (4 times per year). This can be achieved by means of a DGPS survey along the shellfish line;
- annual bathymetry with investigation of structural integrity. Suited technology are the multi-beam bathymetry or a net of bathymetric profiles spaced 20 m cross shore and intersecting 5 long-shore profiles, at least one of which crossing the barrier; gaps should be accurately monitored;
- annual characterisation of sediment distribution.

The collected information should provide a feedback to the maintenance programme. Evaluation of the annual loss in the protected area, related to the sediment distribution, gives sufficient information of the amount of required nourishment and of the morphological behaviour of the defence structure, also in view of possible design modification.

12.6.6. Recommendations for construction phase

The structure can be built by pontoon. Bottom should be preliminary flattened, in order to supply sufficient depth to allow the placement of both armour and filter.

The filter should be accurately mixed, and in absence of a proper technology, the bigger fraction (> 100 mm) may be placed separately in three layers, on top of the mix.
Both the filter and the geotextile are not entirely reliable due to construction problems: during placement of the filter the fine material may be washed out or may not be sufficiently mixed to the coarser part; conversely the geotextile may be removed or folded by waves before being anchored by the stones.

Waves should be $H_{rms} < 0.10$ m (maximum 0.25 m) during placing of geotextile and of first part of filter layer. Stability is much dependent on a proper realisation of the filter and geotextile. Possible over dimension of the armour (D_{n50a}) is not dangerous provided that $D_{n50a} < 4\, D_{n85f}$, where subscript «*f*» refer to the filter.

12.7. CONCLUSIONS

This Chapter presented the application of integrated design approach for the selection of a coastal defence scheme in Lido di Dante. In the example application it is assumed that at the initial (hypothetical) design stage the coast was defended only by three groynes, and as a consequence subject to great erosion which justify an intervention for better protection of the beach and the related human activities.

The preliminary investigation of European directives, environmental constraints and site characteristics allowed identification five design alternatives: pure nourishment; a submerged barrier; emerged barriers parallel to the shore; prolongation of the two external existing groynes; a submerged barrier with submerged connectors to the existing groynes.

The inputs for the integrated design consisted of available data on climate, environmental conditions, habitat and species, preferences of visitors; tools (see Chapter 13) for establishment of design wave climate, selection structure type and their lay-out and geometries; tools for simulating waves and currents induced by the structures and the consequent morphological changes.

Engineers would have selected emerged barriers or submerged barrier with connectors as preferred schemes for beach defence; ecologists would have preferred submerged barriers for minimising ecological impacts or the prolongation of groynes for maximising species biodiversity and natural resources; socio-economists would have chosen submerged structures mainly for aesthetic reasons but also for water quality. The global evaluation of design alternatives resulted in the selection of the submerged barrier which was then optimised accounting for general multidisciplinary perspectives achieved within DELOS.

The analysis performed and the results presented for this site emphasized the strict interactions among LCS construction, habitat changes, hydrodynamics, beach erosion, water quality and thus beach value; it appears therefore necessary to follow general LCS design guidelines to account for the multiple effects of LCS on the littoral environment and thus promote an effective and environmentally sustainable defence scheme.

Part III

Tools

CHAPTER 13

Design tools related to engineering

13.1. SITE CONDITION PARAMETERS

This Section provides a description of the most important site condition parameters related to the design of LCSs.

13.1.1. Bathymetry and morphology

(Burcharth, AAU)

The bathymetry of the sea bed, the beach and the adjacent coastal land formations must be known, not only at the location of the LCS scheme but also for the neighbouring stretches along the coast because of potential distant effects of the structures. On charts for navigation purposes the sea bed level is most often defined relative to the chart datum, commonly taken as the lowest astronomic spring tide level. The coastal profile is very important for the assessment of the wave regime and its impact on morphology and the structure itself.

Morphological impact due to seabed erosion and sedimentation causes the bathymetry to vary with time. One storm can impose significant changes as can seasonal variations in storm intensities. On eroding coasts such short-term bathymetric modifications appear as fluctuations on top of the long-term retreat of the coastal profile. For the design of LCSs it is important to know the lowest seabed level at the position of the structure, bearing in mind that the structure impose local changes if scour occurs.

The rate of seabed morphological changes depends on the divergence of the sediment transport. Large gradients are generally related to situations with high sediment transport, i.e. under conditions of storm waves and strong currents. With the exception of tidal currents, there is a strong correlation between waves and currents, which again under storm conditions in shallow water are correlated to the local water depth due to depth limitation of the waves. The water depth is determined not only by the seabed level but also by the water level, which, with respect to the storm surge component, is strongly correlated to the waves.

The complicated interaction between the morphological changes and the hydrographic conditions makes prediction of changes in coastal profiles difficult and rather uncertain (see Section 13.10). Historical data on seabed and shoreline changes therefore becomes of great importance for the understanding of coast dynamics as a basis for design of LCS schemes.

13.1.2. Water levels, waves and currents

(Burcharth, AAU)

Prediction of water level is very important in shallow water as it determines the water depth and thereby the upper limit for wave heights. Changes in water level are due to astronomical tide and storm surge, the latter being the effect of barometric pressure variations and set-up caused by wind and waves.

Most LCSs are constructed in shallow water on coasts with mildly sloping seabeds. For such coastlines, the storm surge can be significant, say a rise in water level up to approximately 2-3 metres. Tropical storms can generate much higher storm surges. Storm surge is then dominating on coasts with small astronomic tide as for example in the Mediterranean Sea. Storm surge is strongly correlated to wind and waves.

Water level changes are of importance for the design of LCSs. Generally it is easier to optimize LCSs with respect to crest level when only small water level variations occur, because the distance from the crest to the still water table determines largely the wave energy that can be transmitted over the structure. Very few LCSs are built on coasts with large tidal ranges although it is certainly possible to design for such conditions.

Large water level variations give high exchange of the water which helps maintaining good water quality. On coasts with small tidal range, long periods with warm and calm weather and consequently no storm surge conditions might result in stagnant water of poor quality. Closed-cell LCS-schemes should then be avoided.

The mean water level (MWL) is known with high accuracy on European coastlines. It can be determined with good accuracy by measurements over a period of some months.

The change in water level, Za, caused by atmospheric pressure variations can be estimated at equilibrium as:

$$Za = 0.01\ (1013 - pa) \tag{13.1}$$

where pa is the pressure at sea level in mbar or hPa. Za is water level change in metres, positive for rise in water level. A common low pressure of 960 mbar causes a rise of 0.53 m.

Wind generated shear stress on the water surface causes a tilt of the water surface in shallow water in the continental shelf. Onshore winds then generate a rise in water level on the coast termed wind set-up. For long straight coasts with a mild sloping seabed with shore-parallel depth contours and a constant onshore wind field the rise in sea level S at a distance F from deep water can be roughly estimated as:

$$S = \frac{\rho_a}{\rho_w} f \frac{U_{10}^2}{g(D_1 - D - S)} \ln\left(\frac{D_1}{D + S}\right) F \tag{13.2}$$

where f is the air-water friction coefficient ($1 \cdot 10^{-3} - 3 \cdot 10^{-3}$), ρ_a and ρ_w are the mass density of air and water respectively ($\rho_a / \rho_w \cong 1/800$), and U_{10} is the average onshore directed wind velocity at 10 m height. D_1, D, S and F are explained in Figure 13.1.

Wind set-up is sensitive to the alignment of the coastline. Bays result in relatively large set-up at the shoreline whereas wind set-up is usually marginal on convex coastlines.

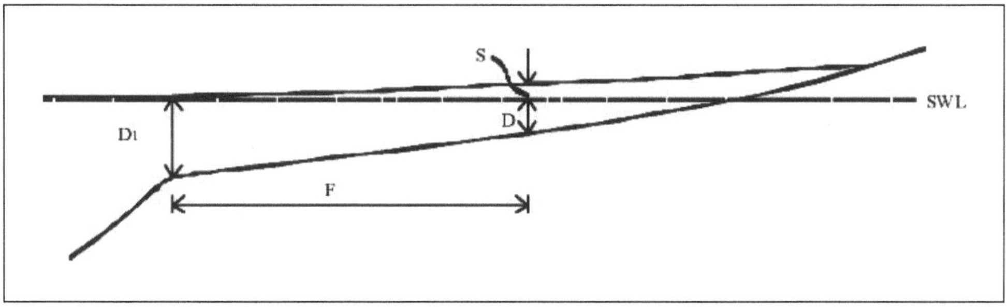

Figure 13.1. Definition of geometrical parameters for calculation of wind set-up.

Waves impose the largest impacts on open coasts. Related to evaluation of the morphological effect of LCS-schemes it is important to know the yearly average nearshore wave climate in terms of combined statistics of wave heights, wave periods, and wave direction as well as the correlation to water levels and currents. For the structural design of the LCSs the waves imposing the most damaging effect on the location of the structures must be identified.

LCS-schemes are generally located in shallow water where the larger waves break before reaching the coastline. Open littoral coasts with limited tidal range have bars on which the storm waves break. The number and the positions of the bars changes with time resulting in changes in waves as well as in currents at given locations. However, the yearly average conditions at a location vary only slowly.

As the waves approaches from deeper water into shallow they are refracted resulting in a turn of the wave crest to be parallel to the seabed depth contours. As water depth diminishes, shoaling (steepening) of the waves takes place resulting in wave breaking when the wave height exceeds approximately 80% of the water depth. The wave height reduces as energy is dissipated by breaking. The shoaling process is influenced by the seabed slope. The wave breaking and wave transformation is described in detail in Section 13.2.

Breaking waves approaching the coastline cause a raise in water level termed wave set-up due to changes in the radiation stress (wave thrust). For waves approaching perpendicular to a straight coastline with a plane sloping seabed, the water level set-up at the shoreline can be approximated in excess by

$$S = \frac{3}{8} H_b^2 \frac{1}{D_b} \cong 0.25 H_b \tag{13.3}$$

where H_b and D_b are wave height and water depth, respectively, at the breaker line.

This value, which is the theoretical maximum, is practically never reached as irregularities in coastline alignment and seabed topography cause generation of compensating return flows. For oblique waves, only the coast-perpendicular component of the radiation stress generates wave set-up.

Astronomical tide water level variations are well known along practically all coastlines as they can be calculated. Astronomical tide is not correlated to storm surge. Storm surges are normally correlated to large offshore waves whereas tide is uncorrelated to offshore waves. However, in the shallow water coastal zone both types of water level variations influences the nearshore waves due to depth limitation of wave heights.

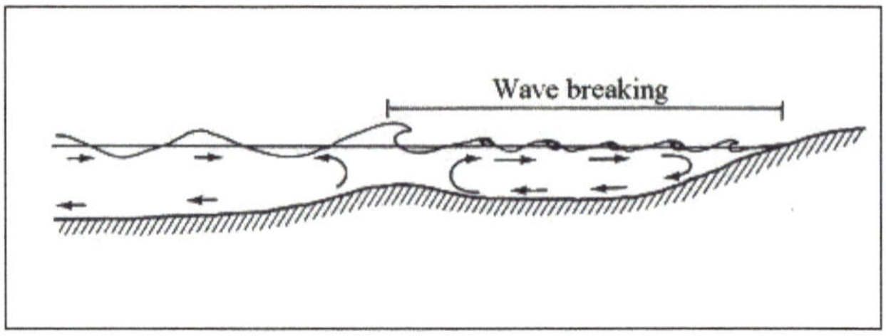

Figure 13.2. Sketch of net circulation patterns due to wave breaking.

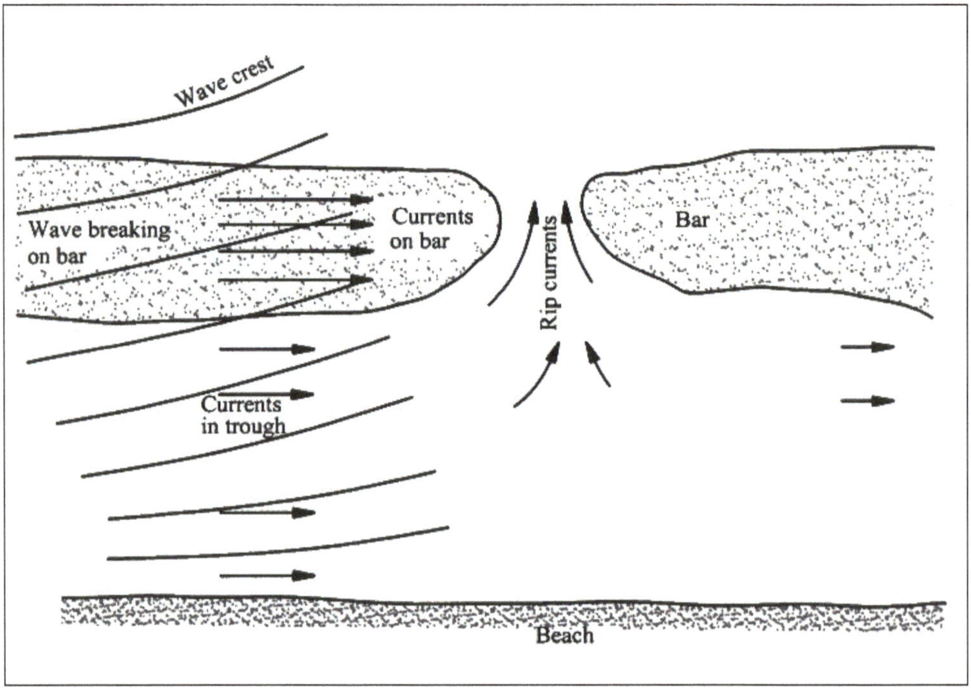

Figure 13.3. Wave induced currents in case of oblique waves.

Coastal currents are generated by tides, by changes in water levels due to storm surge and by breaking waves. Tidal currents in the nearshore zone are mainly shore parallel on straight coastlines, but more complex patterns are generated around, and especially in the lee of, protruding headlands or structures or other irregularities along the coastline. This includes estuaries. Tidal currents can be predicted quite accurately if the seabed topography is known.

Longshore storm surge generated currents are caused by water level gradients along the coast and can be predicted if the gradients are known. Like for tidal currents, more complex local patterns are caused by irregularities along the coastline. Storm surge also generate cross shore currents which together with wave generated currents can result in complex patterns.

The most dominant wave generated currents are those caused by breaking waves. On a plane coast with shore parallel depth contours and perpendicular waves, the seaward undertow is the most significant current, see Figure 13.2.

Oblique waves on a barred coast create complicated patterns as illustrated on Figure 13.3, dominated by strong longshore currents in the breaker zone on the bars and return flows as rip-currents to compensate for the net-inflow of water over the bars.

13.1.3. Extreme events analysis

(Lamberti, Archetti, UB)

Extreme value theory is used in storm, flood, wind, sea waves and earthquake estimation, according to the theory of extreme values: the largest or smallest value from a set of independent identically distributed random variables, tends to an asymptotic distribution form that only depends on the tail of the distribution of the parent variable.

Obviously if the sign of the variable is changed, the order of the order statistics is reversed, maximum is changed into minimum and the distribution function values are changed into their complement to 1. The theory of extreme distributions is normally presented for the maxima but can be easily translated to minima.

Let X be a random variable and $X_1, X_2, X_3, \ldots, X_n$ an independent sample from it, i.e. a set of n random variates with a common distribution $F_X(x)$, where x is the current value of the variable and n is the sample size. Let also $X_{(1)}, X_{(2)}, \ldots, X_{(n)}$ represent the ordered set of the same variables, or order statistics, with $X_{(1)} \leq X_{(2)} \leq \ldots \leq X_{(n)}$, the distribution of $X_{(i)}$ (or $X_{(i;n)}$ when the sample size is emphasised) is given by:

$$F_{X_{(i;n)}}(x) = \sum_{j=1}^{n} \binom{n}{j} [1 - F_X(x)]^{n-j} [F_X(x)]^j \tag{13.4}$$

In particular for $i = n$, $F_{X_{(i;n)}}$ provides the distribution of the maximum as:

$$F_{X_{(i;n)}}(x) = [F_X(x)]^n.$$

As n increases indefinitely, the distribution of the standardized maximum $Y_n = (X_{(n)} - b_n)/a_n$ converges to a limit distribution, where $a_n > 0$ denotes a scale parameter and b_n a location parameters both of which may depend on sample size n but in a very simple way.

The limiting distribution must be one of the following types: where γ denotes a positive constant and Y is the asymptote of Y_n.

a) Gumbel or Type I extreme distribution, applicable when the parent cumulative distribution has an exponential upper tail of asymptotic form $1 - \exp\left(-\frac{x-b}{a}\right)$:

$$F_Y(y) = \exp(-e^{-y}) \qquad -\infty, < y < +\infty; \qquad a_n = a, \qquad b_n = b + a \ln n \tag{13.5}$$

b) Frechet or Type II extreme distribution, applicable when the parent cumulative distribution has an upper tail of the form $1 - \left(\frac{x-b}{a}\right)^{-\gamma}$:

$$F_Y(y) = \begin{cases} \exp\left(-y^{-\gamma}\right) & y > 0 \\ 0 & y \le 0 \end{cases}; \quad a_n = a \cdot n^{1/\gamma}, b_n = b \qquad (13.6)$$

c) Weibull or Type III extreme distribution, applicable when the parent distribution is upper bounded with cumulative distribution near the bound of the form $1 - \left(\frac{x-b}{a}\right)^{\gamma}$:

$$F_Y(y) = \begin{cases} \exp\left[-(-y)^{\gamma}\right] & y < 0 \\ 0 & y \ge 0 \end{cases}; \quad a_n = a \cdot n^{-1/\gamma}, b_n = b \qquad (13.7)$$

The probability density functions and the cumulative density functions of the three type of distributions are plotted in Figure 13.4.

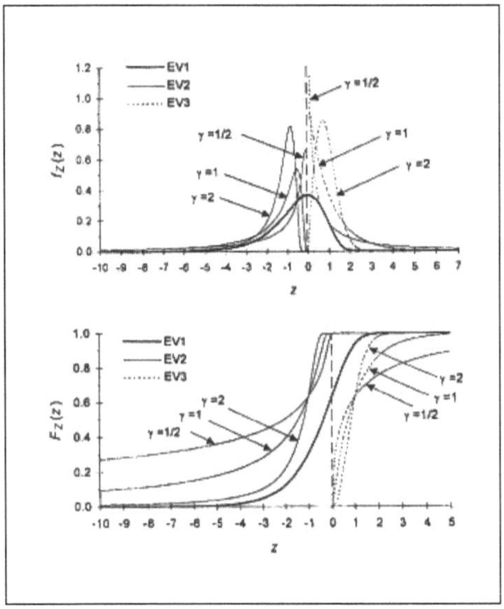

Figure 13.4. pdf and cdf of distributions EV1, EV2 and EV3.

The three distributions are referred as EV1, EV2, and EV3; they can be represented by a single distribution function named the Generalized Extreme Value (GEV) distribution.

$$F_{X\max} = \exp\left\{-\left[1 - \frac{k(x-\varepsilon)}{\alpha}\right]^{1/k}\right\} \qquad (13.8)$$

where α denotes a scale parameter, ε a location parameter and k the shape parameter.

Note that for negative k the GEV represents an EV2, in the opposite case, i.e. $k > 0$, this model becomes EV3; the case $k = 0$ corresponds to the Gumbel distribution (EV1) with scale parameter α and location parameter ε.

13.1.3.1. Generalized Extreme Value moments

The mean and the variance of the GEV distribution are given by:

$$\begin{aligned} E\left[X_{\max}\right] &\equiv m_1 = \varepsilon + \frac{\alpha}{k}\left[1 - \Gamma(1+k)\right] && \textit{for} \quad k > -1 \\ Var\left[X_{\max}\right] &\equiv m_2 - m_1^2 = \left(\frac{\alpha}{k}\right)^2 \left[\Gamma(1+2k) - \Gamma^2(1+k)\right] && \textit{for} \quad k > -1/2 \end{aligned} \qquad (13.9)$$

respectively, therefore the mean diverges for $k < -1$ and the variance for $k < -1/2$.

The coefficient of skewness is given by:

$$\gamma_{1,Xmax} = sign(k)\frac{-\Gamma(1+3k)+3\Gamma(1+k)\Gamma(1+2k)-2\Gamma^3(1+k)}{\left[\Gamma(1+2k)-\Gamma^2(1+k)\right]^{3/2}} \quad for\ k > -1/3 \quad (13.10)$$

13.1.3.1.1. GEV L-moments

Moments are very sensitive to extreme values of the distribution and to outliers, that with high probability will fall among extremes; the *L* moments, here described are expected to be less prone to adverse sampling effects (introducing outliers).

Let $X_{i,n}$ be the ith largest observation in a sample of size *n*, then the second and third *L* moments are defined as:

$$\begin{aligned} L_1 &= E(X) \\ L_2 &= \frac{E(X_{2:2} - X_{1:2})}{2} \\ L_3 &= \frac{E(X_{3:3} - 2X_{2:3} + X_{1:3})}{3} \end{aligned} \quad (13.11)$$

The first *L* moment is the mean; the second and third are measures of dispersion and skewness.

For any distribution, the *L* moments can be given in terms of the probability-weighted moments:

$$\begin{aligned} L_1 &= M_0 \\ L_2 &= 2M_1 - M_0 \\ L_3 &= 6M_2 - 6M_1 + M_0 \end{aligned}$$

where $M_n = \int x[1-F(x)]^n dF(x)$ is a probability weighted average.

The parameter of the GEV distribution are related to the first three *L* moments as follows:

$$\begin{aligned} L_1 &= \varepsilon + \frac{\alpha}{k}\left[1-\Gamma(1+k)\right] \\ L_2 &= \frac{\alpha}{k}\left(1-2^k\right)\Gamma(1+k) \\ \frac{L_3}{L_2} &= \frac{2\left(1-3^{-k}\right)}{\left(1-2^{-k}\right)} - 3 \end{aligned} \quad (13.12)$$

13.1.3.2. Estimation of parameters

13.1.3.2.1. Method of moments

The method of moments is a long established procedure for finding point estimators. When fitting a parametric distribution to a set of data by this method, we equate the sample moments to those of the fitted distribution in order to estimate the parameters. For example, in the case of the GEV distribution if the first moments of X_{max} exist and are known, the values of the three parameters α, k and ε can be determined from the mean, the variance and the skewness coefficient of the data.

The 3 first sample moments are evaluated (giving to any value in the sample probability 1/n) and from these the sample variance and skewness.

The parameter k depends only on the skewness coefficient for $k > -1/3$, so it can be found by solving Eq. (13.10), substituting in it the sample skewness coefficient, or by using the plot in Figure 13.5; after some substitutions the other two coefficients can be determined by:

$$\alpha = \sqrt{\frac{k\,\sigma^2}{\Gamma(1+2k) - \Gamma^2(1+k)}} \tag{13.13}$$

where the sample variance is substituted for $\sigma^2 = \mathrm{Var}[X]$. Finally the location parameter is computed from:

$$\varepsilon = \mu - \frac{\alpha}{k}\left[1 - \Gamma(1+k)\right] \tag{13.14}$$

Where the sample mean is substituted for μ.

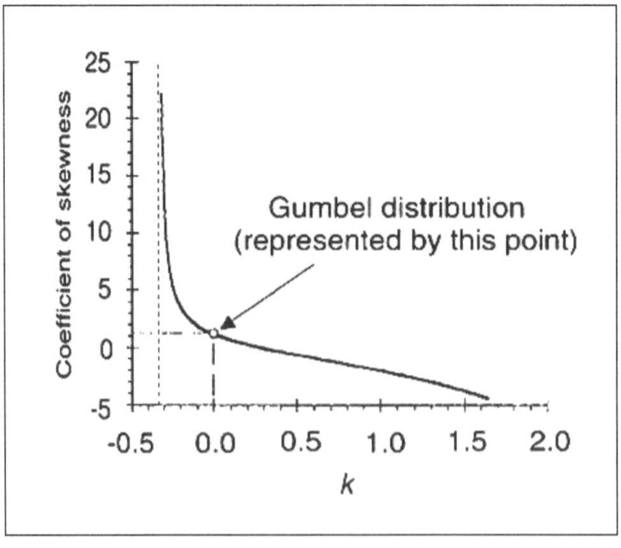

Figure 13.5. Coefficient of skewness versus the exponent (shape parameter) of the GEV distribution.

13.1.3.2.2. Method of maximum likelihood

A consistent estimator for the parameters of the GEV distribution is given by Maximum Likelihood (ML) method.

The maximum likelihood procedure, or ML, is an alternative to the method of moments. For a random variable X with a known pdf $f_X(X)$ and observed values $x_1, x_2, x_3, \ldots, x_n$, in a random sample of size n, the likely function of the set of unknown parameters θ, is defined as:

$$L(\vartheta) = \prod_{i=1}^{n} f_X\left(x_i \middle| \vartheta\right) dx_i \tag{13.15}$$

The objective is to maximize $L(\theta)$ with respect to θ for a given data set. This is easily done by taking m partial derivatives of $L(\theta)$, where m is the number of parameters, and equating them to zero. We then obtain the maximum likelihood (ML) estimators of the parameter set θ from the solution of the equations. In this way the greatest probability is given to the observed set of events, provided that we know the true form of the probability distribution (Kottegoda and Rosso, 1997).

The ML is the only presented method that can easily provide through Fisher's information matrix (defined as the expected value of the squared gradient of minus the log-likelihood function) the estimation of the errors, see Ibragimov & Has'minskii (1981) and for applications the Matlab Statistics Toolbox.

13.1.3.2.3. Method of L moments

As moment and ML estimators perform poorly when the distributions of the observations deviates significantly from the fitted distributions, the alternative method of L-moments is suggested (LM). The LM are expected to be less prone to adverse sampling effects (presence of outlyers) as they give a probability weight to the moments.

After the sample values of L_1, L_2 and L_3 are estimated from the data, associating to each ordered variate probability $1/n$ the cdf value provided by a proper formula (Hazen formula $F_i(i-0.5)/n$ is appropriate) one can solve for k the last equation.

An approximate solution of Eq. (13.12) (3rd equation) is:

$$k = 7.859\left(\frac{2L_2}{L_3+3L_2} - \frac{\ln 2}{\ln 3}\right) + 2.95554\left(\frac{2L_2}{L_3+3L_2} - \frac{\ln 2}{\ln 3}\right)^2 \tag{13.16}$$

Then, the estimate of α is obtained as:

$$\alpha = \frac{kL_2}{\left(1-2^{-k}\right)\Gamma(1+k)} \tag{13.17}$$

finally the location parameter is:

$$\varepsilon = L_1 - \frac{\alpha}{k}\left[1 - \Gamma(1+k)\right] \tag{13.18}$$

13.1.3.3. Suggestions

Among the methods presented for the estimation of the parameters all are valid. When we are sure of the data source and we are sure that the data set has been cleaned from outliers or erroneous data, the method of moments (Eq. (13.10), Eq. (13.13) and Eq. (13.14)) is the simplest to use with hand calculation.

When some outliers can be present in the data set we suggest to use the method of L moments because the parameters are easily estimated through Eq. (13.16), Eq. (13.17) and Eq. (13.18).

The ML method is the only one that gives an estimate of the parameter error. It requires automatic computation and the absence of outliers should be checked.

Whenever estimates provided by the three methods are significantly divergent the guess made on the parent distribution is probably wrong, for instance because of the presence of erroneous data in the data set.

13.2. TRANSFORMATION OF WAVES FROM DEEP TO SHALLOW WATER

(Martinelli, Zanuttigh, Clementi, UB)

This Section briefly describes the wave transformation processes, such as shoaling, refraction, diffraction, breaking and energy dissipation, and presents consolidated models to be solved, in the general case, by means of numerical modelling. For coastlines with straight and parallel isobaths, simplified equations (e.g. Snell's law) or diagrams (Goda, 1985; CUR/CIRIA, 1991) are reported.

Notations

b = distance between adjacent wave rays
b_o = rays distance in deep water
C = wave celerity ($L/T=\omega/k$)
C_o = wave celerity in deep water
C_g = group wave celerity
C_{go} = group wave celerity in deep water
E = wave energy density
f = bottom friction coefficient
g = gravitational acceleration
H = wave height
H_b = breaking wave height
H_d = diffracted wave height
H_i = incident wave height
H_{m0} = spectral wave height
H_o = wave height in deep water
H_{rms} = root mean square wave height
H_s = significant wave height
H_{Tr} = transitional wave height
$H_{x\%}$ = wave height of percentile $x\%$
h_o = offshore water depth
K_d = diffraction coefficient
K_r = refraction coefficient
K_s = shoaling coefficient
k = wave number ($2\pi/L$)
L = wave length
L_b = breaking wave length
L_o = wave length in deep water
L_{op} = L_o related to the peak frequency
m = beach slope
m_0 = zero spectral moment
n = energy flux parameter
R_c = crest freeboard (positive if structure is submerged)
T = wave period
u_b = wave velocity at the bottom
a = wave amplitude
q = wave direction
r = water density

$H_{1/3}$ = average of 1/3 higher waves
h = water depth
h_b = breaking water depth
ω = wave angular frequency ($2\pi/T$)
η = surface elevation

13.2.1. Basic concepts

The simplest way to describe a wave, propagating along the x direction is:

$$\eta(x, t) = a \cos (kx - \omega t) \tag{13.19}$$

In linear theory, wave length $L = 2\pi/k$ is related to the local water depth, h, and period, $T = 2\pi/\omega$, by the dispersion relationship:

$$\omega^2 = gk \tanh kh = gk_0 \tag{13.20}$$

Period and water depth are usually given and wave numbers (or length) is obtained.

Wave length decreases as the wave propagates from deep to shallow water, assuming the value of $L_o = gT^2/2\pi = 1.56\,T^2$ (SI units) is deep water and $L = \sqrt{gh}T$ in shallow water. Wave celerity is defined as $C = L/T$.

If the wave is propagating in an arbitrary direction, water elevation is expressed by:

$$\eta(x, y, t) = a \cos [(k \cos \theta)x + (k \sin \theta)y - \omega t + \chi_0] = a \cos \chi(x, y, t) \tag{13.21}$$

where $\chi(x, y, t)$ is the phase function for given L, T and χ_0. The wave crest is the line formed by points with maximum elevation (where $\chi = 2n\pi$, $n = 0, 1, 2,..$).

Wave energy is proportional to the square of wave amplitude and travels in wave direction at group celerity C_g which may differ from wave celerity C:

$$C_g = nC = \frac{1}{2}\left(1 + \frac{2kh}{\sinh 2kh}\right)C \tag{13.22}$$

n is defined by Eq. (13.22) itself and is 1/2 in deep water and 1 in shallow water, where the group and wave celerity become function of depth only (not dispersive conditions).

Waves at sea can be considered as the superposition of many (infinite) small waves with different period and direction and random phase. A time history of real waves appears indeed as an irregular record, with elevation crossing a mean value (zero) alternatively downward and upward. Single waves may be identified extracting the record between two consecutive zero up- or down-crossing, and the set of periods and heights may be statistically described in an easy way: periods are usually concentrated around a mean value; the statistical distribution of wave heights in deep water tends to the Rayleigh one, which is function of a single parameter, e.g. H_{rms} or H_s.

13.2.2. Energy conservation

Conservation of wave energy in stationary conditions and in absence of currents is expressed by:

$$\nabla(EC_g) = 0 \tag{13.23}$$

where $E = \frac{\rho g H^2}{8}$ for regular waves and $E = \frac{\rho g H_{rms}^2}{8}$ for irregular waves.

During propagation in absence of energy dissipation, three physical phenomenon may be recognised: shoaling, refraction and diffraction, which are described by separate factors K:

$$\frac{H}{H_o} = K_s K_r K_d \tag{13.24}$$

Directional spreading has usually a significant effect on refraction and diffraction. In the following, waves are considered to be long-crested (i.e. monodirectional) for sake of simplicity, but influence of spreading must be considered in practice. This can be easily done by subdividing the spectrum in different directional classes and applying wave transformation to each class.

13.2.2.1. Wave shoaling

Shoaling is the modification of the wave specific energy E induced by group celerity variations. Eq. (13.25) describes the shoaling effect when waves propagate along a straight line and gives:

$$\frac{H}{H_0} = \sqrt{\frac{C_{go}}{C_g}} = \sqrt{\frac{1}{2n \tanh kh}} = K_s \tag{13.25}$$

$K_s(h)$ is equal to 1 in deep waters, it has a minimum of 0.91 in intermediate waters and then rises to infinity as the water depth approaches zero. In practice waves do not grow to infinity since they are limited by breaking.

13.2.2.2. Wave refraction

Refraction is a change of wave direction associated to the modification of celerity. It is encountered typically by waves approaching obliquely a sloping beach, in which case water depth, and therefore wave celerity, decreases along the front, and the wave bends toward the shore. By simple geometrical considerations, it is seen that:

$$\frac{\sin\theta}{c} = \frac{\sin\theta_o}{c_o} = \text{constant (Snell's law for a long-shore uniform bathymetry)} \tag{13.26}$$

Refraction on non-uniform bathymetries may be obtained solving numerically:

$$\nabla \times \vec{k} = \frac{\partial(k \sin\theta)}{\partial x} - \frac{\partial(k \cos\theta)}{\partial y} = 0 \tag{13.27}$$

As effect of refraction, the distance among wave rays changes and the wave height varies accordingly (decreases). The wave height variation which may be specifically attributed to refraction is given by the conservation of energy flux in case of constant group celerity:

$$K_r = \frac{H}{H_o} = \sqrt{\frac{b_o}{b}} = \sqrt{\frac{\cos\theta_o}{\cos\theta}} \text{ (for a large-shore uniform bathymetry)} \tag{13.28}$$

In general, offshore contours are irregular and vary along the coast, so that a solution for θ and b can not be found as easily as in Eq. (13.26) and Eq. (13.28) and numerical modelling is required. Ray tracing techniques, described for instance in Dean and Dalrymple (1992), were specifically developed to solve refraction and shoaling following wave-path.

13.2.2.3. Wave diffraction

Wave diffraction is the process by which wave energy spreads perpendicularly to the dominant direction of wave propagation.

Wave diffraction is specifically concerned with sudden changes in boundary conditions such as at breakwater roundheads, where wave energy is transferred into the shadow zone by diffraction. For uniform water depth, Helmholtz equation can be used to describe diffraction and obtain K_d:

$$\Delta\phi - k^2\phi = 0 \tag{13.29}$$

where $\phi(x, y)$ is the unknown horizontal variation in velocity potential Φ, i.e. $\Phi(x, y, z, t) = \phi(x, y)\, Z(z) \cos(\omega t)$.

The above equation is obtained solving the Laplace condition over the wave field $\Delta\Phi = 0$ considering $Z(z)$ a known exponentially decreasing function of uniform depth.

In general a different equation, instead of Eq. (13.29), is used, which is valid for (mild) sloping bottoms and accounts for diffraction, shoaling and refraction:

$$\nabla \cdot (CC_g \nabla\Phi) + \omega^2 \left(\frac{C_g}{C}\right)\Phi = 0 \text{ (Berkhoff, 1972)} \tag{13.30}$$

For irregular waves, Eq. (13.30) is evaluated for each class of the directional spectrum.

The diffraction coefficient K_d is found in literature for typical cases also in presence of directional spreading (Goda, 2000).

13.2.3. Wave energy dissipation

During wave propagation, in particular approaching the shoreline, some dissipative phenomena occur, such as wave breaking and bottom dissipation. In these cases the energy flux convergence is equal to the energy dissipation rate D:

$$\nabla \cdot (EC_g) = -D \tag{13.31}$$

13.2.3.1. Wave breaking criteria

Breaking conditions occur when the horizontal particle velocity u at the crest of the wave equals or exceeds the wave celerity C, or when the vertical acceleration of the particles at the surface exceeds gravity, causing an instable free surface. In practice we can predict breaking when wave height exceeds a certain fraction of water depth or of wave length. In these cases the wave breaks, producing turbulence, dissipating energy and causing a rapid reduction in wave height.

Breaking position or point is defined as where the wave front becomes vertical and it is determined when weves in their propagation reach breaking wave height (H_b, see below).

Breakers have different shapes, which are usually grouped into 3 classes (a 4th class, «collapsing», refers to conditions between surging and plunging) and may be predicted on the basis of the *surf similarity parameter*:

$$\xi_b = \frac{m}{\sqrt{H_b/L_o}} \begin{cases} \xi_b > 3.3 & \text{Surging} \\ 0.5 < \xi_b < 3.3 & \text{Plunging breakers} \\ \xi_b > 0.5 & \text{Spilling breakers} \end{cases}$$

The following subparagraphs present consolidated models for the evaluation of breaking wave height and the consequent energy dissipation in case of regular and irregular waves.

13.2.3.1.1. Breaking wave height

Waves break when they reach the upper wave height limit, H_b, which is function of depth h, wave length L and bottom slope m.

In the following, 5 models to estimate H_b are presented. Models 1 to 3 are related to regular waves, models 4 and 5 are related to irregular waves.

1) McCowan (1894) introduces the *breaker depth index* γ_b:

$$\gamma_b = \frac{H_b}{h_b} = 0.78 \tag{13.32}$$

to be applied in shallow water conditions (depth limited waves).

2) Miche criterion (1944):

$$\frac{H_b}{L_b} = 0.14 \tanh(kh) \quad \text{or} \quad kH_b = 0.88 \tanh(kh) \tag{13.33}$$

which becomes: $H_b = 0.14L_b$ in deep water and $H_b = 0.88h_b$ in shallow water.

3) Weggel (1972) introduces the influence of the foreshore slope m:

$$\frac{H_b}{h_b} = \left(c_1 - c_2 \frac{H_b}{gT^2}\right) \tag{13.34}$$

where:

$$c_1 = \frac{1.56}{1+e^{-19.5m}}$$
$$c_2 = 43.75(1-e^{-19m})$$

Note that for long waves as the beach slope approaches zero, the breaker index tends to 0.78; as the beach slope approaches infinity this index tends to 1.56 (sum of the incident and perfectly reflected wave component).

4) Kamphuis (1991) proposes the following extensions to the practical case of irregular waves; the limit shall be imposed to H_s : $H_s \leq H_b$ where:

$$H_b = 0.095e^{4.0m}\tanh\left(\frac{2\pi h_b}{L_{bp}}\right)L_{bp} \text{ for steepness limited breaking}$$

$$H_b = 0.56e^{3.5m}h_b \text{ for depth limited breaking} \tag{13.35}$$

5) Hur *et al.* (2003) describe the breaking over a submerged permeable breakwater, far from the edges breaking limit is:

$$\frac{H_b}{L_{os}} \approx (0.095 \div 0.106)\tanh\left(\frac{2\pi R_c}{L_{os}}\right) \tag{13.36}$$

with L_{os} being thre off-shore wave height relative to the significant wave period. It was found that multidirectionality of waves has little effect.

13.2.3.1.2. Energy loss due to breaking

Three models are summarised in the following.

1) Battjes and Janssen (1978) describe the energy dissipation per wave on the basis of the bore analogy:

$$D = \frac{1}{4}\alpha Q_b \rho g H_b^2 f \tag{13.37}$$

where: $\alpha \cong 1$ is the dissipation coefficient, Q_b is the fraction of breaking waves and f is wave arrival frequency.

If waves are Rayleigh distributed, Q_b can be derived from:

$$(1 - Q_b)/\ln(Q_b) = (H_{rms}/H_b)^2$$

where H_b is obtained by $kH_b = 0.88\tanh(\gamma_b kh/0.88)$ with $\gamma_b = 0.5 + 0.4\tanh(33\, H_{rms}/L_{op})$.

2) Dally, Dean and Dalrymple (1985) describe the dissipation in shallow water,

assuming that beyond the breaking point breaking waves continue to dissipate energy until a stable wave height is reached:

$$D = \frac{\kappa}{h}(EC_g - (EC_g)_s) \tag{13.38}$$

where: k expresses the rate at which wave height decays, $(EC_g)_s$ is the energy flux associated with a stable wave height, $H_e = \gamma_e h$. For regular waves, $0.1 < \kappa < 0.275$ and $0.35 < \gamma_e < 0.475$; for irregular waves, $\kappa = 0.15$ and $\gamma_e = 0.4$. Different values of the coefficients are suggested in the case wave set up is not considered: $\kappa = 0.17$ and $\gamma_e = 0.5$.

Wave height in the surf zone can be predicted on the basis of this model for dissipation D, by solving equation Eq. (13.31).

3) Goda (1985) defines indirectly a criterion for evaluation of energy decay giving the wave height distribution after the breaking process.

Waves with height from H_2 to H_1 have a probability to break which varies linearly from zero to 1, so that no wave higher than H_1 may exist. After breaking, waves are assumed to be distributed in the range of wave heights 0 - H_1, with a probability proportional to the distribution of unbroken waves.

For given wave period water depth and foreshore slope, the various breaking wave heights are provided by:

$$\frac{H_b}{L_o} = A\left[1 - \exp\left\{-1.5\frac{\pi h}{L_o}\left(1 + 15m^{4/3}\right)\right\}\right] \tag{13.39}$$

where $A = \begin{cases} 0.17 & \textit{for the unique limit in case of regular waves} \\ 0.18 & \textit{for the upper breaking limit in case of irregular waves} \quad (H_1) \\ 0.12 & \textit{for the lower breaking limit in case of irregular waves} \quad (H_2) \end{cases}$

13.2.3.2. Energy dissipation over rough bottom

The energy rate dissipated by bottom friction in absence of currents is

$$D = \langle \tau_b \cdot u_b \rangle = \langle 1/2\, f\rho u_b^2 |u_b| \rangle$$

where < .. > denotes time averaging. When the boundary layer is turbulent (high waves and/ or rough bottom) the dissipation becomes:

$$D = \frac{\rho f (u_{b\max})^3}{6\pi} = \frac{\rho f}{6\pi}\left(\frac{H\omega}{2\sinh kh}\right)^3 \tag{13.40}$$

The decay with distance of a regular wave height can be obtained from the energy balance:

$$\frac{d(Ec_g)}{dx} = -D \quad \rightarrow \quad \frac{1}{8}\rho g C_g \frac{DH^2}{dx} = -\frac{\rho f}{48\rho} \frac{\omega^3}{\sinh^3 kh} H^3 \tag{13.41}$$

and therefore assuming constant friction along a flat bottom (starting from $x = 0$, where H_o is given), integration of Eq. (13.41) gives:

$$H(x) = \frac{H_o}{1 + \frac{f}{3\pi} \frac{k^2 H_o}{(2kh + \sinh_2 kh)\sinh kh} x} = K_f H_0 \tag{13.42}$$

13.2.4. Technical methods for irregular wave decay

13.2.4.1. Goda (2000)

This consolidated method accounts for shoaling and breaking under the hypothesis of Rayleigh distributed waves. Refraction and diffraction, if present, should be assessed separately considering the directional spreading.

Figure 13.6 presents the non-linear shoaling factor K_s. The dotted lines in the figure for the different bed slope separate the regions of breaking and non-breaking waves. When the intersection of the relative water depth (h/L_o) and the equivalent deepwater steepness (H_o'/L_o) falls in the region of the dotted lines, the structure is subjected to the action of breaking waves.

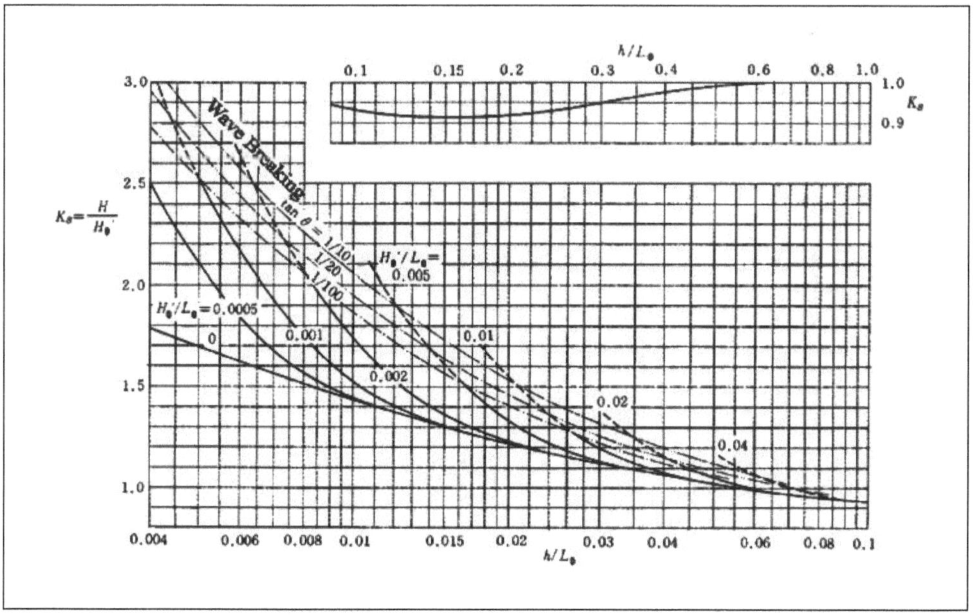

Figure 13.6. Diagram of non linear wave shoaling.

Wave height within the surf zone can be expressed as follows:

$$H_{1/3} = \begin{cases} K_s H'_o & h/L_o \geq 0.2 \\ \min\{(\beta_o H'_o o + \beta_1 h), \beta_{\max} H'_o, K_s H'_o\} & h/L_o < 0.2 \end{cases} \quad (13.43)$$

$$H_{\max} \equiv H_{1/250} = \begin{cases} 1.8\, K_s H'_o & h/L_o \geq 0.2 \\ \min\{(b_o^* H'_o o + b_1^* h), b_{\max}^* H'_o,\ 1.8\, K_s H'_o\} & h/L_o < 0.2 \end{cases} \quad (13.44)$$

where $H'_o = K_f K_d K_r (H_{1/3})_o = H_s/K_s$ is the equivalent deep water wave height corresponding (in a wave flume) to the local significant wave height and the coefficient β_o, β_1, ... are listed in Table 13.1.

Table 13.1. Coefficients for approximate estimation of wave heights within the surf zone.

Coefficients for $H_{1/3}$
$\beta_0 = 0.028(H'_0/L_0)^{-0.38} \exp[20 \tan^{1.5} \theta]$ $\beta_1 = 0.52 \exp[4.2 \tan \theta]$ $\beta_{\max} = \max\{0.92, 0.32(H'_0/L_0)^{-0.29} \exp[2.4 \tan \theta]\}$
Coefficients for H_{max}
$\beta_0^* = 0.052(H'_0/L_0)^{-0.38} \exp[20 \tan^{1.5} \theta]$ $\beta_1^* = 0.63 \exp[3.8 \tan \theta]$ $\beta_{\max} = \max\{1.65, 0.53(H'_0/L_0)^{-0.29} \exp[2.4 \tan \theta]\}$

13.2.4.2. CUR/CIRIA (1991)

This method is based on design curves for the combined effect of shoaling and breaking on uniform foreshore slopes. These graphs were obtained from the ENDEC model (Van der Meer, 1990a,b), which makes use of the Battjes and Janssen (1978) energy dissipation model.

Input data are off-shore peak wave length and steepness, local water depth and foreshore slope; the output consists of the local ratio H_{m0}/h.

The graphs (Fig. 13.7) are provided for wave steepnesses in the range 0.01 - 0.05; a couple of similar graphs are available accounting also for the obliquity of the incident wave (Fig. 13.8).

13.2.5. Wave height distribution in shallow water

13.2.5.1. Glukhovskiy (1966)

In shallow water, the Rayleigh distribution significantly underestimates the lower wave heights, and overestimates the highest. Several works deals with semiempirical adaptation to the Rayleigh distribution to allow for the effect of shallow water and breaking. Glukhovskiy (1966) proposed a Weibull type distribution that accounts for depth-limited

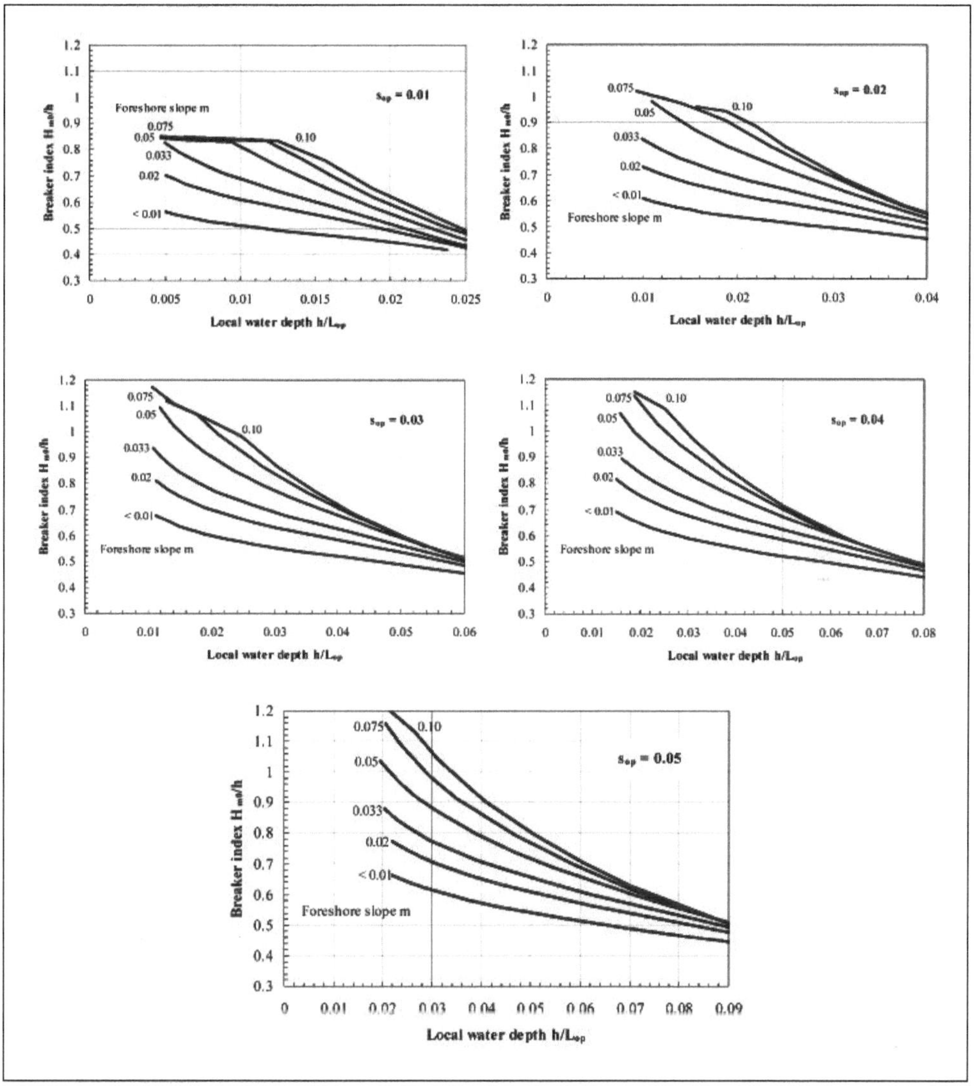

Figure 13.7. Diagrams of breaker indices for different wave steepness (increasing from top to bottom) as function of local water depth and foreshore slope.

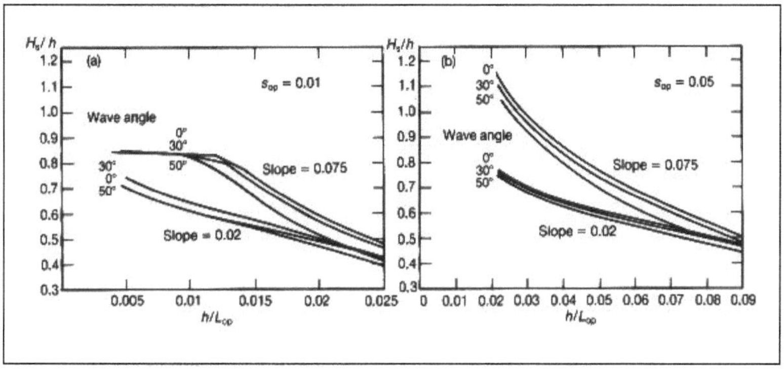

Figure 13.8. Diagrams of breaker indices accounting for wave obliquity.

breaking by making the exponent K in Eq. (13.45) an increasing function of wave height to depth ratio:

$$F(H) = 1 - \exp\left(-A\left(\frac{H}{H_{rms}}\right)^{k}\right) \qquad (13.45)$$

To assure consistency, the second moment of the distribution has to equal H_{rms}^2; this yields the following relation between the coefficient A and the exponent k:

$$A = \left(\Gamma\left(\frac{2}{k}+1\right)\right)^{\frac{k}{2}} \qquad (13.46)$$

According to Klopman (1996) formulation, the exponent k is assumed to be a function of the ratio H_{rms}/h.

$$k = \frac{2}{1-\beta\dfrac{H_{rms}}{h}} \qquad (13.47)$$

Klopman assumes the relation between H_{rms} and m_o to be as for a narrow-banded Gaussian process:

$$H_{rms} = \sqrt{8m_0}$$

From fitting of laboratory data, the optimal value of β is found to be 0.7.

13.2.5.2. Battjes & Groenendijk (2000)

Battjes & Groenendijk (2000) suggest another method for wave height distribution on shallow foreshores. Their model consists of an appropriate combination of two Weibull distributions, to represent a linear trend for the lower heights and a downward curved relation for the higher waves, limited by breaking. The distributions match at the transition wave height H_{Tr}, given by:

$$H_{Tr} = (0.35 + 5.8\tan\theta)\,h \qquad (13.48)$$

The resulting characteristic waves $H_{1/3}$, $H_{1/10}$, $H_{2\%}$, $H_{1\%}$, $H_{0,1\%}$ are tabulated in the quoted paper, normalized with H_{rms}:

$$H_{rms} = \left(2.69 + 3.24\sqrt{m_0}/h\right)\sqrt{m_0} \qquad (13.49)$$

Table 13.2. Characteristic dimensionless wave heights.

H_{Tr}/H_{rms}	$H_{1/3}/H_{rms}$	$H_{2\%}/H_{rms}$
0.05	1.279	1.548
0.50	1.280	1.549
1.00	1.324	1.603
1.20	1.371	1.662
1.35	1.395	1.717
1.50	1.406	1.778
1.75	1.413	1.884
2.00	1.415	1.985
2.50	1.416	1.978
3.00	1.416	1.978

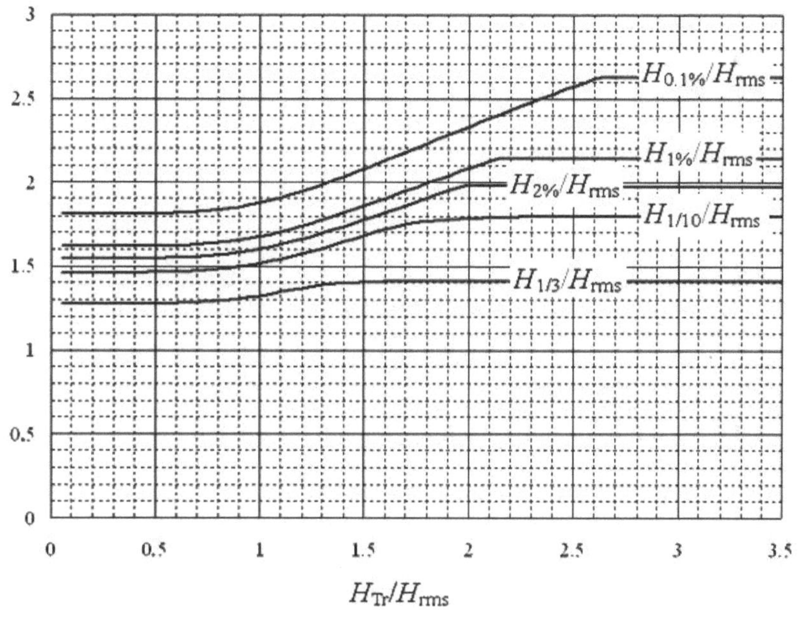

Figure 13.9. Characteristic waves $H_{1/3}$, $H_{1/10}$, $H_{2\%}$, $H_{1\%}$, $H_{0,1\%}$, for given H_{rms} and H_{Tr} , according to Battjes & Groenendijk (2000) distribution.

Tab. 13.2 reports some normalised values of $H_{1/3}$ and $H_{2\%}$ corresponding to H_{Tr} in the range 0.05 - 3.00. A plot of characteristic wave distributions is given in Fig. 13.9.

13.3. WAVE TRANSFORMATION BY STRUCTURES

(Van der Meer, INF)

Waves coming from deep water may reach a structure after refraction and breaking, see the previous section on wave transformation. As soon as waves reach a structure, such as an

LCS, a lot of processes start. The waves may break on the structure, overtop it, generate waves behind the structure and reflect from the structure. Another effect is wave penetration through openings between structures and diffraction around the head of structures. Both wave penetration and diffraction do not depend on the fact whether the structure is low-crested or not and, therefore, one is referred to handbooks for these items (CEM, 2001; Massie, 1986).

13.3.1. Wave transmission

The main effect of an LCS is that energy can pass over the crest and generate waves behind the structure. The main parameters describing wave transmission are given in Figure 13.10, here for a rubble mound structure. These are:

H_i = incident significant wave height, preferably H_{m0i}, at the toe of the structure
H_t = transmitted significant wave height, preferably H_{m0t}
T_p = peak period
s_{op} = wave steepness, $s_{op} = 2\pi H_i/(gT_p^2)$
R_c = crest freeboard
H_c = structure height
K_t = transmission coefficient H_t/H_i
ξ_{op} = breaker parameter $\xi_{op} = \tan\alpha/(s_{op})^{0.5}$

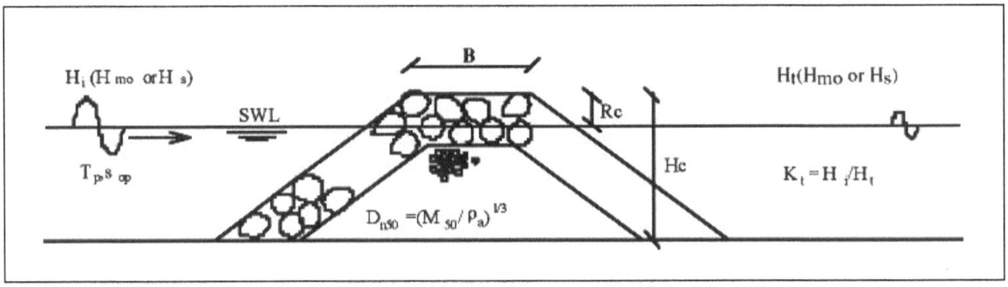

Figure 13.10. Governing parameters for wave transmission.

13.3.1.1. Rubble mound low-crested structures

An extensive database on wave transmission was gathered in the DELOS project. This database was analysed to come up with the best formulae describing wave transmission. The full analysis is given in Briganti *et al.* (2003). The gathered database, made up of 2337 tests, include the data by Van der Meer and Daemen (1994) and by d'Angremond *et al.* (1996) on rock and tetrapod structures (old database); Calabrese *et al.* (2002) with large scale tests on shallow foreshores (GWK); Seabrook and Hall (1998) on submerged structures with very wide crests; Hirose *et al.* (2000) on Aquareef blocks with very wide crests; and Melito and Melby (2000) on structures with corelocs. Within the DELOS project, tests were performed at the University of Cantabria (UCA) and the Polytechnic University of Catalonia (UPC), both in Spain. Table 13.3 gives the datasets with the number of tests and ranges tested.

The main conclusion by Briganti *et al.* (2003) is that, if submerged rubble mound structures with very wide crests are considered, two formulae should be considered, one for relatively narrow crested structures and one for very wide and submerged structures. The formulae are given by:

Table 13.3. Overall view of extensive database on wave transmission at rubble mound structures.

Database	*Armour type*	R_c/H_i	B/H_i	B/L_{op}	ξ_{op}	H_i/D_{n50}	H_i/h	s_{op}	*Test #*
Old database	various	– 8.7 4.0	0.37 43.48	0.009 0.51	0.7 8.26	0.3 6.62	0.03 0.62	$2{\cdot}10^{-4}$ 0.06	398
UCA	rubble mound	– 1.5 1.53	2.67 30.66	0.04 0.4	3.97 12.98	0.84 2.42	0.1 0.37	0.002 0.02	53
UPC	rubble mound	– 0.37 0.88	2.66 8.38	0.07 0.24	2.69 3.56	2.65 4.36	0.17 0.33	0.02 0.034	24
GWK	rubble mound	– 0.76 0.66	1.05 8.13	0.02 0.21	3 5.21	1.82 3.84	0.31 0.61	0.01 0.03	45
M & M	core locks	– 8.2 8.9	1.02 7.21	0.02 0.13	2.87 6.29	0.68 4.84	0.05 0.5	0.01 0.054	122
Seabrook	rubble mound	– 3.9 0	1.38 74.47	0.04 1.66	0.8 8.32	0.78 3.2	0.11 0.58	0.01 0.06	632
Aquareef	aquareef	– 4.77 – 0.09	1.24 102.12	0.02 2.1	1.78 5.8	0.59 4.09	0.1 0.87	0.01 0.08	1063

$$K_t = -0.4\frac{R_c}{H_{si}} + 0.64\left(\frac{B}{H_{si}}\right)^{-0.31}(1 - e^{-0.5\xi}), \qquad B/H_i < 10 \tag{13.50}$$

$$K_t = -0.35\frac{R_c}{H_{si}} + 0.51\left(\frac{B}{H_{si}}\right)^{-0.65}(1 - e^{-0.41\xi}), \qquad B/H_i > 10 \tag{13.51}$$

Eq. (13.50) is the original formula of D'Angremond *et al.* (1996), which proved to be applicable to the dataset with the restriction given on crest width. For wider crests, Eq. (13.51) was derived with a similar structure. Both formulae shall be limited by plausible lower and upper bounds. These are 0.07 and 0.80 for narrow crests; for wide crests, 0.05 and:

$$K_{tu} = -0.006\frac{B}{H_i} + 0.93 \tag{13.52}$$

the transition between Eq. (13.50) and (13.51) is not continuous. If a continuous transition is required, it is suggested to use Eq. (13.50) for $B/H_i \leq 8$ and Eq. (13.51) for $B/H_i \geq 12$. For $8 < B/H_i < 12$ one should interpolate between the values for $B/H_i = 8$ and 12.

A comparison of calculated and measured transmission coefficients is given in Figure 13.11. The results show quite some scatter. The performance of Eq. (13.50) and Eq. (13.51) + Eq. (13.52) may be evaluated in terms of root mean square error (RMSE) and R^2. They show an RMSE of 0.072 and 0.082 and R^2 equal to 0.91 and 0.90, respectively.

The DELOS project gave also results with regard to oblique wave attack and transmission, see Van der Meer *et al.* (2003). The main conclusion on the effect of angle of wave attack

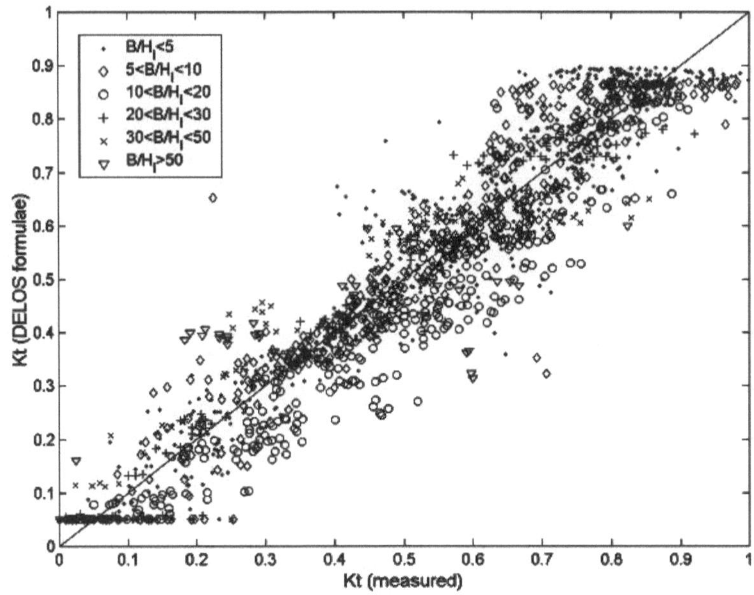

Figure 13.11. Calculated (Eq.s (13.50), (13.51), (13.52)) and measured transmission coefficients on rubble mound structures (Briganti *et al.* 2003).

was that there was none to marginal influence on wave transmission up to a wave angle of 70° (0° is perpendicular wave attack). This conclusion means that Eq. (13.50) to Eq. (13.52), developed for perpendicular wave attack, can also be used for oblique wave attack, up to 70°.

Another question with regard to oblique wave attack is whether the transmitted wave angle is similar to the incident wave angle. The same research showed that this was not the case, the transmitted wave angle is consequently smaller than the incident one:

$$\beta_t = 0.80\ \beta_i \text{ for rubble mound structures} \tag{13.53}$$

where β_t = the angle of transmitted waves and β_i = the incident wave angle.

13.3.1.2. Smooth low-crested structures

Not all low-crested structures are of the rubble mound type. Sometimes smooth and impermeable structures exist, for example low-crested structures covered with asphalt or armoured with a block revetment. Often the slope angles of the structure are gentler (1:3 or 1:4) than for rubble mound structures, mainly for construction reasons.

Wave transmission over smooth low-crested structures is completely different from rubble mound structures. First of all, the wave transmission is larger for the same crest height, simply because there is no energy dissipation by friction and porosity of the structure. Furthermore, the crest width has less or even no influence on transmission, as also on the crest there is no energy dissipation, which is completely different from rubble mound structures. Only for very wide (submerged) structures there could be some influence of the crest width, but this is not a case that will often be present in reality as asphalt and block revetments are mainly constructed in the dry and not under water. The presence of tide or storm surges make it possible to construct these kind of structures above water.

As smooth structures are different from rubble mound structures, also different formulae will be given for the transmission coefficient and the influence of oblique wave attack. The wave transmission can be calculated by, see Van der Meer *et al.* (2003):

$$K_t = [-0.3\, R_c/H_i + 0.75[1 - \exp(-0.5\xi_{op})]]\, \cos^{2/3}\beta \tag{13.54}$$

with as minimum $K_t = 0.075$ and maximum $K_t = 0.8$ and limitations:

$$1 < \xi_{op} < 3 \quad 0° \le \beta \le 70° \quad 1 < B/H_i < 4$$

Eq. (13.54) already includes the effect of oblique wave transmission by the term $\cos^{2/3}\beta$. It was very clear from the experiments that wave transmission decreases with increasing obliquity. Figure 13.12 show this dependency, where on the vertical axis the measured transmission coefficient is given as a ratio to Eq. (13.54), without the cosine part.

Oblique wave attack has also influence on the transmitted wave angle and in a different way than for rubble mound structures. Up to 45° the transmitted wave angle is similar to the incident one. Beyond 45° the waves jump along the structure and generate consequently a transmitted wave angle of 45°, regardless of the incident angle. Thus:

$$\begin{aligned} \beta_t &= \beta_i && \text{for } \beta_i \le 45° \\ \beta_t &= 45° && \text{for } \beta_i > 45° \qquad \text{for smooth structures} \end{aligned} \tag{13.55}$$

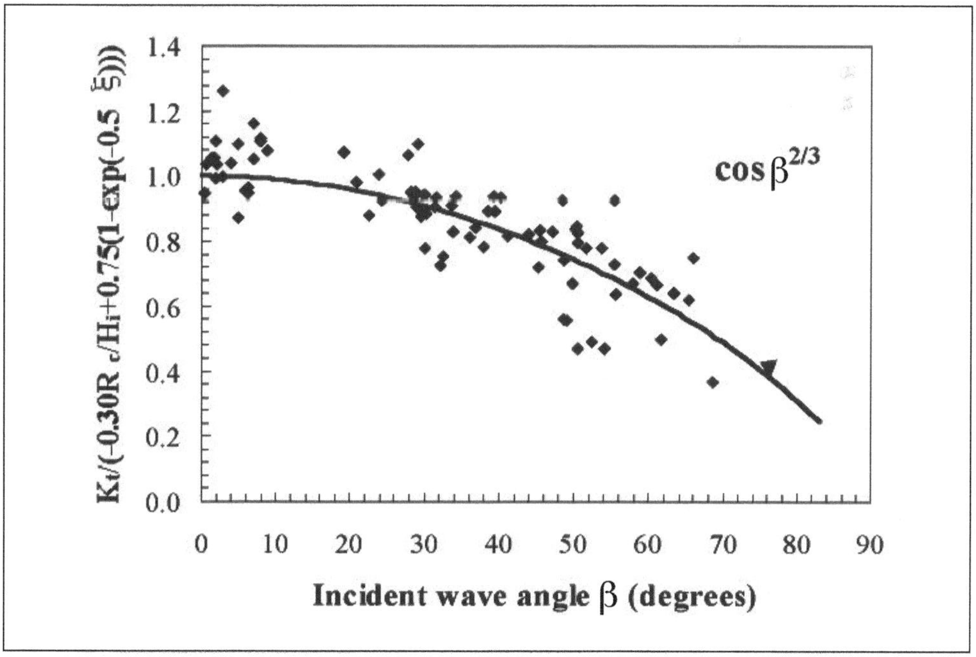

Figure 13.12. Influence of angle of wave attack on wave transmission for smooth structures.

13.3.1.3. Application of a neural network

It is clear in Figure 13.11 that quite some scatter still exists if formulae are based on various investigations and a large dataset. One of the main drawbacks of empirical formulae is that, in order to keep the application fairly simple, a reduced number of parameters are taken into account.

A neural network is a tool which has proven its usefulness if a process is difficult to describe and if a large dataset is available. In fact this is the case for wave transmission at rubble mound low-crested structures. In Panizzo *et al.* (2003) a neural network was made with the DELOS dataset as described in Table 13.3. Figure 13.13 gives the structure of the neural network and also the input parameters. The number of input parameters is larger than in Eq. (13.50)-Eq. (13.52). The parameters in the formulae are R_c/H_i; B/H_i; and ξ_{op} (in Figure 13.13 given as *Ir*). For the neural network also H_i/D_{n50}; B/L_{op}, and H_i/h were added. This gives the added effect of the rock size, another effect of the wave length than only the breaker parameter, and the effect of wave height to water depth.

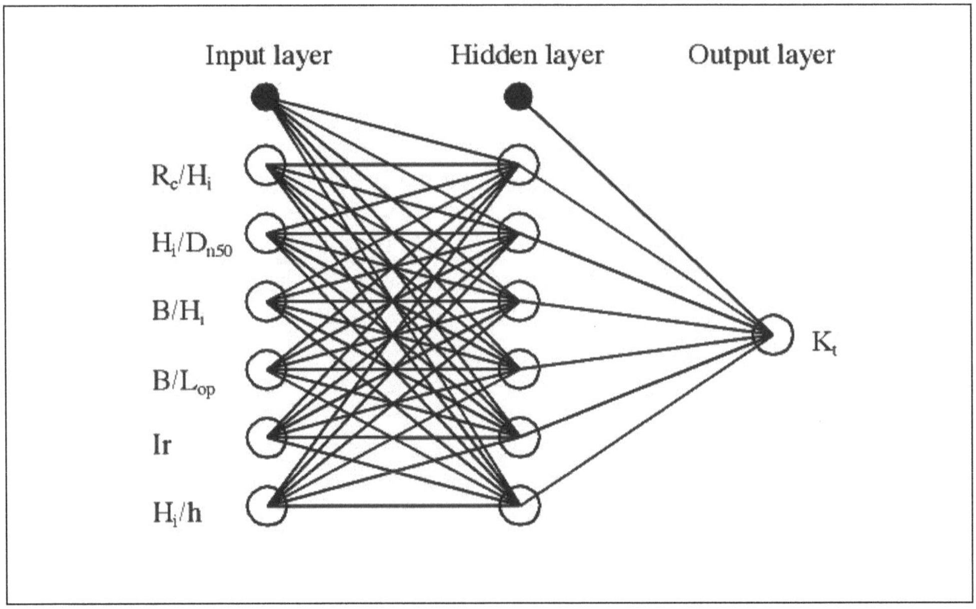

Figure 13.13. Structure of the neural network with the input parameters used.

The results of the neural network are given in Figure 13.14 as predicted versus measured wave transmission coefficients. This should be compared with Figure 13.11 and it is clear that, due to the presence of an extensive dataset, the neural network performs much better than the empirical Eq. (13.50)-Eq. (13.52).

The drawback of a neural network is that an equation is not available. The method can only be used with direct access to the neural network, which is not publicly available for the wave transmission prediction.

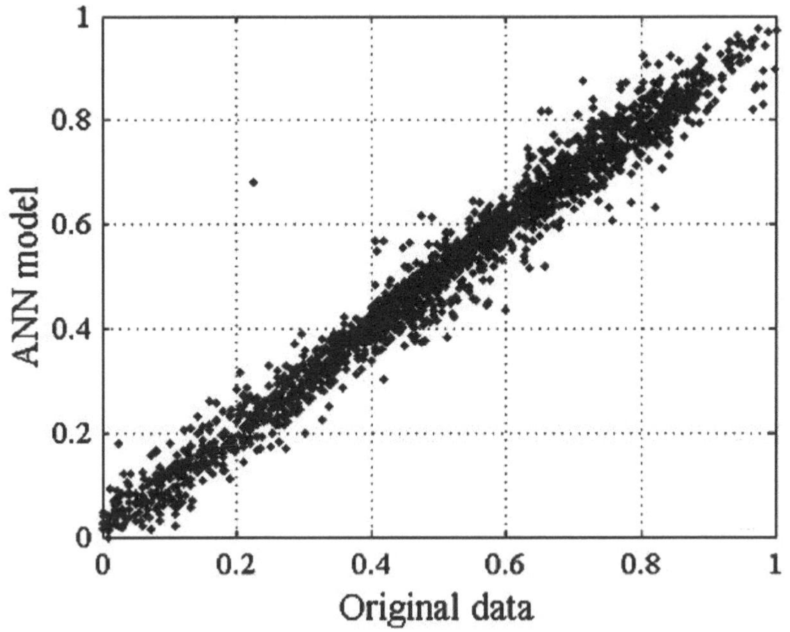

Figure 13.14. Comparison of wave transmission predicted by the neural network and measured.

13.3.1.4. Spectral change due to wave transmission

Transmitted spectra are often different from incident spectra. Waves breaking over a low-crested structure may generate two or more transmitted waves on the lee side. The effect is that more energy is present at higher frequencies than for the incident spectrum. In general the peak period is quite close to the incident peak period, but the mean period may decrease considerably. A first analysis on this topic can be found in Van der Meer *et al.* (2000).

The wave transmission coefficient only contains information about the wave heights behind the structure. It is the spectrum which contains wave period information. Very often information is required on both wave heights and periods, for example for wave run-up or overtopping at structures behind a low-crested structure, or for calculation of morphological changes.

Figure 13.15 shows an example of a transmitted spectrum for a smooth structure and gives clearly the picture that energy is present more or less at a similar level up to high frequencies. Based on this, a simple and crude model was developed by Van der Meer *et al.* (2000), which is shown in Figure 13.16. In average 60% of the transmitted energy is present in the area of $< 1.5 f_p$ and the other 40% of the energy is evenly distributed between $1.5 f_p$ and $3.5 f_p$.

The division of energy in 60%/40% parts and the frequency of $f_{max} = 3.5 f_p$ were only based on a limited number of tests. The assumptions by Van der Meer *et al.* (2000) were refined with new data of the DELOS project, see Briganti *et al.* (2003) and Van der Meer *et al.* (2003).

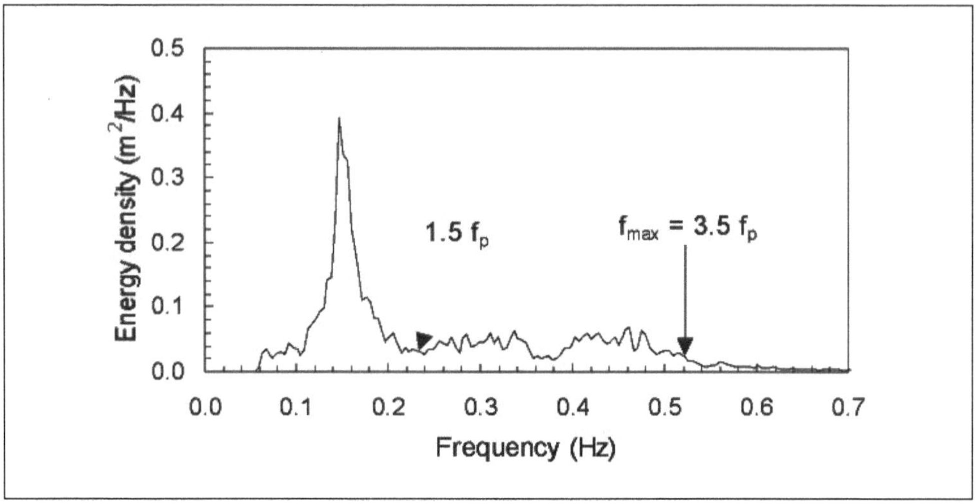

Figure 13.15. Example of transmitted spectrum with energy at high frequencies.

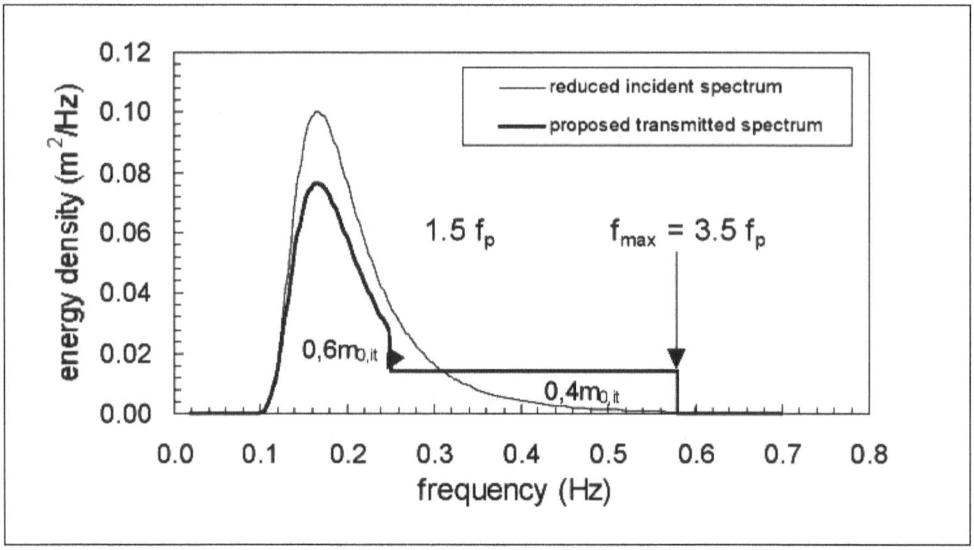

Figure 13.16. Proposed method by Van der Meer *et al.* (2000) for transmitted spectrum.

The conclusion was that overall results are similar to the proposed method in Figure 13.16, although rubble mound structures give a little smaller values than smooth structures. Briganti *et al.* (2003) analyzed this a little further and concluded that rubble mound and smooth structures do not give a similar behaviour. The method is also applicable to submerged rubble mound structures, *but not to emerged ones*. In the latter case much less energy goes to the higher frequencies and f_{max} may become close to $2.0f_p$. More research is needed to improve the method as described above.

13.3.2. Wave reflection

As far as wave transformation over low-crested structures is concerned, the DELOS project focused on wave transmission only. Wave reflection was not considered to be an important aspect and was only treated at the end of the project. Preliminary results are given here for rubble-mound structures.

Wave reflection at non-overtopped structures is described in the Rock Manual (CUR/CIRIA, 1991). For rock structures the data source is: Van der Meer (1988) and Allsop and Channel (1989). The most simple prediction formula given in the Rock Manual is:

$$K_r = 0.14\,\xi_{op}^{\;0.73} \text{ for } \xi_{op} < 10 \tag{13.56}$$

This formula, together with the original data, is shown in Figure 13.17. A more elaborated formula for rock slopes in the Rock Manual is:

$$K_r = 0.071\,P^{-0.82}\cot\alpha^{-0.62}\,s_{op}^{\;-0.46} \tag{13.57}$$

In this formula the slope angle has a little larger influence than the steepness, compared to the relationship in the breaker parameter ξ_{op}. Also the permeability has a small influence, see Van der Meer (1988). In the case of overtopped structures, the P-value will often be close to $P = 0.4 - 0.6$ and the influence of the slope angle will reduce if the structure becomes more submerged. Therefore the simple Eq. (13.56) was taken for comparison.

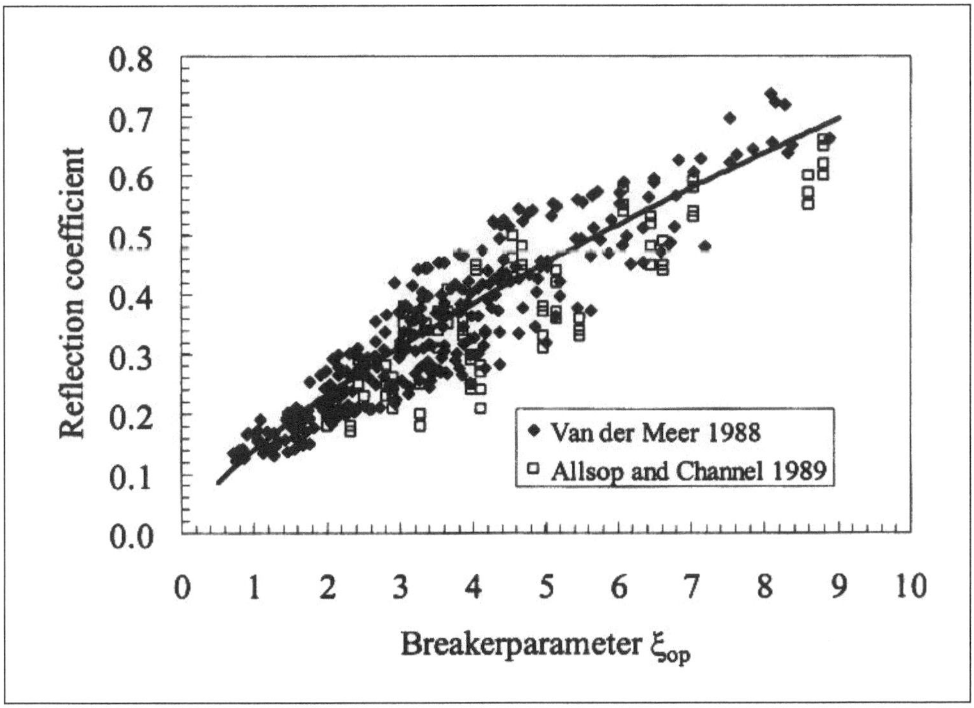

Figure 13.17. Reflection on non-overtopped rock slopes, CUR/CIRIA (1991).

It is expected that (very) submerged structures will have smaller reflection than non-overtopped, due to the fact that more energy will go over the structure. It is also expected that the relative crest height R_c/H_s has the main influence on a possible reduction of the reflection coefficient. The crest width will have no influence as waves reflect from the seaward side only.

Within the DELOS project there are 4 data sets with low-crested structures:

- UPC: Large scale 2D tests at the Polytechnic University of Catalonia, Spain. In total 63 tests.
- UCA: Small scale 2D tests at University of Cantabria, Spain. In total 53 tests.
- UB: 3D tests at Aalborg University, Denmark by University of Bologna. In total 28 tests (random waves, lay-out 1).
- INF: 3D tests at Aalborg University by Infram. In total 19 tests (rubble mound structure, perpendicular attack).

Comparison of reflection coefficients with Figure 13.17 showed, for various reasons, quite some scatter. But it was clear that lower structures gave indeed lower reflection. In order to reduce the scatter and to come to a conclusion about the reduction in reflection by low-crested structures, the averages of groups of similar data points were taken. Furthermore, it was assumed that for the highest structures tested ($R_c/H_i > 0.5$), the influence on the reflection would be very small or not existing.

Based on these assumptions a reduction in average reflection coefficients was determined for data groups of the four mentioned data sets. Figure 13.18 gives the final graph, which still must be considered as a preliminary result.

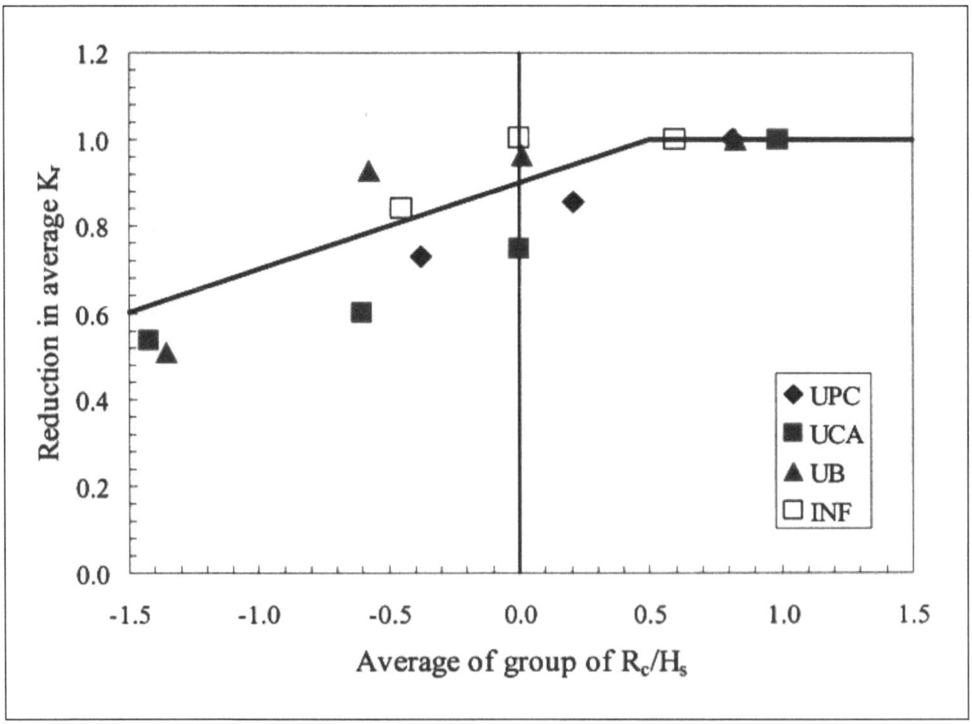

Figure 13.18. Reduction in reflection coefficient for low-crested rubble mound structures.

The reduction factor f_r on K_r for LCSs is:

$$f_r = 0.2\, R_c/H_s + 0.9 \qquad \text{for } R_c/H_s < 0.5$$
$$f_r = 1 \qquad \text{for } R_c/H_s \geq 0.5 \qquad (13.58)$$

The reduction factor f_r in Eq. (13.58) can be applied to reflection coefficients determined by Eq. (13.56) or by other existing equations for wave reflection. Eq. (13.58) is valid for rubble mound structures. There is no method for smooth structures other than using also Eq. (13.58), but now applied to a prediction formula for smooth non-overtopped structures. Such prediction formulae can be found in the Rock Manual.

13.4. HYDRODYNAMIC NUMERICAL MODELS TO PREDICT LOCAL HYDRODYNAMICS IN THE VICINITY OF THE STRUCTURES

(de Vries, WL-DH; Zyserman, DHI; Losada, UCA; Gonzalez-Marco & Arcilla, UPC)

13.4.1. Introduction and concepts

For the design of hydraulic structures, the hydraulic design data (*e.g.*, water levels, waves and currents) need to be assessed. To achieve this, use is often made of measurements and numerical modelling. The hydraulic design data are used as input for the design of the coastal protection structures. The conceptual design of these structures is often based on empirical formulae. These formulae have a limited range of validity, and for some cases do not provide sufficiently accurate estimates. For instance, the geometry of the structure may be different from those structures on which the empirical formulae were based, leading to unacceptable uncertainties in the predictions of hydraulic interactions and structural response. For this reason, there is a need for additional information that can be obtained from measurements or numerical modelling. In this section, some basic aspects of numerical modelling related to hydraulic structures consisting of rock are discussed. The numerical models provide a useful tool in the *pre-design* phase, but for the final design of the coastal protection structures, verification in physical scale models are in some cases indispensable.

13.4.2. Types of models and modelling

Hydraulic phenomena can be represented physically, in physical or scale models, or numerically, in numerical or mathematical models. The latter type of modelling is discussed in this Section. For a discussion of physical or scale models, the reader is referred to Section 13.12. Processes and phenomena relevant to low-crested structures which may be subject to modelling are water levels, currents, waves, wave reflection, wave run-up, wave overtopping, wave transmission. Scour, forces and the stability of stones is typically a topic for study in physical models.

13.4.2.1. Mathematical models

Mathematical models are based upon descriptions of physical phenomena through (a set of) mathematical equations. The equations are then solved numerically for the parameters of interest by a numerical model, usually in a computer program.

In many numerical models for hydraulic applications, such programs solve the equations of continuity and momentum or energy. These numerical models simulate for instance the motion of water, or the interaction of water with hydraulic structures. Another type of numerical models is built around analytical solutions and/or empirical formulae describing a phenomenon. Examples are the formulae for stability of rock. Also models exist based on processing a large amount of available data to obtain estimates of relevant design parameters, *e.g.* artificial neural network modelling.

13.4.2.2. Phase-resolving versus phase-averaged (spectral) modelling

For obtaining hydraulic design data from numerical wave model simulations, there are several options. The main choice is between phase-resolving and phase-averaged models. Phase-resolving models can be both time-domain models (for example solving the Boussinesq equations or the hyperbolic approximation of the mild-slope equation, MSE) and stationary conditions models (based on the fully-elliptic MSE or on the parabolic approximation of the MSE). Phase-averaged models are the so-called spectral models; these can integrate the equation of energy in the time-domain or solve boundary value problems achieving stationary conditions. A further category of models currently applied in the nearshore areas are the so-called flow models: these take as input data the wave field predicted by a separate model and simulate the wave-induced currents and long waves.

The choice of the most appropriate numerical model to be employed in practical applications depends on the required accuracy of the wave conditions near the dikes, the dominant physical phenomena to be reproduced, the available budget and time for obtaining these conditions, the available data, etc. Also possible developments in the future have to be taken into account. Not only the applied hardware (PC, workstations, network) will improve, but also the models themselves. New insight into physics will result in improved parameterizations and more reliable wave predictions. Furthermore, the numerical models may speed up significantly by improving numerical techniques.

Phase-resolving models can provide a very accurate prediction of the wave field in the vicinity of structures, as they can simulate wave-shoaling, refraction, diffraction and reflection. By using *ad hoc* techniques it is also possible to include a description of the dissipative effects due to the wave breaking and to the bottom friction. Time-domain models such those based on the Boussinesq equations can also simulate the propagation of irregular waves and most of the nonlinear phenomena that occur in the nearshore aresas, like wave-wave interaction, long waves and currents generation. Stationary conditions models are mostly based on linearized governing equations, simulate monochromatic waves propagation and cannot take into account the generation of long waves and currents; these models can however, be run for each spectral component of a random sea state and the total wave field can be reconstructed by linear superposition of the results. Typically phase-resolving models require several computational grid nodes per wave-length (about 10 for MSE models and more than 20 for Boussinesq models); the number of time interval required for integrating the governing equations depends on the local wave celerity and in the case of nonlinear models can be extremely high.

Phase-averaged models (spectral models) solve the energy equation for each component of an irregular sea state and can describe the wave field over wide geographical areas, while are not so accurate in proximity and especially in the lee of the structures. These models can simulate wave-shoaling and refraction, while can simulate in a very approximate manner the wave diffraction. Wave breaking, bottom friction and wind forcing can also be included in the governing equations. In principle the computational grid nodes can be spaced in order to obtain a reasonable description of the wave field over the area of interest, since there are not mathematical constraint in this case.

As far as flow models are concerned, these take as input data the wave field calculated by a separate model (usually a MSE or a spectral model) and simulate the wave-induced currents and long waves. The advantage of decoupling the simulation of short-waves and currents is that separate computational grids can be used. More specifically flow models can be applied over wide areas, since they do not need very fine grids. Flow models are based on depth-integrated equations and in principle provide a single value of the hydrodynamic parameters (flow velocity in the two horizontal directions, mean water level set-up) at each computational point; however in the last decades several advanced formulations have been proposed that can partially take into account for the vertical structure of the currents, so that nowadays these models are commonly referred to as quasi-3D models.

Nowadays it is common practise to use spectral wave models, such as SWAN, or models based on the mild-slope equation, like MIKE 21 PMS, to predict the wave field in the vicinity of structures. Spectral wave models can rather accurately predict the wave motion inside tidal basins or outside the surf zone. However, in very shallow regions, such as tidal flats and surf zones, the accuracy decreases. Spectral models describe the wave motion in a statistical way. The wave parameters such as significant wave height and wave period are averaged measures, which are used to assess the safety of sea defences.

Alternatively, time-domain wave prediction models can be used. Nowadays, Boussinesq-type wave models are appropriate to determine the wave conditions in the vicinity of coastal structures. If the model includes a description of wave breaking, simultaneous computation of the wave-induced flow field is possible. In the future (say within 10 years from now) non-hydrostatic flow models may also form an alternative. A disadvantage is that time-domain models require significantly more computational time compared to spectral wave models for computing the wave motion in the same domain. Therefore, time domain models are restricted to smaller domains. On the other hand, if the focus is on the wave conditions near the sea defences, it is not necessary to consider the whole wave field offshore. If proper «offshore» boundary conditions (which are not necessarily deep-water conditions) are available, for instance from a phase-averaged wave model, time domain models can be used to determine the hydraulic boundary conditions. The offshore boundary for the time-domain model is located inside the larger domain of interest. The boundary conditions can be obtained from measurements or from a spectral wave model describing the wave motion in somewhat deeper water.

The pros and cons of phase-averaged and time-domain models are often complementary and can be combined. Time domain models provide accurate wave predictions in the region near the sea defences, whereas the wave field in the rest of the tidal basin can be obtained with a spectral model. Consequently, by coupling the two types of models accurate results can be obtained.

13.4.2.3. Points to be considered

Improper schematizations and choice of computational grids may introduce numerical effects. Some are easily recognised, but others may be hard to discover. Instability problems, for instance, are obvious and can be remedied by adjusting the grid and/or time step. However, tracing of model inaccuracies is possible, for example, by varying the conditions or by comparison with similar cases, but generally requires special expertise.

Generally, a mathematical model is designed for a restricted number of phenomena (tide, flow, waves, wave run-up, wave overtopping and morphology). The following criteria must be met to obtain reliable results:

- mathematical description of the relevant phenomena is correct (equations);
- numerical accuracy (to limit the differences between the mathematical equations and the discretised equations);
- boundary conditions must be sufficiently accurate;
- schematisations of bathymetry, structure geometry, boundaries (friction, porosity) sufficiently accurate;
- the post-processing and interpretation of results should be correct;
- the numerical model should be calibrated correctly;
- the numerical model should be validated sufficiently.

A wide variety of numerical models with a wide variation in quality exist. To develop a reliable numerical model is however complex and requires expertise from various backgrounds. Often numerical models that have not been sufficiently validated are applied in design processes. Also adequately validated numerical models exist, but also those are often applied outside their range of validity. Care should be taken to correctly analyse and interpret the results to obtain suitable information from numerical models.

13.4.2.4. Selection of a suitable model

Scale and mathematical models are used for different types of problems. Which type of model is the most suitable one depends on various factors (nature of the problem, size of model, complexity of set-up of model, accuracy of model, scale effects, schematisation effects, numerical effects, time required per condition, 2D or 3D effects, turbulence, etc). In some cases several types of models can be used, then an adequate selection has to be made. In some other cases a combination of two or three models is used to obtain the required information. For instance, an overall mathematical model of a large area delivers boundary conditions for a detailed scale model of a smaller area. From the small area much more detailed information is obtained from the scale model than the mathematical model can provide. This is for instance often the case if hydraulic wave conditions near coastal structures are obtained based on numerical modelling, while the analysis of the stability of the structure is modelled in a physical scale model.

Advantages of physical scale models include the possibility of direct (audio-) visual observation and registration, that 3-D effects are represented, relatively limited schematisation effects, and that the stability of rock slopes can be modelled more accurate than in numerical models. Advantages of many numerical models include that larger regions can be modelled and that many computations for various situations can often be made relatively fast. Therefore, numerical models are mostly applied in the pre-design phase, whereas scale models can be used for the final design of hydraulic structures.

For all types of modelling, interpretation of the results is of vital importance for a proper use of the results and this requires knowledge of the processes involved.

Models also require that the accuracy is tested in some way, in order to improve the reliability of predictions. A clear distinction has to be made between *calibration* and *verification* of a model.

Calibration of a model implies adjusting the model (e.g. by means of field measurements) in such a way that the model data fit the prototype data sufficiently. The model is then reproducing a specific, known, situation in the prototype.

Verification of a model implies hindcasting of another known situation without adjusting the model parameters anymore. In fact, verification is a must because calibration alone is not a sufficient guarantee for reliability.

A calibrated and verified model can be considered operational for delivering forecasts of future changes as a result of hydraulic engineering works. However, it will never represent all physical phenomena exactly, but only the most important aspects selected by the designer.

It leaves the designer with the responsibility to select the suitable model for the problem to be solved. The availability of accurate field data also plays a role in the process of the ultimate selection of a model. Selection is based on (and thus requires knowledge of) data on for instance:

- the phenomena to be quantified (including possible interactions between the structure and the phenomena of concern);
- data (boundary conditions), which are available or to be acquired (from existing files or from measurements);
- the limitations of available tools ranging from simple design formulae to existing models;
- the accuracy of available tools (range of validity, and uncertainties within the range of validity);
- extent and accuracy of information needed for the purpose of design and construction.

Finally, the designer should be capable to make a good interpretation of the model results to be used in the design process.

13.4.3. Numerical modelling systems available for engineering applications

Mathematical modelling tools are nowadays available as commercial software from major hydraulic laboratories and universities. In the following Sub-sections, model tools are divided into tree groups, namely (a) flow models, (b) wave models and (c) fluid dynamics (CFD) models. The main characteristics of these models are summarised in the Tables 13.4 and 13.5, which provide information on the output quantities generated by different type of numerical models and their limitations and suitability for different applications.

13.4.4. Flow modelling tools

13.4.4.1. Delft3D modelling framework (Delft Hydraulics)

Delft3D-FLOW is applied to simulation of 2- and 3D hydraulics in lakes, estuaries, bays, coastal areas and seas. WL Delft Hydraulics has developed a fully integrated modelling framework for a multi-disciplinary approach and 3D computations for coastal, river, lake and estuarine areas. It can carry out simulations of flows, sediment transports, waves, water quality, morphological developments and ecology. It has been designed for experts and non-experts alike. The Delft3D framework is composed of several modules, grouped around a mutual interface, while being capable to interact with one another.

Delft3D can switch between the 2D vertically averaged and 3D mode simply by changing the number of layers. This feature enables to set up and investigate the model behaviour in 2D mode before going into full 3D simulations.

Table 13.4. Functionalities of models (a).

Model	*Dim.*	*Spatial scale [m]*	*Time scale*	*Output quantities*				
				Engineering parameters$^{(*)}$				*Impact parameters*
				H [m]	*T [s]*	*q [m²/s]*	$H_{2\%}$ *[m]*	*Velocity at bottom [m/s]*
Flow								
COPLA	2DH	$O(10^2-10^6)$	hours-months					x
D3D-FLOW	2DH/3D	$O(10^3-10^7)$	hours-months					x
MIKE 21 HD	2DH	$O(10^2-10^6)$	hours-months					x
SHORECIRC	2DH/3D	$O(10^2-10^4)$	hours-months					x
LIMCIR	Q3D	$O(10^2-10^6)$	hours-months			x		x
Wave								
BMV	1DH	$O(10_1-10_3)$	minutes	x	x	x	x	x
DELFT- TRITON	1-2DH	$O(10^2-10^3)$	minutes-hours	x	x		x	x
MIKE 21 BW	2DH	$O(10^2-10^3)$	minutes-hours	x	x	x	x	
MIKE 21 PMS	2DH	$O(10^2-10^3)$	days-months	x	x		x	
OLUCA-SP	2DH	$O(10^2-10^3)$	stationary conditions	x	x		x	
REF-DIF	2DH	$O(10^2-10^3)$	stationary conditions	x	x			
LIMWAVE	2DH	$O(10^2-10^3)$	stationary conditions	x	x			
CFD								
COBRAS	2DV	$O(10^1-10^2)$	minutes	x	x	x	x	x
DELFT-SKYLLA	2DV	$O(10^1-10^2)$	minutes	x	x	x	x	x
NS3	3D	$O(10^1-10^2)$	minutes	x	x	x	x	x
Other								
Breakwat	–	–	–			x	x	
LIMORPH	Q3D	$O(10^1-10^3)$	minutes-weeks					

$^{(*)}$ Engineering parameters indicated in the columns are wave height, wave period, wave overtopping discharge per unit length and wave run-up (expressed, e.g. in terms of $H_{2\%}$).

Modular setup

Delft3D is composed of a number of modules, each addressing a specific domain of interest, such as flow, near-field and far-field water quality, wave generation and propagation, morphology and sediment transport, together with pre-processing and post-processing modules. All modules are dynamically interfaced to exchange data and results where process formulations require. In the following chapters these modules are described in more detail.

Table 13.5. Functionalities of models (b).

Model	*Availa-ble for end users*	*Suitability for pre-or detailed design*		*Limitations*	*Geographical domain of application*		
		Pre-design	*Detailed design*		*offshore*	*near-shore*	*near structure*
COPLA	yes (**)	x		depth-averaged flow velocities and set-up	x	x	x
D3D-FLOW	yes	x		waves only in combination with WAVE module	x	x	x
MIKE 21 HD	yes (*)	x		depth-averaged flow velocities only	x	x	x
SHORECIRC	yes	x		Quasi 3D flow velocities and surface elevation		x	x
DELFT-TRITON	no	x		accuracy decreases for very short waves		x	x
MIKE 21 BW	yes	x		computing time		x	x
MIKE 21 PMS	yes	x		stationary conditions		x	x
OLUCA-SP	yes (**) (free)	x		Stationary conditions		x	x
REF-DIF	yes (free)	x		Stationary conditions		x	x
BMV	no	x	x	Suitable for nearshore hydrodynamics (shallow waters waves)		x	x
COBRAS	no	x		computing time			
DELFT-SKYLLA	no	x		computing time			x
NS3	no	x	x	computing time			x
Breakwat	yes	x		only suitable for design of structure; no computation of wave propagation			x
LIMCIR	no	x		short boundary conditions		x	x
LIMWAVE	no	x		energic model with only first reflection considered	x	x	x
LIMORPH	no	x		short boundary conditions for water and sediment fluxes		x	x

(*) Commercial license. (**) Spanish and French version available. English version to be completed. User-friendly interface included with permission of the Spanish Ministry of the Environment granted through UC.

Delft3d-FLOW

The hydrodynamic module, Delft3D-FLOW, is a multi-dimensional hydrodynamic simulation program that calculates non-steady flow and transport phenomena resulting from tidal and meteorological forcing on a curvilinear, boundary-fitted grid. In 3D simulations, the hydrodynamic module applies the so-called sigma co-ordinate transformation in the vertical, which results in a smooth representation of the bottom topography. It also results in a high computing efficiency because of the constant number of vertical layers over the whole computational domain.

Module description

The hydrodynamic module is based on the full Navier-Stokes equations with the shallow water approximation applied. The equations are solved with a highly accurate unconditionally stable solution procedure. The supported features are:

- three co-ordinate systems, i.e. rectilinear, curvilinear and spherical in the horizontal directions and a sigma co-ordinate transformation in the vertical;
- domain decomposition both in the horizontal and vertical direction;
- tide generating forces (only in combination with spherical grids);
- simulation of drying and flooding of inter-tidal flats (moving boundaries);
- density gradients due to a non-uniform temperature and salinity concentration distribution (density driven flows);
- for 2D horizontal large eddy simulations the horizontal exchange coefficients due to circulation's on a sub-grid scale (Smagorinsky concept);
- turbulence model to account for the vertical turbulent viscosity and diffusivity based on the eddy viscosity concept;
- selection from four turbulence closure models: k-ε, k-L, algebraic and constant coefficient;
- shear stresses exerted by the turbulent flow on the bottom based on a Chézy, Manning or White-Colebrook formulation;
- enhancement of the bottom stresses due to waves;
- automatic conversion of the 2D bottom-stress coefficient into a 3D coefficient;
- wind stresses on the water surface modelled by a quadratic friction law;
- space varying wind and barometric pressure (specified on the flow grid or on a coarser meteo grid), including the hydrostatic pressure correction at open boundaries (optional);
- simulation of the thermal discharge, effluent discharge and the intake of cooling water at any location and any depth in the computational field (advection-diffusion module);
- the effect of the heat flux through the free surface;
- online analysis of model parameters in terms of Fourier amplitudes and phases enabling the generation of co-tidal maps;
- drogue tracks;
- advection-diffusion of substances with a first order decay rate;
- online simulation of the transport of sediment (silt or sand) including formulations for erosion and deposition and feedback to the flow by the baroclinic pressure term, the turbulence closure model and the bed changes;
- the influence of spiralling motion in the flow (i.e. in river bends). This phenomenon

is especially important when sedimentation and erosion studies are performed;
- modelling of obstacles like 2D spillways, weirs, 3D gates, porous plates and floating structures;
- wave-current interaction, taking into account the distribution over the vertical;
- many options for boundary conditions, such as water level, velocity, discharge and weakly reflective conditions;
- several options to define boundary conditions, such as time series, harmonic and astronomical constituents;
- online visualisation of model parameters enabling the production of animations.

Applications

Delft3D-FLOW is for example applied to the following related problems:
- harbours-wave disturbance, seiches, breakwater alignment, ship motion;
- sediment erosion, transport and deposition;
- salt intrusion in estuaries;
- fresh water river discharges in bays;
- thermal stratification in lakes and seas;
- cooling water intakes, heat and salt recirculation and waste water outlets;
- sediment transport including feedback on the flow;
- transport of dissolved material and pollutants;
- storm surges, combined effect of tide and wind/typhoon;
- bottom vanes, spurs, groynes, bridges, weirs and levees.

More references to Delft3D models: http://www.wldelft.nl/soft/d3d

13.4.4.2. MIKE 21 Modelling System (DHI Water & Environment)

MIKE 21 is a professional engineering software package containing a comprehensive modeling system for 2D free-surface flows. MIKE 21 is applicable to the simulation of hydraulic and related phenomena in lakes, estuaries, bays, coastal areas and seas where stratification can be neglected.

MIKE 21 provides the design engineer with a unique and flexible modeling environment using techniques which have set the standard in 2D modeling. It is provided with a modern user-friendly interface facilitating the application of the system. A wide range of support software for use in data preparation, analysis of simulation results and graphical presentation is included.

MIKE 21 utilises some of the most modern computer hardware and software and is available for PCs. MIKE 21 is compiled as a true 32-bit application implying that it can only be executed under Windows 98, NT, 2000 and XP.

MIKE 21 is the result of more than 20 years of continuous development and is tuned through the experience gained from thousands of applications worldwide. DHI continues to use MIKE 21 in its own studies, thus giving a valuable symbiosis between development and application.

Modular Construction

MIKE 21 is constructed in a modular manner around the four main application areas:
- coastal hydraulics and oceanography
- environmental hydraulics

- sediment processes
- waves

Applications

MIKE 21 can be used to study a wide range of phenomena related to hydraulics. Examples are:

- tidal exchange and currents
- storm surge
- heat and salt recirculation
- water quality
- harbours-wave disturbance, seiche, breakwater alignment, ship motion, sediment erosion, transport and deposition.

For additional references on MIKE 21, see http://www.dhisoftware.com/mike21/

MIKE 21 HD

MIKE 21 HD is the basic module of the entire MIKE 21 system. It provides the hydrodynamic basis for the computations performed in most other modules, for example the Advection-Dispersion and Sediment Transport modules.

MIKE 21 HD simulates the water level variations and flows in response to a variety of forcing functions in lakes, estuaries, bays and coastal areas. The water levels and flows are resolved on a rectangular grid covering the area of interest when provided with the bathymetry, bed resistance coefficients, wind field, hydrographic boundary conditions, etc.

MIKE 21 HD is applicable to a wide range of hydraulic phenomena such as tidal exchange and currents, storm surges, secondary circulations, eddies and vortices, harbour seiching, dam breaks, tsunamis, wave-driven currents (eventually combined with tidal and/or wind-driven currents), etc.

The hydrodynamic module of MIKE 21 solves the vertically integrated equations of continuity and conservation of momentum in two horizontal dimensions. The following effects are accounted for:

- convective and cross momentum
- wind shear stress at the surface
- barometric pressure gradients
- Coriolis forces
- momentum dispersion
- sources and sinks for mass and momentum
- evaporation.

The instantaneous water levels and fluxes are obtained from the solution of the continuity and momentum equations:

$$\frac{\partial \zeta}{\partial t} + \frac{\partial p}{\partial x} + \frac{\partial q}{\partial y} = S - e \qquad (13.59)$$

$$\frac{\partial p}{\partial t}+\frac{\partial}{\partial x}\left[\frac{p^2}{h}\right]+\frac{\partial}{\partial y}\left[\frac{p\cdot q}{h}\right]+gh\frac{\partial \zeta}{\partial x}$$
$$+\frac{g\sqrt{\frac{p^2}{h^2}+\frac{q^2}{h^2}}\cdot\frac{p}{h}}{c^2}-fVV_x-\frac{h}{r_W}\cdot\frac{\partial p_a}{\partial x}$$
$$-\Omega q-\left[\frac{\partial}{\partial x}\left[E_x\cdot h\cdot\frac{\partial u}{\partial x}\right]+\frac{\partial}{\partial y}\left[E_y\cdot h\cdot\frac{\partial u}{\partial y}\right]\right]=S_{ix} \qquad (13.60)$$

$$\frac{\partial q}{\partial t}+\frac{\partial}{\partial y}\left[\frac{q^2}{h}\right]+\frac{\partial}{\partial x}\left[\frac{p\cdot q}{h}\right]+gh\frac{\partial \zeta}{\partial y}$$
$$+\frac{g\sqrt{\frac{p^2}{h^2}+\frac{q^2}{h^2}}\cdot\frac{q}{h}}{c^2}-fVV_y-\frac{h}{\rho_W}\cdot\frac{\partial p_a}{\partial y}$$
$$+\Omega p-\left[\frac{\partial}{\partial x}\left[E_x\cdot h\cdot\frac{\partial v}{\partial x}\right]+\frac{\partial}{\partial y}\left[E_y\cdot h\cdot\frac{\partial v}{\partial y}\right]\right]=S_{iy} \qquad (13.61)$$

$\zeta(x, y, t)$ is the instantaneous water surface above datum, $p(x, y, t)$ and $q(x, y, t)$ are the flux densities in x- and y- directions, $h(x, y, t)$ is the total water depth, S is a source magnitude per unit horizontal area, S_{ix} and S_{iy} are sources for impulse in x- and y-directions (for example, gradients in radiation stress field), e is the evaporation rate, g is gravitational acceleration, c is Chezy's resistance number, f is wind friction factor, V, V_x and V_y are wind speed and its components in x- and y-directions, p_a is barometric pressure, ρ_w is density of water, Ω is Coriolis coefficient, $E(x, y)$ is the momentum exchange coefficient (eddy viscosity), x, y are space co-ordinates and t is time.

The equations are solved by implicit finite difference techniques with the variables defined on a space-staggered rectangular grid. A «fractioned step» technique combined with an Alternating Direction Implicit (ADI) algorithm is used in the solution to avoid the necessity for iteration. Second-order accuracy is ensured through the centring in time and space of all derivatives and coefficients. The ADI algorithm implies that at each time step a solution is first made in the x-direction using the continuity and x-momentum equations followed by a similar solution in y-direction.

The implicit scheme is used in MIKE 21 HD in such a way that stability problems do not occur provided that the input data is physically reasonable, so that the time step used in the computations is limited only by accuracy requirements.

The following basic input is required by MIKE 21 HD:

- bathymetry data
- time step and length of simulation
- bed resistance
- momentum dispersion coefficients
- wind friction factor
- initial conditions (water surface level and flux densities in x- and y-directions)

- boundary conditions (water levels or flow magnitude, flow direction)
- wind speed and direction
- radiation stress fields
- source/sink discharge magnitude and speed.

The following output can be obtained from MIKE 21 HD:
- time series of water depth maps
- time series of 2D maps of x- and y-components of flux (p and q).

Variables such as surface elevation, current speed and direction, x- and y-velocity components may be derived from the basic output by use of MIKE 21 pre- and post-processing tools.

13.4.4.3. SHORECIRC (C.A.C.R., University of Delaware)

SHORECIRC is a numerical model developed at C.A.C.R., University of Delaware, able to reproduce currents and long waves forced by wind and short waves.

The model is quasi-3D since it is able to approximately reproduce the vertical variation of the current flow, which decisively contributes to the horizontal exchange of momentum known as «lateral mixing». This is done by using an analytical solution for the 3D current profiles in combination with a numerical solution for the depth-integrated 2D horizontal equations. The theoretical background for SHORECIRC is described in Putrevu and Svendsen (1999) which is an extension of Svendsen and Putrevu (1994).

SHORECIRC is coupled with the numerical model REF-DIF which calculates short-wave quantities that are provided as input to the model by means of the radiation stresses. SHORECIRC solves the depth integrated continuity and momentum equations, providing information about the total depth integrated volume fluxes and the surface elevations. The vertical variation of the current velocities are calculated as well in the process and the effect of this variation is taken into account through the dispersive mixing coefficients. Several types of boundary conditions can be used on the computational grid boundary, in order to match the user's needs. More specifically it is possible to impose specific fluxes, periodicity conditions, no flux/straight wall, absorbing/generating conditions, and no flux following still water line.

A detailed description of the model, the user's manual and the program source codes (FORTRAN) are distributed, after registration, by the Authors of the model at the official SHORECIRC web page
http://chinacat.coastal.udel.edu/~kirby/programs/shorecirc/shorecirc.html

13.4.4.4. LIMCIR (Universitat Politècnica de Catalunya)

The LIMCIR code is an advanced Q-3D circulation model, developed at the Universitat Politècnica de Catalunya (Cáceres, 2004), solving the depth and time averaged continuity and momentum equations while recovering a depth averaged undertow. The resulting partial differential equations are solved with a staggered grid and an Alternating Direction Implicit method that allows, at the end of each iteration, to obtain a centered scheme in space and time.

The closure sub models are based on state of the art formulations.
- Bed shear stresses are obtained according to Madsen (1994) in the presence of waves.
- Roller model is based on Dally and Brown (1995).
- Eddy viscosity is evaluated based on Nielsen (1985) formulation to consider the

bottom turbulence and Osiecki and Dally (1996) to consider the roller turbulence. It can also employ the Smagorinsky model.

- Wave induced mass flux can be obtained from De Vriend and Stive (1987) or Fredsoe and Deigaard (1992).
- Wind stress is considered using the Yelland and Taylor (1996) formulation.
- The overtopping term can be obtained following Owen (1980), Hedge and Reis (1998), Van der Meer and Janssen (1995), or Allsop *et al.* (1995) considering sloping or vertical structures.

13.4.5. Wave modelling tools

13.4.5.1. BMV, Boussinesq model with vorticity (C.A.C.R., University of Delaware, U.S.A.; University of Roma TRE, University of Genova, University of Catania, Italy)

BMV is a one-dimensional numerical model based on the Boussinesq-type equations. It was originally developed at C.A.C.R., University of Delaware by Veeramony and Svendsen (1999, 2000) and then extended within the framework of the DELOS Project by a group of researchers from three Italian Universities (Rome TRE, DSIC; Genova, DIAm; Catania, DICA).

The Boussinesq-type model equations were derived without making the assumption of irrotational flow; coupling with the vorticity transport equation allows for taking into account horizontal axis, vorticity induced by wave-breaking. On the basis of the experimental study of Svendsen *et al.* (2000) a physically sound description of wave-breaking is introduced into the model, by applying at the lower edge of the surface roller a vorticity distribution similar to that measured in weak turbulent hydraulic jumps.

The main advantage of the present approach in comparison with standard Boussinesq models is that BMV can provide a very accurate description of the flow in the surf zone: although it is based on depth-integrated equations coupling with the vorticity transport equations allows modeling of non self-similar velocity profiles over the depth and therefore allows reproduction of the undertow currents.

Within the framework of the DELOS Project the model was extended in order to give a more accurate description of the flow in the swash zone, developing new shoreline boundary conditions (Bellotti and Brocchini, 2001 and 2002). Further developments were aimed at incorporating into the model a more physically sound description of turbulence, allowing the eddy viscosity to vary over the water depth; since the original model by Veeramony and Svendsen (1999, 2000) used a semi-analytical method to solve the vorticity transport equation that did not allow for vertical varying eddy viscosity, a full numerical solution to this equation was included, by coupling to the Boussinesq solver a further module that solves the vorticity transport equation with arbitrary values of the eddy viscosity at each computational point; see Briganti *et al.* (2004) for more details.

13.4.5.2. TRITON (Delft Hydraulics)

Application

Wave propagation in shallow water plays an important role both physically and economically in, e.g., coastal regions and harbour areas. Due to the existence of relatively large waves in shallow water non-linear effects are significant in these regions, especially when compared to wave propagation in deep water. A second important process in these regions is frequency

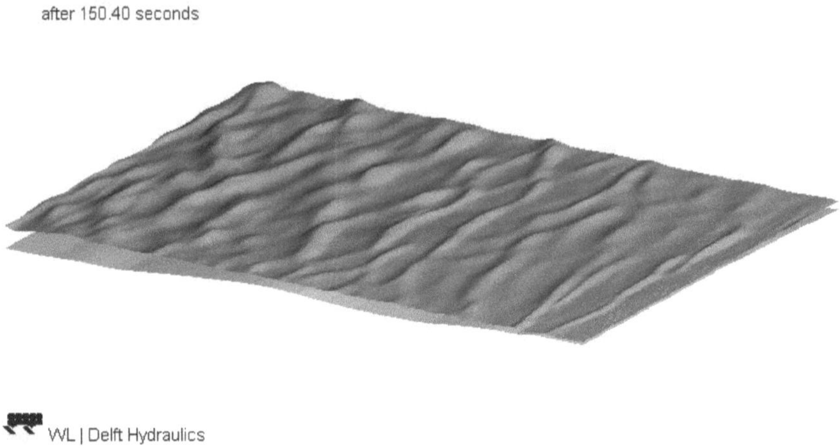

Figure 13.19. Refraction interference pattern of waves propagating over a 2D shoal on a sloping bed.

dispersion, i.e., the physical phenomenon that wave components of different frequencies propagate at different speeds. Standard shallow-water models, that are only valid for very long waves, do not take frequency dispersion into account. In the two-dimensional time-domain Boussinesq-type model TRITON both non-linear wave behaviour and frequency dispersion are represented, making the model suitable to be applied in coastal regions and harbours to provide hydraulic boundary conditions for coastal structures, coastal morphology and harbours.

Model description

TRITON is a two-dimensional Boussinesq-type model with improved linear- and non-linear behaviour (Borsboom *et al.*, 2000). The model has been extended with the implementation of a 2D wave breaking model based on a combination of the eddy viscosity concept and the surface roller concept (Borsboom *et al.*, 2001).

The combination has a number of features that makes it suitable for near-shore applications. Mass and momentum are strictly conserved while the wave breaking model only dissipates energy, which is in agreement with physical laws. The results and the comparison with experiments under very different wave conditions demonstrate the good performance of the model.

TRITON accounts for the following physics:

- wave propagation in time and space: shoaling, refraction due to depth variations, frequency dispersion and diffraction;
- non-linear wave-wave interactions;
- wave breaking;
- wave reflection.

Coupling with other models

The TRITON model is boundary driven, which implies that at the model boundaries the incident waves in terms of surface elevation as function of space and time should be prescribed. Both regular and irregular waves can be imposed at the boundary of the model. The latter are either based on a parametric spectrum or on a user-defined time signal. An interface with the spectral model SWAN, a third generation wave model developed at Delft University of Technology, has also been implemented to allow for boundary conditions based on spectra computed by SWAN. The shoreward boundaries can be fully absorbing, partially or fully reflective. TRITON calculates the instantaneous flow solution, i.e. the surface elevation and the depth-integrated velocities. These quantities can be generated as output on a grid covering the whole computational domain, along a ray or at singular locations. The model has been validated based on physical model tests and field measurement.

In addition to the regular boundary types, the boundary conditions for TRITON may also be obtained from observations or from other sources such as other numerical models. TRITON has been succesfully coupled to the spectral model SWAN, and the 3D potential flow model RAPID, which has been developed at MARIN. The latter allows for studies on ship-induced waves (Raven, 1996).

13.4.5.3. MIKE 21 BW (DHI Water & Environment)

MIKE 21 BW is a state-of-the-art numerical modelling tool for studies and analysis of wave disturbance in ports, harbours and coastal areas. MIKE 21 BW can be used for the analysis of operational and design conditions of coastal structures and within ports and harbours. Through the inclusion of surf and swash zone dynamics, the application range is extended further into the coastal engineering.

The model is capable of reproducing the combined effects of most wave phenomena of interest in port, harbour and coastal engineering. These include:

- shoaling and refraction;
- diffraction;
- bottom dissipation;
- partial reflection and transmission;
- non-linear wave-wave interactions;
- frequency spreading;
- directional spreading.

MIKE 21 BW is based on the numerical solution of the time domain formulations of Boussinesq type equations, Madsen and Sørensen (1991, 1992). The Boussinesq equations are solved using a flux-formulation with improved frequency dispersion characteristics. The enhanced Boussinesq type of equations make the model suitable for simulation of the propagation of directional wave trains travelling from deep to shallow water. The maximum depth to deep-water wavelength is $h/L_0 \approx 0.5$ (or $kh \approx 3.1$, where kh is the relative wave number) for the Boussinesq dispersion coefficient $B = 1/15$. For the classical Boussinesq equations ($B = 0$) the maximum depth to deep-water wavelength is $h/L_0 \approx 0.22$ (or $kh \approx 1.4$).

The Boussinesq equations solved by MIKE 21 BW are expressed in terms of the free surface elevation, ξ, and the depth-integrated velocity-components, P and Q.

The equations have been extended into the surf zone by inclusion of wave breaking and moving shoreline according to Madsen *et al.* (1997a,b), Sørensen and Sørensen (2001) and Sørensen *et al.* (2004).

The Boussinesq equations read:

Continuity

$$n\frac{\partial \xi}{\partial t}+\frac{\partial P}{\partial x}+\frac{\partial Q}{\partial y}=0 \qquad (13.62)$$

x-momentum

$$\begin{aligned}&n\frac{\partial P}{\partial t}+\frac{\partial}{\partial x}\left(\frac{P^2}{h}\right)+\frac{\partial}{\partial y}\left(\frac{PQ}{h}\right)+\frac{\partial R_{xx}}{\partial x}+\frac{\partial R_{xy}}{\partial x}+\\&n^2gh\frac{\partial \xi}{\partial x}+n^2P\left[\alpha+\beta\frac{\sqrt{P^2+Q^2}}{h}\right]+\frac{gP\sqrt{P^2+Q^2}}{h^2C^2}+n\Psi_1=0\end{aligned} \qquad (13.63)$$

y-momentum

$$\begin{aligned}&n\frac{\partial Q}{\partial t}+\frac{\partial}{\partial y}\left(\frac{Q^2}{h}\right)+\frac{\partial}{\partial x}\left(\frac{PQ}{h}\right)+\frac{\partial R_{xx}}{\partial x}+\frac{\partial R_{xy}}{\partial x}+\\&n^2gh\frac{\partial \xi}{\partial y}+n^2Q\left[\alpha+\beta\frac{\sqrt{P^2+Q^2}}{h}\right]+\frac{gP\sqrt{P^2+Q^2}}{h^2C^2}+n\Psi_2=0\end{aligned} \qquad (13.64)$$

where the dispersive Boussinesq terms Ψ_1 and Ψ_2 are defined by

$$\begin{aligned}\Psi_1 \equiv &-\left(B+\frac{1}{3}\right)d^2\left(P_{xxt}+Q_{xyt}\right)-nBg\,d^3\left(\xi_{xxx}+\xi_{xyy}\right)\\&-dd_x\left(\frac{1}{3}P_{xt}+\frac{1}{6}Q_{yt}+nBgd\left(2\xi_{xx}+\xi_{yy}\right)\right)\\&-dd_y\left(\frac{1}{6}Q_{xt}+nBgd\,\xi_{xy}\right)\\\Psi_2 \equiv &-\left(B+\frac{1}{3}\right)d^2\left(Q_{yyt}+P_{xyt}\right)-nBg\,d^3\left(\xi_{yyy}+\xi_{xxy}\right)\\&-dd_y\left(\frac{1}{3}Q_{yt}+\frac{1}{6}P_{xt}+nBgd\left(2\xi_{yy}+\xi_{xx}\right)\right)\\&-dd_x\left(\frac{1}{6}P_{yt}+nBgd\,\xi_{xy}\right)\end{aligned} \qquad (13.65)$$

Subscripts x, y and t denote partial differentiation with respect to space and time, respectively. P is the flux density in the x-direction ($m^3/m/s$), Q is the flux density in the y-direction ($m^3/m/s$), B is Boussinesq dispersion coefficient (–), h is the total water depth ($= d + \xi$), d is the still water depth (m), g is gravitational acceleration ($= 9.81\ m/s^2$), n is the

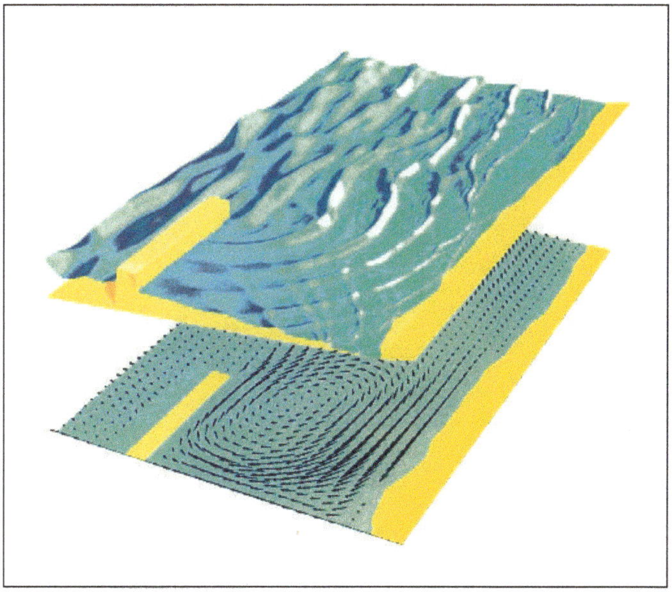

Figure 13.20. Wave and depth-averaged flow fields around a shore-parallel breakwater calculated by MIKE 21 BW.

porosity (–), C is Chezy resistance number ($m^{0.5}/s$), α is the resistance coefficient for laminar flow in porous media (–), β the resistance coefficient for turbulent flow in porous media (–) and ξ is the water surface elevation above datum (m).

The incorporation of wave breaking (available in the 1DH model) is based on the concept of surface rollers, where the terms denoted R_{xx}, R_{xy} and R_{yy} account for the excess momentum originating from the non-uniform velocity distribution due to the presence of the surface roller. R_{xx}, R_{xy} and R_{yy} are defined by

$$R_{xx} = \frac{\delta}{1-\delta/h}\left(c_x - \frac{P}{h}\right)^2$$

$$R_{xy} = \frac{\delta}{1-\delta/h}\left(c_x - \frac{P}{h}\right)\left(c_y - \frac{Q}{h}\right) \qquad (13.66)$$

$$R_{yy} = \frac{\delta}{1-\delta/h}\left(c_y - \frac{Q}{h}\right)^2$$

Here $\delta = \delta(t, x, y)$ is the thickness of the surface roller and c_x and c_y are the components of the roller celerity.

Model Input Data

The necessary input data to the two models in MIKE 21 BW can be divided into the following groups:

Basic data:
- bathymetry
- type of model and equations
- numerical parameters
- type of boundaries
- time step and length of simulation

Calibration data:
- initial conditions
- boundary data
- internal wave generation data
- wave breaking
- moving shoreline
- bottom friction
- partial wave reflection/transmission
- wave absorbing

Output data:
- deterministic output
- statistical output
- moving shoreline output

Model Output

Two types of output data can be obtained from the model:
- Deterministic data
- Statistical data

Deterministic output data consists basically of e.g. time series of surface elevations and depth-integrated velocity components. Statistical output data is obtained by user defined time-integration of derived variables.

13.4.5.4. MIKE 21 PMS (DHI Water & Environment)

MIKE 21 PMS is based on a parabolic approximation to the mild-slope equation governing the refraction, shoaling, diffraction and reflection of linear water waves propagating on a gently sloping bathymetry. The parabolic approximation is obtained by assuming a principal wave direction (x-direction), neglecting diffraction along this direction and neglecting backscatter. Neglection of backscatter means that modelling of wave conditions in the vicinity of reflecting structures by use of MIKE 21 PMS should be avoided. In addition, improvements to the resulting equation allow the use of the parabolic approximation for waves propagating at large angles to the assumed principal direction.

An additional feature of MIKE 21 PMS is the ability to simulate directional and frequency spreading of the propagating waves by use of linear superposition.

MIKE 21 PMS can be applied to any water depth on a gently sloping bathymetry, and it is capable of reproducing phenomena, such as shoaling, refraction, dissipation due to bed friction and wave breaking, forward scattering and partial diffraction, which makes it suited for application to the range of problems considered in the present study. The numerical solution is based on a single marching procedure from the offshore boundary to the coastline.

MIKE 21 PMS can be used to determine wave fields in open coastal areas, in coastal areas with structures where reflection and diffraction along the x-direction are negligible, in navigation channels, etc. Furthermore, MIKE 21 PMS can produce the wave radiation stresses required for the simulation of wave-induced currents.

The parabolic mild-slope equation applied in MIKE 21 PMS is:

$$\frac{\partial A}{\partial x} + \frac{\sigma_1}{\omega C_g}\frac{\partial}{\partial y}\left(CC_g\frac{\partial A}{\partial y}\right) + \frac{\sigma_2}{\omega C_g}\frac{\partial^2}{\partial y \partial x}\left(CC_g\frac{\partial A}{\partial y}\right) + \left[i\left(k_o - \beta_1 k\right) + \frac{1}{2C_g}\frac{\partial C_g}{\partial x} + \frac{\Omega}{2C_g}\right]A = 0 \tag{13.67}$$

where

$$\sigma_1 = i\left(\beta_2 - \beta_3\frac{k_o}{k}\right) + \beta_3\left(\frac{1}{k^2}\frac{\partial k}{\partial x} + \frac{1}{2kC_g}\frac{\partial C_g}{\partial x}\right) \qquad \sigma_2 = -\beta_3/k \tag{13.68}$$

$A(x, y)$ is the slowly varying complex wave amplitude, C is the phase velocity, C_g is the group velocity, k is wave number, k_0 is average wave number in y-direction, β_1, β_2 and β_3 are coefficients in the parabolic approximation, ω is the angular wave frequency, Ω is a complex dissipation coefficient due to bed friction and wave breaking, i is the imaginary unit and x, y are Cartesian co-ordinates.

For the parabolic approximation, three different techniques are implemented via the coefficients of the rational approximation β_1, β_2 and β_3:

- simple approximation (also known as (1,0) Padé approximation) ($\beta_1 = 1$, $\beta_2 = -1/2$ and $\beta_3 = 0$);
- (1, 1) Padé approximation ($\beta_1 = 1$, $\beta_2 = -3/4$ and $\beta_3 = -1/4$);
- minimax approximation for different apertures (10, 20, ..., 90 deg). Each aperture width has a set of coefficients.

The formulation of bed friction is based on the quadratic friction law. The description of the dissipation due to wave breaking is based on the expressions given by Battjes and Janssen, (1978).

The parabolic mild-slope equation in MIKE 21 PMS is solved using the Crank-Nicolson finite difference techniques with variables defined on a rectangular grid.

In MIKE 21 PMS, the following basic input data is required:

- bathymetry data
- bed friction data (optional)
- wave breaking parameters (optional)
- boundary conditions.

For monochromatic unidirectional waves, the incoming wave conditions are specified by the wave height, wave period and wave direction. For irregular and/or directional waves, the incoming wave conditions are given by the directional-frequency wave energy spectrum, prepared using the MIKE 21 pre-processing program *m21spc*.

MIKE 21 PMS produces four main types of output:
- integral wave parameters: the significant wave height, the peak wave period, the mean wave direction (MWD);
- 2D map of instantaneous surface elevation;
- 2D map of vector components $H_s \cdot \cos(\text{MWD})$ and $H_s \cdot \sin(\text{MWD})$;
- 2D map of radiation stresses.

13.4.5.5. OLUCA, part of the University of Cantabria (UC) Coastal Modelling System

The Coastal Modelling System (SMC) is a user-friendly software package developed by the University of Cantabria for the Dirección General de Costas (Spanish Ministry of the Environment). SMC encloses some numerical models for the application in coastal projects of the methodologies and formulations proposed in several manuals elaborated for the Ministry. The SMC is structured in five modules: (1) A pre-process module which generates all of the input data for the short- medium- and long-term numerical models. This module obtains (for any location along the Spanish coast including the islands) the bathymetry, wave directional regimes and the littoral flooding risk. (2) The short-term module includes numerical evolution morphodynamic models for monochromatic and irregular input waves, in a process on a scale of hours to days. (3) The medium- and long-term module allows the analysis of the medium-term processes (seasonal changes) and long-term response of the system on a scale of years. (4) The bathymetry renovation module permits easy updating of the actual bathymetry including different elements (sand fills in equilibrium beaches: plan and profile, coastal structures, etc.) in order to evaluate the different alternatives proposed using the numerical models.

The input files on bathymetry, wave climate and flooding risk have been also developed for other countries such as Colombia, Costa Rica and Tunisia and is currently under development for other countries. However, the user-friendly interface allows the use of input files for any bathymetry or wave information and therefore, makes the system applicable at any coastal site.

The Spanish Ministry of the Environment has delivered free versions of SMC to Spanish consultants and administrations and signed agreements with other countries to develop new ad hoc versions. SMC has been consistently applied to hundreds of real cases in Spain and in other countries, especially in Latin America.

The most relevant hydrodynamic modules for the application to LCS design are:
- OLUCA-SP and
- COPLA.

For further reference please visit http://www.smc.unican.es.

OLUCA-SP (University of Cantabria)

OLUCA-SP is a wave propagation model based on the parabolic approximation of the mild-slope equation. In essence it is equivalent to other models such as REF-DIF (University of Delaware) and MIKE 21 PMS (DHI Water & Environment).

OLUCA-SP is able to model most of the wave propagation processes but is limited to the restrictions inherent to linear wave theory and the parabolic approximation. The equations solved in OLUCA-SP, Kirby (1986), is able to include the effect of currents.

For spectral wave conditions the model input is based on a frequency spectrum that can be read directly from a file or a TMA spectrum together with a directional spreading function.

Figure 13.21. Altafulla Beach. Mediterranean Spanish Coast. H_s based computed by OLUCA-SP. The incident wave climate is defined by directional spectrum consisting of a TMA frequency spectrum with the following characteristics $H_s = 2.5$ m, $h = 10$ m, $T_p = 10$ s, $\gamma = 7$; number of components 5 and a directional spreading function, $\theta_m = 0^0$; $\sigma = 20^0$; number of components 5.

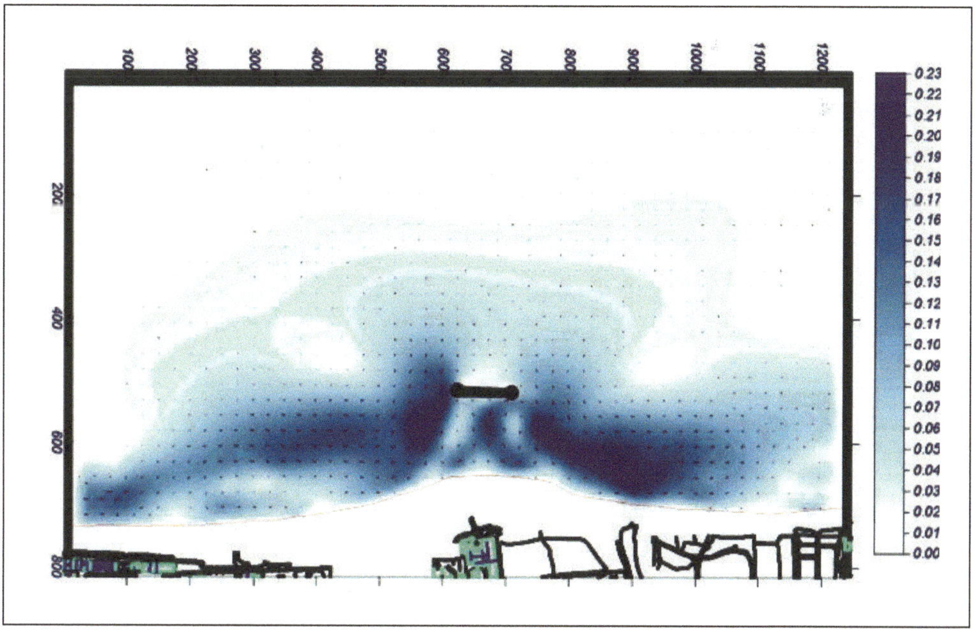

Figure 13.22. Circulation system around the LCS at Altafulla for the same incident conditions current intensities in blue scale and directions in scale vectors.

Wave dissipation includes laminar and turbulent boundary layers, bottom permeability and wave breaking. Wave breaking may be considered based on different models. OLUCA-SP includes the following options: Battjes and Janssen (1978), Thornton and Guza (1983) and Winyu and Tomoya (1998).

The model is appropriate to determine the wave field in open areas even in the presence of structures. However, it has to be pointed out that results in zones with high reflection or with large angles deviating from the principal direction of wave propagation should be discarded.

The model is also useful to evaluate radiation stresses and therefore to drive nearshore circulation models such as COPLA-MC/SP.

COPLA-MC/SP (University of Cantabria)

COPLA-MC/SP provides the circulation and water level variations in the nearshore as a response to wave forcing. It solves the vertically and time-averaged continuity and momentum equations in two horizontal dimensions (2DH model). The currents are driven thanks to the radiation stress gradients calculated from the COPLA-MC/SP model.

The model accounts also for convective and cross-momentum, turbulent diffusion and bottom friction than can be expressed in terms of a Chezy coefficient.

13.4.5.6. REF-DIF (C.A.C.R., University of Delaware, U.S.A.)

REF-DIF is a numerical model developed at C.A.C.R., University of Delaware, U.S.A. The model solves the parabolic approximation of the mild-slope equation and can simulate the effect of wave shoaling, refraction, wave-breaking and bottom friction and approximate diffraction; wave reflection cannot be reproduced by the model. The wave height and wave direction at each computational grid node are the output of the model; on the basis of these results the radiation stresses to be provided to flow models can be easily calculated. REF-DIF considers monochromatic waves but random sea states can be reproduced by using linear superposition of each component.

The model is provided as it is by C.A.C.R. for free. More details, as well as a detailed user's manual, the FOTRAN source code and some compiled version of the model can be obtained, after registration on the web site, at
http://chinacat.coastal.udel.edu/~kirby/programs/refdif/refdif.html

13.4.6. Fluid dynamics modelling tools

13.4.6.1.COBRAS (Cornell University/University of Cantabria)

Application

COBRAS is a 2DV numerical model that allows the simulation of wave-induced motion around coastal structures including the most relevant processes: shoaling, reflection, transmission, overtopping, porous flow, wave breaking, run-up, nonlinear effects and turbulence generation and transport in the fluid and permeable regions.

The model is able to reproduce complicated geometries and multi-layered structures from deeply submerged to emerged.

The model has been extensively validated against analytical solutions and laboratory experiments of flow around LCS's, wave breaking on impermeable and permeable slopes and wave interaction with other types of structures. Comparisons have included free surface transformation, pressure fields around and inside the structures, velocity fields and turbulence.

The input required is incident wave conditions, water depth, structure geometry and some characteristic coefficients of the permeable material of the different layers for multi-layer permeable structures.

As an output the model can provide directly: free surface, pressure and mean velocity time records at any point of the fluid domain; turbulence intensity and vorticity. Based on this information further magnitudes can be obtained: forces, moments mean flow, mass flux, shear stresses, overtopping discharges, etc.

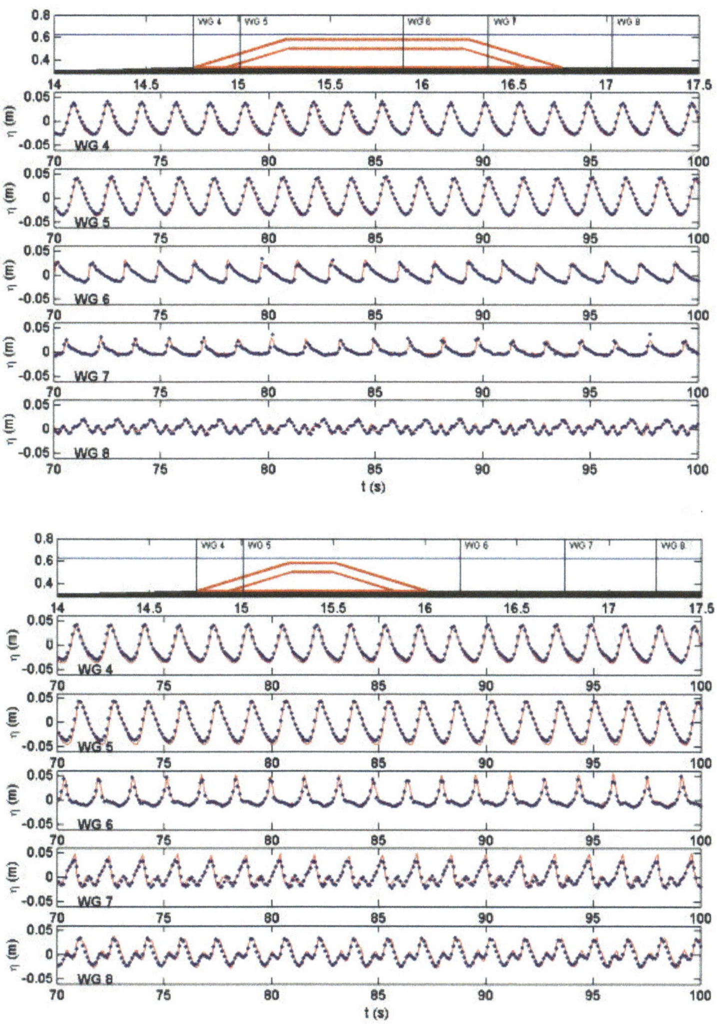

Figure 13.23. Comparison of free surface time series at different locations, for two different LCS built of two different permeable layers. ($h = 40$ cm, $T = 1.6$ s, $H = 10$ cm). Solid lines: experimental data. Dots: numerical results.

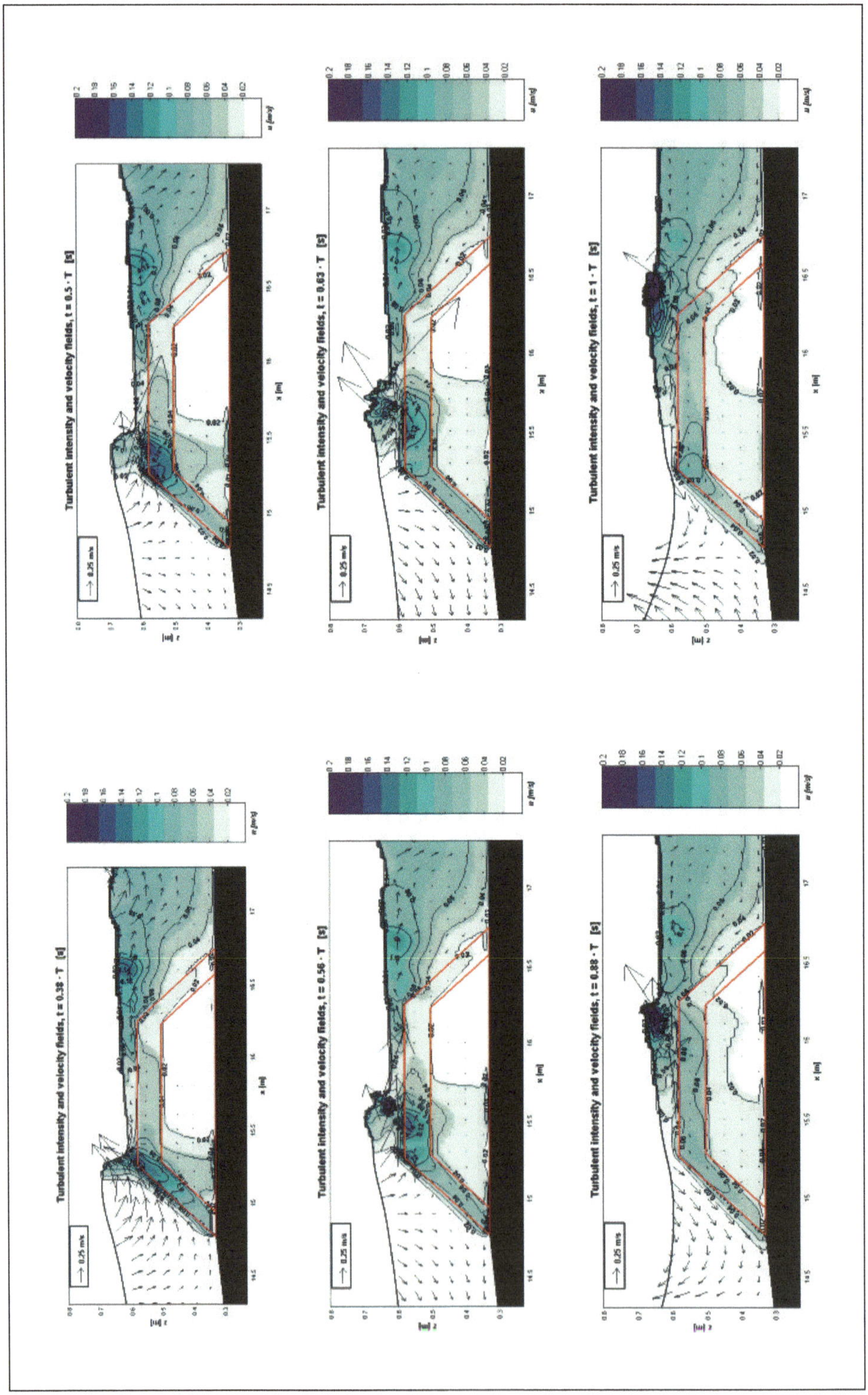

Figure 13.24. Simulated turbulent intensity and velocity fields around a LCS. Contour lines of turbulent intensity have intervals of 0.02 m/s.

Model description

The COBRAS model (Lin and Liu, 1998; Liu *et al.*, 1999, 2000; Hsu *et al.*, 2002) solves the 2DV Reynolds Averaged Navier-Stokes (RANS) equations, based on the decomposition of the instantaneous velocity and pressure fields into mean and turbulent components. Reynolds stresses are closed with an algebraic nonlinear k-ε turbulence model that can solve anisotropic-eddy-viscosity turbulent flows. The flow in the porous structure is described in the COBRAS model by the Volume-Averaged Reynolds Averaged Navier-Stokes (VARANS) equations, obtained by integration of the RANS equations in a control volume larger than the pore structure but smaller than the characteristic length scale of the flow (Hsu *et al.*, 2002). A new set of k-ε equations equivalent to those of the fluid region are obtained by volume averaging and used to model turbulence production-dissipation within the porous media.

The movement of the free surface is tracked by the volume of fluid (VOF) method as described by Hirt and Nichols (1981) which satisfies both the kinematic and dynamic free surface boundary conditions for the mean flow is imposed no-slip boundary condition at the solid boundaries. With respect to the turbulence field, a log-law distribution of the mean tangential velocity in the turbulent boundary layer is considered near the solid boundary, where the values of k (turbulent kinetic energy) and ε (dissipation rate of turbulent kinetic energy) can be expressed as functions of the distance from the solid boundary and the mean tangential velocity outside the viscous sublayer. On the free surface, the zero gradient boundary conditions for both k and ε are based on the assumption of no turbulence exchange between the water and air. The initial condition consists of a still water situation, with no wave or current motion.

Regular and irregular waves can be generated at the right boundary of the domain based on a source function. Also currents can be superimposed to the waves.

A detailed description of the governing equations, boundary conditions and numerical integration can be found in Lin and Liu (1998); Liu *et al.* (1999, 2000) and Hsu *et al.* (2002).

13.4.6.2. SKYLLA (Delft Hydraulics)

Application

The wave model SKYLLA simulates wave motion on coastal structures such as dikes and breakwaters. The two-dimensional numerical model can simulate breaking waves because use is made of the powerful «Volume Of Fluid» (VOF) method. This method is used to solve the well known Navier Stokes equations. The model is able to simulate very complex shapes of the free surface like those occurring in breaking waves and can be applied to compute pressures on a slope caused by breaking waves (Doorn and Van Gent, 2003). Furthermore, the model can simulate porous media flow (laminar and turbulent flow) to enable simulations of waves on and inside permeable coastal structures. In addition the model has been verified using analytical solutions and physical model tests (Petit *et al.*, 1994 and Van Gent, 1995a).

Model description

The numerical model SKYLLA allows for detailed modelling of the free surface flow near structures. The modelling of the flow is based on the incompressible Navier-Stokes

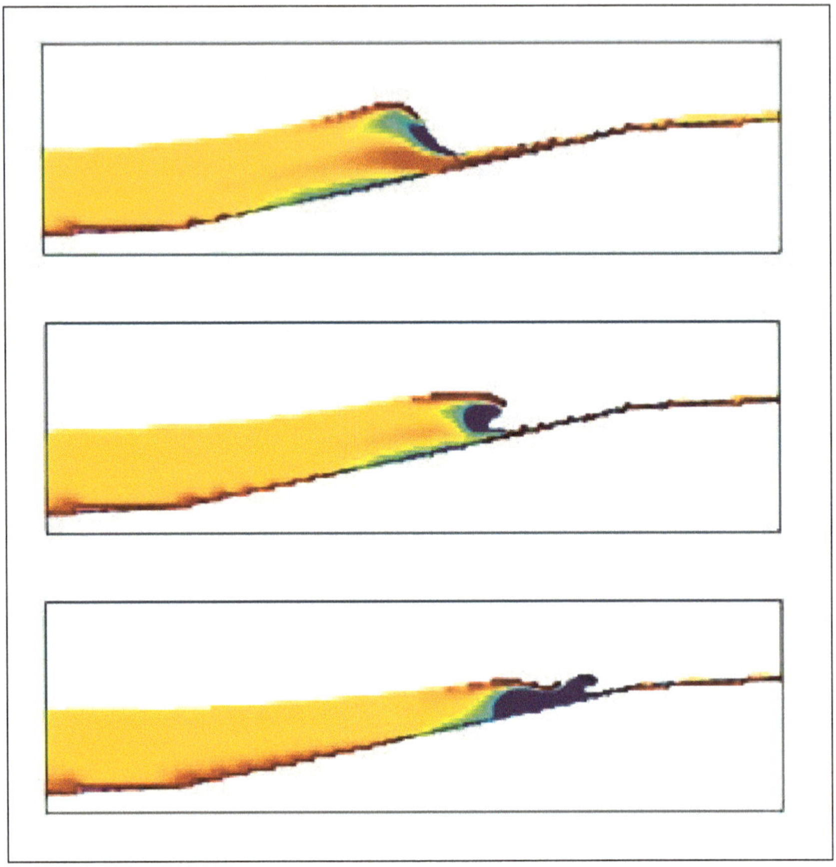

Figure 13.25. Breaking wave on a slope, computed by SKYLLA.

equations, that are solved by means of a pressure correction method; the free-surface is modelled by means of a VOF method. The model SKYLLA can combine detailed modelling of free-surface wave motion with porous media flow (Van Gent, 1995b).

Structures can be specified in detail because cells can be filled with impermeable material or can be permeable. Inside the structure, regions of different porosity and permeability can be specified. Impermeable slopes as well as combinations of impermeable parts with permeable parts can be modelled. This allows to model wave motion on coastal structures for a wide range of configurations.

The computational grid is such that smaller cells can be used in regions where the flow field is expected to become relatively complex, for instance in regions where overturning waves occur. Cells are assigned a specific porosity n that is equal to 1.0 in the region of the external wave motion and a different porosity in regions where porous media flow will be simulated.

The left and right boundary of the computational domain can be open, in which case these boundaries act as weakly reflecting boundaries. Regular/monochromatic or irregular/random waves can be imposed at these boundaries while reflected waves can leave the computational domain here.

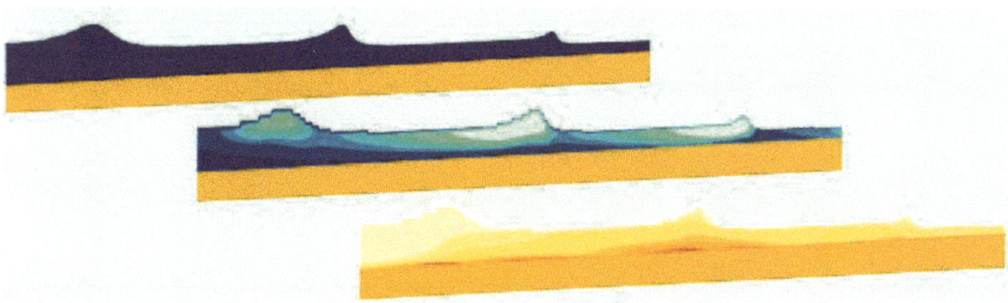

Figure 13.26. The top panel shows the shape of the breaking waves in the surf zone, the second indicates the turbulence intensities and the lowest the sediment concentrations under breaking waves.

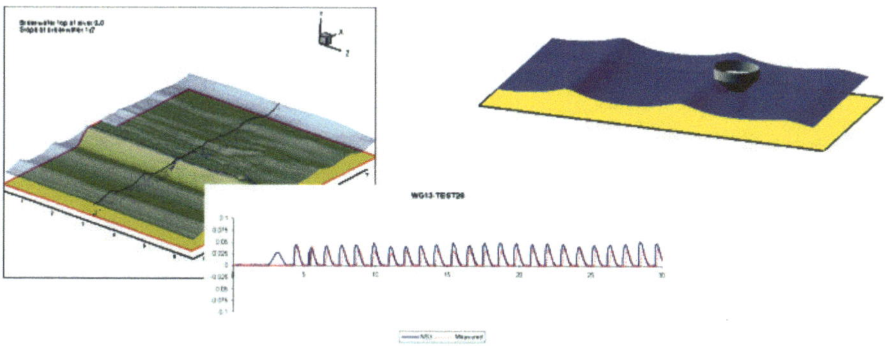

Figure 13.27. Two examples of the use of NS3 for studying coastal structures. On the left, wave overtopping over a submerged breakwater is studied. A comparison between measured and modelled wave heights on the front top of the breakwater shows good agreement. The right figure is an example of waves hitting the foundation of an offshore wind turbine.

Coupling with other models

Up to now the SKYLLA model has not been applied coupled to other numerical models. However, it is possible to impose timesignals with surface elevations computed with other numerical models, such as, e.g., TRITON.

13.4.6.3. NS3 (DHI Water & Environment)

NS3 is an advanced numerical Navier-Stokes solver for the computation of three-dimensional flows and sediment transport, and has been developed by DHI with focus on the free-surface description and adaptive grid technology, see Mayer *et al.* (1998) for further references.

The model features a flow adaptive curvilinear grid, which allows for moving boundaries, Volume of Fluid (VOF) representation of free surfaces, multi-block formulation, which allows for complex layouts, and advanced turbulence models. To improve the computational speed, parallel methods have been implemented. Therefore it is now possible to run large full three-dimensional computations on multiprocessor computers.

The model has been applied to calculate the forces and moments exerted on structures by the combination of currents and non-linear waves, run-up and green water effects, sedimentation in waves and currents, wave-breaking and associated sediment transport in the surf-zone, and sediment transport near reflective structures.

The VOF-method was applied to simulate the free surface for the detailed study of sediment transport under spilling breakers in the surf zone. A k-ε turbulence model was used for the production, transport and dissipation of turbulent kinetic energy. This was combined with a model for the sediment transport.

Wave overtopping and wave induced forces on coastal structures can easily be studied using the refined flow model NS3. As the figure below illustrates, the analyses include full three-dimensional intra-wave simulation of the wave-structure interaction.

13.4.7. Other modelling tools

13.4.7.1. Breakwat (Delft Hydraulics)

Application

For more than 10 years earlier versions of BREAKWAT have been widely used as a tool to guide and assist in the design of many types of breakwaters. In these 10 years new developments in the technical aspects of breakwater design as well as developments in the user-friendliness of computer programs in general have taken place.

With the newest version, BREAKWAT 3.0, a conceptual design can be made for statically stable structures, like rubble mound breakwaters with an armour layer of rock or concrete units, as well as for dynamically stable structures, like berm breakwaters, reef type structures and near-bed structures. It is also possible to make calculations for vertical (caisson) structures.

Model description

BREAKWAT 3.0 uses modern design formulae to perform calculations to the hydraulic response:

- wave height distribution
- wave run-up
- wave overtopping
- wave transmission

or to the structural response:

- rock stability of armour layer and toe berm
- stability of concrete armour units

of several types of structures:

- statically stable structures (rubble mound breakwaters)
- dynamically stable structures
- vertical (caisson) breakwaters.

BREAKWAT 3.0 is a Windows based product. It is programmed in the Visual Basic 6.0 program language. The main general features of BREAKWAT 3.0 are:

- flexible set-up, easy to implement new modules and formulae

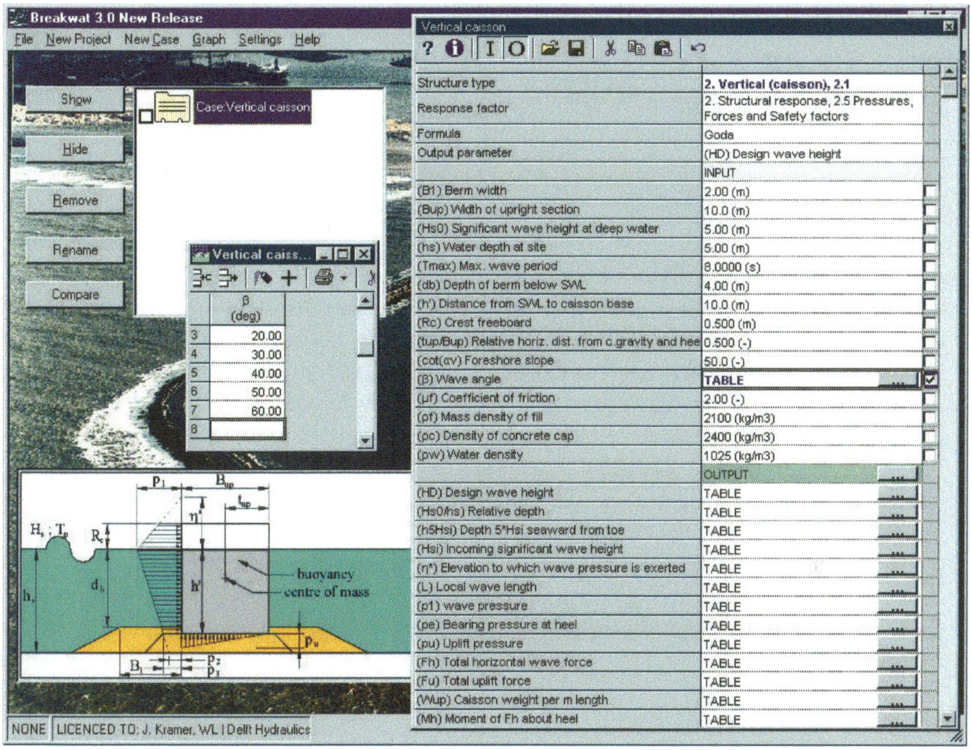

Figure 13.28. BREAKWAT user interface for case with vertical caisson and different wave angles.

- report-ready graphical presentation of results
- ability to work with input and output files
- possibility to calculate and compare more than one scenario at one time
- ability to copy data to and from clipboard
- «hard» and «soft» limits to validity of formulae
- extensive digital help function.

Coupling with other models

BREAKWAT is the last link in a modelling chain, starting with the modelling of the offshore wave field and ending with the modelling of the wave impact on the structure. This wave impact, in terms of wave overtopping or wave run-up, is computed by means of analytical solutions and empirical formulae. Although the model uses input from other wave models, to be exact: the wave height and wave period, the model cannot directly be coupled to these wave simulations programs.

For further information please visit http://www.wldelft.nl/soft/chess/breakwat/

13.5. PREDICTION OF WAVE INDUCED WATER FLOW OVER AND THROUGH THE STRUCTURE, OF SET-UP AND RIP-CURRENTS

(Lamberti, Martinelli, Zanuttigh, UB)

13.5.1. Introduction

13.5.1.1. LCS peculiarities

For LCSs in contrast to emergent structures, the flow rate over the barrier is high and related to the piling-up at the rear. Overtopped water accumulates behind the structure, establishing a higher mean water level, or piling-up, which drives return flows along different paths.

In case of impermeable structures, water may return off-shore through gaps or, if the crest is submerged, over the barrier itself. In this case, the flux over the barrier during the wave cycle is alternately directed inshore and offshore, driven by waves and piling-up.

LCS, however, are typically made of permeable rubble mound so that filtration takes place; the average flow is driven by the unbalance between piling-up and wave thrust due to breaking waves; the first is usually dominant causing a return flow through the structure. A fraction of the volume of water overtopping the structure offshore edge percolates through the crest, causing flows directed partially inshore and partially offshore.

The flow within the rubble matrix is dominated by the oscillatory wave flow and can usually be assumed fully turbulent with an average component much smaller than the oscillation amplitude.

13.5.1.2. Flow description

For an emerged structure, overtopping wave crests pile up water inshore of the structure until a level is reached that forces return flows (through the structures and through gaps) globally equal to the overtopping discharge. The value of piling-up depends on flow resistance of all return paths acting in parallel: it is maximum for laterally confined conditions as in a wave flume with no recirculation, where the net mass flux across the structure is zero; it is significantly lower in presence of gaps, that make up easy return paths and induce a horizontal recirculation.

For a submerged structure, water can return offshore also over the berm. The net inshore flux over the berm is the difference between the flow associated to overtopping crests and the return flow at troughs. The net flux may have an effect on the breaking process and wave transmission.

In both cases, emerged and submerged, the offshore directed flux through the gap/s compensate the net inshore water flux across the barrier/s, including net flux over the berm and through the structure.

13.5.1.3. Dynamics

For emerged structures the overtopping process (wave crests topping over the structure crest) is not significantly influenced by piling-up and return flows. It can be assumed an imposed flow, on which piling-up and the other return flows do depend.

For submerged structures, wave crests breaking on the structure berm (submerged structure crest) release their momentum to the water mass they merge with.

This momentum release is the cause of an increase of the water level across the structure, similar to wave set-up on a beach. It is still named piling-up, because, due to the significant structure slope, local wave conditions are much more related to incident waves than to local water depth and therefore the relation among incident waves, structure profile and wave set-up is quite different from the one holding for a beach.

Piling-up and net flow over the structure are in this case strictly related to each other as well as to incident waves. In particular, an inshore directed mean flow reduces momentum released by breaking crests (reduced number of breaking waves and velocity difference) and induces resistance to flow; both effects cause a significant reduction of piling-up.

The accentuated oscillatory character of velocities strongly affects flow resistance over and within the structure; the resulting mean head loss is not proportional to the square of the mean velocity but is rather proportional to the product of the mean velocity and the amplitude of the oscillating component.

13.5.1.4. Wave pumping

The head losses associated to rip currents can be represented by a return flow characteristic curve and the relation between piling-up and net mass flux across the structure can be similarly described by a barrier pumping curve.

The system operational point at equilibrium may be obtained as the intersection between the two curves: one representing the piling-up versus net overtopping discharge relation and the other representing a similar relation for all the remaining return flows.

The pumping curve for the barrier has been experimentally investigated in wave flumes equipped with a recirculation system and it was found to be approximately linear by Ruol *et al.* (2004) and Cappietti *et al.* (2004).

The curve can therefore be described by two points, for instance the two extremes: the net mass flux at zero piling-up $Q_0 = Q_{net}|_{P=0}$ and piling-up for zero mass flux (i.e. in absence of recirculation) $P|_{Qnet=0}$. Even when the relation is not linear, these two point represent two peculiar conditions of the pumping system.

13.5.1.5. Structure of the section

Overtopping, piling-up and return flows, presented respectively in Sub-sections 13.5.2, 13.5.3 and 13.5.4 are indeed strictly correlated, due to the water balance condition and to the specific relations between the common head difference and the flow through each path, so that the quantification of each process can be given only for fixed and precise conditions of the others. Therefore special attention is paid in the text to the relations between piling up and return flows for different flow paths: over the barrier crest, through the porous matrix and through gaps. In Sub-section 13.5.5 it is finally presented and verified how the actual piling-up and circulation can be determined in a wave flume and in 3-D conditions.

13.5.2. Wave mass flux, overtopping

The oscillatory nature of waves induces positive mass and momentum fluxes; the divergence of the latter is balanced by water level gradients, water acceleration and friction on the bed.

13.5.2.1. Wave mass flux

Outside the surf zone, mass flux is a second order effect and momentum flux has null divergence. The mass transfer per unit width, given by the vertical integration of the velocity, is concentrated, according to the Eulerian 1st order description, in the region bounded by water level excursion. For horizontal bottom, it is given by $\rho g < \eta^2 > /C$. In practice, a certain volume of water is cyclically pushed forward by propagating waves.

The pumped water volumes are far greater where the oscillation pattern is very asymmetric (and the 1st order approximation is not satisfactory), like in case of breaking or broken waves, or where some obstacle prevents the flow to return offshore at trough, like in presence of a screen/barrier with the crest around mean water level.

While propagating across the structures, the waves break in conditions which are obviously affected by the structure freeboard. Breakers occur on the structure slope for emerged structures and on the crest in submerged conditions.

For rubble mound structures, the up-rushing tongue that would form over the crest if the breakwater was impermeable is partially transmitted into the porous medium. In the case of an emerged rubble mound, the water volumes periodically transmitted behind the structure are mainly transferred through the structure itself and are thus much lower than in the low crest case where overtopping is significant.

13.5.2.2. Overtopping frequency, volumes and discharge

Formulations are available essentially for emerged structures and irrelevant piling-up. In this subsection discharge shall be interpreted as overtopping discharge in absence of piling-up.

Overtopping can be estimated as an average discharge or in greater detail as the sum of the volumes overtopped by the single waves; some waves do not overtop (zero volume), the others (P_{ot}) produce overtopping volumes (V_{ot}) variable from wave to wave.

Overtopping discharge per unit width q_{ot} can be therefore represented as:

$$q_{ot} = P_{ot} E\left(V_{ot}\right)/T_m \qquad (13.69)$$

where T_m is the mean period of incident waves, P_{ot} is the overtopping probability and $E(V_{ot})$ is the mean volume of overtopping crests. The fraction P_{ot}/T_m is the occurrence frequency of overtopping events.

Volume statistics can be directly estimated or can be assessed in relation to run-up R_u of each wave.

For regular waves, P_{ot} is equal to 0.0 if $R_u \le R_c$ and equal to 1.0 if $R_u > R_c$.

For irregular waves, P_{ot} is equal to the probability that the Weibull distributed run-up exceed the crest freeboard R_c

$$P_{ot} = \exp(-(R_c/b)^c) \quad \text{for } R_c \cdot 0 \qquad (13.70)$$

where van der Meer (1992) suggests: $b = 0.4 H_{si} s_{om}^{-0.25} \cot g\alpha^{-0.2}$, with s_{om} = mean wave steepness and α – mean offshore slope; c is 3.0 $\xi_m^{-0.75}$ for plunging waves ($\xi_m < 2.5$) and 0.52 $P^{-0.3} \xi_m^{\ P} \sqrt{\cot\alpha}$ for surging waves ($\xi_m > 2.5$), where ξ_m is the Irribarren number based on mean wave period and P is structure notional permeability.

CEM (2001) suggests that the run-up distribution is Rayleighian ($c = 2$ in Eq (13.70)) and provides an expression for the rms run-up value b for any structure profile.

Pilarczyk (1990) evaluates the overtopping volume V_{ot} through the empirical relation:

$$V_{ot} = 0.1 \cdot \left(\cot\alpha\right)^{1/5} \left(R_u - R_c\right)^2 \qquad (13.71)$$

obtained for high banks with mild slopes (cot α = 3 ÷ 5). Eq. (13.71) describes the volume of water running over the structure, which has the form of a prism with angle dependent on the seaward slope angle.

According to Van der Meer and Janssen (1995) the overtopping volume distribution is well approximated by a Weibull distribution F_{Vot} with a fixed shape parameter (3/4):

$$F_{Vot} = 1 - \exp(-(V_{ot}/a)^{3/4}) \tag{13.72}$$

The scale parameter is related to the mean overtopping volume $a = 0.84 \cdot E(V_{ot})$.

In practice the average overtopping rate per unit width q_{ot} is directly investigated and the mean overtopping volume and the scale parameter are obtained by reversing Eq. (13.69), e.g. $E(V_{ot}) = q_{ot}\,T_m/P_{ot}$.

13.5.2.3. Empirical overtopping formulae

Van der Meer and Janssen (1995) provide different formulae for the overtopping discharge due to plunging and surging waves, Eq (13.73) and Eq (13.74) respectively. The reported regression coefficients are adopted by the TAW code, based on van der Meer *et al.* (1998), and are valid only for emerged structures.

$$\frac{q_{ot}}{\sqrt{gH_s^3}} = \frac{0.067}{\sqrt{\tan\alpha}}\gamma_b \xi_{op} \exp\left(-5.2\frac{R_c}{H_s \xi_{op} \gamma_b \gamma_f \gamma_b \gamma_v}\right) \quad \text{(for plunging waves)} \tag{13.73}$$

$$x_{op} = \frac{\tan\alpha}{\sqrt{s_{op}}}\gamma_b, \qquad s_{op} = \frac{2\pi H_s}{g T_p^{\,2}}$$

$$\frac{q_{ot}}{\sqrt{gH_s^3}} = 0.2\exp\left(-2.6\frac{R_c}{H_s \gamma_f}\right) \quad \text{(for surging waves)} \tag{13.74}$$

where s_{op} is the deep water peak wave steepness, ξ_{op} is the Iribarren or surf-similarity parameter, γ_b is the reduction factor for berms, γ_f is the reduction factor for slope roughness and $\tan\alpha$ is the structure slope. The γ factors may be considered in first approximation equal to 1. For more details see CEM (2001) or the quoted paper.

In case of LCSs, waves can be generally assumed of the surging-type. Kofoed and Burcharth (2002), on the basis of their tests and including the dataset from van der Meer and Janssen (1995) and Oumeraci *et al.* (1999), suggest the following reduction factor for the overtopping discharge obtained from Eq. (13.74):

$$\gamma_{R_c} = 0.6 + 0.4\sin\left(2\pi\frac{R_c}{3H_s\gamma_f}\right) \quad \text{for} \quad \frac{R_c}{H_s\gamma_f} \le 0.75$$

$$\gamma_{R_c} = 1.0 \text{ otherwise} \tag{13.75}$$

Schüttrumpf and Oumeraci (2001) suggest the following expression on the basis of a dataset including different structures, ranging from zero freeboard to quite emerged:

$$\frac{q}{\sqrt{2gH_s^3}} = 0.038\, C \exp\left(-b\frac{R_c}{CH_s}\right) \quad \text{for } C < 2$$

$$\frac{q}{\sqrt{2gH_s^3}} = \left(0.096 - \frac{0.160}{C^3}\right) \exp\left(-b\frac{R_c}{CH_s}\right) \quad \text{for } C \geq 2 \qquad (13.76)$$

with $C = R_{u2\%}/H_s$ and $b = -3.67$.

Overtopping rate can be also described with a weir model: instantaneous discharge is proportional to the 3/2 power of the water elevation above the structure crest and can be integrated within the wave period assuming a fixed wave form (e.g. sinusoidal). If the water level in front of the structure does not exceed the crest freeboard, overtopping is trivially zero. Assuming this weir approach, Hedges and Reis (1998) re-analyzed the data by Owen (1980) with the aim of improving the predictions in the vicinity of the physical boundaries (large freeboards and freeboard close to zero), obtaining the following expression:

$$\frac{q}{\sqrt{g(CH_s)^3}} = A_2\left(1 - \frac{R_c}{CH_s}\right)^{B_2} \qquad (13.77)$$

where A_2 and B_2 are regression coefficients and C is the ratio between maximum run-up and the significant incident wave height ($C\,H_s = R_{max}$) see Tab. 13.6. Suitable expressions suggested for the significant run-up are:

$$R_s/H_s = 1.35 \cdot \xi_p \text{ if } \xi_p \leq 2 \text{ or } R_s/H_s = 3.00 - 0.15 \cdot \xi_p \text{ if } \xi_p \geq 2.$$

For Rayleigh distributed run-up, $R_{max,p}/R_s = (0.5 \cdot (\ln N - \ln(-\ln p)))^{0.5}$ and therefore, for wave records of 100 waves as in Owen (1980) dataset, the most probable maximum $R_{max,37\%} = 1.52\, R_s$ and the extreme one $R_{max,99\%} = 2.15\, R_s$. Discharge is null for $R_c > C\, H_s$.

Table 13.6. Coefficients for Hedges and Reis (1998) model.

		$R_{max} = 1.52\, R_s$ *(crest frequently overtopped)*	$R_{max} = 2.15\, R_s$ *(crest almost never overtopped)*
Slope 1:1	A_2 B_2	0.00703 3.42	0.00515 6.06
Slope 1:3	A_2 B_2	0.00753 4.17	0.00542 7.16
Slope 1:4	A_2 B_2	0.0104 6.27	0.00922 10.96

Overtopping has been mainly investigated for long-crested perpendicular waves. Only few tests examined the effect of spreading or oblique wave attack. For oblique waves, the incident energy per unit length of the structure is reduced. Banyard and Herbert (1995) suggest a reduction factor on overtopping discharge $\gamma_\beta = 1 - 0.00015\ \beta^2$, β being wave obliquity in degrees.

Van der Meer and Janssen (1995) suggest, in case of long-crested waves, a further reduction factor on run-up in Eq. (13.74) $\gamma_\beta = \cos(\beta - 10°)$ with a lower limit of 0.6; short crestedness is accounted either decreasing the angle of a fixed amount (10°) or using a different law, $\gamma_\beta = 1 - 0.0033\ \beta$.

For submerged structures, the overtopping process is different and can not be properly described by available formulae. Mass flux over the barrier during the wave cycle is alternately directed inshore and offshore, driven by waves and piling-up. Sub-section 13.5.4.2 analyses the return flows over the structure.

13.5.3. Piling-up

13.5.3.1. Introduction

Natural beaches are usually rather uniform along shore and characterized by mild slopes. Sub-section 13.5.3.2 describes wave set-up for such simple reference conditions. For a defended beach, conversely, the barrier and the beach behind it may vary significantly along the shore; the barrier moreover has never a mild slope. Set-up, in this case also named piling-up, is affected by the rapid variation of water depth and by a wide range of possible paths overtopping water can follow to return offshore.

Evaluation of set-up behind the barriers should then consider the specific degree of confinement. Mass balance, applied to the area protected by the structures, requires that overtopping discharge, which (per unit length) is described in Sub-section 13.5.2, equals the sum of all returning flows, described in Sub-section 13.5.4. In general, piling-up is the forcing of all return flows and is eventually established at the value that satisfies water mass balance equation behind the structure.

For example, let us consider a beach protected by an indefinitely long parallel emerged structure. Mass balance requires that the seaward directed filtration equals overtopping and the piling-up is thus influenced by the structure permeability. These lateral constraints determine the maximum piling-up. Should part of the overtopping water be recirculated off-shore through gaps, piling-up would decrease; in the theoretical limit case of infinite conductivity, piling-up would decrease down to zero.

Wave flume experiments carried out by Ruol *et al.* (2003), relative to low emerged structures, and repeated by Cappietti *et al.* (2004), who also tested zero freeboard and submerged structures, quantitatively analyze the effect of lateral conditions (which are schematized by different degrees of recirculation) on piling-up. Piling-up reaches its maximum in absence of recirculation and decreases to zero when the overtopping discharge is totally recirculated; the relation is approximately linear, see Figure 13.29.

In case of emerged or zero freeboard structures, the overtopping discharge can be determined on the basis of equations given in Sub-section 13.5.2, and piling-up can be obtained by imposing that inshore and off-shore directed flows are equal, piling-up being the unknown.

In case of submerged structures, see Sub-section 13.5.2.3, formulae describing accurately inshore mass flux due to overtopping are not yet available: the extrapolation of existing

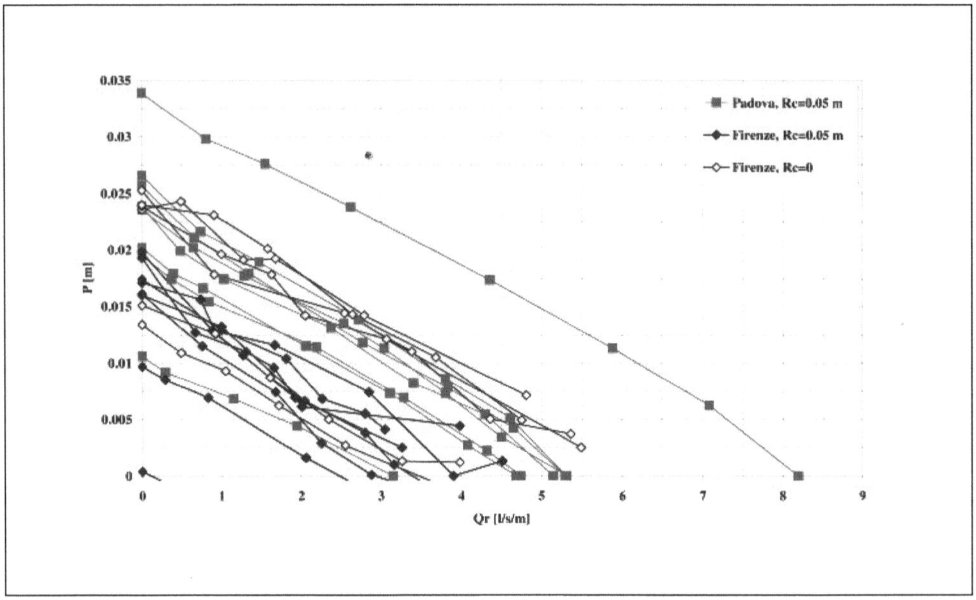

Figure 13.29. Pumping curves for similar LCSs under varied wave conditions.

empirical formulae leads to overestimate the overtopping discharge, interpreted as Q_0, by a factor 2-4; their use therefore cannot be recommended.

Wave crests transport a certain mass of water shoreward; for any positive piling-up a return flow over and through the structure is generated; the first can be schematically described by the «weir» analogy: the average offshore directed flow is related to piling-up by Eq. (13.88), or an equivalent one including flow resistance, and shall be subtracted to wave crest transport (overtopping discharge) providing a net discharge over the structure. Flow through the structure can be similarly related to piling-up and subtracted to obtain the net discharge across the structure.

The maximum piling-up P_0 (piling-up at zero net discharge across the barrier) can be directly described based on momentum balance, and then flow resistance induced by the net flow over the barrier can be estimated and the induced head drop subtracted to P_0. These methods are described in Sub-sections 13.5.3.3 and 13.5.3.4, and compared to recent experimental data in 13.5.3.5.

13.5.3.2. Momentum balance for mild slope bottom

Considering the propagation of a progressive wave with angle θ to the x_1 direction, the average momentum excess caused by waves in the water column is the radiation stress tensor (sum of vertically averaged pressure and momentum flux per unit width)

$$\begin{bmatrix} S_{11} & S_{12} \\ S_{21} & S_{22} \end{bmatrix} = \frac{E}{2}G\begin{bmatrix} 1 & 0 \\ 0 & 1 \end{bmatrix} + \frac{E}{2}(1+G)\begin{bmatrix} \cos^2\theta & \cos\theta\, sin\theta \\ \cos\theta\, sin\theta & sin^2\theta \end{bmatrix} \tag{13.78}$$

where $G = \dfrac{2kh}{\sinh(2kh)}$ is 1 in shallow water and 0 in deep water.

The average momentum balance can be written

$$\rho(h+\overline{\eta})\left(\frac{\partial}{\partial t}+U_j\frac{\partial}{\partial x_j}\right)U_i+\rho g(h+\overline{\eta})\frac{\partial\overline{\eta}}{\partial x_i}+\frac{\partial}{\partial x_i}\left(S_{ij}+S'_{ij}\right)+\tau_i^b=0 \tag{13.79}$$

where $\overline{\eta}$ is the average water level above datum, U_i the current (mean velocity) vector, S'_{ij} is the depth integrated Reynolds stress tensor and τ_i^b is the average shear stress on the bed.

Waves propagating outside the surf-zone, e.g. in non-breaking conditions, do not induce currents (nor average shear stress or turbulence) but only a small set down, increasing as waves shoal on the beach and reaching a maximum of about 4% of the breaking depth h_b.

Inside the surf zone, the cross-shore and long-shore wave thrust (divergence of the radiation stress) originated by breakers are substantially balanced by set-up in the cross shore direction and by bottom shear stress related to long-shore currents in the long-shore direction.

Eq. (13.79) is integrated inshore the breaking line under the following hypotheses:

- waves propagate approach the beach with a small angle ($\cos\theta \cong 1$) and with constant wave height to depth ratio (constant breaker index $\gamma \cong 0.6$ at mild slope beaches);
- mean cross-shore velocities, bottom friction and turbulent stresses are negligible.

The derived set-up is given by

$$\overline{\eta}-\overline{\eta}_b=\frac{3}{8}\gamma^2\Big/\left(1+\frac{3}{8}\gamma^2\right)\cdot\left(h_b-h\right)\cong 0.12\cdot\left(h_b-h\right) \tag{13.80}$$

This value shall be incremented due to the effect of wave and breaker drift near the water surface and of the compensating under-tow (approximately + 20%). For mild slope profiles, the maximum set-up value at the shoreline is about 10% of h_b.

The breaker index value increases significantly with bed slope (as well as set-up at the shoreline) and can not be considered constant in particular when depth suddenly changes due to the presence of a barrier. Waves almost preserve at breaking the height they can have on the foreshore depth. For submerged LCSs waves break on the berm, where water depth is small, breaking continue a while inshore the barrier crest and cause a set-up far greater than at a mild slope beach (Eq. 13.80). The phenomenon is qualitatively not so different from the one described earlier, but is more intense, therefore we shall use a different term «piling-up» and symbol P to represent it.

The term refers probably to the case of an emerged barrier, where overtopping induces a water accumulation inshore the barrier (not related to wave thrust: force balance is assured by the structure reaction) to which the term piling-up seems most appropriate.

Since there is a smooth transition between submerged and emerged structures, the term and the symbol P shall be used for both cases.

13.5.3.3. Piling-up behind submerged barriers

The piling-up for zero net inshore discharge can be determined for instance by the CVB method, described in Calabrese *et al.* (2003, 2005).

This method considers the momentum balance across the barrier under the following assumptions:

- uniform alongshore conditions;
- orthogonal waves;
- negligible flow through the structure;
- breaking on the seaward slope, continuing all over the berm;
- mean water level linearly varying across the structure.

In the surf zone (control volume in Fig. 13.30) the gradient of the mean hydrostatic pressure is constant and the resultant pressure force on the control volume surface Π is equal to the volume times the pressure gradient $\Pi = -\rho g P h_m$ where h_m is the average water depth from the breaking point to breaking end. Let h_{m0} be the average water depth in absence of piling-up, in presence of piling-up the average depth is increased by $P/2$; when, for instance, breaking ends near the berm inshore edge the depths are:

$$h_m = h_{m0} + P/2 = h - \left\{ h_c - \frac{(h_b + R_c)}{2(B + x_b)} x_b \right\} + P/2,$$

where R_c is freeboard (negative for submerged structures), B is the crest width, h_c is the structure height, h is water depth at structure toe, h_b is the breaking depth and x_b is the distance between the breaking point and the seaward crest edge.

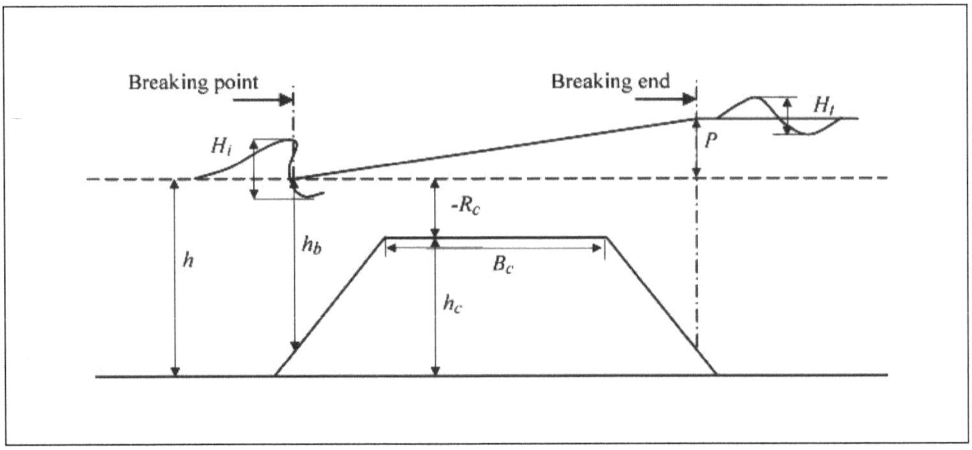

Figure 13.30. Control volume for momentum balance.

When a regime is reached this force is balanced by:

- the resultant of radiation stress (momentum excess due to waves) through the offshore and inshore boundaries of the surf zone S,
- friction force on the barrier R,
- net momentum excess due to currents C.

Let A be the resultant of these forces $A = S + R + C$, one can easily obtain piling-up from the momentum balance $\Pi + A = 0$:

$$P = \sqrt{h_{m0}^2 + 2A} - h_{m0} \cong A/h_{m0} \tag{13.81}$$

Under the following simplifying hypotheses:

- radiation stress can be calculated according to the linear wave theory 1/16 $\rho g H_s^2$ (1/2 + G),
- negligible average flow and shear stress on the berm,
- the simplified expression for the «static» piling-up results:

$$P_0 = \frac{\left(H_{si}^2 - H_{st}^2\right)}{16 \cdot h_m}\left(\tfrac{1}{2} + G\right) \tag{13.82}$$

where H_{si} and H_{st} are the incident and transmitted significant wave heights.

Eq. (13.82) is approximately explicit when submergence is not small compared to incident wave height; otherwise Eq. (13.82) becomes a second degree equation, whose solution is given by (13.81) above.

If the actual P is lower than P_0, the water momentum balance is not reached and an inshore average flux q is originated until the related shear stress on the barrier surface compensates the unbalance.

For a permeable structure, piling-up induces a return flow through the porous matrix and an equal inshore flow over the berm; shear stress on the crest contrasts wave action. Piling-up is therefore somewhat overestimated by Eq (13.82).

Any recirculation is associated to a force imbalance. The formula given in Eq. (13.82) assumes that there is no flow across or through the structure nor friction on it. The CVB formula accounts for the mass drift carried inshore by the wave motion and, in the latest version (Calabrese *et al.*, 2005), for the roller, by representing the resultant shear stress on the structure contrasting undertow; it is therefore more accurate. Both do not consider the flow through the porous structure: when such filtration is offshore directed, as in the experiments by Loveless and Debski (1997), a piling-up higher than predicted is observed (see Fig.13.31).

13.5.3.4. Empirical formulae

In the following, literature formulations of piling-up obtained for particular conditions are given. Wave piling-up is predicted by Diskin *et al.* (1970) based on tests on structures with small permeability (stone size 0.4 m at prototype scale) and regular waves:

$$P_0/H_i = 0.6\exp\left[-\left(0.7 - \frac{R_c}{H_i}\right)^2\right] \tag{13.83}$$

For an emergent and truly impermeable structure, overtopping water is piled up inshore until it returns offshore over the structure, therefore P_0 is for this hypothetical structure always greater than the crest level. The maximum scaled piling-up and the associated crest elevation are, according to Eq. (13.83), 0.6 and 0.7, making clear the modest permeability of the tested barrier. The structure of Eq. (13.83) (the bell shaped expression) reflects the concept that piling-up is small both for well emerged structures, for which overtopping is rare, and for deeply submerged structures over which the return flow may reach overtopping discharge under a small piling-up.

A similar formula was proposed by Loveless *et al.* (1988):

$$\frac{P_0}{B} = \frac{1}{8gD_{n50}}\left(\frac{H_i L}{hT}\right)^2 \cdot \exp\left[-20\left(\frac{R_c}{h+R_c}\right)^2\right] \tag{13.84}$$

Basically Eq. (13.84) treats piling-up as the hydraulic head necessary to return offshore the volume of each wave crest ($HL/2\pi$) in one wave period by an essentially turbulent flow through the structure. Stone size at prototype scale is 0.7-1.0 m.

Diskin formula may be used to predict piling-up only for submerged structures, for which the weir mechanism is efficient and predominant over filtration. In case of emerged structures, the Diskin formula may be used only for almost impermeable ones. Loveless formula points out the effect of filtration. Both should be used mainly for regular waves.

13.5.3.5. Comparison of available formulae with experimental data

Eq.s (13.82) and (13.83) are compared (Fig. 13.31 and 13.32) to experimental measurements of piling-up in case of null recirculation.

The data set used for the comparison is derived only by wave flume tests under irregular wave conditions:

- Bristol tests, described in Loveless and Debski (1997) tests on irregular waves (in order to reduce the scatter, tests with small piling-up, close to the measuring accuracy, are not graphed);

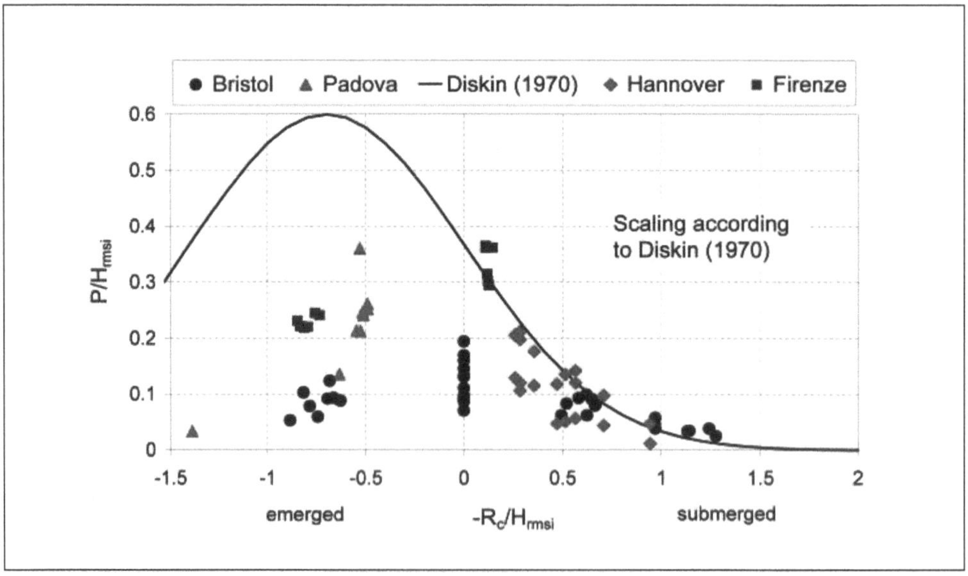

Figure 13.31. Set-up in confined conditions following Diskin *et al.* (1970) non-dimensionalisation, Eq. (13.83). Submerged structures appear at the right side of the plot.

- Padova tests, described in Ruol and Faedo (2002) and Ruol *et al.* (2004);
- Hannover tests, performed at the GWK, described in Calabrese *et al.* (2005);
- Firenze tests, described in Cappietti *et al.* (2004), Clementi *et al.* (2006) and Ruol *et al.* (2006);

In some cases (Firenze and some of Padova tests), carried out in a recirculating flume, overtopping and piling-up were measured for different net discharge across the barrier.

Fig. 13.31 presents piling-up compared to Diskin (1970) formula. It is known (Loveless *et. al.*, 1998) that whenever structure permeability is greater that in Diskin experiments, a smaller piling-up is obtained. Nevertheless, even data from the same tests (and therefore same permeability), appear quite scattered with the proposed scaling.

For tests with irregular waves in submerged conditions, Fig. 13.32 presents the comparison between measured P and the prediction given by CVB formula (Calabrese *et al.*, 2005). Since tests correspond to quite different scales, a variable roughness is used and good calibration was obtained using a Manning-Strickler coefficient $C = 26\ k_s^{1/6}$ with $k_s = 2D_{n50}$. The Stokes drift is not reduced and the random sea state is described as a train of regular waves with $H = H_{rmsi} = H_{si}/1.4$.

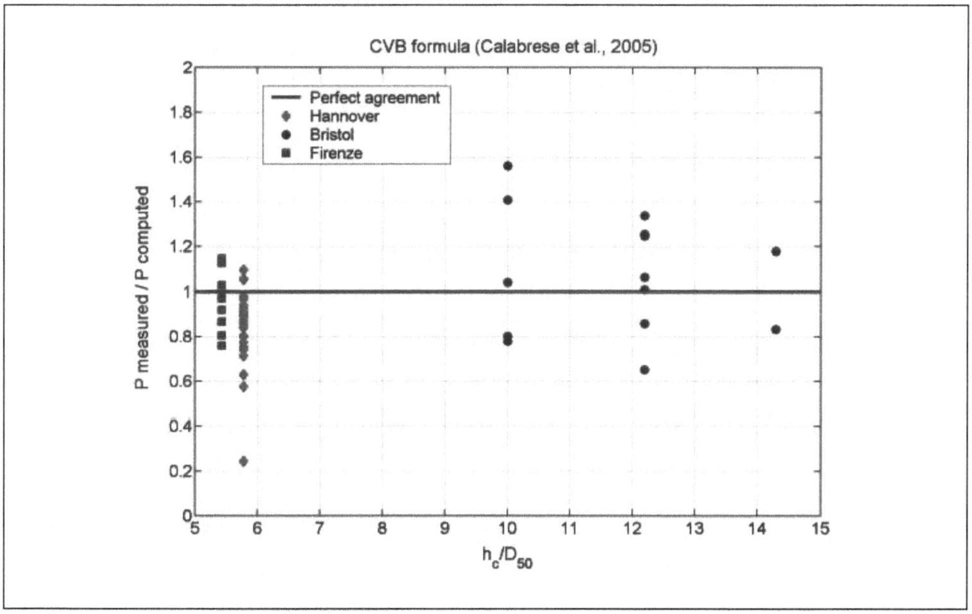

Figure 13.32. Piling-up in confined conditions: computed values are derived using CVB formula, Eq. (13.82). The impermeable structure scheme is satisfactory near to design wave conditions.

13.5.4. Return Flows

13.5.4.1. Filtration

In presence of waves and currents on/through the structure, the wave averaged momentum equation consists of the balance of three terms: divergence of radiation stresses, mean pressure gradient and friction force exerted on the porous medium. For emerged structures in absence of mean filtration (and mean friction force), the momentum released from waves causes an «equilibrium» piling up P_e in the mound (Zanuttigh and Lamberti, 2006). For zero freeboard and submerged structures and zero net inshore flow, water flows inshore over the structure and offshore in the barrier, and mean filtration velocity drop to zero for an almost zero piling-up. The unbalance of actual P and P_e causes filtration through the structure (or is balanced by the friction force).

An estimate of P_e for emerged permeable structures can be obtained from momentum balance. Neglecting wave transmission and assuming shallow water conditions for the sake of simplicity of the formula, momentum balance equation is

$$1/16\, H^2_{si}(1/2 + G) = P_e(h + P_e/2)$$

from which, assuming $P_e << h$, one obtains

$$P_e = 1/16\, H^2_{si}(1/2 + G)/h \cong 0.07\, H_{si}$$

The Forchheimer equation (see for instance van Gent, 1993) may be used to predict friction slope and flow through a rock structure for a given hydraulic gradient or head difference per unit length I. This equation can be written as

$$I = \frac{Xu}{D_{n50}^{2}} + \frac{Yu|u|}{D_{n50}} + Z\frac{\partial u}{\partial t} = \frac{\alpha_f \nu}{g}\left(\frac{1-n}{nD_{n50}}\right)^2 \frac{u}{n} + \frac{\beta_f}{g}\frac{1-n}{nD_{n50}}\frac{u|u|}{n^2} + \frac{n + C_m(1-n)}{g}\frac{\partial u}{n\partial t} \tag{13.85}$$

where u is bulk velocity through the porous medium, C_m is the added mass coefficient and α_f, β_f are constants depending on flow shape in pores (KC number, rock grading, element shape, marginally porosity); X, Y, Z depend also on porosity n, since it controls the average pore radius $n/(1-n)\cdot D_{n50}/6$. The third term in the right hand-side is zero in average and when extreme flow conditions are reached.

The mean hydraulic gradient is therefore evaluated as

$$\langle I \rangle = \frac{X}{D_{n50}^2}\left\langle(\bar{u} + \tilde{u})\right\rangle + \frac{Y}{D_{n50}}\left\langle(\bar{u} + \tilde{u})\cdot|\bar{u} + \tilde{u}|\right\rangle \tag{13.86}$$

where $\bar{u}$ is the mean seepage velocity and $\tilde{u}$ is the oscillating velocity component.

Values of α_f and β_f are around 1000 and 1 respectively. For more details the original papers of Burcharth and Christensen (1991), Burcharth and Andersen (1995), van Gent (1992), Garcia *et al.* (2004) should be consulted.

The mean hydraulic gradient $\langle I \rangle$ can be expressed as the net piling-up $P - P_e$ over the average width B of the submerged part of the barrier, which is evaluated at 1/3 of the seepage depth (structure height for submerged and water depth for emergent structures), to account for the greater filtration in the upper part of the structure. The average quadratic term in Eq. (13.86) can be evaluated approximately as $k\,\bar{u}\,\tilde{u}_{rms}$ whenever $|\bar{u}| < \tilde{u}_{rms}$, where $\tilde{u}_{rms}$ denotes the root mean square of the oscillatory velocity component. The coefficient k is equal to 1.8 for a sinusoidal fluctuation, whereas in the extreme case of a Gaussian fluctuation it is 1.6

and 2.0 for fluctuations jumping between equiprobable values. In the following, $k = 1.8$ is adopted.
Considering wave conditions that contain a significant number of breaking waves, wave piezometric slope is an order of magnitude higher than mean piezometric slope and Eq. (13.86) can be rewritten as

$$\frac{P-P_e}{B}=\left(\frac{X}{D_{n50}^2}+1.8\frac{Y\cdot\tilde{u}_{rms}}{D_{n50}}\right)\cdot\frac{q_f}{\min(h,h_c)} \tag{13.87}$$

from which mean off-shore filtration discharge q_f can be derived, if wave velocity is estimated. The laminar flow term in (13.87) results an order of magnitude smaller than the other, therefore scale considerations presented below account only for the second term. Under breaking waves, the instantaneous friction slope is limited by some finite value below 1; $\tilde{u}_{rms}$ is therefore more or less constant depending on structure permeability and submergence. This is the reason why the relation between piling-up $P-P_e$ and seepage discharge in literature appears to be linear for a given structure and variable incident waves (Ruol and Faedo, 2002; Cappietti *et al.*, 2004), see Fig. 13.29.
Quoted experiments suggest that $\tilde{u}_{rms}$ can be obtained from the relation

$$Y\cdot\tilde{u}_{rms}^2/D_{n50}\approx 0.1\div 0.2$$

depending on structure submergence (the lower value is for zero freeboard, the greater for emerged structures).

Zanuttigh and Lamberti (2006) clearly show that the filtration process is different for emerged and submerged or zero-freeboard structures, as it has been already observed by Debski and Loveless (1997), but additionally prove that it is possible to identify a unique curve also for emerged structures, showing some scatter for the lowest P over B values.

For zero-freeboard and submerged structures the water mass exchanges over the barrier crest and the vertical percolation inside the barrier play the most relevant role. For emerged structures, for lower P over B values, waves build up pressure inside the structure and filtration may result in-shore directed; with increasing P above the «threshold» P_e, i.e. when piling-up becomes predominant over wave generated head in the porous structure, off-shore directed filtration occurs.

13.5.4.2. Return flow over a submerged structure

In case of submerged structures an additional return path acting in parallel with filtration may be considered: the offshore flow over the crest q_o.

An estimate of the discharge can be obtained by applying a weir model, with flow seaward directed. In case of small submergence, critical depth may be reached on the weir, whereas for significant submergence the weir may result drowned. When the crest is wide, friction losses along the crest shall be accounted for reducing the effective head.

Calabrese *et al.* (2003, 2005) considered the friction due to undertow with a Gauckler-Strickler formula; the undertow discharge q_u compensates Stokes drift

$$\frac{1}{8}H_{rms}^2\sqrt{\frac{g}{h}}$$

(Calabrese *et al.*, 2003) and, in addition, the roller mass flow

$$\frac{A_r}{T} = \frac{0.06HL}{T}$$

(Calabrese *et al.*, 2005) for breaking waves. In 2005 they suggest calibrated values for the coefficients: 0.02 (substituting 1/8) in the drift term and 6 $m^{1/3}$ s^{-1} for the berm roughness.

The same approach may be followed, further assuming that:
- the oscillatory component $\tilde{u}$ prevails on the average glow u;
- the outlet head losses (or current momentum) are described as in a channel.

The corresponding resistance term in the momentum balance equation is:

$$R = \frac{1}{2}\rho \cdot f \cdot \frac{q_o + q_u}{h_m} 1.8 \cdot \tilde{u}_{rms} B_c + \rho \frac{|q_o|}{h_m} q_o \lambda_m \tag{13.88}$$

where λ_m is a calibration factor considering the velocity distribution, $\tilde{u}_{rms}$ for breaking waves is given by $\tilde{u}_{rms} = 0.2 \div 0.4\sqrt{gH_{rms}}$, $f = 0.25 \div 0.35$ and

$$q_u = \lambda_{qu1} \tfrac{1}{8} H_{rms}^2 \sqrt{g/h} + \lambda_{qu2} 0.9 H_{rms}^2 / T;$$

H_{rms} is here an average value along the berm; λ_{qu1} and λ_{qu2} are calibration factors, i.e. may differ from 1.

13.5.4.3. Return flow through gaps

Interest in rip currents is motivated by their importance for near shore processes such as off-shore sediment transport, shoreline evolution and pollutant transport; public interest in rip currents is due to beach safety issues and beach erosion.

If the beach is protected by a multi-structure, most of the return flow is concentrated at gaps: the actual discharge depends on the gap to structure length ratio, structure porosity and freeboard.

A simple way to evaluate the velocity at gap derives by the application of the generalised Bernoulli theorem (Mei, 1989, p. 472), along the return flow pattern.

The first point for the balance is placed inshore the barrier centre, where piling-up is maximum and velocity is almost null due to symmetry; the second point is the gap centre, where the gap velocity is unknown. Along the pattern between these two points, head losses (ΔH) due to bed friction should be considered.

The balance equation is:

$$H_1 - \Delta H = H_2 \tag{13.89}$$

where the head H in presence of waves is given by the sum of piling-up P, the current kinetic energy due to mean velocity u and wave pressure excess height:

$$H = P + \frac{u^2}{2g} + \frac{\eta_{rms}^2 k_s}{\sinh(2k_s h)}$$

The mean flow head losses ΔH can be calculated as: $\Delta H = j_1 l_1 + j_2 l_2$ with $j = \tau/(\gamma h)$, h = water depth at the structure toe and $\tau = (1/2)\,\rho f <| \tilde{u}_{rms} |> u$.

Eq. (13.89) allows to relate the velocity at gap u_2 with piling-up P_1 in the protected area since all the other variables may be assessed at least in first approximation.

- velocity u_1 can be considered null due to symmetry;
- η_{rms1} can be derived from the transmitted wave height calculated as in Sub-section 13.3.1;
- piling-up P_2 can be assumed null;
- η_{rms2} can be assumed equal to the incident wave amplitude;
- l_1 is the long-shore distance between points 1 and 2, i.e. one half the sum of barrier and gap length;
- l_2 is the cross-shore distance between points 1 and 2, i.e. half the distance of the barrier from the shoreline;
- f is the bottom friction coefficient, due to presence of waves and currents; its value is in the range 0.01 (smooth bed) - 0.1 (rough and rippled bed) see Nielsen (1992);
- $\tilde{u}_{rms}$ is the wave velocity at the bottom for *rms* wave height, $\tilde{u}_{rms} = \omega H_{rms} / [2 \sinh(k_s h)]$;
- k_s is the significant wave number.

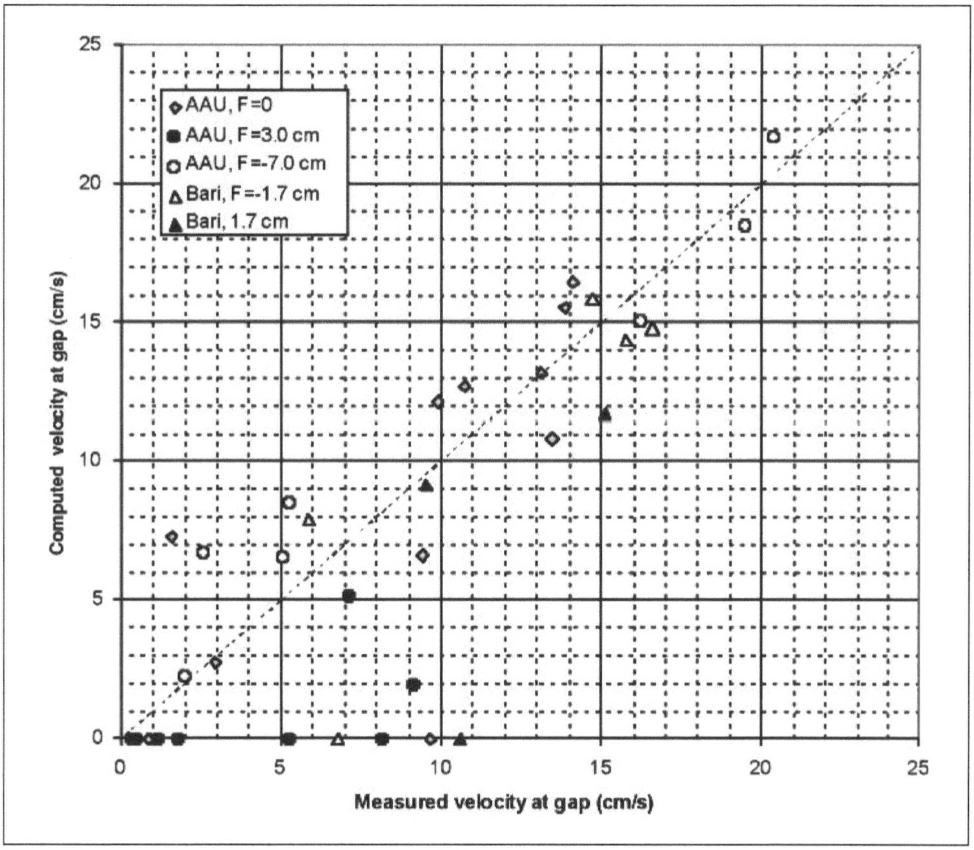

Figure 13.33. Comparison between measured velocities at gaps and velocities derived from piling-up using Eq. (13.89).

Fig. 13.33 shows the results obtained applying Eq. 13.89 to experimental tests performed on fixed (AAU, Zanuttigh and Lamberti, 2006), and mobile beds (Bari, Martinelli *et al.*, 2006). Values for u_1, η_{rms1}, P_1, u_2, η_{rms2}, P_2, h were measured; f is assumed equal to 0.02 and 0.05 for fixed and mobile bed respectively. The experimental results show that the wave term is not negligible. Some cases exist in which it was not possible to compute velocity through Eq. (13.89), i.e. square root of negative values, and are reported in the graph associated to zero computed velocity. These points are possibly affected by higher measurement errors, since they are characterised by lower wave energy.

13.5.5. Verification of the circulation model

The global LCS circulation can be obtained by the combination of the equations reported above.

For both emerged and submerged structures, filtration can be estimated with eq. 3 and velocity at gap can be derived from the balance Eq. (13.89).

Overtopping discharge is evaluated from the Eq.s 13.74 and 13.75 for emerged structures, whereas for submerged structures, flux over the crest is function of P and is computed by solving Eq. (13.81) for q_0.

13.5.5.1. Confined conditions

This model was applied to the data set described in Sub-Section 13.5.3.5, limitedly to cases for which piling-up and overtopping discharge were contemporary measured. 14 tests refer to emerged ($0 < R_c/H_{si} < 1$), 14 to zero freeboard conditions and 8 to submerged LCSs ($-1 < R_c/H_{si} < 0$).

The objective of calibration is to obtain an accurate pumping relation and is checked comparing experimental and model values of P_0 (piling-up for no net discharge across the barrier) and q_0 (discharge across the barrier that reduces piling-up to zero). The calibration parameters were the friction factor f ($f = 0.2$ is obtained), and a minimum wave height to mean water depth ratio over the barrier crest ($h > H_s/4$).

The average wave condition on barrier crest are assumed equal to the armonic mean among incident and transmitted wave height.

The width of the structure in Eq. (13.87) is the structure width at 2/3 h_c.

All predicted and measured values of P_0 and q_0 do not differ more than a factor of 2. The inter-quartile range of the predicted to measured ratios are [0.90-1.30] for discharge and [0.85-1.30] for piling-up.

13.5.5.2. 3-D conditions

The following analysis is based on data acquired in the Bari wave basin (Martinelli *et al.*, 2006). The structure consisted of two horizontal layers, the foundation (D_{n50} = 3.0 cm) and the structure itself (D_{n50} = 4.5 cm), which was 11.0 cm high and 33.3 cm wide at the crest level; foreshore slope was 1:200. Tested freeboards were in the range +/–1.7 cm. Irregular waves were generated, with H_{si} ranging from 3.5 to 7.5 cm and steepness ranging from 0.02 to 0.045.

Experimental results are shown in Fig. 13.34. They provide a quantification of phenomena described in Sub-section 13.5.4 and can be based on the given model.

For constant wave conditions, Fig. 13.34 shows that piling-up in the channel (in confined conditions) is quite greater than for a multi-structure with narrow gaps (L_g/L_b = 1/4, with L_g = gap width and L_b = barrier length), and even greater when compared to a multi-structure with wide gaps (L_g/L_b = 1). Indeed the overall return flow resistance decreases with increasing ratio L_g/L_b, and consequently the piling-up required to drive all the return flows is smaller.

For constant wave conditions and variable crest freeboard, Figure 13.35 shows piling-up behind the barrier centre and mean overtopping discharge across the barrier measured during the experiments and derived as the crossing point between the barrier pumping curve and the return resistance relation. The comparison shows that the evaluation procedure provides reasonable results and that, even for a gap to barrier length ratio equal to 1/4, the actual operating point is near to the extreme zero piling-up condition and far from the zero net overtopping discharge.

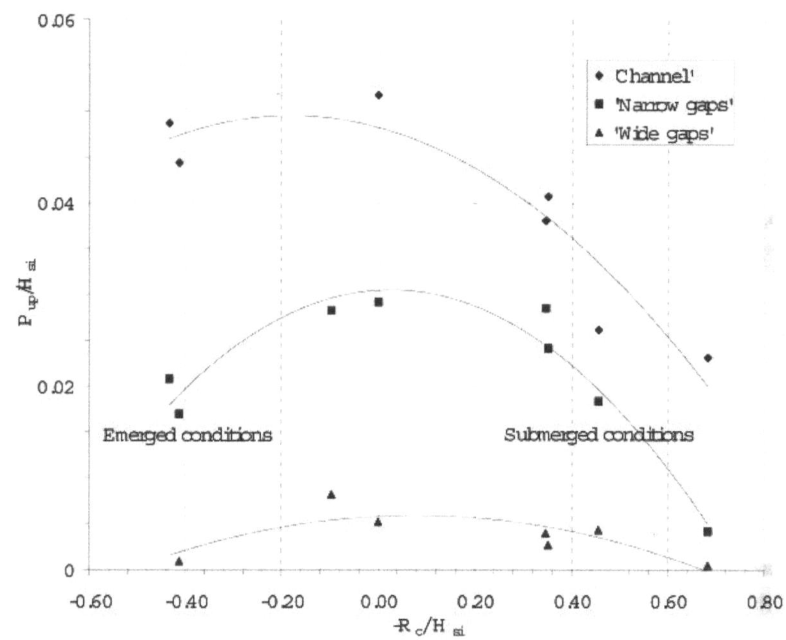

Figure 13.34. Piling-up P for different confinement conditions and relative submergence R_c/H_s, from Martinelli *et al.* (2006). Tests are characterised by a peak wave steepness in the range 0.042-0.054.

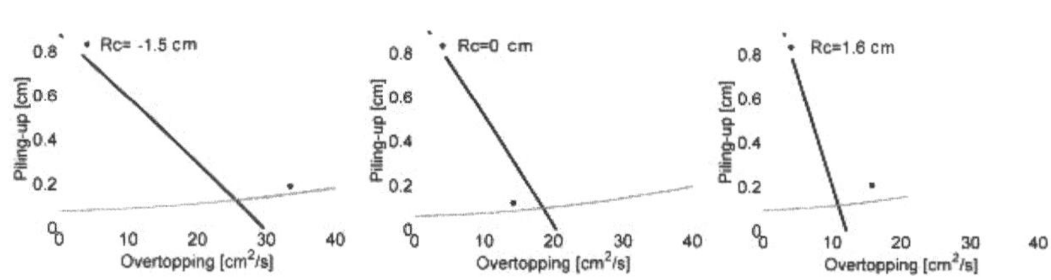

Figure 13.35. Pumping curves for different submergences and comparison between couples, piling-up behind the barrier - overtopping discharge, obtained from the proposed evaluation scheme and measured in Bari experiments for the narrow gap case. From left to right tested conditions are respectively: H_{si} = 5.28, 4.20, 4.40 cm; T_p = 1.03 s; K_t = 0.48, 0.44, 0.27; h = 12.5, 11.0, 9.4 cm.

13.6. CROSS-SHORE EQUILIBRIUM PROFILE

(Vidal, UCA)

13.6.1. Introduction

Various expressions have been proposed over the years for the equilibrium profile (see González *et al.* 1997 as a general reference). The most widely used formulation, very simple and easy to apply, is the 2/3-power profile shape proposed by Bruun (1954) and Dean (1977). Both authors concluded that the beach profile shape could be adequately represented by:

$$h = Ax^{2/3} \tag{13.90}$$

where h is the total water depth, A is a dimensional shape parameter that depends on the grain size, see Figure 13.36, and x is the horizontal distance from the shoreline.

Dean (1977) found that the 2/3 profile could be obtained considering that the time-averaged energy dissipation rate per unit volume across the beach, caused by wave breaking D^*, was held constant and dependent on beach grain size:

$$\frac{1}{h}\frac{dF}{dx} = D^* \tag{13.91}$$

The influence of a coastal structure on the equilibrium profile can be evaluated if a proper energy dissipation model and wave height variation across the profile is provided for the energy flux balance. In the case of two-dimensional, submerged breakwaters, the water depth in the leeside of the structure, h_i, can be obtained if the breakwater is inside a surf zone and the transmission coefficient over the structure, K_t, is known:

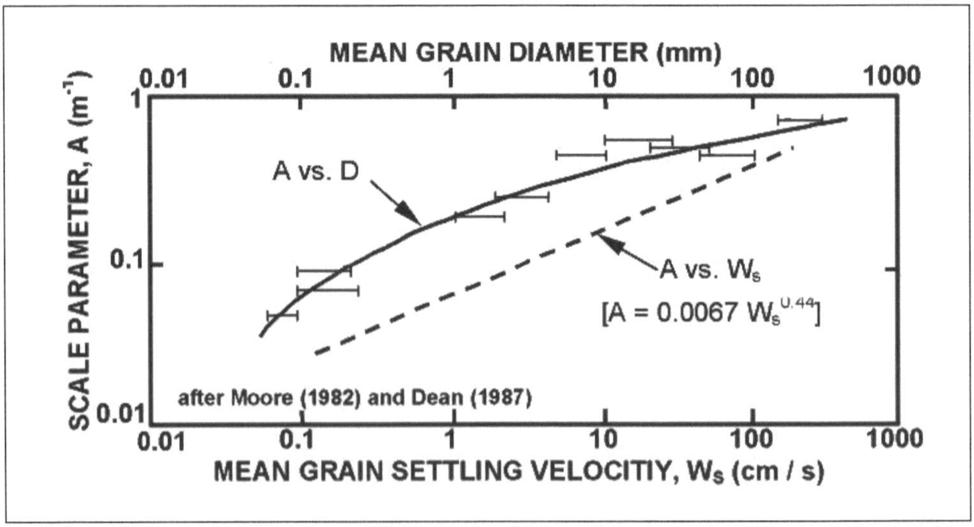

Figure 13.36. Dependence of the A coefficient in the Bruum/Dean profile Eq. (13.90) on the mean grain diameter or settling velocity of the beach sand. From Komar, (1998).

$$h_i = \frac{H_i}{\gamma} = \frac{H_e K_t}{\gamma} \tag{13.92}$$

where H_e in Eq. (13.92) is the incident wave height and γ a constant wave height to water depth ratio in the surf zone. Once the water depth in the leeside is known, the beach profile Eq. (13.90) can be applied. There are two cases when the 2D profile can be obtained using the energetic approach developed above:

- perched beaches: where a narrow-crest submerged breakwater situated in the surf zone modifies wave transmission due mainly to wave reflection on the structure;
- reef-protected beaches: where a wide submerged breakwater allows the wave breaking to stabilize to a bore in equilibrium with the water depth over the crest.

13.6.2. Perched beaches

A perched beach is characterised by a profile shifted in the off-shore direction with respect to its original one (see Fig. 13.37); such change is produced by a reduction of the incident wave energy, generally due to a dissipation caused by an artificial structure.

In Fig. 13.37 the original and the perched profile are represented schematically by a $h = x^{2/3}$ curve. It is expected that the profile significantly deviates from such curve, both at the shoreline, where a milder slope is more appropriate, and at the barrier, where the effect of breakers may induce erosion or deposition depending on the barrier width.

The perching amount can be derived from the hypothesis of constant water depth to local breaking wave height ratio in the surf zone. In such simple case, an artificially induced dissipation reduces the incident wave height and proportionally the equilibrium depth.

This 2D conceptual model considers the morphology to be determined by the steadiness of the dissipation rate and it may fail in 3D environments with circulations inducing different morphological mechanisms, as those described in Section 13.10.

Incident energy is divided between transmitted and reflected energy and the reflected fraction is theoretically evaluated.

The energy flow balance on both sides of the structure is:

$$\Gamma_i - \Gamma_e = \Gamma_r \tag{13.93}$$

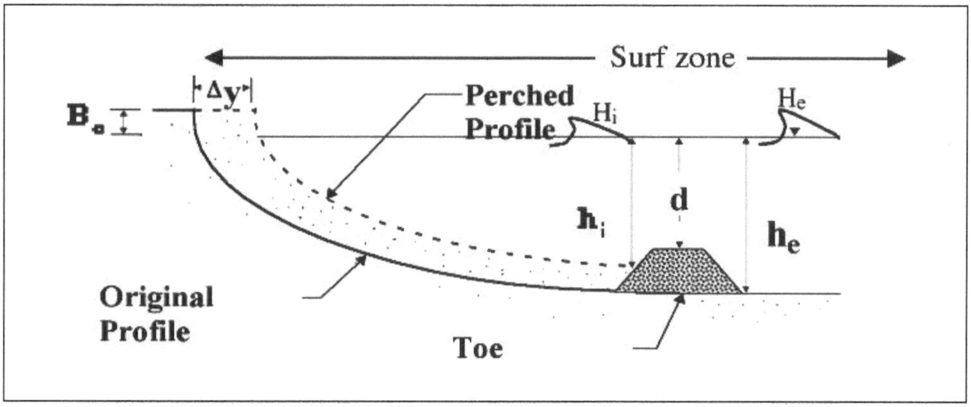

Figure 13.37. Definition sketch of a perched beach.

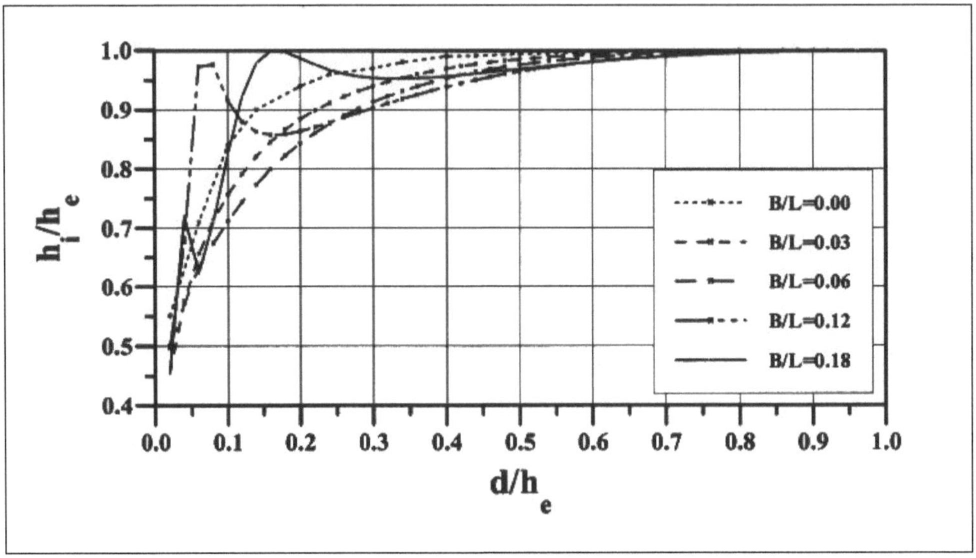

Figure 13.38. Relative depths h_i/h_e versus relative crest submergence, d/h_e for different relative width, B/L.

If shallow water conditions are assumed in the surf zone, using equations Eq. (13.92) and Eq. (13.93), the water depth in the leeside, h_i, can be obtained:

$$h_i = h_e K_t^{2/5} \tag{13.94}$$

where K_t is the transmission coefficients.

Using the above mentioned procedure, Gonzalez *et al.* (1999) evaluated the water depth ratio h_i/h_e versus the dimensionless water depth, d/h_e, for different breakwater crest widths, B/L, see Figure 13.38. From Figure 13.38 it can be concluded that for relative submergence d/h_e greater than 0.5 minor benefits are achieved with the construction of a submerged breakwater ($h_i \sim h_e$). A considerable reduction in h_i/h_e is obtained for $d/h_e < 0.1$.

Gonzalez *et al.* (1999) used laboratory data from Chatham (1972) and Sorensen and Beil (1988) and field data from Dean *et al.* (1977) and Ferrante and Franco (1992) to validate Eq. (13.91). The proposed model fitted well in all cases.

13.6.3. Reef-protected beaches

13.6.3.1. Introduction

Gourlay (1994) demonstrated that on a reef, the breaking process will take a distance (one or two wave lengths) to reduce this wave energy flux to a stable value. This result agrees with Muñóz *et al.* (1998) field data, which showed that for a natural reef-protected beach to exist, the reef width must exceed three wave lengths.

If the man-made offshore structure is wide enough it will resemble the effect of a natural reef. It is well known that the spilling-wave breaking assumption with a constant wave height to water depth ratio, γ, is not adequate for waves breaking on a shelf. Horikawa and Kuo (1966), computed theoretical curves that have a consistent agreement with experimental data in the case of wave transformation on a horizontal bottom. The ratio between the local wave height and the mean water depth decreases from 0.8, at the initial wave breaking point, to become almost constant, about 0.5, in the inner zone.

From the above can be concluded that the wave height, H_{rp}, that reaches the sandy beach toe, which is located at the depth h_r, see Figure 13.39, is lower than the wave height, H, that would reach that particular depth in a beach without the hard shelf. Consequently, the total amount of energy that has to be dissipated by the sandy profile is minor.

13.6.3.2. Energy Flux Balance

A simple relationship between the shape parameter for reef-protected beaches, hereafter denoted as A_{rp}, and non-reef-protected beaches, A, can be obtained considering that the energy flux Ec_g at h_r must be dissipated along the beach profile in both cases:

$$(EC_g)_{h_r} = \int D^* \; h \, dx \tag{13.95}$$

Assuming linear shallow wave theory and Eq. (13.90) valid along the entire profile, it yields:

$$W_{rp} = W \cdot \left(\frac{\Gamma}{\gamma}\right)^2 \tag{13.96}$$

where Γ is the breaker-to-depth ratio for a reef-protected beach and γ is the breaker-to-depth ratio in a non-reef-protected beach. For a wide shelf ($l \approx \infty$), typical values of Γ range between 0.55 to 0.35 (Nelson, 1994). Values of γ depend on beach slope and wave steepness, and have a wider range of variability. Kaminsky and Kraus (1993) compiled a large database of wave breaking parameters and showed that for typical field beach slopes (1/30 to 1/80) most of γ values are encountered in the range 0.65 to 1.1 with an average value of 0.79.

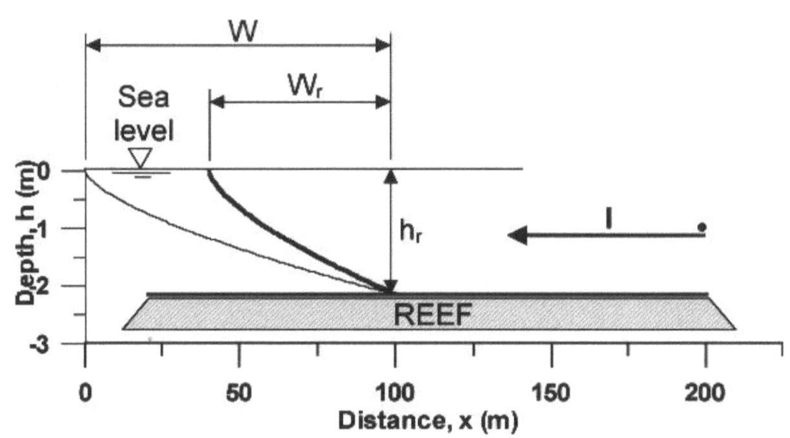

Figure 13.39. Definition sketch of parameters for the reef-protected beach.

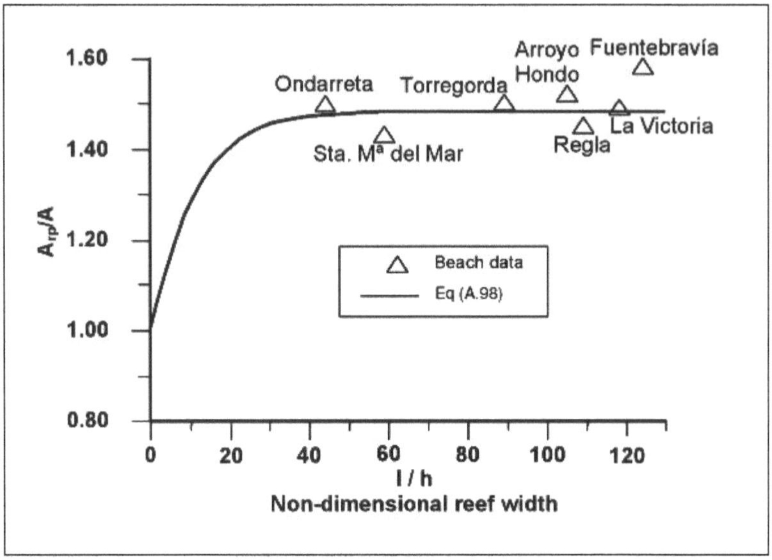

Figure 13.40. Non-dimensional shape parameter A_{rp}/A.

Introducing Eq. (13.90) in Eq. (13.96), a relationship between the shape parameters can be found as:

$$\frac{A_{rp}}{A} = \left(\frac{\gamma}{\Gamma}\right)^{\frac{4}{3}} \tag{13.97}$$

where A_{rp} is the shape parameter for the reef-protected beach and A is the non-reef-protected beach shape parameter.

Using the set of field data compiled by Gomez-Pina (1995), Muñoz *et al.* (1998) verified the above described model. Over 50 profiles from seven beaches were used. The predicted values of A_{rp} using Eq. (13.97) and the best-fitted values are compared in Figure 13.40. The predicted values are computed using Fredsoe and Deigard's (1992) model for Γ. It is seen in Figure 13.40 that Eq. (13.97) provides a good representation of the beach shape parameter A_{rp}. The asymptotic best fit for a wide shelf ($l/h > 60$) is $A_{rp} = 1{,}48$ A which corresponds to a value of $W_{rp} = 0.56\ W$.

13.7. CROSS-SHORE SEDIMENT TRANSPORT

(Zyserman, DHI)

In nature, the profile of a sandy beach changes continuously in response to gradients in cross-shore transport. These gradients may be quite large, causing the beach profile to vary considerably even during the course of a single storm. Therefore, reliable calculation of cross-shore sediment transport rates is a pre-requisite to simulating the development of the beach profile in response to the incident wave forcing.

Several mechanisms are active in connection with cross-shore transport outside and in the surf zone (Fredsøe and Deigaard, 1992). Streaming in the wave boundary layer, non-linearity of the shoaling waves and Lagrangian drift are the main transport mechanisms under non-breaking waves. The surf zone is characterised by strong energy dissipation due to wave breaking; the high levels of turbulence are capable of keeping significant concentrations of sediment in suspension. The water carried shoreward by the surface rollers returns below wave-trough level in the form of offshore-directed undertow. Sediment transport within the surf zone is strongly related to the undertow and, as such, directed mainly offshore.

A number of empirical models for computation of cross-shore transport are available from the literature. Among them, the models by Madsen and Grant (1976), Shibayama and Horikawa (1980), Sawamoto and Yamashita (1986), Sleath (1978) and Trowbridge and Young (1989) may be mentioned. All these models have been developed or calibrated/ validated using specific data sets. Thus, application of the models should be restricted to similar conditions as found during the experiments.

Bailard (1981) developed a total-load transport model based on an energetic approach. It calculates the depth-integrated suspended and bed-load transport rates on the basis of near-bed velocity moments, for arbitrary angles between the direction of wave propagation and the depth-averaged flow velocity. This model is widely used because it is easy to apply, especially in the form of a computer program. A limitation of this model is that it does not include transport mechanisms related to energy dissipation due to wave breaking in the surf zone. This shortcoming is usually overcome by use of calibration factors (the so-called «efficiency factors»).

The local cross-shore sediment transport rate reads according to Bailard (1981):

$$\langle i_x \rangle = \rho c_f u_m^3 \left\{ \frac{\varepsilon_B}{\tan\phi} \left[\psi_1 \cos\alpha + \delta_u^3 + \delta_u \left(\frac{1}{2} + \cos^2\alpha + \delta_v^2 \right) + \delta_v \sin\alpha \cos\alpha - \frac{\tan\beta}{\tan\phi} (u_3)^* \right] + \frac{u_m}{W} \varepsilon_s \left[\psi_2 \cos\alpha + \delta_u (u_3)^* \right] - \frac{u_m^2}{W^2} \varepsilon_s^2 \tan\beta (u_5)^* \right\} \tag{13.98}$$

where $\langle\ \rangle$ indicates time averaging over the wave period, c_f is a drag coefficient, ϕ is the internal friction angle of the bed sediment, W is its fall velocity, ε_B and ε_S are the efficiency factors for bed and suspended load transport, $\tan\beta$ is the seabed slope, α is the angle of wave propagation measured respect to the beach normal, θ is the angle between the steady current $\bar{u}$ and the beach normal.

The oscillatory wave-induced near-bed velocity (above the wave boundary layer) is expressed as $\tilde{u} = u_m \cos\sigma t + u_{2m} \cos 2\sigma t + ...$ where $\sigma = 2\pi/T$ is the wave frequency.

The relative steady current strengths δ, δ_u and δ_v are defined as:

$$\delta = \frac{\bar{u}}{u_m} \qquad \delta_u = \frac{\bar{u}}{u_m}\cos\theta \qquad \delta_v = \frac{\bar{u}}{u_m}\sin\theta$$

The velocity moments ψ_1 and ψ_2 are defined as $\psi_1 = \langle \tilde{u}_3 \rangle / u_m^3$ and $\psi_2 = \langle |\vec{u}_3\ \tilde{u}| \rangle / u_m^4$ and the integrals $(u_3)^*$ and $(u_5)^*$ are evaluated as:

$$\left(u_3\right)^* = \frac{1}{T}\int_0^T (\delta^2 + 2\delta\cos(\theta-\alpha)\cos\sigma t + \cos^2\sigma t)^{3/2}\,dt$$

$$\left(u_5\right)^* = \frac{1}{T}\int_0^T (\delta^2 + 2\delta\cos(\theta-\alpha)\cos\sigma t + \cos^2\sigma t)^{5/2}\,dt$$

Stive and Battjes (1984) developed a model in which the offshore-directed sediment transport was found from the product of an offshore-directed depth-uniform velocity and the near-bed concentration of suspended sediment. Deigaard *et al.* (1988) followed a similar approach, but taking into account the vertical structure of the cross-shore flow (the undertow) and the suspended sediment concentration when calculating the offshore transport.

Most advanced cross-shore sediment transport models applied today follow this approach, namely to compute separately the vertical structure of the flow and the suspended sediment, and then compute the suspended load transport by integrating the product of both along the vertical. A drawback of these model's complexity is that they cannot be expressed through a formula.

Roelvink and Brøker (1993) gave a review of cross-shore model concepts and presented an intercomparison of the most important models.

More recently, quasi-3D transport models based on the three-dimensional structure of the shear stress outside and within the surf zone (Deigaard, 1993) have been developed. Q3D models allow simultaneous computation of cross-shore (both onshore and offshore directed) and longshore transport rates taking into account the vertical structure of the concentration of suspended sediment and the time-averaged flow. Application of such a model is presented in Elfrink *et al.* (2000).

In the last few years fully 3D models of hydrodynamics, sediment transport and morphological change have become available and have been applied to realistic design problems (e.g. Lesser *et al.*, 2003; Roelvink *et al.*, 2002).

13.8. LONG-SHORE SEDIMENT TRANSPORT (AMOUNT AND DISTRIBUTION OVER THE COASTAL PROFILE)

(Zyserman, DHI)

Longshore sediment transport is closely related to the longshore current that is generated when waves break obliquely to the coast. The yearly littoral drift associated with the waves will often be the dominant factor in the sediment budget for an exposed coastline.

The idea that longshore sediment transport is mainly driven by the incident waves rather than by tides and ocean currents became generally accepted early in the 20th century.

Therefore, formulas and models for the computation of littoral drift (either total or local transport rates) have been developed since 1938 based on this idea (Fredsøe and Deigaard, 1992). An usual assumption is that sediment is stirred and brought into suspension by the waves and then transported by the littoral current.

One of the most-widely used methods for calculating the total (i.e. integrated across the surf zone) longshore transport is the CERC formula (Komar and Inman, 1970) which relates the transport rate to the longshore component of the wave energy flux at the breaker line:

$$Q_1 = \frac{K}{\rho g(s-1)} P_{ls} \tag{13.99}$$

where Q_1 is the rate of total longshore sediment transport measured as solid volume, P_{ls} is the so-called longshore energy flux factor, K is a constant (= 0.77), ρ is the density of water, s is the relative sediment density and g is the acceleration of gravity. P_{ls} is evaluated as

$$P_{ls} = \frac{1}{16} \rho g H_{rms,b}^2 c_{g,b} \sin 2\alpha_b \tag{13.100}$$

where the subscript «*b*» indicates values at the point of breaking, α_b is the angle between the waves and the coast at the breaker line, c_g is wave group celerity and H_{rms} is the root-mean-square wave height. If the significant wave height is used instead of H_{rms} to evaluate P_{ls}, then the value of the constant K has to be adjusted accordingly.

Kamphuis (1991) presented a formula to compute the total rate of longshore transport based on dimensional analysis. Later on, Kamphuis (2002) used recent data to validate the expression he derived in 1991.

None of the above models permits to compute the variation of longshore transport along the beach profile. This feature became available when Longuet-Higgins (1970) developed a model for the longshore current based on the concept of radiation stresses.

Bijker's (1971) made the first detailed longshore sediment transport model, using the littoral current model of Longuet-Higgins (1970) for a beach of constant slope together with a sediment transport model for wave and currents.

Most models used nowadays in coastal engineering practice combine a module to compute wave transformation due to refraction, shoaling and breaking with a module that calculates the cross-shore variation of the longshore current velocity; these parameters are then used as input to a sediment transport model capable of computing local sediment transport rates.

The already mentioned Bailard's model also allows to compute the longshore component of the local sediment transport rate:

$$\langle i_y \rangle = \rho c_f u_m^3 \left\{ \frac{\varepsilon_B}{\tan\phi} \left[\psi_1 \sin\alpha + \delta_v^3 + \delta_v \left(\frac{1}{2} + \sin^2\alpha + \delta_u^2\right) + \delta_u \sin\alpha \cos\alpha \right] + \frac{u_m}{W} \varepsilon_s \left[\psi_2 \sin\alpha + \delta_v (u_3)^* \right] \right\} \tag{13.101}$$

The variables involved were already described in the previous Section 13.7.

Formulas for the calculation of local sediment transport rates that are frequently cited in the literature are Bailard (1981), Dibajnia and Watanabe (1996), and Soulsby and Van Rijn and derived models (Soulsby, 1997; van Rijn, 2000), among many others. Again, it should be kept in mind that these models have been developed or calibrated/validated using specific data sets. Thus, application of the models should be restricted to similar conditions as used for their derivation.

The Soulsby-van Rijn formula applies to total load transport in combined waves and currents on horizontal and sloping beds, and it is intended for ripple-covered beds. The formula reads:

$$q_t = (A_{sb} + A_{ss})\overline{U}\left[\left(\overline{U^2} + \frac{0.018}{C_D}U_{rms}^2\right)^{½} - \overline{U_{cr}}\right]^{2.4}(1 - 1.6\tan\beta) \qquad (13.102)$$

where

$$A_{sb} = \frac{0.005h(d_{50}/h)^{1.2}}{\left[(s-1)gd_{50}\right]^{1.2}}$$

$$A_{ss} = \frac{0.012d_{50}D_*^{-0.6}}{\left[(s-1)gd_{50}\right]^{1.2}}$$

$$C_D = \left[\frac{0.40}{\ln(h/z_0) - 1}\right]^2 = \text{drag coefficient due to current alone,}$$

$\overline{U}$ = depth-averaged current velocity, U_{rms} = root-mean-square wave orbital velocity, $\overline{U}_{cr}$ = critical current velocity, β = bed slope in current direction (positive uphill), h = water depth, d_{50} = median grain diameter, z_0 = bed roughness $\simeq$ 0.006 m, s = relative density of sediment and

$$D_* = \left[\frac{g(s-1)}{\nu^2}\right]^{1/3} d_{50}$$

with ν = kinematic viscosity of water.

Deigaard *et al.* (1986b) developed a model to calculate local rates of total-load sediment transport. The model includes a longshore current model for arbitrary coastal profiles. Calculation of local rates of total sediment transport were performed using the deterministic sediment transport model for combined current and waves developed by Fredsøe *et al.* (1995) and extended to include surf-zone waves by Deigaard *et al.* (1986a). The sediment transport model solves the wave boundary layer in an intra-wave fashion to compute instantaneous flow profiles, and the diffusion equation for suspended sediment to determine the instantaneous concentration of suspended sediment. Instantaneous suspended load transport is found by integration of the product of both variables along the vertical. Being deterministic, this model is not limited to a range of input variables, and can be applied to a wide range of conditions including breaking/unbroken waves propagating at an arbitrary angle to the current, horizontal or sloping seabed, plane or ripple-covered bed, uniform or graded bed sediment, etc. A drawback of this model is its complexity, which does not allow to specify it through one or more simple formulas.

Lately, more advanced deterministic models including a quasi-3D description of flow and sediment transport have become available, see e.g. Elfrink *et al.* (2000). These models allow simultaneous computation of the longshore and cross-shore components of the local sediment transport rates along a given beach profile or over a selected area.

13.9. EMPIRICAL DIAGRAMS/FORMULAE FOR PREDICTION OF FORMATION OF SALIENTS AND TOMBOLOS

(Vidal, UCA; Sánchez-Arcilla, UPC)

13.9.1. Introduction

Static equilibrium shoreline models, are used to predict tombolo and salient formations for both natural and man-made coastal structures. Offshore breakwaters are generally shore-parallel structures that effectively reduce the amount of wave energy reaching a protected stretch of shoreline. One of the main problems in the design of these coastal structures is the prediction of the shoreline response.

The empirical approach requires an a priori assumption of the shape of the shoreline. Empirical analyses have been carried out by a number of researchers based on beach equilibrium concepts, e.g. Noble, (1978); Gourlay, (1980); Nir, (1982); Dally and Pope, (1986); Suh and Dalrymple, (1987); Hsu and Silvester, (1990); Ahrens and Cox, (1990); McCormick, (1993); González and Medina, (2001) and on small-scale models and field observations, see Rosati, (1990), and ASCE, (1994), as general references.

This section is divided into two parts. In the first part, the methodology proposed by González and Medina (2001) for testing or designing «static equilibrium beaches» is presented. It is based on the equilibrium beach concept (combining shoreline and cross-shore profile) and a semiempirical model. The proposed methodology includes existing equilibrium profile models and a modified static equilibrium plan form formulation. This methodology has been applied to some natural and man-made beach cases, showing the capability for the design of new nourishment projects. In the second part, the semi-empirical approach presented by González and Medina (1999) predicting the shoreline response behind an offshore breakwater is described.

13.9.2. Proposed methodology for emerged breakwaters

There are in the literature many simple rules for prediction of salient and tombolo formation. Tables 13.7 and 13.8 give a summary of those rules. Table 13.8 gives some conditions for minimal shoreline response.

In Tables 13.7, 13.8 and 13.9, L_B means the breakwater length, Y_B is the distance from the breakwater to the undisturbed shoreline, and G_B is the gap aperture in the case of multiple breakwaters.

González and Medina (2001) carried out analytical and empirical approaches in order to develop a modified methodology for testing or designing static equilibrium shorelines (*SES*). Using an analytical expression of *SES* and 26 fully-developed equilibrium bay beaches along the Atlantic and Mediterranean coasts of Spain, the «downcoast» limit, P_0, was defined (see Figure 13.41). The point P_0 defines the starting point where the parabolic model (Hsu and Evans, 1989) is applicable, and it is a function of the angle α_{min} and the

Table 13.7. Summary of rules for tombolo formation.

Condition	*Comments*	*Reference*
$\frac{L_B}{Y_B} > 2$	Double salient	Gourlay (1981)
$\frac{L_B}{Y_B} > 0.67$ to 1.0	Tombolo (shallow water)	Gourlay (1981)
$\frac{L_B}{Y_B} = 2.5$	Periodic tombolo	Ahrens and Cox (1990)
$\frac{L_B}{Y_B} > 1.5$ to 2.0	Tombolo	Dally and Pope (1986)
$\frac{L_B}{Y_B} > 1.5$	Tombolo (multiple breakwaters)	Dally and Pope (1986)
$\frac{L_B}{Y_B} > 1.0$	Tombolo (single breakwater)	Suh and Dalrymple (1987)
$\frac{L_B}{Y_B} > 2\frac{G_B}{L_B}$	Tombolo (multiple breakwaters)	Suh and Dalrymple (1987)

distance from the «control point» to the prolongation of the straight alignment downcoast of the beach, Y. Furthermore, the angle α_{min} is a function of the dimensionless distance of the beach to the length wave Y/L_s, where L_s is the wave length. This scaling wavelength, L_s, was calculated using the mean water depth along the wave front close to the control point, h_p, and the mean wave period associated with the wave height exceeded 12 hours per year, $H_{s12,}$ hereafter called, T_{H12}. Figure 13.42 shows the measured α_{min} versus Y/L_s for the selected fully developed Spanish beaches. The variables β and R_0, which are used in Hsu and Evans (1989) equilibrium shape formulation are related to the variables α_{min} and Y as $\alpha_{min} = 90^{\circ} - \beta$ and $R_0 = Y/\cos\alpha_{min}$ (see Figure 13.41). The best fit for α_{min} is given in Figure 13.42.

In order to test the stability of an existing bay beach or to predict the static equilibrium shape for newly designed bay, the following procedure must be carried out.

1) determine the position of the control point, C;
2) determine the orientation of the wave front at the control point, C. This orientation corresponds to that of the mean energy flux of the waves in the area;
3) define one point at the shoreline $P_c(\theta_c > \beta, R_c)$ as shown in Figure 13.41.
   – To test stability of an existing beach: select any point along the static equilibrium shoreline, taking into account that this point must not be affected by any other local diffraction.
   – To design a new bay beach: select one point in the bay of the future shoreline. In the selection of this point it must be taken into account that the beach profile should

Table 13.8. Summary of rules for salient formation.

Condition	*Comments*	*Reference*
$\frac{L_B}{Y_B} < 1.0$	No tombolo	SPM, Shore Protection Manual (1984)
$\frac{L_B}{Y_B} > 0.4$ to 0.5	Salient	Gourlay (1981)
$\frac{L_B}{Y_B} = 0.5$ to 0.67	Salient	Dally and Pope (1986)
$\frac{L_B}{Y_B} > 1.0$	Salient (single breakwater)	Suh and Dalrymple (1987)
$\frac{L_B}{Y_B} < 2\frac{G_B}{L_B}$	No tombolo (multiple breakwaters)	Suh and Dalrymple (1987)
$\frac{L_B}{Y_B} < 1.5$	Well-developed salient	Ahrens and Cox (1990)
$\frac{L_B}{Y_B} < 0.8$ to 1.5	Subdued salient	Ahrens and Cox (1990)

Table 13.9. Simple rules for minimal shoreline response.

Condition	*Comments*	*Reference*
$\frac{L_B}{Y_B} \leq 0.17$ to 0.33	No response	Irman and Fautschy (1966)
$\frac{L_B}{Y_B} \leq 0.27$	No sinuosity	Ahrens and Cox (1990)
$\frac{L_B}{Y_B} \leq 0.5$	No deposition	Nir (1982)
$\frac{L_B}{Y_B} \leq 0.125$	Uniform protection	Dally and Pope (1986)
$\frac{L_B}{Y_B} \leq 0.17$	Minimal impact	Noble (1978)

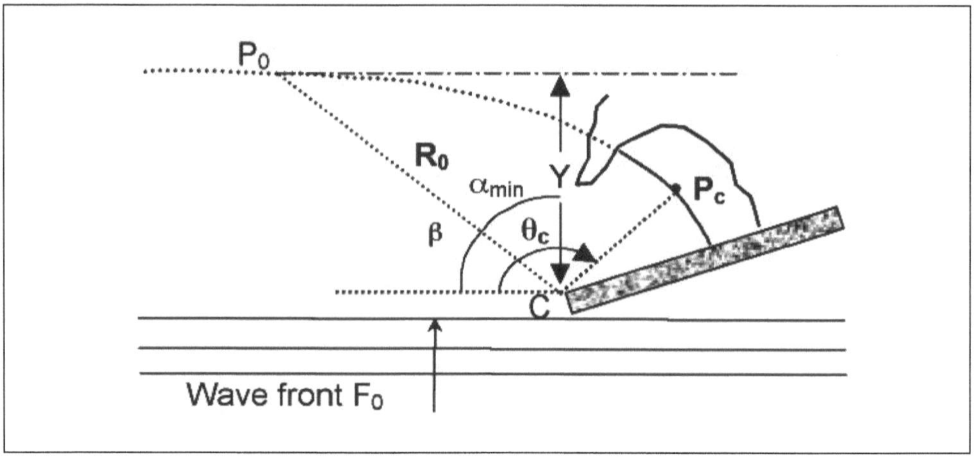

Figure 13.41. Definition Sketch.

be contained between the lateral boundaries of the beach. This condition should be checked at the end of this procedure.

4) Define the scaling wave length near the control point, $L_s = f(h_p, T_{H12})$, being h_p the mean water depth along the wave front close to the diffraction point and the mean wave period associated with the wave height exceeded 12 hours per year, H_{s12}.
5) Define de distance Y (see Figure 13.41). In the case of the design of a new beach, the straight alignment downcoast does not exist and the distance Y must be assumed taking into account that the beach downcoast of point P_0 should be nearly parallel to the incident wave height at the diffraction point. The validity of this assumption will be checked at the end of this procedure.

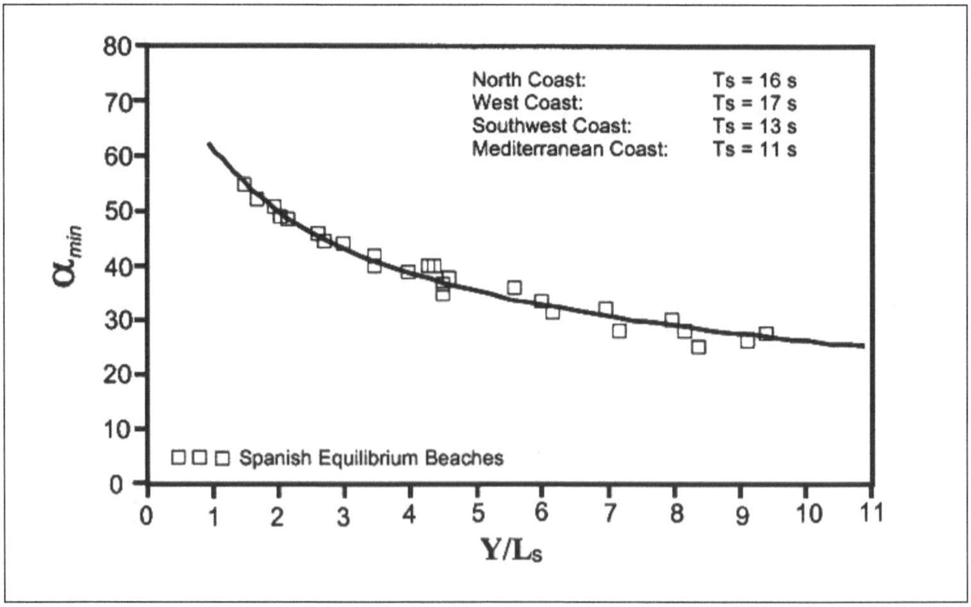

Figure 13.42. Best fit for α_{min} versus Y/L_s for several fully-developed Spanish beaches.

6) Evaluate the angle β using $\alpha_{min} = f(Y/L_s)$, Figure 13.42.

$$\beta = 90° - \alpha_{min} \tag{13.103}$$

7) Define the point P_0. This point can be defined evaluating R_0 from the parabolic model of Hsu and Evans (1989) as:

$$R_0 = \frac{R_c}{C_0 + C_1\left(\frac{\beta}{\theta_c}\right) + C_2\left(\frac{\beta}{\theta_c}\right)^2} \tag{13.104}$$

with C_0, C_1 and $C_2 = f(b)$ can be obtained from Shu and Evans (1989) (see Table 13.10). R_c and q_c where defined previously in step (3) by the point P_c.

8) Recalculate $Y = R_0 \cos \alpha_{min}$; if Y' is far from the initially supposed Y value, go back to step (5).

9) Using Hsu and Evans' (1989) parabolic formulation, the radii, R, can be obtained for different angles q, yielding the equilibrium shape:

$$\frac{R}{R_0} = C_0 + C_1 \cdot \left(\frac{\beta}{\theta}\right) + C_2 \cdot \left(\frac{\beta}{\theta}\right)^2 \tag{13.105}$$

The above-mentioned methodology has been applied to several beaches throughout the world for both high- and low-tide shoreline with very good results. Some applications have been presented by González and Medina (2002).

Table 13.10. Hsu and Evans (1989) parabola's coefficients.

β^o	C_0	C_1	C_2
20	0.054	1.040	– 0.094
22	0.054	1.053	– 0.109
24	0.054	1.069	– 0.125
26	0.052	1.088	– 0.144
28	0.050	1.110	– 0.164
30	0.046	1.136	– 0.186
32	0.041	1.166	– 0.210
34	0.034	1.199	– 0.237
36	0.026	1.236	– 0.265
38	0.015	1.277	– 0.296
40	0.003	1.322	– 0.328
42	– 0.011	1.370	– 0.362
44	– 0.027	1.422	– 0.398
46	– 0.045	1.478	– 0.435
48	– 0.066	1.537	– 0.473
50	– 0.088	1.598	– 0.512
52	– 0.112	1.662	– 0.552
54	– 0.138	1.729	– 0.592
56	– 0.166	1.797	– 0.632
58	– 0.196	1.866	– 0.671
60	– 0.227	1.936	– 0.710
62	– 0.260	2.006	– 0.746
64	– 0.295	2.076	– 0.781
66	– 0.331	2.145	– 0.813
68	– 0.368	2.212	– 0.842
70	– 0.405	2.276	– 0.867
72	– 0.444	2.336	– 0.888
74	– 0.483	2.393	– 0.903
76	– 0.522	2.444	– 0.912
78	– 0.561	2.489	– 0.915
80	– 0.600	2.526	– 0.910

13.9.3. Tombolo and salient prediction for emerged breakwaters

Using the relationship $\alpha_{min} = f(Y/L)$ obtained in the previous section and the static equilibrium shoreline shape formulation given by Hsu and Evans (1989), it is possible to determine the morphological characteristics of the shoreline response due to an offshore breakwater, González and Medina, (1999): (1) tombolo, (2) salient and (3) double salient (DS) (Figures 13.43, 13.44 and 13.45).

13.9.3.1. Tombolo case

If the distance from the breakwater to the shoreline is close enough, and the breakwater is long with respect to the length of the incident waves, sand will accumulate behind the breakwater until a tombolo forms; that is, the shoreline continues to build seaward until it connects with the breakwater. The variables governing the equilibrium shape are (Figure 13.43): the length of the breakwater, *2B*, the distance from the breakwater to the shoreline, *Y*, and the wavelength, *L*, which defines α_{min}. The unknown variables, namely, the shoreline length affected by the breakwater, $2B_1$, and the attachment width at the breakwater, B_k, can easily be obtained from Hsu and Evans' (1989) parabolic-shaped formulation and the α_{min} expression (Figure 13.42). The solutions for these variables are presented in Figure 13.46, see González and Medina, (1999) for details about the formulations.

13.9.3.2. Salient case

When the breakwater is far from the shoreline and its length is short with respect to the length of the incident waves, the shoreline will build a salient seaward. The governing variables involved in the equilibrium shape of the salient are the same as in the case of the tombolo, namely: the length of the breakwater, *2B*, the distance from the breakwater to the shoreline,

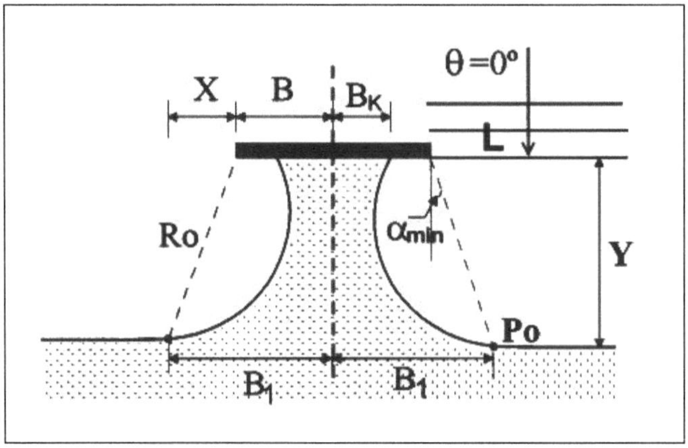

Figure 13.43. Definition sketch of a Tombolo. The typical unknown variables, when designing a tombolo are the shoreline length affected by the breakwater, $2B_1$, and the attachment width at the breakwater, B_k.

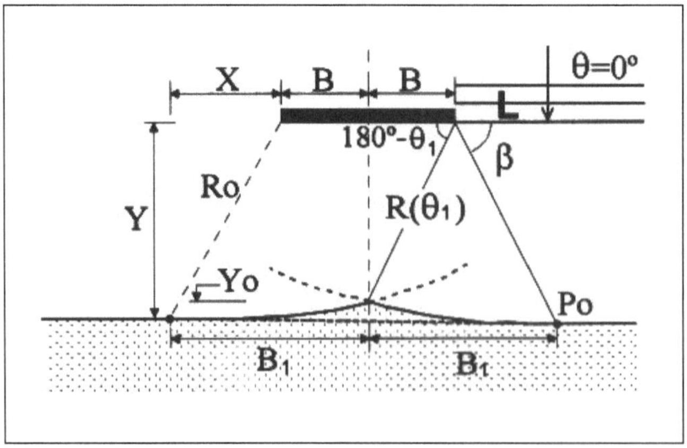

Figure 13.44. Definition sketch of a theoretical Salient. The typical unknown variables, when designing a salient are the salient apex, Y_0, and the shoreline length affected by the breakwater, $2B_1$.

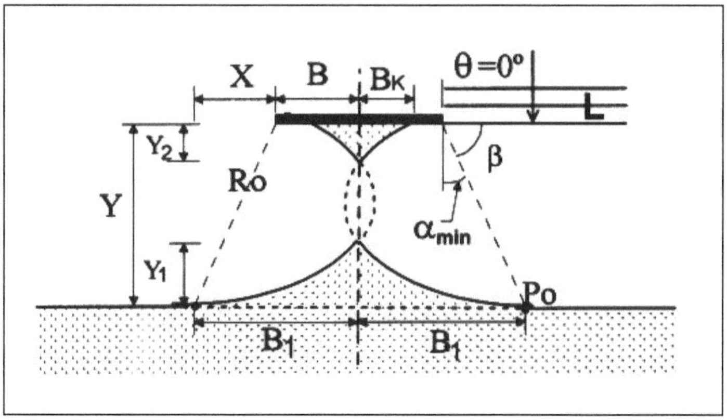

Figure 13.45. Definition sketch of a theoretical Double Salient (DS). The typical unknown variables, when designing a DS are a combination of the above parameters for Tombolo and Salient.

Y, and the wavelength, L, which defines α_{min}. The unknown variable is the salient apex, Y_0 (see Figure 13.44). As in the tombolo case, the unknown variable can easily be obtained from Hsu and Evans' (1989) parabolic-shaped formulation and the α_{min} expression (Figure 13.42).

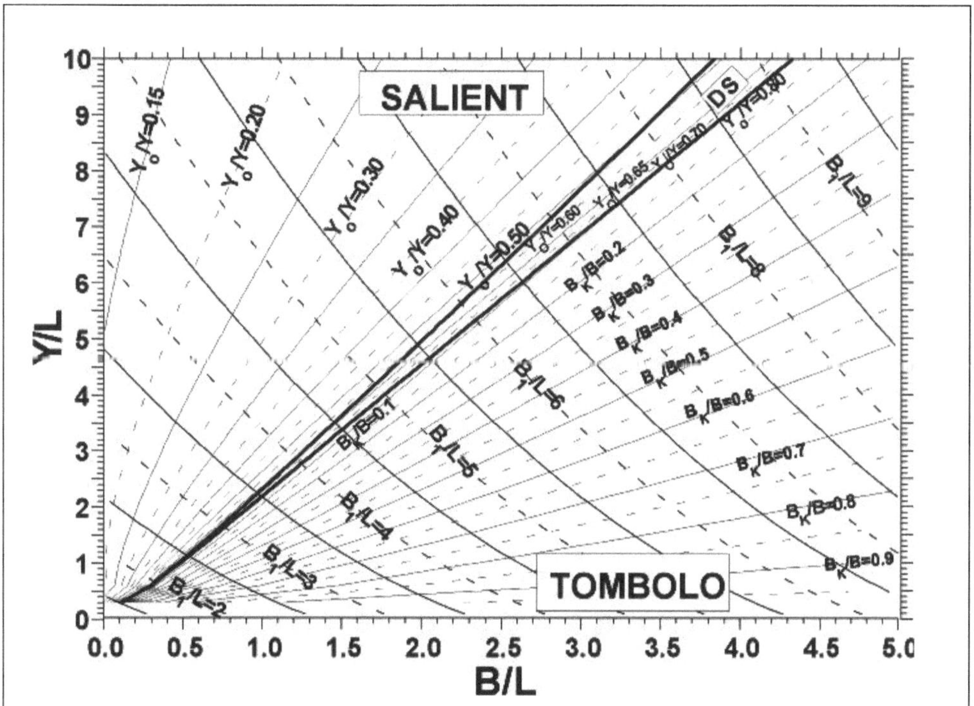

Figure 13.46. Variation of the non-dimensional equilibrium plan form parameters: for Tombolo: (B_k/B, B_1/L) Salient (Y_0/Y, B_1/L) and Double Salient (Y_0/Y, B_1/L) for different values of the length of the breakwater, 2*B*, the distance from the breakwater to the shoreline, *Y*, and the wavelength, L (see Figures 13.43, 13.44 and 13.45 for a definition sketch of the different variables).

The solution for this variable is also presented in Figure 13.46 (see González and Medina (1999) for details about the formulations).

13.9.3.3. Double Salient

Double salient can be interpreted as an intermediate case between the tombolo and the salient. In this case the sand is accumulated both at the lee side of the breakwater and at the coastline as it is shown in Figure 13.45. The variables governing the equilibrium shape are the same as in the case of the tombolo and Salient. In addition, Y_2 is the distance from the land spit at its apex, measured from the equilibrium shoreline, as shown in Figure 13.45 (see González and Medina (1999) for details about the formulations).

The proposed equilibrium shape model is able to adequately represent the equilibrium shoreline in cases where the beach is affected only by one diffracting point. These include the cases of tombolos and of salients formed by T-Groins, where each side of the salient is affected by only one tip of the offshore breakwater. Only in these cases, the salient apex, Y_0, given in Figure 13.44, applies. In general diffraction at the two breakwater tips affects both sides of a salient yielding an apex length, Y_s, shorter than Y_0, (see Figure 13.45). Hsu and Silvester (1990) proposed an empirical formulation which defines the apex position, Y', ($Y' = Y - Y_s$) as a function of the ratio of the distance, S, from the original shoreline to the breakwater and the breakwater length, $2B$. As stated previously, a constant value of the angle β was assumed in their work ($\beta = 40°$).

Since the range of the available data for B/L and Y/L *is* too small for separating the influence of the wavelength in β, a single curve, valid for $0.3 < B/L < 1.5$ and $2.0 < Y/L < 4.0$ is proposed. The relationship obtained is similar to the one proposed by Hsu and Silvester (1990) and is plotted in Figure 13.48.

$$\frac{Y'}{2B} = 0.50\left(\frac{2B}{Y}\right)^{-\sqrt{2}} \qquad (13.106)$$

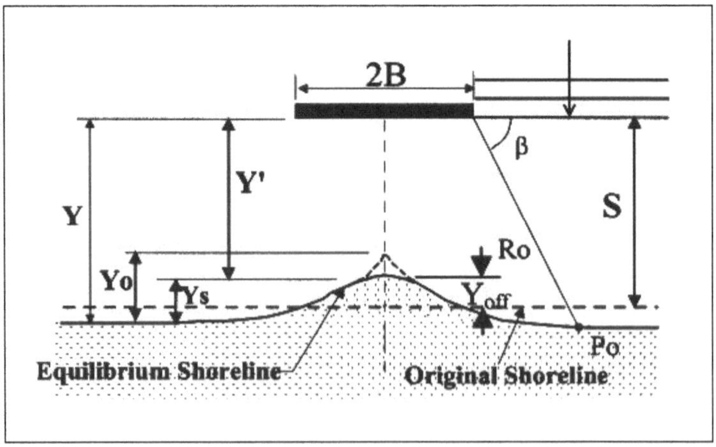

Figure 13.47. Definition sketch of a salient. In this figure both the theoretical (dashed line) and the actual salient shape (solid line) are graphed.

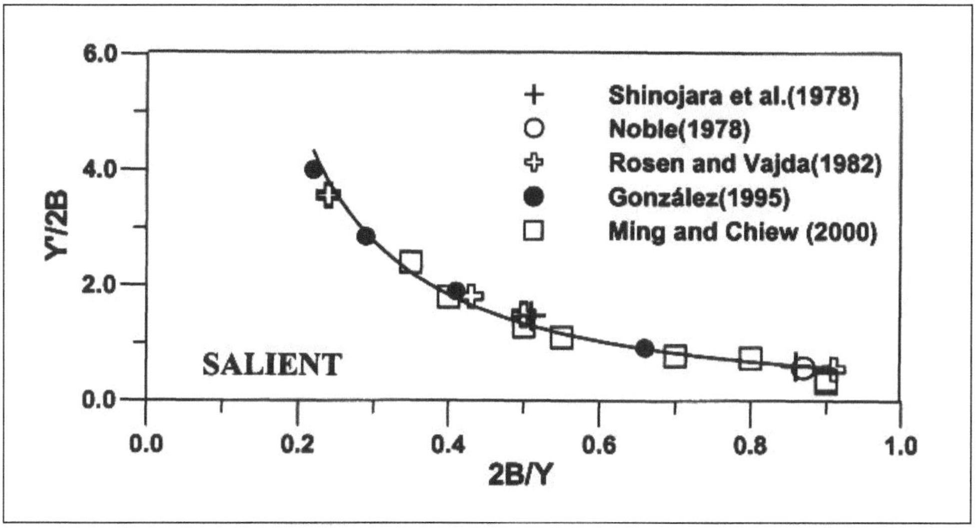

Figure 13.48. Relationship between the theoretical and the actual salient shape expressed in terms of Y *and* Y′ (see Figure 13.47). Dimensionless parameters, Y′/*2B* and 2B/Y *are* used to fit the field and laboratory data. Eq. (13.106) is also graphed (solid line).

González and Medina (2001) analysis can be applied for the design of the shoreline response due to a single offshore breakwater. It has been shown that if Y, B and L are known, it is possible to determine: (1) the kind of response (tombolo, salient or double salient), (2) the beach shape and (3) the affected area, $2B_1$, and therefore, the sand needed.

13.9.4. Submerged Breakwaters

Simple rules for prediction of tombolo or salient formation in the case of submerged breakwaters are given in Pilarczyk (2003), see Figure 13.49:

Salient formation when:

$$\frac{B}{S} < \frac{1}{1-K_t} \tag{13.107}$$

Salient for multiple breakwaters:

$$\frac{G \cdot S}{B^2} > 0.5(1-K_t)$$

where:

K_t is the wave transmission coefficient;
G, the gap distance between breakwaters.

Following Black and Andrews (2001a) salients form in the lee side of submerged offshore breakwater when:

$$\frac{B}{S} < 2 \tag{13.108}$$

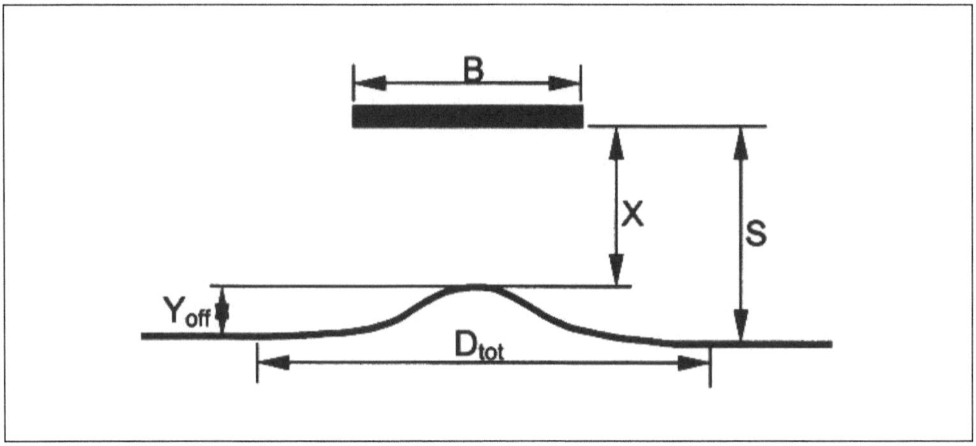

Figure 13.49. Definition sketch for Black and Andrews (2001) salient formation behind a submerged reef.

where, see Figure 13.49, B is the breakwater length and S is the distance to the original shoreline. If $B > 2\ S$ the shoreline continues undisturbed.

The distance from the tip of the salient and the breakwater, X, see Figure 13.49, is given by:

$$\frac{X}{B} = 0.498\left(\frac{B}{S}\right)^{-1.268} \tag{13.109}$$

and the length of the shoreline affected by the salient, D_{tot} (or the width of the salient) is given by:

$$\frac{Y_{off}}{D_{tot}} = 0.125 \pm 0.02 \tag{13.110}$$

where $Y_{off} = S - X$ is the salient amplitude, measured from the undisturbed shoreline see Figure 13.49.

The shape of the salient is best described by a sigmoid function.

13.10. COMBINED HYDRODYNAMIC AND MORPHOLOGIC NUMERICAL MODELS TO PREDICT SHORT AND LONG-TERM SPATIAL AND TEMPORAL EFFECTS

(Roelvink, WL-DH; Vidal, UCA; Zyserman, DHI; Arcilla, UPC)

13.10.1. Processes under simulation

The hydrodynamic and sediment transport processes around LCSs are usually very non-uniform both in the horizontal and vertical direction.

The following processes impact on the morphology of beaches located behind and adjacent to LCSs:

- wave shoaling, refraction and breaking;
- longshore current, with peaks over sand bars and near the shore, or with a single peak in case of a monotonic profile; in case of sharp gradients this current is often unstable, leading to «shear waves» that propagate with the flow;

- cross-shore flow, with weak onshore near-bed currents outside the breaker zone and strong offshore currents inside it;
- rip current patterns that can be seen as perturbations of the uniform situation, with shallow shoals and narrow rip channels;
- longshore sediment transport governed by the longshore current and the combined mixing by orbital motion, the longshore current and breaker-induced turbulence;
- cross-shore transport composed of counteracting components by return flow, wave skewness and asymmetry effects, bed slope effects and long wave/short wave coupling;
- long waves associated with wave groups, which can be cross-shore leaky modes or alongshore propagating edge waves.

There are then processes which are particularly related to LCSs like:

- abrupt wave breaking on the LCS, with wave transmission dependent on the freeboard, the crest width, the incident wave height and the breakwater material;
- strong deceleration of the longshore flow as enters a sheltered area, with non-equilibrium flow and sediment transport profiles, and acceleration as the longshore current picks up downstream of the structures;
- strong horizontal circulations induced by the waves breaking over the LCS. This drives an onshore current over the LCS, while set-up differences in turn drive the flow away of the LCS-sheltered area. In LCSs with gaps, this can lead to strong offshore flows and associated sand losses. It is important to note that the circulation cells at the ends of LCSs have an opposite flow direction than those near emerged breakwaters, so that they generally lead to transport away from the structure;
- vertical velocity profiles that are very non-uniform due to sharp gradients in forcing by wave breaking and set-up differences;
- effect of spiral flow or «helical motion» in the strongly curved circulation patterns, where the near-bed flow is turned towards the inside of a cell and the near-surface flow is towards the outside.

13.10.2. Model classification

There is not a universal model for analysing and predicting beach evolution and its governing processes on all time and length scales involved. Instead, depending on the nature of the problem and project objectives, there is a wide range of models available, each focusing on the problem from a specific standpoint. The work by Hanson *et al.* (2003) gives a good summary of the different models available in terms of time and length scale covered (see Figure 13.50).

a) Analytical models

These models are linear approximations of the equation of shoreline or profile change, often with schematised geometry, boundary and wave conditions, Larson *et al.* (1997). Analytical solutions serve mainly as a means to identify characteristic trends in beach change through time and to investigate basic dependencies of the change on the incident waves and water levels as well as the initial and boundary conditions. As a result, analytical models typically have a longer time perspective than their numerical counterparts.

Typical length and time scales of application are on the order of tens of km and decades, respectively.

b) Morphological state models

These models predict the evolution of a small number of parameters that describe the coastal profile. In the case of beach state models (Lippmann and Holman, 1990; Wright and Short, 1984), beach states are described subjectively, based on visual observations. Predictions based on empirical relationships between observed states and measured forcing parameters have been shown to be pretty accurate, Larson *et al.* (2003).

This approach resolves time and length scales ranging from 1 month to several years and bar length to maximum surf zone width, respectively.

c) Equilibrium based models

These models assume that both the equilibrium profile shape and equilibrium shoreline orientation are known (see Section 13.9).

Profile evolution models, see Swart (1975), have the property that the chronology of the hydrodynamic forcing has negligible effects, provided that the forcing is allowed to act long enough for equilibrium to occur. Kriebel and Dean (1985) predict beach evolution as a result of cross-shore transport while longshore processes are omitted or described in a schematised fashion (Larson and Kraus, 1989; Steetzel, 1993).

This type of models is quite successful in predicting short-term events as the erosive impact of storms; however, applications for medium- and long-term predictions have been limited because of difficulties in formulating sediment transport formulas that produce reliable and robust profile evolution at these time scales.

One-line shoreline evolution models have demonstrated their predictive capabilities in numerous projects, Hanson *et al.* (1988), Hanson and Kraus (1989).

Changes in shoreline position are assumed to be produced by temporal evolution of spatial differences in the total longshore sand transport rate. Thus, this type of model is best suited to situations where there is a systematic trend in long-term change in shoreline

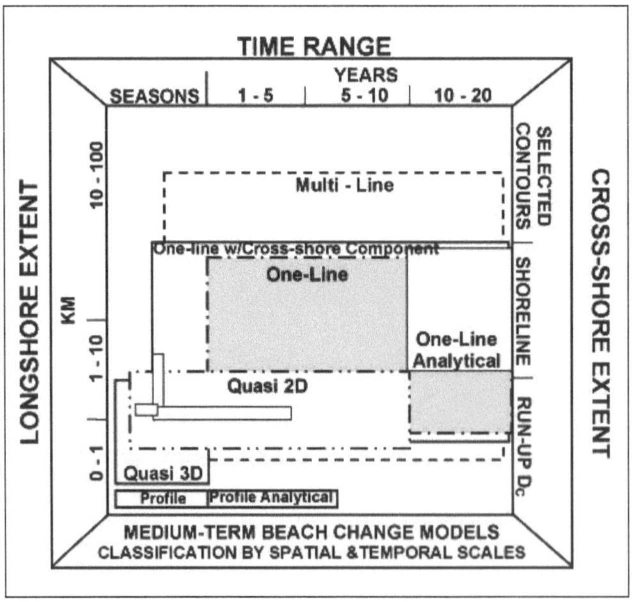

Figure 13.50. Classification of beach change models by spatial and temporal scales. From Hansom *et al.* (2003).

position, such as is the case after a LCS construction. In these models, cross-shore transport effects are assumed to cancel over a long enough simulation period, or are accounted for through external calculation. These models are well introduced in the engineering practice.

Typical time and length scales are or the order of years to decades and hundred of meters to tens of km, respectively.

Multiline models take into account the cross-shore transport schematising the profile with a sequence of mutually interacting layers, Bakker (1969), Perlin and Dean (1979).

Some recent developments have substantially increased its applicability, see Steetzel and Vroeg (1999), Hanson and Larson (1999).

The typical time and length scales of these models range from seasons to centuries and from hundred of meters to hundred of km, respectively. However, these approaches have not yet found their way into the engineering practice.

d) Process-based models

This class of models basically simulates hydrodynamics and sediment dynamics on the actual scale of the forcing, although mostly averaged over the short wave period time scale. In principle, bed updating is done on the same scale. These models account for strongly nonlinear internal dynamics, so that both effects of chronology and effects of inherent morphological behaviour may be expected.

Process-based profile evolution models have been applied in coastal engineering practice since the late 1980's, Roelvink and Brøker (1993) Schooness and Theron (1995). The application of these models is still restricted to relatively short time scales because while it seems that the first order dynamics are reasonably described by the models, there are the more subtle higher order effects which are responsible for the bed profile evolution, which becomes especially relevant when trying to simulate on longer time scales.

First procedures to apply *Process-based, beach shape models* to medium term scales have been reported in the mid 1990's, de Vriend *et al.* (1993). While initially based on depth-averaged models (2-DH) the necessity of including depth-varying effects (such as those included in profile models) has lead to quasi-3D (Q3D) models. Recent developments in the introduction of depth-varying effects, such those due to flow curvature, have led to attempts for fully 3D approaches. These models are typically composed of wave, average flow, sediment transport and bed modules. Wave and flow modules of various modelling systems do not differ significantly, but in search of a classification of transport models, the dimension of the flow model (2DH/Q3D/3D) and the dimension of the transport model have to be distinguished. Transport models of lower dimension can be applied in one context with flow models of equal or higher dimension, the other way round being rather unlikely.

13.10.3. 2DH and Q3D models

A 2DH or Q3D approach may be sufficient to adequately simulate most of the processes described in Sub-section 13.10.1; however, in order to capture the effects of long waves, a time-dependent wave- and roller-energy balance must be included in the model suite, rather than a stationary wave model as it is used more often.

Two-dimensional, depth-averaged (2DH) schemes have been developed over the past twenty years or so, see Fleming and Hunt (1976), Latteux (1980), Coeffe and Pechon (1982), Yamaguchi and Nishioka (1984), Watanabe (1985), O'Connor and Nicholson (1989), Andersen *et al.* (1991), Wang *et al.* (1992), de Vriend *et al.* (1993), Tanguy and Zhang (1994), Sato *et al.* (1995), Leont'yev (1999). Nicholson *et al.* (1997) present a comparison of the performances of different 2DH numerical models applied to a schematic configuration.

Later on, quasi-three-dimensional (Q3D) and three-dimensional (3D) schemes have been implemented, see de Vriend and Stive (1987), Briand and Kamphuis (1993), Roelvink *et al.* (1994) and more recently Zyserman and Johnson (2002) and Lesser *et al.* (2004). A review of these schemes is reported by de Vriend (1996). These Q3D approaches include a description of the vertical strtucture of the flow and the suspended sediment transport.

Coastal area morphological models integrate the waves, flow and sand transport models in order to compute the time-evolution of bed level changes at a given coastal area. The iterative procedure is well represented by Fig. 13.51 that shows the scheme of the Coastal Area Morphological Shell (CAMS) developed by DHI Water & Environment. Some of the morphodynamic models available in the literature are briefly described in the following.

MIKE 21 CAMS, developed by DHI Water & Environment, is built around standard modules of the MIKE 21 model suite (wave and current module already described in Section 13.4) and is based on an explicit forward-time integration scheme for bathymetry evolution (Zyserman and Johnson, 2002; Zyserman *et al.*, 2005). Execution is controlled by a shell, which also ensures the flow of information among the components of the modelling system.

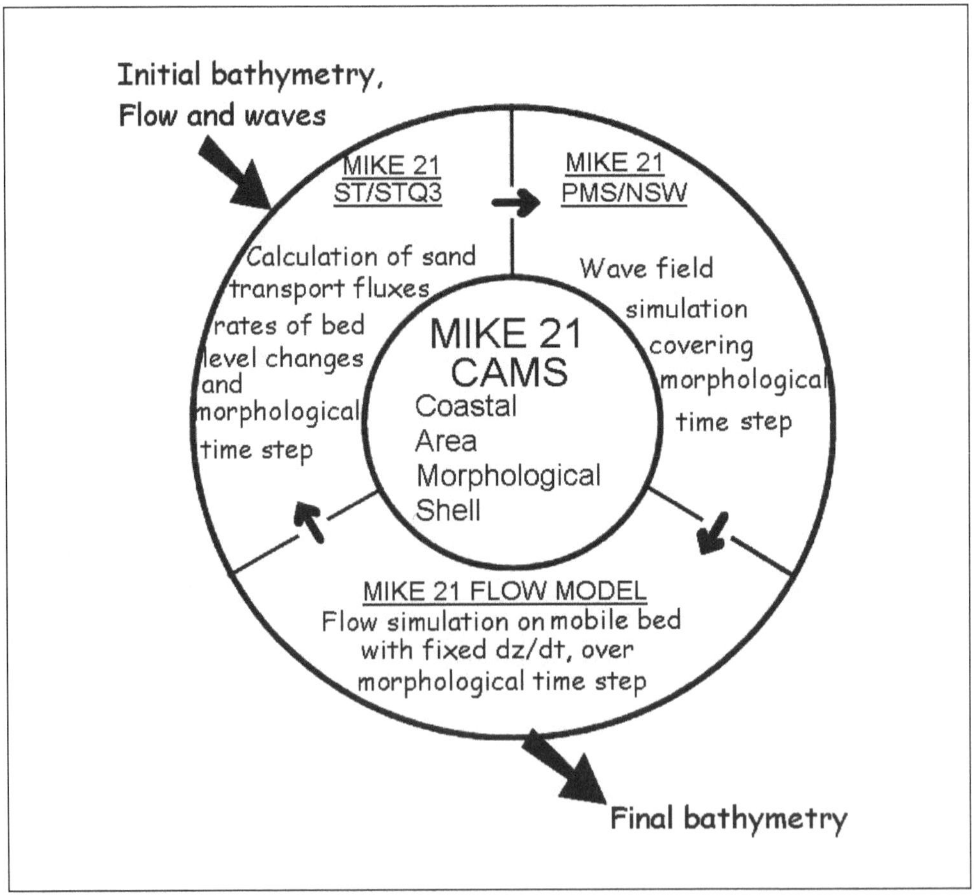

Figure 13.51. Iterative procedure of morphodynamic models.

The evolution of the model bathymetry under a number of forcing processes can be simulated as the wave, current and sediment transport fields are calculated on the updated bathymetry. The sediment transport module ST-Q3 calculates the rates of non-cohesive sediment sand transport using a Q3D approach for combined waves and current situations; it implements a deterministic algorithm based on the model of Deigaard *et al.* (1986a, b) and evaluates separately bed load transport and suspended load.

The DELFT3D package, developed by WL-Delft Hydraulics in close cooperation with Delft University of Technology, is a model system that consists of a number of integrated modules which together allow the simulation of hydrodynamic flow (under the shallow water assumption), computation of the transport of water-borne constituents (e.g., salinity and heat), short wave generation and propagation, sediment transport and morphological changes, and the modelling of ecological processes and water quality parameters. At the heart of the DELFT3D modelling framework is the FLOW module (already described in Section 13.4) that performs the hydrodynamic computations and simultaneous calculation of the transport of salinity and heat. The large number of processes included in DELFT3D-FLOW (wind shear, wave forces, tidal forces, density-driven flows and stratification due to salinity and/or temperature gradients, atmospheric pressure changes, drying and flooding of intertidal flats, etc.) mean that DELFT3D-FLOWcan be applied to a wide range of river, estuarine and coastal situations. The online sediment version allows calculation of morphological changes due to the transport, erosion, and deposition of both cohesive (mud) and non-cohesive (sand) sediments in conjunction with any combination of the above processes.

The LIMOS model, developed at the Universitat Politècnica de Catalunya (Alsina, 2005), consists of a complex formulation to obtain the sediment transport rates as a function of different flow regime considerations, and the sediment mass conservation equation to compute bottom update. The sediment transport formulation is based on Bailard's (1981) sediment transport model which was developed from the energetic arguments proposed by

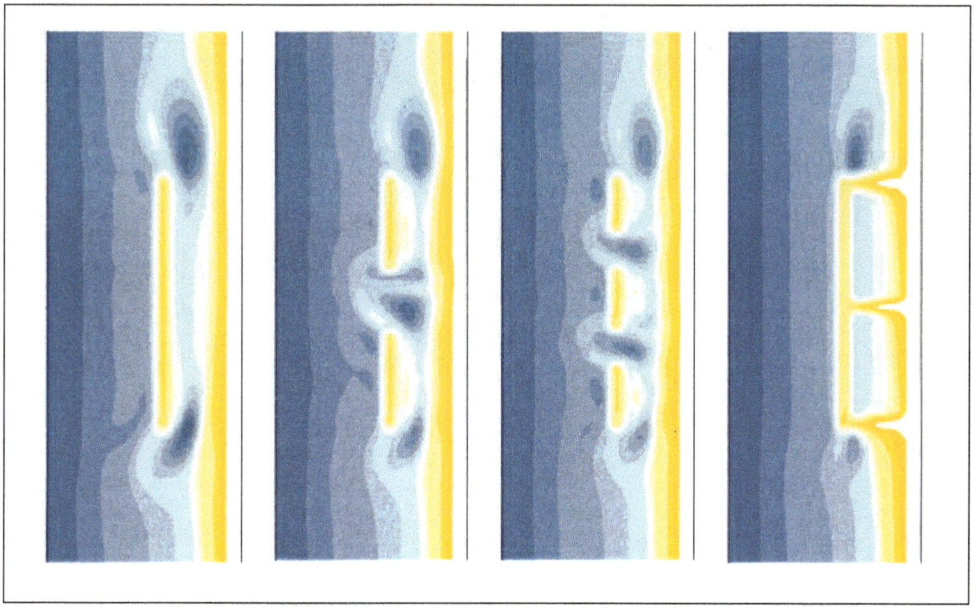

Figure 13.52. Effect of some LCS schemes on morphology after 1 year, from Lesser *et al.*, (2003).

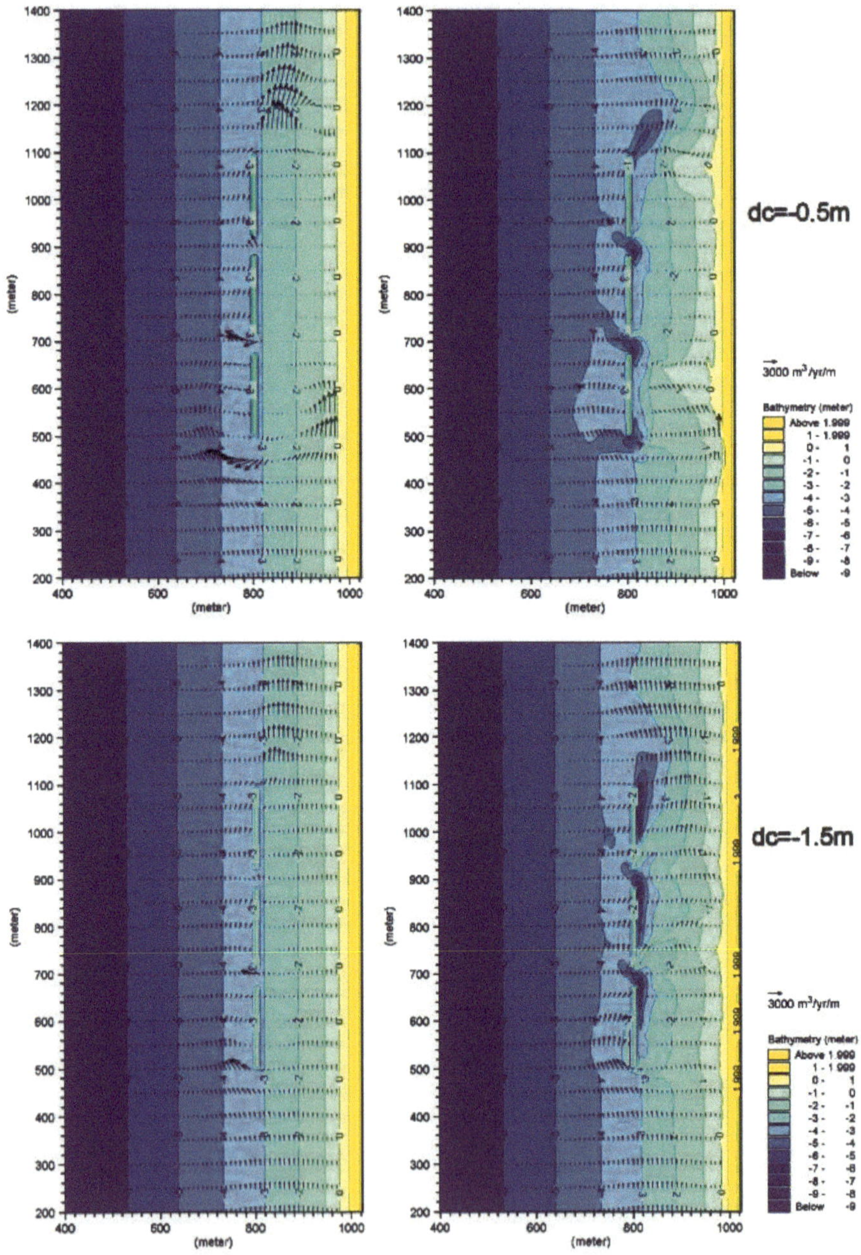

Figure 13.53. Initial bathymetries (left panel) and simulated bathymetries after 28 days morphological simulation (right panel). Vectors represent the sand transport fluxes. Still water depth above breakwater crest equals 0.5 m (top) and 1.5 m (bottom). From Zyserman *et al.* (2005).

Bagnold (1966). The general approach establishes that the work done in transporting sediment is a fixed proportion of the total energy dissipated by the flow. The code takes into account: bed-load and suspended-load transport; waves and currents, including the effects of wave asymmetry, bed slopes in arbitrary directions, among others.

Examples of the application of 2DH or Q3D models to study prototype cases can be found in the literature, e.g.: Damgaard *et al.* (2002) and Ranasinghe *et al.* (2004) examine rip currents and bar evolution at Palm Beach (Australia) and compare numerical results to data derived from video images; Cayocca (2001) performs a long-term simulation of the tidal Arcachon inlet in France; Lesser *et al.* (2004) analyze the evolution of the sea bed and adjacent coast at IJmuiden, The Netherlands.

In Lesser *et al.* (2003) and Roelvink *et al.* (2002) the DELFT 3D model was applied to analyse the sediment bypassing and sand budget of various submerged breakwater schemes over a period of one year after construction (Fig. 13.52).

Zyserman *et al.* (2005) used MIKE 21 CAMS to investigate the influence of structure freeboard on the calculated erosion patterns around submerged detached breakwaters (Fig. 13.53).

Detailed simulations of field cases are time demanding, due to both field data collection and computational time (of the same order of the simulation period) reasons; due to these reasons, more or less all the quoted works come to simplified assumptions on the wave climate or in the bathymetry used for simulations. Only a recent paper (Elias *et al.*, 2006) analyses Texel tidal inlet dynamics by running a three month simulation on a surveyed bathymetry with the morphodynamic Delft 3D code using measured waves, winds, tide and water levels as forcing.

Coastal area morphological models are thus most suitable for medium-term morphological investigations (several weeks to months) over a limited coastal area. The typical dimensions are about 10 km in the alongshore direction and 2 km in the offshore direction. The computational effort can become quite large for long-term simulations (several years), or for larger areas.

13.10.4. One and Multi-line models

One and multi-line models are useful to evaluate the long-term coastline evolution under a large number of wave/current and human intervention scenarios.

Among several models available in the literature, LITPACK, the one-line model developed by DHI Water & Environment, is described including sample results. LITPACK is composed by several modules:

- LITSTP calculates the local rates of non-cohesive sediment transport in combined waves and currents.
- LITDRIFT simulates the cross-shore distribution of wave height, set-up and longshore current for an arbitrary coastal profile. It provides a detailed deterministic description of the cross-shore distribution of the longshore sediment transport for an arbitrary bathymetry for both regular and irregular sea states. LITDRIFT calculates the net/gross littoral transport for a section of coastline over a specific design period (Fig. 13.54). Important factors, such as the linking of the water level and the profile to the incident sea state, are included.
- LITLINE simulates coastline evolution along a quasi-uniform coastline by solving a continuity equation for the sediment in the littoral zone; the influence of structures, sources and sinks are included.

– LITPROF simulates cross-shore profile evolution for oblique waves by solving the bottom sediment continuity equation, based on the sediment transport rates calculated by sediment transport model STP_Q3. LITPROF, being a time-domain model, includes the effects of changing morphology on the wave climate and transport

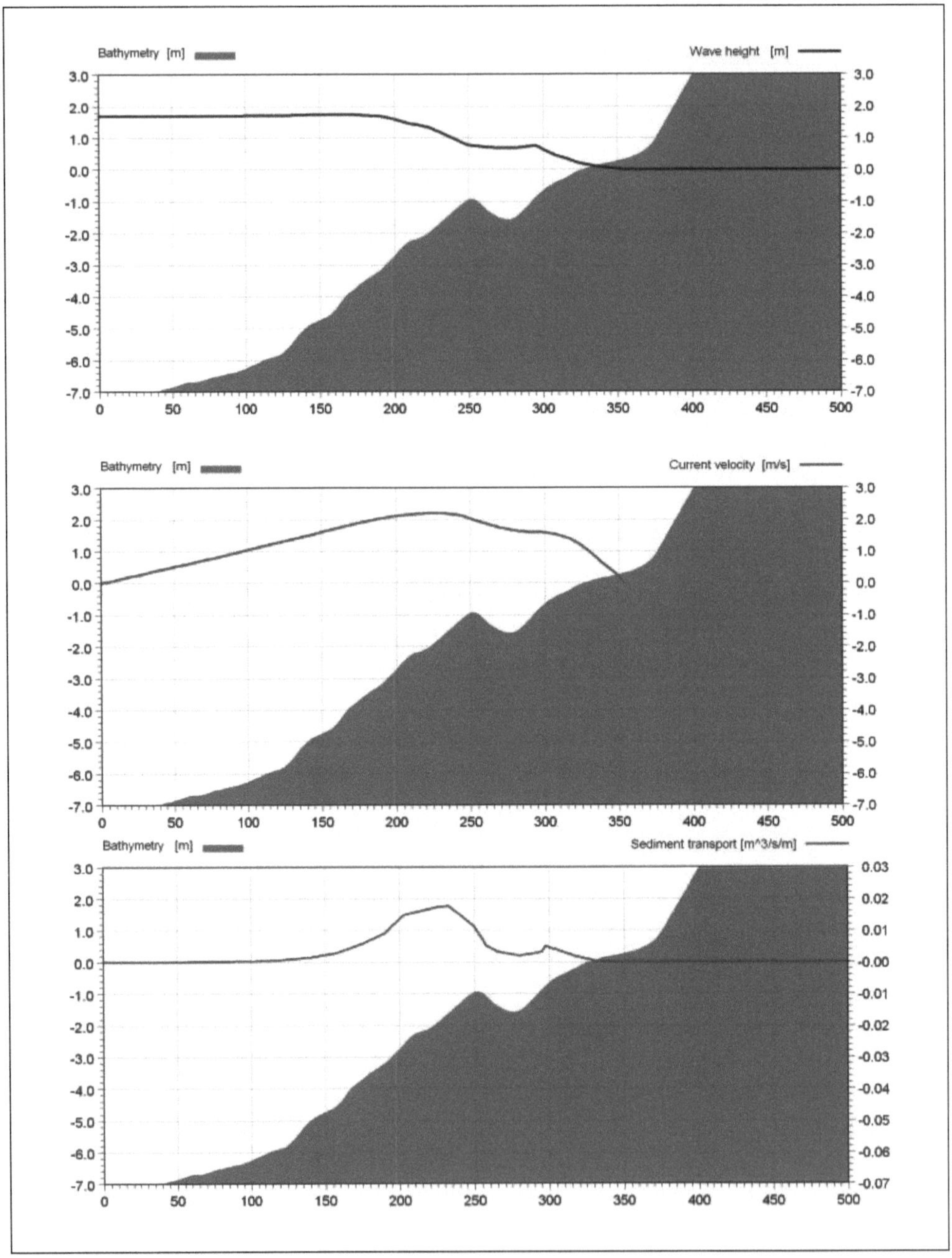

Figure 13.54. Results of simulations with LITDRIFT, from top to bottom: longshore sediment drift, longshore current velocity, cross-shore profile evolution with indication of wave height and water level.

regime; this enables a simulation of profile development for a time-varying incident wave field.

- LITTREN finds applications in areas where the suspended load is not in equilibrium with the local hydrodynamics, for example channel back-filling and intake intrusion problems. LITTREN simulates trench sedimentation accounting for non-equilibrium sediment transport in combined waves and currents; full morphological feed back between bed level change, waves, currents and sediment transport; current and wave refraction over the channel.

13.11. FORMULAE FOR STRUCTURAL STABILITY

13.11.1. Hydraulic armour layer stability

(Kramer & Burcharth, AAU)

13.11.1.1. Introduction

For conventional breakwaters only a small amount of energy is allowed to pass over or through the structure. Damage will therefore mainly happen to the front slope. For LCS wave energy can pass over the structure resulting in exposure also of the crest and the rear side. However, LCSs are generally more stable than the conventional type. Consequently smaller rubble stones can be used in the armour layer. The waves are generally depth limited and therefore higher waves occur when the water level is high, e.g. during high tide or in case of storm surge. Water level variation compared to water depth is usually relevant, therefore the worst condition for the stability of LCSs should be evaluated for all possible combinations of waves and water levels.

In 2D hydraulic model stability tests on LCSs it is very important that the set-up in the leeward side of the structure is well controlled. If not controlled overtopping waves will

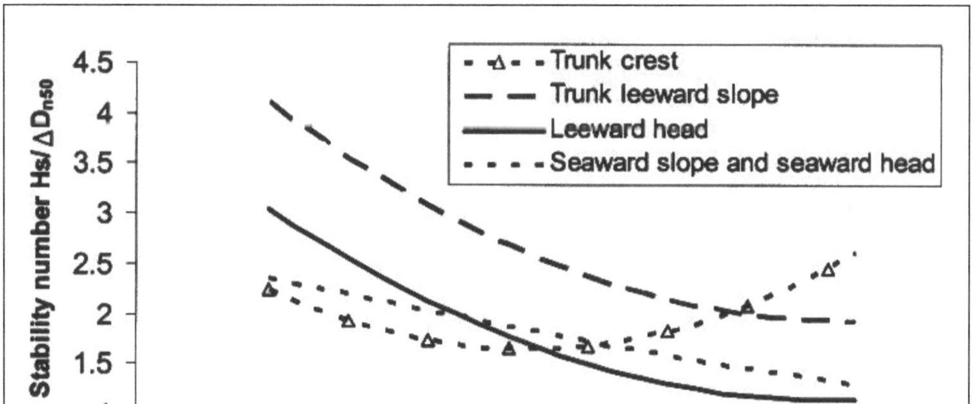

Figure 13.55. Design diagram for LCS armour stability, initiation of damage. Vidal *et al.* (1992, 1995). Non-depth limited waves.

accumulate water behind the breakwater, which will cause a backward flow over the crest and through the structure if permeable. This effect can influence the damage directly and indirectly by changing the wave breaking on and in front of the structure. Thus it should be made clear for which set-up levels the model tests are performed. In 3D test in wave basins the set-up is usually negligible due to the unhindered return flow around the heads.

13.11.1.1.1. Earlier trunk and roundhead stability tests

Several researchers have investigated trunk armour layer stability of LCSs; see e.g. Powell and Allsop (1985), Givler and Sorensen (1986), Ahrens (1987), Van der Meer (1988), and Loveless and Debski (1997). However, the most extensive work was performed by Vidal *et al.* (1992), Burger (1995) and Kramer and Burcharth (2003), which is described in more detail in the following.

Vidal *et al.* performed laboratory experiments on a complete 3D structure to investigate trunk and roundhead damage. The experiments and elaboration of results are given in Vidal *et al.* (1992), (1995) and (2000). The cross section had slopes 1:1.5 on both seaward and landward sides and a crest width of 6 D_{n50}. The waves were non-depth-limited and perpendicular to the trunk. Vidal showed that the trunk crest was the least stable part of the structure in case of submerged structures, and that the leeward part of the head was the least stable part under emergent conditions, see Figure 13.55 (parameters in the figure are defined subsequent in Sub-section 13.11.1.2).

Vidal *et al.* (1992) divided the structure into several sections in order to study the distribution of the damage. It should be noted that the definition of crest in these tests contained the upper parts of the two slopes. A steel frame was covering the surface of the structure along the sections, and a steel mesh was covering the parts where damage was not measured. Damage interactions among the sections were thereby not possible, e.g. damage to the crest section could not influence damage to the seaward slope section and visa versa. Further the steel frame restricted stones from movements along the boundaries within the sections. These effects most probably stabilized the stones making the sections in the experiments more stable than what would be the case for real structures. Vidal *et al.* (1992) also studied the response of a complete trunk section without steel mesh covering. The results are implemented in Figure 13.56.

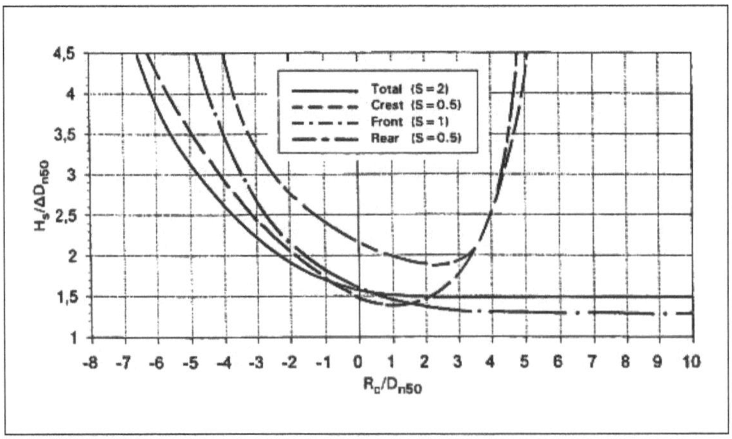

Figure 13.56. Design diagram for trunk armour stability for initiation of damage, based on tests by Delft (1988) and NRC (1992). Burger (1995). Non-depth-limited waves.

Burger (1995) performed new laboratory experiments on trunk stability and re-analysed the existing tests reported by Van der Meer (1988) and Vidal *et al.* (1992). The cross sections of Van der Meer and Burger had slope 1:2 at the seaward side and slope 1:1.5 at the landward side. The crest width was 8 D_{n50}. The waves were non-dept-limited and perpendicular to the trunk. The analysis is described in detail in Burger (1995) and is summarized in Van der Meer *et al.* (1996). The trunk was divided in seaward slope, crest and leeward slope. Related to initiation of damage stage the stability was reported both for each sector and for the total trunk sector, see Figure 13.56. From the figure it is seen that the crest is the least stable part of the trunk under submerged and slightly emergent conditions. For more emergent conditions the seaward slope is the least stable part.

13.11.1.1.2. New model tests within DELOS

The DELOS stability tests on LCSs (mainly roundhead but also trunk) were performed to supplement existing tests in order to identify the influence on rubble stone stability of:

- obliquity of short crested waves including depth limited conditions;
- wave height and steepness including depth limited conditions;
- crest width;
- freeboard.

A detailed report about the tests is available in the deliverables for the DELOS project, see Kramer *et al.* (2003). An overview of the experimental layout can be found in Kramer *et al.* (2005). In Kramer and Burcharth (2003) some recommendations for design were given. They are repeated in the following.

Table 13.11. Model characteristics for NRC, Delft and AAU tests.

Parameter	*Test facility and year*			
	NRC 1992	*Delft 1988 (trunk)*	*Delft 1995 (trunk)*	*AAU 2002*
Armour unit size D_{n50} [m]	0.025	0.034	0.035	0.033
Structure height H/D_{n50}	16.0	8.7, 11.6, 15.3	19.1	9.1
Crest width B/D_{n50}	6.0	8.0	–	3.0 and 7.6
Freeboard R_c/D_{n50}	– 2, 0, 0.8, 1.6, 2.4	– 2.9, 0, 3.6	2.0	– 3.1, – 1.5, 0, 1.5
Structure slope	1:1.5	1:2, leeward 1:1.5	1:2, leeward 1:1.5	1:2
Foreshore slope	Horizontal	1:30	Horizontal	1:20
Type of waves	2D irregular(*)	2D irregular(*)	2D irregular(*)	3D irregular
Wave direction	Head on (0º)	Head on (0º)	Head on (0º)	– 20º to + 20º
Reference	Vidal *et al* 1992	Van der Meer (1988) and Burger (1995)	Burger (1995)	Kramer *et al.* (2003)
(*) Non-depth limited waves				

The data sets described in Table 13.11 were compared in Kramer *et al.* (2003). Structure geometries, wave basin/flume layouts, stone characteristics and types of waves generated were different in all the datasets. However, when the differences are kept in mind, Kramer *et al.* (2003) concluded that all data sets are in reasonable agreement.

Major results of Kramer *et al.* (2003) are summarised in the following points.

Wave direction. All parts of the trunk are slightly more stable under oblique wave attack than under normal incidence wave attack. The stability of the roundhead sections in case of oblique waves < 0° (a large part of the head exposed to direct wave attack) is the same as for normal incidence waves. The stability of the leeward and middle part of the roundhead in case of oblique waves > 0° (when a large part of the head is in lee of direct wave attack) is the same as for normal incidence waves, but the area of damage shifts towards the middle part of the head. During the experiments it was experienced that wave breaking tends to focus at the roundhead forming a jet of water slamming down on the top part of leeward head. This effect shifted towards the middle head in case of oblique waves making the middle head more prone to damage.

Wave steepness. The investigation showed that the damage data for s_{0p} = 0.02 and s_{0p} = 0.035 were fairly close. However, the series with s_{0p} = 0.02 (long waves) tend to give slightly more damage than series with s_{0p} = 0.035 (short waves) meaning the structure is more stable for s_{0p} = 0.035.

Crest width. No significant difference in response could be identified for the tested crest widths indicating that for the tested range the influence of crest width was small.

Freeboard. The tests showed that stability is highly influenced by freeboard.

Structure slopes. Only a structure slope of 1:2 was tested in the DELOS tests. The results

Table 13.12. Sections prone to damage. Filled black areas indicate exposed stones.

Freeboard	*Damage to trunk*	*Damage to roundhead*
$R_c > 0$ Slightly emergent crest	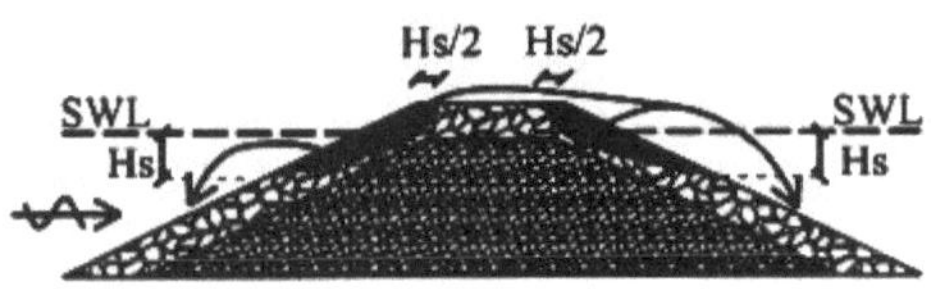	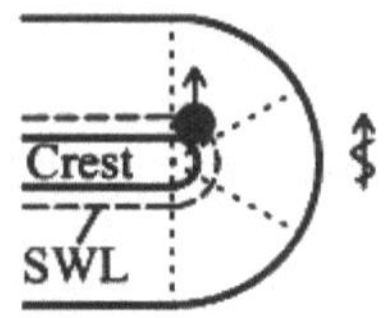
$R_c = 0$	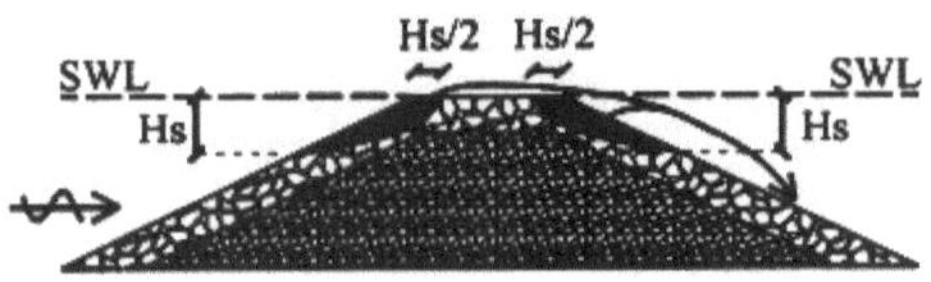	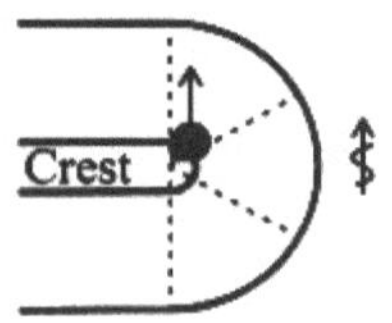
$R_c < 0$ submerged crest	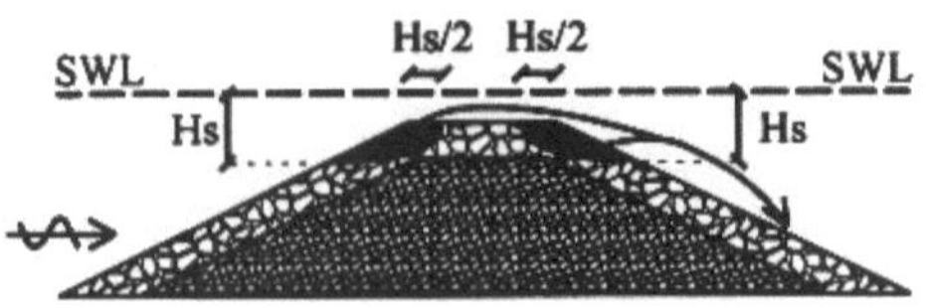	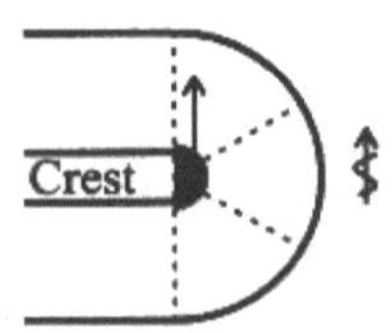

can therefore only be applied for structures with slopes 1:2. There were too many other differences between the NRC 1992 tests (slope 1:1.5) and the AAU 2002 tests (slope 1:2) to assess the influence of the slope.

Kramer and Burcharth (2003) described the exposed areas of the breakwater as given in Table 13.12. The information in the table is important if there is a wish for optimization by using different stone sizes in the different parts of the armour layer.

In the AAU 2002 tests the trunk and the roundhead were divided in different sections and damage was measured within each section, see Figure 13.57. Narrow LCSs built in shallow water are only a few stones high and wide. One stone removed from the edge of the crest will cause a large hole in the cross-section. When one section reached the initiation of damage stage it was therefore chosen to define the whole structure to be in this stage. In Figure 13.58 (left) a line representing the lower limit of the test results is given. This line represents the least stable part of the structure. The function for the line is given below by Eq. (13.111). If the highest waves are depth limited then the significant wave height can be replaced by the approximation $H_s = 0.6 \cdot h$ (h is water depth). By inserting in Eq. (13.111) $\rho_r = 2.65$ t/m^3 corresponding to $\Delta = 1.6$, and $H_s = 0.6 \cdot h$ the curves in Figure 13.58 (right) are obtained. Under breaking wave conditions, increasing water level increases wave load and the damage to the structure, until submergence reaches condition $R_c = -0.36 \cdot H_c$. Further water level increase will cause a dominant self protection of the structure by submergence. The R_c/H_c relation is used in Eq. (13.111) to calculate the required D_{n50} and the following rule of thumb is found: $D_{n50} = 0.3 \cdot H_c$.

If the saturation values $H_s/h \neq 0.6$, a similar procedure can be applied. Eq. (13.111) together with $\Delta = 1.6$ is used to evaluate the worst water level condition. The relative freeboard R_c/H_c is strongly dependent on the chosen saturation value. An increase in this

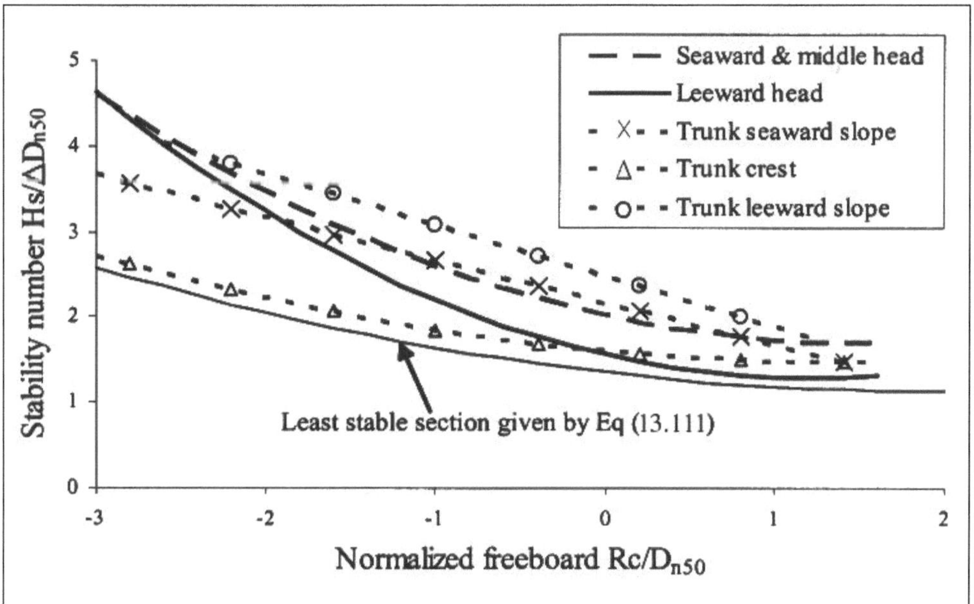

Figure 13.57. Design diagram for LCS armour stability, initiation of damage. Kramer *et al.* (2003). Depth limited waves.

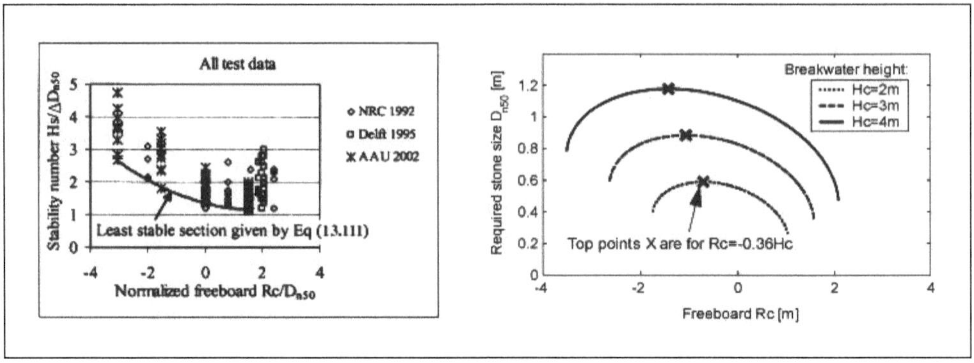

Figure 13.58. Design graphs for stability of low crested breakwaters corresponding to initiation of damage. Test results (left) and formula in case of depth limited waves (right).

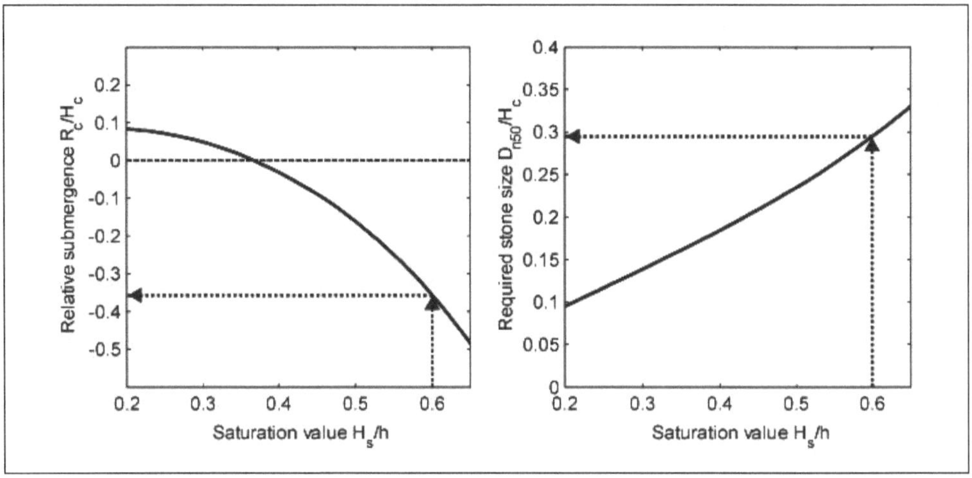

Figure 13.59. Design graphs according to Eq. (13.111). The arrows indicates depth-limited conditions with $H_s/h = 0.6$. Left: relative submergence corresponding to minimum stability. Right: required stone sizes corresponding to minimum stability.

value will allow higher waves in shallow water giving minimum stability for a larger submergence. This effect is shown in Figure 13.59 (left). The required stone size corresponding to the worst relative submergence can be found from Figure 13.59 (right).

13.11.1.1.3. Comparison of new and existing design curves

The AAU 2002 experiments showed basically the same overall behaviour as the NRC 1992 tests, i.e. the trunk crest was the least stable part under submerging conditions, and the leeward part of the roundhead was the least stable part in case of emergent conditions. If the same stone type is used in all sections the following rules for design can be given.

- $R_c \leq 0$, submerged conditions. The crest is the least stable part, the more submerging the more stable. Existing 2D tests and formulae for trunk armour layer stability of LCSs can be used in the design of the armour layer for the whole structure.
- $R_c > 0$, emergent conditions. Leeward part of the roundhead is the least stable, the more emergent the less stable. It is therefore on the safe side to design the roundhead

according to existing knowledge about stability of roundheads for non-overtopped breakwaters.

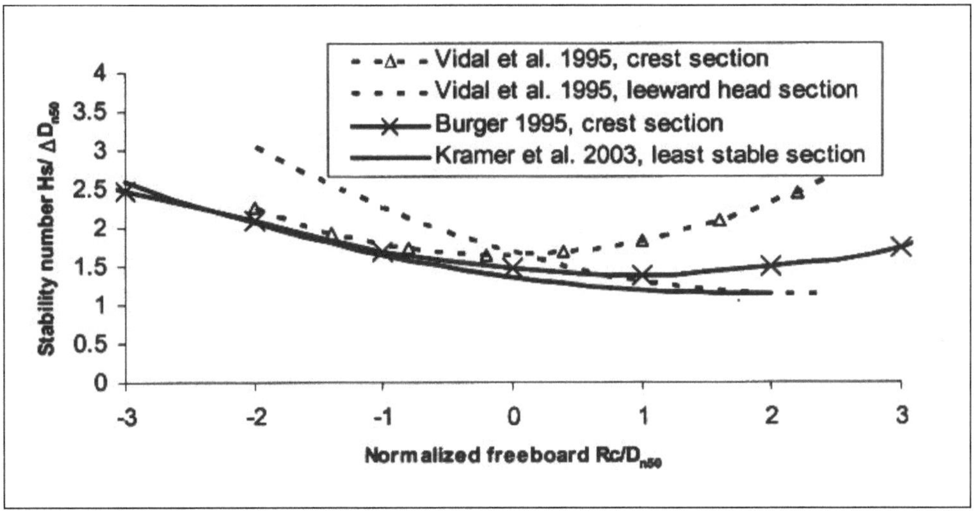

Figure 13.60. Comparison of design curves for armour damage, initiation of damage.

The design curves for the least stable sections given by Vidal *et al.* 1995 (design curves for leeward head and crest given in Figure 13.55), Burger 1995 (design curve for crest damage shown in Figure 13.56), and Kramer *et al.* 2003 (design curve for least stable section given in Figure 13.58) are compared in Figure 13.60.

The design curves shown in Figure 13.60 are in good agreement. For submerging conditions ($R_c/D_{n50} < 0$) the design curves given by the 3 researchers for the crest follows each other giving the same stability number for a certain freeboard. Under emergent conditions ($R_c/D_{n50} > 0$) the curves for the leeward head by Vidal *et al.* (1995) and Kramer and Burcharth (2003) gives approximately the same stability number. Design by the single formula provided by Kramer and Burcharth (2003) will therefore be safe.

13.11.1.2. Recommendations for design of armour layer

It is recommended to choose a crest width at least equal to the largest significant wave height. The crest width should correspond to at least three stones. If the structure is expected to be exposed to oblique wave attack the same rock type should be applied in the whole roundhead. Anyway, for LCSs it is usually chosen to use stones in the trunk and the roundhead of the same size. In this case design can be done according to Eq. (13.111) or Eq. (13.112). If it is chosen to use only one stone size (no core, i.e. homogeneous cross-section) design by Eq. (13.111) and Eq. (13.112) given below will be conservative. As LCSs are low the use of fairly gentle slopes does not increase the total required quantity of material significantly. It is therefore recommended to use 1:2 slopes or even gentler slopes. For gentler slopes the structure will be more stable than given by Eq. (13.111) or Eq. (13.112).

13.11.1.2.1. Rock shape and grading

Burger (1995) and Van der Meer *et al.* (1996) investigated the influence of rock shape and grading on the stability of a slightly emerged low-crested breakwater and concluded that the influence was very small, especially for low damage levels. A rock type with relatively many elongated/flat rocks showed a similar stability as more uniform rock types. No influence was found for gradings D_{85}/D_{15} smaller than about 2, but it was recommended not to use gradings with $D_{85}/D_{15} < 2.5$. The conclusion was further to release customary strict restrictions on shape or grading of armour material during construction.

13.11.1.2.2. Required stone size in shallow water waves

When designing a low crested breakwater the highest significant wave heights must be calculated for different water depths caused by tide and storm surge. The corresponding necessary stone sizes for each of these water depths can then be found from the Figures 13.55 to 13.60. In this way the «worst condition» will be the water depth giving the largest stone size. It is recommended to choose the stone size according to the lower line shown in Figure 13.58 (left) given by Eq. (13.111).

$$\frac{H_s}{\Delta D_{n50}} = 0.06\left(\frac{R_c}{D_{n50}}\right)^2 - 0.23\frac{R_c}{D_{n50}} + 1.36, \quad for\ -3 \le R_c / D_{n50} < 2 \qquad (13.111)$$

In Eq. (13.111) H_s is the significant wave height, R_c is the freeboard (negative if submerged), D_{n50} is the mean nominal diameter of the armour, and $\Delta = (\rho_r - \rho_w)/\rho_w$, where ρ_r and ρ_w are the densities of rock and water, respectively. An example of the use of Eq. (13.111) is shown in Figure 13.58 (right) and Figure 13.59.

The validity of the formula is examined through all the parameters involved.

Freeboard. The formula is only valid for relatively low freeboards given by the ranges in Eq. (13.111). For more emergent structures design according to the upper limit of Eq. (13.111) is most likely sufficient, or existing formulae for roundhead stability of non overtopped breakwaters can be used. The upper limit of Eq. (13.111) is $R_c/D_{n50} = 2$ corresponding to a stability number of $H_s/\Delta D_{n50} = 1.14$, which in terms of stone size is $D_{n50} = H_s/1.14\Delta$.

Wave obliquity. The formula is safe to apply also in case of oblique wave attack. The tests by Kramer *et al.* (2003) showed that wave directions in the range – 20º to + 20º leads to a slightly larger stability. However, the increase did not justify for a reduction in the necessary rock size within the tested range of obliquities.

Wave steepness. The formula is tested for fairly long waves ($s_{op} = 0.02$) and rather short waves ($s_{op} = 0.035$). If extremely long waves are expected design by Eq. (13.111) may underestimate the necessary stone size.

Stone-type. The formula is only valid for armour material consisting of quarry rock.

Layers. A two-layer fairly permeable rubble mound structure was tested. However, it is safe to use the formula for design of homogeneous structures. For multilayered or impermeable rubble mound structures caution should be taken if Eq. (13.111) is used to design the armour.

Slopes. The breakwater should be built with slopes not steeper than 1:2. Breakwaters with less steep slopes are more stable and design by Eq. (13.111) will therefore be safe.

Crest-width. The formula is developed for narrow-crested breakwaters (crest widths less than approximately $10D_{n50}$).

Trunk/roundhead differences. The formula is based on the assumption that the same stone size and type will be used in all armouring parts of the breakwater. If there is a wish for optimizations by using different stone sizes in the different outer sections of the breakwater, design can be done according to the Figures 13.55, 13.56, and 13.57. In this case important information about the location of the most exposed areas can be seen in Table 13.12.

13.11.1.2.3. Required stone size in depth limited waves

If the highest waves are depth limited and regular rock are used then Kramer and Burcharth (2003) showed that submerging conditions are the most critical. In this case Eq. (13.111) is reduced to Eq. (13.111) and the required D_{n50} can be estimated by the following rule of thumb:

$$D_{n50} = 0.3 \cdot H_c,\ H_c \text{ is the structure height} \qquad (13.112)$$

The rule of thumb is valid for breaking wave conditions with $H_s/h = 0.6$. According to Eq. (13.112) the structure height will be no more than 3 to 4 D_{n50}, which is very typical for existing LCSs. For other H_s/h values Figure 13.59 can be used in the design.

If the structure is emerged under design conditions the upper limit of Eq (13.111), corresponding to $D_{n50} = H_s/1.14\,\Delta$, is most likely sufficient for design. By inserting $\rho_r = 2.65$ t/m^3 corresponding to $\Delta = 1.6$ and the approximation $H_s = 0.6 \cdot h$, the required stone size is $D_{n50} = 0.33 \cdot h$.

Table 13.13. Design conditions.

Structure height	*Freeboard at MSL*	*Design freeboard and water depth*	*Design waves*	*Design tool*
$H_c < \sim 4$ m	Slightly emerged to slightly submerged	Worst condition is for $R_c/H_c \cong -0.3$ if obtainable. Typically the highest design water depth is the worst condition.	Depth limited	Rule of thumb (13.112)
	Very submerged ($R_c/H_c < -0.4$)	Worst condition is for $R_c/H_c \cong -0.3$ if obtainable. Typically a frequently occurring low water level or even the lowest design water depth is the worst condition.	Depth limited	Rule of thumb or if very submerged Eq. (13.111)
$H_c > \sim 4$ m	Very emerged structures	*Not a low crested structure*		
	Slightly emerged to slightly submerged	Worst condition is usually for the highest design water level.	The design waves may not be fully depth limited ($H_s/h < 0.6$)	Eq. (13.111)
	Very submerged ($R_c/H_c < -0.4$)	*Structure does not exist. However* Eq. (13.111) *may still be used for design, e.g. artificial reefs.*		

13.11.1.2.4. Design conditions: waves and water levels

Table 13.13 is based on the knowledge about existing structures (see Table 13.14), the behaviour of Eq. (13.111) and the rule of thumb (13.112).

Table 13.14. Existing EU breakwater designs. RoT is «Rule of Thumb». From Burcharth *et al.*, (2006).

Breakwater	*Armour size D_{n50} [m]*	*Structure height H_c [m]*	*Freeboard R_c (MSL) [m]*	*Water depth h (MSL) [m]*	$\frac{H_c}{D_{n50}}$	*Satisfies*	
						RoT	*Eq. (13.111)*
DK, Lønstrup	0.80	2.3	+ 1.3	1.0	2.9	√	√
DK, Skagen	0.71	2.0	+ 1.0	1.0	2.9	√	√
GR, Lakopetra	1.00	4.0	+ 0.7	3.3	4.0	÷ [(1)]	√
GR, Alaminos	1.10	3.5	+ 0.5	3.0	3.1	√	√
GR, Paphos	1.40	4.5	– 0.3	4.8	3.2	√	√
UK, Elmer	1.45	6.0	+ 4.3	1.7	4.1	÷ [(2)]	√
UK, Monk's Bay	1.31	3.7	+ 2.2	1.5	2.8	√	√
ES, Altafulla	1.31	4.5	+ 0.5	4.0	3.4	√	√
ES, Comin	0.87	3.0	+ 0.5	2.5	3.4	√	√
ES, Postiguet	0.57	2.0	– 2.0	4.0	3.5	√	√
ES, Palo	0.91	2.8	– 1.5 to – 2.0	4.3 to 4.8	3.1	√	√
IT, Punta Marina	0.90	2.8	– 0.2	3.0	3.1	√	√
IT, Lido di Dante	0.80	2.5	– 0.5	3.0	3.1	√	√
IT, Cesenatico	0.90	2 to 2.5	– 0.5	2.5 to 3.0	2.2 to 2.8	√	√
IT, Ostia (1990)	0.65	2.5	– 1.5	4.0	3.9	÷	÷ [(3)]
IT, Ostia (2003)	0.90	3.0	– 1.0	4.0	3.3	√	√
IT, Sirolo	0.90	2.5 to 4.0	– 1.0	3.5 to 5.0	2.8 to 4.4	÷	÷ [(4)]
IT, Scossicci	0.99	4.20	– 1.0	5.20	4.2	÷	÷ [(4)]
IT, Grottammare	0.90	1.6	– 0.9	2.5	1.8	√	√
IT, Bisceglie	1.04	2.55 to 4.15	– 0.15	2.7 to 4.3	2.5 to 4.0	(÷) [(5)]	√
IT, Nettuno	0.86	2.5	– 0.5	3.5	2.9	√	√
IT, Amendolara	1.36	2.3	– 0.5	2.8	1.7	√	√
IT, Pellestrina	0.76	2.5	– 1.5	4.0	3.3	√	√

Notes:

(1) GR, Lakopetra: $H_{s,\ design}$ = 2.4 m occurring during the design water depth $h \cong 4$ m corresponding to approximately zero freeboard. For this event Ns=1.4, which satisfies equation (13.111).

(2) UK, Elmer: Extreme high water depth h=5.4m corresponding to freeboard $R_c = +\ 0.6$ m. The maximum significant wave height is estimated as $H_s = 0.6 * h = 3.2$ m corresponding to $N_s = 1.4$. This is slightly more than the stability number calculated by equation (13.111). The Elmer structures have gentle slopes of 1:2.5 and wider roundheads, which makes the structures more stable than calculated by (13.111).

(3) IT, Ostia: Over a decade (1990-2003) reshaping was experienced resulting in crest lowering of about 0.5 m. Damage to the structures was in the range 4% to 25%. In 2003 the structures were therefore recharged and raised to $R_c = -\ 1.0$ with larger rocks. The 1990 breakwaters did not satisfy the rule of thumb.

(4) IT, Sirolo and Scossicci: Damage to some structures experienced. Some structures have been rebuilt. The breakwaters does not satisfy the rule of thumb and equation (13.111).

(5) IT, Bisceglie. $H_{s,\ design}$ = 2.8 m occurring during the design water depth $h = 5.1$ m corresponding to freeboard $R_c = -\ 1.0$. For this event $N_s = 1.6$, which satisfies equation (13.111).

High structures (slightly emerged to slightly submerged) cannot get a large relative submergence ($H_c/D_{n50} < -0.3$) and the rule of thumb does not apply. Instead the equation (13.111) should be used.

13.11.1.3. Validation of stability formulae with prototype experience

The rule of thumb and Eq. (13.111) have been validated with information about the breakwaters described in Table 13.14 and a good agreement was found. All breakwaters in the DELOS inventory for which the required parameters were available have been included in the list. For further information about the DELOS inventory see Lamberti *et al.* (2005). In three cases armour damage was experienced (Table 13.14: *IT Ostia 1990* (slope 1:5), *IT Sirolo, IT Scossicci*). This is in agreement with the formulae as these three cases do not satisfy Eq. (13.111).

When no notes about damage are given the structures have not showed any sign of damage.

For the low structures ($H_c < 4$ m) the same rock type, crest width and slopes are used in trunk and roundhead sections. Design condition is depth limited waves under submerged conditions, which in most cases corresponds to the highest design water level. For the submerged ($R_c \leq -1$ m) and very low ($H_c \leq 3$ m) structures the design water depth is during normal water level conditions or even for the lowest design water level. This is for example the case for *ES Paolo*, for which $h_{\text{design, lowest}} = 3.8$ m.

For the high structures ($H_c \geq 4$ m) wider crests and/or less steep slopes are used in the roundhead. This is the case for *UK Elmer, GR Lakopetra*, and *GR Paphos*. At *ES Altafulla* a wider roundhead with larger rocks were used.

13.11.1.4. Residual stability and damage development

The following formulae were based on laboratory tests with 2D-irregular, head-on waves. Real LCSs will usually be designed for depth-limited 3D-waves, which are more damaging to the structure. The following formulae are therefore expected to underestimate the required rock-size, and caution should therefore be taken if the formulae are used for design in such conditions. However, the formulae are very useful to evaluate the residual stability if some reshaping and crest-lowering of the breakwater is allowed.

The damages experienced to the Ostia breakwaters in Italy (see Table 13.14) are in agreement with the predictions by the formula by Van der Meer (1991), see Lamberti *et al.* (2005).

13.11.1.4.1. Van der Meer (1990) formula, reef breakwaters

The formula was established for the trunk of low-crested reef homogeneous breakwaters. The formula was based on laboratory tests with 2D-irregular, head-on waves.

The equilibrium height of the structure (irregular, head-on waves) is:

$$h_c = \sqrt{\frac{A_t}{\exp(aN^*_s)}} \text{ with a maximum of } H_c \tag{13.113}$$

where A_t area of initial cross section of structure
h water depth at toe of structure
H_c initial height of structure

N_s^* = spectral stability number, $N_s^* = \frac{H_s}{\Delta D_{n50}} s_p^{-1/3}$

s_p = wave steepness

$$a = -0.028 + 0.045\frac{A_t}{H^2_t} + 0.034\frac{H_c}{h} - 6\cdot 10^{-9}\frac{A^2_t}{D^4_{n50}}$$

Data source: Ahrens (1987), van der Meer (1990).

No ranges of the parameters in Eq. (13.113) were given by Ahrens or Van der Meer. However, Eq. (13.113) seems only to be valid for fairly narrow structures. This is explained further. For structures with wider crests (i.e. larger area A_t) the required stone size is larger, given that the crest lowering is fixed. This is not in agreement with the physics (a wider structure should be at least as stable as a narrow one). Van der Meer tested a structure with $0.5 \leq B/H_c \leq 1$ (B is crest width). It is therefore assumed that the equation is only valid for fairly narrow structures as indicated by the shape of the sketch in Figure 13.61.

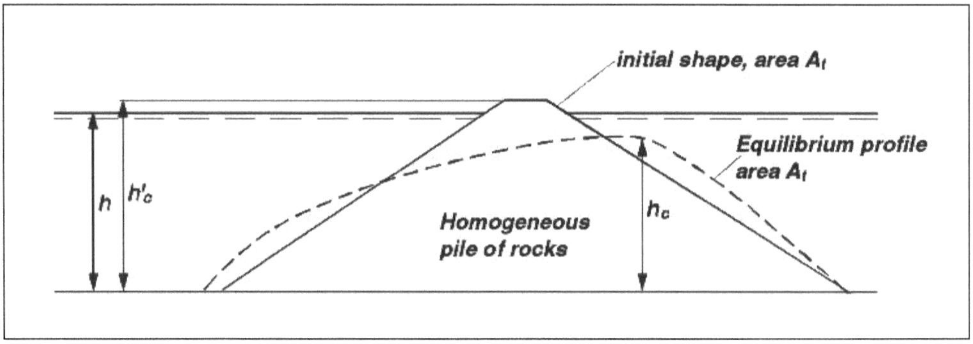

Figure 13.61. Definition sketch for reshaping reef breakwaters.

13.11.1.4.2. Van der Meer (1991) formula, submerged breakwaters

The formula was established for the trunk of submerged breakwaters with two-layer armour. The formula was based on laboratory tests with regular and some 2D-irregular, non depth-limited, head-on waves.

$$\frac{H_c}{h} = (2.1 + 0.1S)\exp(-0.14N_s^*) \qquad (13.114)$$

where h water depth
H_c height of structure over sea bed level
S relative eroded area

Data source: Givler and Sorensen (1986): regular head-on waves, slope 1:1.5
van der Meer (1991): irregular head-on waves, slope 1:2.

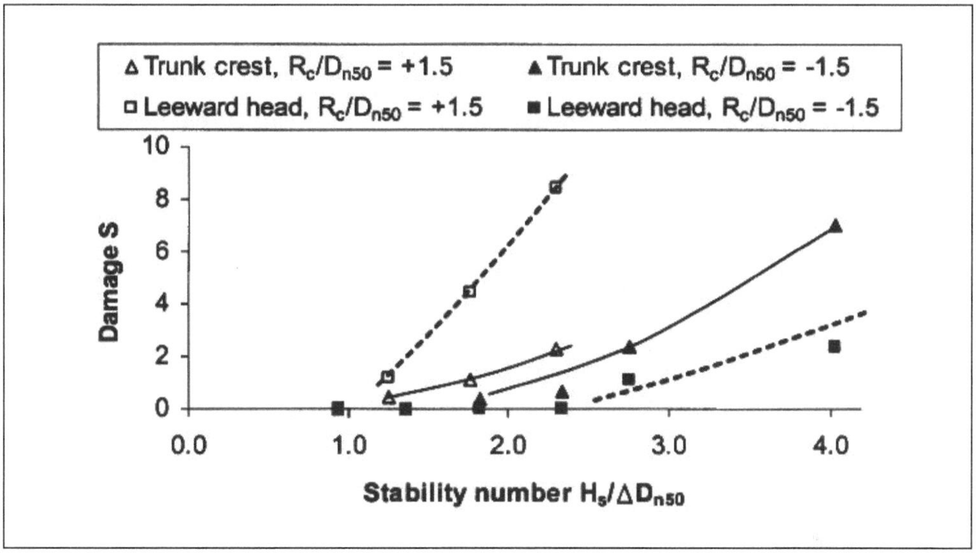

Figure 13.62. Typical example of damage development. Markers are test results. The lines indicate the trend of the data; dashed lines are for leeward head and full lines are for trunk crest. Tests by Kramer *et al.* (2003).

13.11.1.4.3. Typical example of damage development in trunks and roundheads, Kramer et al. (2003)

Kramer *et al.* (2003) showed that the leeward part of the roundhead is the most exposed part of the breakwater for emerged conditions. For submerged conditions the trunk crest is the most exposed part. An example of the test data for emerged conditions (R_c/D_{n50} = +1.5) and submerged conditions (R_c/D_{n50} = −1.5) is shown in Figure 13.62. The test results shown are for head-on 3D waves with s_p = 0.02.

From Figure 13.62 it is seen that the structure is most vulnerable under emerged conditions as the unfilled markers in the figure corresponds to larger damage than the filled markers. Further it is observed that the leeward head is the most exposed part for emerged conditions but the most stable part for submerged conditions. For emerged conditions the progress of the damage of the leeward head is much more rapid than for the trunk crest (the slope of the left line in the figure is much steeper than the others), meaning the difference in stability numbers between initiation of damage and complete destruction is small. For emerged conditions the selection of proper safety margins for the roundhead is therefore important as exceedance may lead to quick destruction. If design condition is for submerged conditions then less strict safety factors are necessary.

The result is well in agreement with the way existing LCSs are designed. From Table 13.14 it was concluded that low regularly overtopped breakwaters have the same rock type, crest width and slopes in trunk and roundhead sections. For the high emerged breakwaters wider crests, larger rocks and/or less steep slopes are used in the roundhead.

13.11.1.4.4. Example of required stone size according to the formulae and diagrams

In Table 13.14 it is seen that the height of a typical LCS cross-section is about H_c = 2 to 4 m. In this example a cross-section height H_c = 3 m, slopes 1:2 and a crest-width of 3 m is used. Rock with submerged density Δ = 1.6 is applied. Two conditions with depth-limited wave attack are investigated:

Table 13.15. Example of required stone size according to armour stability formulae for a typical structure with height $H_c = 3$ m. Depth-limited waves. ID is Initiation of Damage.

Formula	*Damage*	*Required stone size* *Zero freeboard condition* *(h = 3.0 m)*	*Submerged condition* *(h = 4.0 m)*
Rule of thumb	ID	0.90	0.90
Equation (13.111)	ID	0.83	0.88
Burger (1995)	$S = 2$	0.70	0.83
Vdm (1991), formula	$S = 0$ $S = 2$ $S = 5$	0.78 0.70 0.61	0.75 0.69 0.62
Vdm (1990), formula	$h_c = H_c$ $h_c = 0.9\ H_c$ $h_c = 0.8\ H_c$	0.53 0.45 0.38	0.67 0.56 0.47
Vidal (1992), trunk	$S = 1{,}5$ $S = 2{,}5$ $S = 6{,}5$	0.70 0.60 0.45	0.70 0.60 0.45

1) Water depth $h = 3$ m corresponding to zero freeboard.
2) Water depth $h = 4$ m corresponding to freeboard $R_c = -1.0$ m (submerged conditions).

The question is: what is the required stone size according to the formulae to resist the conditions?

The significant wave height is estimated as $H_s = 0.6 \cdot h$, and a wave steepness $s_p = 0.02$ is used in the Van der Meer (1990), (1991) formulae.

From the example given in Table 13.15 the following can be concluded.

- According to the van der Meer 1990 formulae a smaller stone size can be used if a homogeneous cross-section is used.
- If some reshaping resulting in crest lowering is allowed the required nominal stone diameter can be reduced by 20-40%.
- The required stone size by the different methodologies varies significantly. The trend seems to be that formulae developed mainly by use of regular non depth-limited 2D waves gives the smallest required stone size, whereas the formulae developed with 3D irregular depth-limited breaking waves leads to the largest required stone size.

The tests with non depth-limited 2D waves is expected to lead to an underestimation of the required rock size for the conditions in Table 13.15. It is therefore recommended to use the results from Table 13.15 only for comparisons to evaluate residual stability and not for design of LCSs in depth-limited 3D waves.

13.11.2. Bedding layer and geotextiles

(Kramer & Burcharth, AAU)

Subsidence of the armour into the sea bed is prevented by a bedding layer and/or geotextiles. A bedding layer helps to distribute the structure's weight over the underlying base material to provide more uniform settlement. Granulated filters are commonly used as a bedding layer on which a coastal structure rests. It is advisable to place coastal structures on a bedding layer (along with adequate toe protection) to prevent or reduce undermining and settlement. When rubble structures are founded on cohesionless soil, especially sand, a bedding layer should be provided to prevent differential wave pressures, currents, and groundwater flow from creating an unstable foundation condition through removal of particles. Even when a bedding layer is not needed in the completed structure, bedding layers may be used to prevent erosion during construction to distribute structure weight or to retain and protect a geotextile filter cloth.

Placing large armour stones or riprap directly on geotextile filter cloth is likely to puncture the fabric either during placement or later during armour settlement. Placing a bedding layer over the geotextile fabric protects it from damage. In this application there is more flexibility in specifying the bedding layer stone gradation because the geotextile is retaining the underlying soil.

13.11.2.1. Bedding layer design

To prevent loss of the bedding layer by leeching through the cover layer, the so called «piping criterion» given by Eq. (13.115), should be satisfied.

$$\frac{D_{15(\text{cover})}}{D_{85(\text{bedding})}} < (4 \text{ to } 5) \tag{13.115}$$

Adequate permeability of the bedding layer is needed to reduce the hydraulic gradient across the layer. The accepted permeability criterion is:

$$\frac{D_{15(\text{cover})}}{D_{15(\text{bedding})}} > 5 \tag{13.116}$$

If the bedding layer material has a wide gradation, there may be loss of finer particles causing internal instability. Internal stability requires:

$$\frac{D_{60\,(\text{bedding})}}{D_{10\,(\text{bedding})}} < 10 \tag{13.117}$$

Bedding layer thickness should be at least two to three times the size of the larger quarry stones used in the layer, but never less than 30 cm thick to ensure that bottom irregularities are completely covered. Considerations such as shallow depths, exposure during construction, construction method, and strong hydrodynamic forces may dictate thicker layers, but no general rules can be stated. For deeper water the uncertainty related to construction often demands a minimum thickness of 50 cm.

In designs where a geotextile fabric is used to meet the retention criterion, a covering layer of quarry spalls or crushed rock (10 cm minimum and 20 cm maximum) should be placed to protect against puncturing by the overlying stones. Recommended minimum bedding layer thickness in this case is 60 cm, and filtering criteria should be met between the bedding layer and overlying stone layer.

If geotextile is not applied, the bedding layer must, similar to Eq. (13.115) and Eq. (13.116), satisfy the filter rules:

$$\frac{D_{15(\text{bedding})}}{D_{85(\text{in situ soil})}} < (4 \text{ to } 5) \text{ and } \frac{D_{15(\text{bedding})}}{D_{15(\text{in situ soil})}} > 5 \qquad (13.118)$$

The use of Eq. (13.118) is illustrated in Figure 13.63.

Due to the limited structure height of typical LCSs there is not enough space to separate coarse materials from sea bed sand if the conventional filter criteria for stone filter layers should be satisfied. However, the internal stability rule can, at least conceptually, be applied repeatedly if the amount of materials in the bedding layer is sufficiently controlled. This is suggested for instance in Pilarczyk (2000), where the internal stability is ensured by using the rules:

$$\begin{aligned} D_{10} &< 4\,D_{05} \\ D_{20} &< 4\,D_{10} \\ D_{30} &< 4\,D_{15} \\ D_{40} &< 4\,D_{20} \end{aligned} \qquad (13.119)$$

With an appropriate grading, Eq. (13.119) can produce a pore size of the bedding layer ($D_{05}/4$) three orders of magnitude smaller than the size of the larger stones in it.

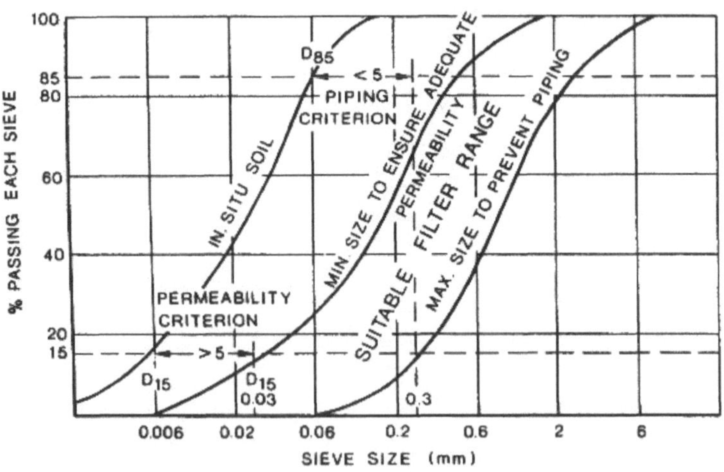

Figure 13.63. Standard design method for granular filters, Pilarczyk (2000).

Satisfying all the conditions mentioned above in the constructed structure may be difficult and requires a careful control of the grading in the prepared mixture and of the placing method.

13.11.2.2. Geotextiles

The main part of the following text is from Pilarczyk (2000). The design of geotextiles in relation to LCSs follows the same procedures as for conventional breakwaters. For in depth guidance on the use and design of geotextiles the reader is referred to standard literature, e.g. Pilarczyk (2000) and PIANC (1992).

The most likely type of damage to the geotextile in LCSs is mechanical damage. Mechanical damage can be prevented by a proper choice of material and a careful execution. Much attention must be paid to the flatness of the surface on which the geotextiles are spread. Danger of puncturing may arise when stones lie under a membrane or when stones are dumped on a membrane. Great differences in tension and deformation lead to the formation of folds. These folds have to be prevented. Damage to the geotextile can be prevented by:

- the application of a load-spreading bedding layer of gravel or light stones (maximum 10 to 60 kg);
- reduction of the height of the fall of rock, by placing the dumping vessel or crane bucket as near to the bedding layer as possible.

In practice, the choice of the strength of the geotextile is very often based on experience. Often, the installation conditions are decisive for design. For example, for bank protection the geotextiles with the unit weight of 200 g/m^2 and tensile strength (in the warp direction) of at least 15 to 20 kN/m^2 are applied. However, in the case of dumped stones, a unit weight of 300 g/m^2 is recommended. In present Dutch practice, the stone classes up to 10/60 kg are dumped directly on geotextiles. For heavier classes the layer of finer stones with a weight of about 200 kg/m^2 is placed first.

Experience shows that often joints, edges, transitions, etc. are the weak points leading to failures. When the subsoil surface is uneven or is compacted insufficiently, or when cyclic loadings appear, there is a great chance of wash-out through the filter and below the filter. Therefore, during design and execution, special attention must be paid to placement methods, and joints and overlaps. The water permeability of a geotextile, especially in overlap zones, may decrease by clogging and blocking. If there is any chance of this, the most suitable geotextile has to be carefully selected, if necessary based upon soil analyses.

A number of precautions must be taken when laying the geotextile. The surface of the subsoil should be a relatively smooth plane, free of obstructions, cavities and soft pockets of material. Cavities in the soil must be filled with compacted material, otherwise the fabric may bridge and tear when the cover layer is placed.

Care must be taken when placing the cover layer. The placing method should avoid damage to the geotextile. With a soft subsoil, the geotextile needs to be able to deform sufficiently to avoid tearing under dumped stone. If the subsoil is rocky, cutting of the geotextile has to be avoided; this can be achieved by using a geotextile with a high tear resistance. It is good practice to insist that the contractor demonstrates that his chosen placing method does not result in damage to the geotextile.

The sea bed level on tidal coasts can vary significantly from season to season and from year to year. It is important that the level of the geotextile is not higher than the predicted lowest level of the sea bed in order to prevent undermining of the structure.

13.11.3. Toe berm stability

(Kramer & Burcharth, AAU)

The function of a toe berm is to support the main armour layer and to prevent damage resulting from scour. Armour units displaced from the armour layer may come to rest on the toe berm, thus increasing toe berm stability. Toe berms are normally constructed of quarryrun, but concrete blocks can be used if quarryrun material is too small or unavailable.

In shallow water with depth-limited design wave heights, support of the armour layer at the toe is ensured either by placing one or two extra rows of main armour units at the toe of the slope or by the use of stones or blocks in the toe that are smaller than the main armour, c.f. examples given in Figures 7.3 and 7.4. These solutions are stable provided that scour does not undermine the toe causing the armour layer to slide. The toe berm must be wide enough to avoid this problem, which will be treated in detail in the chapter subsequent dealing with scour.

Toe berm stability is affected by wave height, water depth at the top of the toe berm, width of the toe berm, and block density. However, wave steepness does not appear to be a critical toe berm stability parameter.

Model tests with irregular waves indicate that the most unstable location is at the shoulder between the slope and the horizontal section of the berm. The instability of a toe berm will trigger or accelerate the instability of the main armour. Lamberti (1995) showed that moderate toe berm damage has almost no influence on armour layer stability, whereas high damage of the toe berm severly reduces the armour layer stability. Therefore, in practice it is economical to design toe berms that allow for little damage.

No model tests dealing especially with toe berm stability of LCSs exist. However, within DELOS a few model tests on LCSs with depth limited waves and wave breaking at the toe showed good agreement with the formula for trunk toe stability of emerging breakwaters given by Eq. (13.120). For LCSs wave energy can pass over the structure making them more stable than the conventional type. Seaward toe berms designed by formulae developed for non overtopped breakwaters will therefore be more stable when used for LCSs. This was confirmed by the model tests performed within DELOS. The tests showed that the seaward toe was more prone to damage than the leeward toe. This indicates that it is safe to apply the same stone type in the leeward toe as used for the seaward toe. Further the DELOS testing showed that oblique wave attack was less damaging than normal incidence wave attack.

13.11.3.1. Toe berm stone sizes in trunk

The formula by Van der Meer *et al.*, (1995) given in Eq. (13.120) may be used to find the required rock size for the toe berm for the trunk. The formula was developed for sloping, emergent rubble mound breakwaters. Stones having a mass density of 2.68 t/m^3 were used, and the berm width was varied.

$$N_s = \frac{H_s}{\Delta D_{n50}} = \left(0.24\frac{h_b}{D_{n50}} + 1.6\right) N_{od}^{0.15} \qquad (13.120)$$

where

H_s Significant wave height in front of breakwater
Δ $(\rho_s/\rho_w)-1$
ρ_s Mass density of stones
ρ_w Mass density of water
D_{n50} Equivalent cube length of median stone
h_b Water depth at top of toe berm
N_{od} Number of units displaced out of the armour layer within a strip width of D_{n50}. For a standard toe size of about 3-5 stones wide and 2-3 stones high:

$$N_{od} = \begin{cases} 0.5 & \text{no damage} \\ 2 & \text{acceptable damage} \\ 4 & \text{severe damage} \end{cases}$$

For a wider toe berm, higher Nod values can be applied.

The formula is valid for:
- Irregular head on waves; nonbreaking, breaking and broken.
- $0.4 < h_b/h < 0.9$, $0.28 < H_s/h < 0.8$, $3 < h_b/D_{n50} < 25$ where h is the water depth in front of the toe berm.

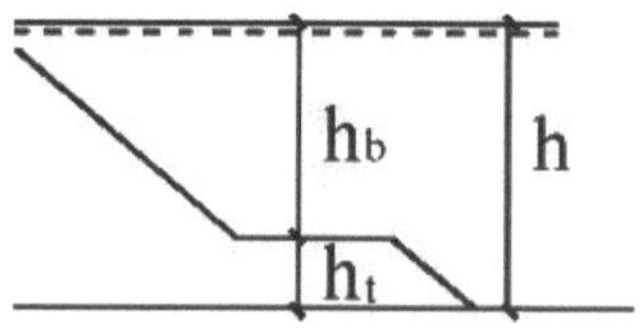

If the highest waves are depth limited then the significant wave height can be replaced by the approximation $H_s = 0.6 \cdot h$. By inserting in Eq. (13.120) $\rho_s = 2.65$ t/m³ corresponding to $\Delta = 1.6$, and $H_s = 0.6 \cdot h$, Eq. (13.120) can be reduced to:

$$N_{od} = 0.5: \begin{cases} D_{n50} = 0.16 \cdot h, & \text{for } h_t = 2 \cdot D_{n50} \\ D_{n50} = 0.20 \cdot h, & \text{for } h_t = 3 \cdot D_{n50} \end{cases} \quad (13.121)$$

$$N_{od} = 2: \begin{cases} D_{n50} = 0.09 \cdot h, & \text{for } h_t = 2 \cdot D_{n50} \\ D_{n50} = 0.11 \cdot h, & \text{for } h_t = 3 \cdot D_{n50} \end{cases}$$

However, if the toe is located in very shallow water and the toe is expected to be very exposed to direct wave action, then the same stone type as used in the armour layer can be applied. This will always lead to a stable conservative design.

13.11.3.2. Toe berm stone sizes in roundheads

For the toe berm in the roundhead no specific recommendations exist. In many situations previous experiences can be used to evaluate the necessary size of the rocks. Rock sizes equal to the sizes by the trunk might be used, but in that case it is recommended to validate the design by the use of model tests. If the LCSs are long and low very large rip currents might occur in the gaps. This might affect the toe stability especially if scour takes place in front of the toe. If model tests are used to design the toe berm it is very important that the rip currents are correctly modelled in the experiments.

If the toe is located in very shallow water and the toe is expected to be very exposed, then the same stone type as used in the main armour layer of the roundhead can be applied. This will always lead to a stable conservative design.

13.11.4. Dimension of scour protection

13.11.4.1. Toe protection

(Sumer, ISVA)

Toe protection layer may be constructed in the form of a protection apron. The apron must be designed so that it will remain intact under wave and current forces, and it should be «flexible» enough to conform to an initially uneven seabed. With this countermeasure, scour can be minimized, but not entirely avoided. Some scour will occur at the edge of the protection layer, and consequently, armour stones will slump down into the scour hole. This latter process will, however, lead to the formation of a protective slope, a desirable effect for «fixing» the scour. The determination of the width of the protection layer is an important design concern. The width should be sufficiently large to ensure that some portion of the protection apron remain intact, providing adequate protection for the stability of the breakwater.

13.11.4.1.1. Toe protection at the trunk section

On the basis of the experiments on scour at LCSs undertaken in DELOS and the experiments conducted in the work of Sumer and Fredsøe (2000) (see pp. 347-365 of Sumer and Fredsøe, 2002), it is recommended that the width of the protection apron (Figure 13.64) be calculated by the following empirical equation

$$W = \frac{L}{4} - mh_b \tag{13.122}$$

where:

- m is the slope of the breakwater (Figure 13.64),
- h the water depth and
- L the wave length of the incident wave.

This is essentially roughly equal to the width of the scour hole measured from the nearest dune crest to the toe of the breakwater in the case of emerged breakwaters, and therefore it is a conservative estimate of the scour-hole extent for submerged breakwaters. It may be noted that Sumer and Fredsøe (2002, p. 362) report that the α value

$$\left(\alpha = 1 - \frac{mh_b}{L/4}\right)$$

measured in the laboratory experiments is 1 for vertical-wall emerged breakwaters, 0.6 for m = 1.2 and 0.3 for m = 1.75 for rubble-mound emerged breakwaters. It should also be mentioned that the preceding relation is valid for shallow waters, the conditions under which experiments were conducted in the DELOS work and in Sumer and Fredsøe (2000), $h/L <$ O (0.1-0.2).

This is for the scour protection at the offshore side of the breakwater. The scour experiments undertaken in DELOS suggest that the same width may be selected for the toe protection apron at the onshore side. Extra precautions must be exercised towards reinforcing

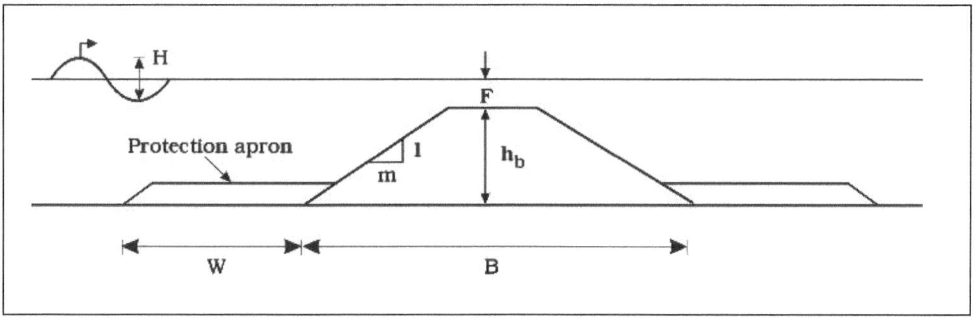

Figure 13.64. Definition sketch.

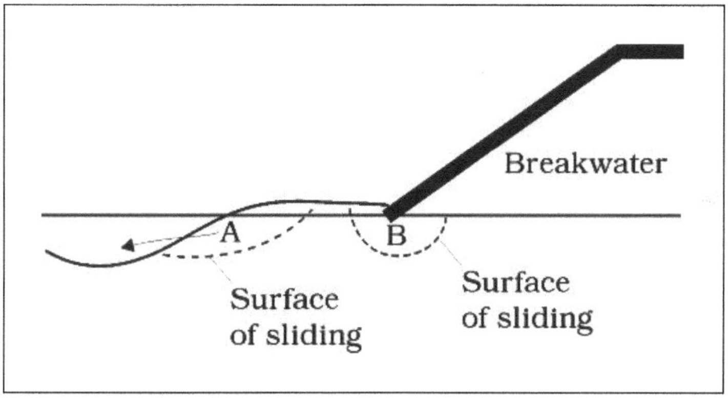

Figure 13.65. Possibility of sand slide in front of breakwater.

the protection layer on this side to protect the protection material against damage caused by wave overtopping.

The volume of the toe berm shall be such that its material is sufficient to protect the scour/erosion hole from further erosion without destabilising the armour layer slope, i.e., its width should be around three times the erosion depth and its thickness at least four times its maximum stone size (SPM, 1984; Burcharth *et al.*, 2006). In this way slided berm stones can form, although dispersed, a stable and continuous slope covering the sand bed.

The equation (13.122) is based on the scour experiments where the mode of sediment transport was in the no-suspension regime. In the case of the suspension-regime sediment transport, from the knowledge of scour at emerged breakwaters, no scour is expected at the toe (at the offshore side of the breakwater), and therefore scour is not an immediate threat to the breakwater. However, soil failure illustrated in Figure 13.65 may be a risk for stability, and hence may need to be considered (Sumer and Fredsøe, 2002).

Furthermore, the preceding equation is for scour protection against the *local* scour caused by the combined effect of steady streaming and phase-resolved stirring of sediment by waves (Sumer and Fredsøe, 2002). Due considerations must be given to *global* scour caused by the far-field flow circulations around the breakwater.

13.11.4.1.2. Toe protection at the head section

It is recommended that the width of the protection apron be calculated by the following empirical equation

$$W = W_e \quad \text{if} \quad \frac{F}{H} < -0.9$$
$$W = \left(-0.29\frac{F}{H} + 0.74\right)W_e \quad \text{if} \quad \frac{F}{H} > -0.9 \tag{13.123}$$

in which

- F Freeboard (Figure 13.64; negative values correspond to slightly or fully emerged breakwaters)
- H Wave height
- W_e Width recommended for «fully» emerged breakwaters, given by $W_e/B = AKC$
- B Diameter of the round head at the bed
- A A is 1.5 for complete scour protection and 1.1 for a scour protection which allows a scour depth of 1% of B
- KC Keulegan-Carpenter number, $KC = (2\pi a)/B$ in which a is the amplitude of the orbital motion of water particles at the bed, and may be calculated using the small-amplitude, linear wave theory.

The above equation is based on the experiments where the breakwater slope was 1:1.5 (i.e., m = 1.5, Figure 13.64). Therefore, for slopes steeper than 1:1.5, the width necessary for protection may be increased, and for slopes milder than 1:1.5, it may be reduced.

Furthermore, the above equation is for scour protection against the *local* scour caused by the combined effect of steady streaming and phase-resolved stirring of sediment by waves (Sumer and Fredsøe, 2002). Due considerations must be given to *global* scour caused by the far-field flow circulations around the breakwater.

Finally, the recommended width is for protection at the offshore side of the head. Experiments show that the implemented widths of the protection layer are able to protect the sand bed against the breaker-induced scour at the onshore side of the head. However, scour (damage) may occur in the protection layer itself due to wave breaking and wave overtopping. Therefore, additional reinforcement is recommended at the onshore side regarding the protection material.

13.11.4.2. Bed protection at gaps

(Martinelli, UB)

In case of submerged structures, rip currents are characterised by great intensity and thus great sediment transport capacity. The erosion induced at gaps can both cause serious problem of structure stability and act as sink for sediments inside the protected area, making them first fall into the hole and then favouring their exit from the gap pushed by the currents.

It is therefore necessary to adequately protect the gaps with a stable and flexible plateau that may follow bottom movements, usually consisting of the same material at the barrier toe.

The objective must be to shift erosion from the structure at such a distance not to compromise structure stability. Gap protection shall be extended more in off-shore than in in-shore direction, although it is not realistic an off-shore protection to the limit of the eroded area. The amount of material must be exceed the strictly necessary quantity in order to fill the holes that inevitably form at the protection boundaries. Maintenance works for restoring toe protection before structure damage occur should be planned.

13.12. MODEL TESTS RELATED TO STRUCTURE DESIGN

(Kramer & Burcharth, AAU)

Physical model experiments are performed when suitable design formulae or numerical models are missing, or are too uncertain. Often model tests are performed to validate a considered design. For large expensive designs model tests should always be performed in order to optimize the design. For example, stability tests should be performed to determine the required armour unit size when existing stability formulae does not cover the preferred structure geometry, the in situ bathymetry or the type of armour unit.

Laboratory tests are generally more expensive than numerical modelling. However the reliability of physical models is generally much better, so far.

Generally, with scale models only some pre-selected phenomena can be well represented, whereas at the same time, other phenomena may not be reproduced correctly and suffer from scale effects. This is a hardly avoidable penalty for not matching all the scale requirements. If, however, the scale effects are considered to be of minor importance for the phenomena of direct concern for the design of a structure, the scale model may provide accurate information. Scale modelling is however complex and requires sophisticated facilities and experimental set-ups. Care should be taken to perform adequate testing (e.g. wave generation techniques, methods to reduce scale effects, analysis techniques) and to correctly analyse and interpret the results to obtain the required information.

When setting up an experiment one should consider the importance of the following:

- scale effects: typically viscous forces are relatively larger in the model than in the prototype;
- laboratory effects: typically the boundaries are different in model and prototype;
- missing conditions: for example neglecting effects of wind shear stresses acting on the free surface, which may lead to neglecting generation of waves and circulation currents leeward of the structure.

In order to make ideal set-ups in the laboratory with respect to different subjects one may distinguish between the following types of tests with LCSs:

Stability tests (typically the stable unit sizes of e.g. armour, core and toe berm are determined).

Hydrodynamic tests (typically wave transmission and reflection characteristics, overtopping, rip-currents and water level set-up in the lee of the structures are investigated).

Morphological tests (typically scour, beach development, and selection of sand for beach nourishment is studied).

An example of the design of model tests related to LCSs can be found in Kramer *et al.*, (2005).

Tests can be performed with either fixed bed (solid boundaries, typically concrete bed) or movable bed (to study sedimentary processes, typically a sandy bed). Some laboratories are specialized in movable bed tests while others only perform fixed bed experiments. Typically fixed bed tests are cheaper and more easily controllable than movable bed tests. Therefore usually only morphological tests are performed with movable bed. In fixed bed tests the bottom bathymetry can be either horizontal, sloping or a certain bathymetry can be modelled (e.g. in concrete). In movable bed tests the bed is typically horizontal at the initiation of the tests. During testing the bed forms and e.g. scour holes develop.

Tests can be performed in wave channels (often referred to as 2D-tests) or in wave basins (often referred to as 3D-tests). Wave channel tests are cheaper than wave basin tests. Phenomena related to perpendicular wave attack on the trunk of the LCS are typically studied in wave channels, while phenomena related to the roundhead and effects of oblique waves and 3-D waves are studied in wave basins.

In order to minimize viscous scale effects the model is typically designed as large as the laboratory limits and the economy permit. If the Reynolds numbers are sufficiently large scaling can be performed solely by Froude's model law. As an example the effect of Reynold numbers on the stability of armour stones have been investigated by various researchers. No scale effects seems present if

$$\text{Reynolds number} = \frac{\sqrt{g \cdot H_s} \cdot D_{n50}}{\nu} > 1.0 \cdot 10^4 \text{ to } 4.0 \cdot 10^4 \qquad (13.124)$$

where g is the gravitation acceleration and ν is the kinematic viscosity.

If for example a significant wave height $H_s = 0.2$ m is generated in the laboratory then a stone size $D_{n50} = 0.03$ m gives a Reynold number $4.2 \cdot 10^4$ (with typical values of $\nu = 10^{-6}\,\text{m}^2/\text{s}$ and $g = 10\,\text{m/s}^2$). According to the limits given, no significant viscous scale effect is present, regarding armour layer response and the scaling can be performed by Froude's law.

For a comprehensive study of physical models and laboratory techniques, see Hughes (1993).

13.13. SAFETY ASPECTS

(Vidal, UCA)

13.13.1. Limit states for maritime structures

Every maritime structure should comply with certain requirements of operationality, functionality and reliability during a specific time interval. One of its purposes is to permit or facilitate a series of economic activities that will have social repercussions as well as impacts on the physical environment. The main objective of the design of the structure is the verification of the fulfilment of these objectives and requirements, repercussions and impacts.

The design of a maritime structure is carried out dividing the project into spatial subsystems and temporal phases. The duration of each project phase the maritime structure undergoes (i.e. construction, operational life, maintenance/repair and dismantling) can be divided into a sequence of project states. The project state defines and describes the behaviour of a subsystem of a structure in a given time interval, for instance the temporary exposed rubble mound foundation during the contruction of a breakwater.

During the occurrence of a project state, the shape, the exploitation of the subsystem and its structural response are assumed to be stationary processes.

The objective of the project design is to verify that the subsystem fulfils the project requirements in each of the project states. In order to simplify the verification of the subsystem, only some of all the possible project states are verified, namely those that represent limit situations of the subsystem from the viewpoint of the structure, its shape, use and exploitation. These states are called limit states, and the verification procedure based on them is called the method of the limit states.

In resume, a limit state is a state in which the combination of project factors produces one or more structural failure or operational breakdown. A failure mode describes the form or mechanism in which the structural failure (or the operational breakdown) of the subsystem or of one of its elements is produced. Three sets of limits states are defined: ultimate, serviceability and operational.

Ultimate limit states are those project states that produce the collapse (unrecoverable state) of the structure usually because of the structural breakdown of some essential and non-repairable part of it. They include all failure modes which may be caused by:

- loss of static equilibrium of the whole structure or relevant part of it;
- excessive deformation, breakage, loss of ability to resist loads in all or part of the structure;
- accumulation of deformation, progressive cracking, fatigue.

Serviceability limit states are those project states that produce a loss of service and functionality in all or part of the structure due to a minor and repairable structural failure. The failure modes related to these limit states are frequently established by functional, environmental or aesthetic legal constraints. These limit states can be reached during the useful life of the structure as a consequence of its use and exploitation, as well as its location in the physical environment. Serviceability limit states include those conditions that reduce or constrain the use and exploitation of the structure and which can signify a reduction of the useful life and the reliability of the residual life of the structure. These states are naturally permanent; repair works become necessary so that the structure can recover its ability to meet the project requirements. They include:

- unacceptable deterioration of the properties of the building materials or soil;
- unacceptable deformations or vibration conditions in the structure for its use and exploitation;
- unacceptable cumulative geometrical changes of the structure for its use and exploitation;
- unacceptable aesthetic damage on the structure.

Operational limit states are those project states in which a structure's use and exploitation is reduced or stopped, due to causes that are external to the maritime structure and its installations, without the existence of structural damage to the structure or any of its elements.

Generally, the operation is stopped in order to avoid this sort of damage to the structure or unacceptable environmental and social consequences. Once the external cause disappears, the structure and its installations totally recover the exploitation requirements of the project. Operational limit states include those failure modes which may be caused by:

- temporary reduction of the reliability and functionality of the maritime structure and its installations;
- temporary unacceptable environmental effects and social repercussions or temporal failure to fulfil environmental legal constraints.

13.13.2. LCS limit states and failure modes

LCS schemes, as any other engineering project, are built to fulfil some functional objectives (described in Chapter 3) during their useful life while maintaining adequate security levels.

Based on the stated limit states established above, the following limit states and corresponding failure modes can be defined for LCS structures.

Ultimate limit states correspond to:

1. loss of the LCS static equilibrium causing the following ultimate failure modes:
   - significant displacement of LCS armour units due to hydrodynamic forces;
   - armour layer sliding due to poor interlocking with filter;
   - displacement of LCS toe berm units inducing significant damage to armour;
   - overall LCS stability failure due to bed scour;
   - overall stability failure due to soil failure;
2. loss of resistance or breakage of LCS units causing the following ultimate failure modes:
   - breaking of armour units due to structural stresses;
   - breaking of armour or filter units do to flaws on the rock;
   - breaking of armour or filter stones do to chemical attack acting on the flaws;
3. deformation of the LCS structure causing the following ultimate failure modes:
   - structure armour dislodging due to filter failure;
   - sinking of the LCs structure or part of it in the sand bed due to filter failure;
   - significant displacement of LCS armour units due to settlement or compactness of the armour.

Serviceability limit states correspond to:

1. unacceptable deterioration of the properties of the building materials or soil causing the following serviceability failure modes:
   - changes in the properties of rock surfaces for its safe use by pedestrian or fishermen;
   - changes in the rock surfaces modifying their ability to sustain attached life;
2. unacceptable cumulative geometrical changes of the structure for its use and exploitation causing the following serviceability failure modes:
   - filling up with sand of the potholes associated to the toe berm modifying the habitat associated to them;
   - filling up of the voids of the structure with attached life and sand, modifying the water interchange in the voids and the associated habitat.

Operational limit states correspond to temporary unacceptable environmental effects and social repercussions or temporal failure to fulfil environmental legal constraints, causing the following operational failure modes:

- excessive wave transmission and/or set-up and mean currents in the sheltered area, affecting beach bathing security conditions;
- insufficient water offshore-inshore interchange through and over the LCS, causing poor water quality conditions for bathing;
- excessive wave transmission and/or set-up and mean currents in the sheltered area, affecting mobile marine life;
- insufficient water offshore-inshore interchange through and over the LCS, causing poor water quality conditions for marine life;

- accumulation of algae and other organic materials in the sheltered area, due low or inappropriate current systems, producing anoxic conditions and bad smells, thus affecting both human usage of the beach and marine life.

The risk analysis of any structural scheme is related to the ultimate, service and operational failures modes and is carried out evaluating the overall probability of failure (*OPF*) and the cost of the consequences (*CC*) of the failure elevated to some power:

The probability of ultimate and service failure during the analysed temporal domain (i.e. the useful life) and the operationality of an LCS depend on how the different failure modes are connected. Sometimes, to simplify the procedure, some principal failure modes are defined, designing the scheme in such a manner that the probability of the occurrence of other failure modes can be assumed negligible. In that case, the overall probability of failure of the LCS depend only on the probability of occurrence of the principal failure modes. To assess the probability of failure of each failure mode, a verification procedure should be established.

The Spanish Recommendations for Maritime Structures, in its document 0.0 (ROM 0.0) provide for instance a set of standards and technical criteria for the design, construction, maintenance, repair and dismantling of maritime and harbour structures of all types and designs, no matter what materials, techniques and elements are used for these purposes. The organization of the ROM 0.0 is indicated in the diagram of Fig. 13.66. ROM 0.0 are difficult to follow step by step and are hardly applicable to LCSs because they are meant for larger structures; they can however provide a general guidance and useful suggestions.

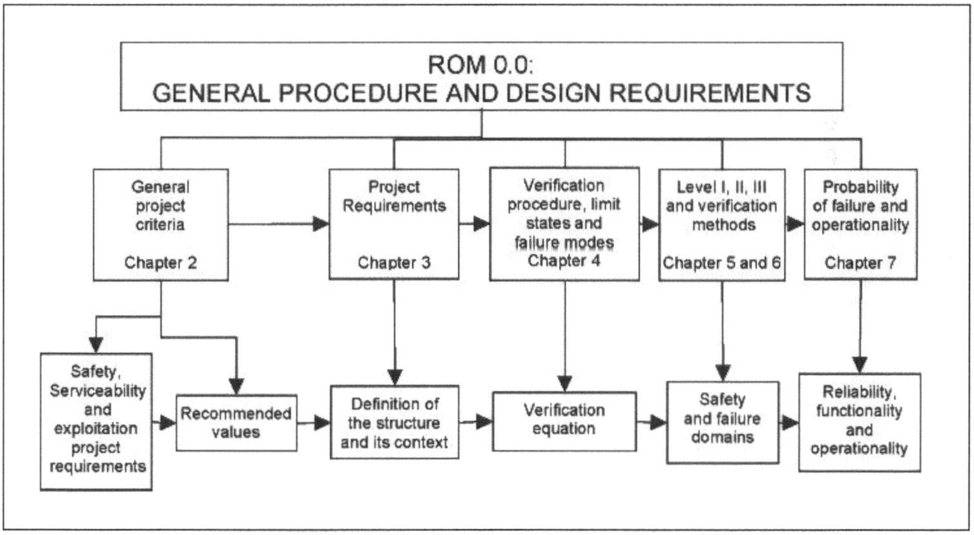

Figure 13.66. ROM 0.0 Organization and contents.

CHAPTER 14

Background knowledge and tools for prediction of ecological impacts

(Moschella, MBA; Abbiati, Airoldi, Bacchiocchi, Bertasi, Bulleri, Ceccherelli, Colangelo, FF; Cedhagen, BIAU; De Vries WL-DH; Dinesen; BIAU; Aberg, Jonsson, Granhag, Sundelöf, UGOT; Gacia, Macpherson, Martin, Satta, CSIC; Frost, Thompson & Hawkins, MBA)

14.1. DEFINITIONS OF MAIN FACTORS INFLUENCING THE DISTRIBUTION AND ABUNDANCE OF SPECIES AND ASSEMBLAGES (BIOTOPES) ON NATURAL SOFT- AND ROCKY BOTTOMS

14.1.1. Broad-scale – Geographic variation

The species pool in a particular locality, is determined by its biogeographic context. This is the result of past events on tectonic/evolutionary time scales (100 million years – 1 million years B.P., e.g. Mediterranean compared to Atlantic) and more recent palaeo-ecological/ geomorphological history (last 20 thousand years e.g. English Channel, North Sea and Irish Sea coastlines).

The evolution of the species pool is a dynamic and ongoing process. Biodiversity patterns on a broad-scale are a function of adaptation, extinctions and speciation. The species pool may also change following introduction of alien species, often through human activities (Stæhr *et al.*, 2000).

Global transfer of species (e.g. Lessepsian migrations via Suez canal) has gathered in importance over the last 200 years.

Broad-scale biodiversity patterns are influenced by major physical factors such as climate, currents, upwelling, tidal elevation, wave climate, salinity, coastal topography and seabed composition, which can all vary with geographical location (e.g. greater waves on Atlantic coast of Ireland versus the more enclosed Irish Sea, salinity in Baltic versus North Sea, tides in Atlantic versus Mediterranean and Baltic).

14.1.2. Mesoscale – Within coastline

The species assemblage found at a specific location is affected by the exchange with neighbouring populations through dispersal, mainly through suspended propagules (e.g. larvae and spores). The spatial distribution of source populations is largely governed by coastal geomorphology that determines the diversity of substrata and hence habitat types in a particular region. Morphodynamics of sediments further affect the coastal-scale distribution of sedimentary habitats. The presence of source populations, however, is not sufficient to ensure exchange between habitats.The dispersal between habitats depends on hydrodynamic transport, although interactions with behavioural responses (or gravitational sinking) may modify dispersal pathways. Hydrodynamic transport includes tidal, wind driven and baroclinic advection (currents) together with turbulent diffusion. Other coastal-scale factors that may influence species assemblages are point sources of nutrients, contaminants, suspended sediment and freshwater (e.g. from riverine discharge). Differences in geomorphology and bathymetry will also cause coastal-scale differences in wave climate that will in turn influence local species distribution.

a) Coastal geology, geomorphology and topography

The topography and geomorphology of the coastline are crucial to the distribution of species. The description of the large-scale distribution of species and assemblages therefore must take account of the characteristics of sediment, natural rock and artificial substrata. The underlying geology of an area can have significant effects on the distribution and abundance of species (Crisp, 1974; Holmes *et al.*, 1997). For example, rock types of differing physical and chemical properties seem to affect the settlement of various barnacle species. Other features of the substratum are also important, such as the surface composition and orientation (Glasby, 2000; Glasby and Connell, 2001). For soft bottom communities this factor is coupled to hydrodynamics, discussed in point c).

b) Localised nutrient supply due to small-scale upwelling, riverine run-off, seawage disposal increasing growth rates (14.1.6d) of algae and frence productivity

Local small-scale upwelling carries nutrients from deeper water to shallow water and changes the local nutrient concentrations. Fresh water run-off can carry nutrients from farmlands and forests via the catchment. Waste discharge may locally increase nutrient availability. Differences in the local concentration of available nutrients will have large impacts on the local species composition (see also 14.1.6d).

c) Hydrodynamic-sedimentary regimes affecting erosion/deposition, disturbance regime, turbidity and long-shore transport

The coastline topography and geomorphology as well as the local bathymetry influence the hydrodynamics regime. Hydrodynamics also determines for the sedimentary regime affecting erosion and deposition of sediments, turbidity, disturbance regimes for the biota and long-shore transport.

Soft-bottom assemblages are greatly affected by changes in the sedimentary regimes (deposition, erosion) and modification of sediment characteristics such as organic matter and granulometry. Turbidity of waters also affects a variety of organisms, including seagrasses, invertebrates and algae by reducing light penetration through the water column.

The factors and processes described above will in turn affect the connectivity of habitats and larval supply – sources and sinks of propagules, recruitment regimes, metapopulation dynamics.

Connectivity of habitats and larval supply can be very important for the large-scale distribution of species and assemblages. In fragmented habitats connectivity is low and the species composition may be affected by chance events. The connectivity and larval supply thus determines colonisation probabilities for species and populations. Low connectivity means low colonisation probability and high connectivity means high colonisation probability. The dynamics caused by extinctions and colonisations is often termed metapopulation dynamics. Post-recruitment events may also control the population survivorship rates and the persistence of recruits is often a more relevant factor in controlling population dynamics than the recruitment itself (Jackson, 1986). Species composition in fragmented habitats is strongly dependent on residual currents. On the other hand, residual currents will be less important for the dispersal of organisms existing in a commonly occurring habitat or where the habitat is narrow but well connected. Assuming a fragmented habitat, the range expansion of species may depend largely on the extreme values of actual water movement, and not the mean residual current.

14.1.3. Local scale – Major abiotic factors and processes

Several abiotic factors affect the distribution of species on a local scale (Lewis, 1964; Stephenson and Stephenson, 1972; Raffaelli and Hawkins, 1996). These include vertical and horizontal patterns of distribution caused by tidal elevation, wave exposure, light penetration and, in sediments, physical and chemical gradients. In addition, local disturbance caused by extreme events such as wave-induced impact, depletion of oxygen and sediment burial can create a mosaic pattern of species occurrence. Some key gradients are summarised below:

a) Tidal elevation/depth.

On macrotidal shores, the time of emersion/submersion and consequently desiccation stresses experienced by intertidal organisms, as well as the time to take up nutrients (algae) and food (invertebrates), markedly depends on the tidal level (Lewis, 1964; Raffaelli and Hawkins, 1996). The distribution of species is affected by tidal level, as physiological tolerance to emersion and desiccation stresses varies between and within species but in general a higher number of species tend to better tolerate lower shore environmental conditions (Lewis, 1964; Newell, 1979; Raffaelli and Hawkins, 1996; Spicer and Gaston, 2000). This pattern is particularly evident on macrotidal shores, where epibiotic assemblages differ markedly between different tidal levels. On microtidal shores, the structure of benthic assemblages changes considerably with increasing depth, from an algal monopolized community to a community dominated by sessile invertebrates. This is mainly due to a decrease in light penetration, which can be further reduced by turbidity (Gaçia *et al.*, 1996; Irving and Connell, 2002).

b) Wave exposure

Wave action plays a major role in the composition of rocky littoral and sub littoral communities shores (Lewis, 1964; Hiscock, 1983; Raffaelli and Hawkins, 1996). On exposed shores, benthic organisms experience greater wave-induced forces and consequently face a higher risk of breakage or dislodgement from the rock and consequently their persistence. Wave action, however, can increase wetting of upper shore species, nutrient supply for algae and suspended food for filter feeders. Foraging times can be both positively and negatively impacted. Conversely, on more sheltered shores, reduced water movement is generally associated with greater sediment deposition and siltation on the rock substratum, which can be cause of disturbance. Species respond differently to this stress gradient (Denny *et al.*, 1988; Denny, 1995); some organisms thrive better and are naturally more abundant in wave swept conditions (e.g. mussels and barnacles), whilst others are adapted to more sheltered conditions (e.g. the macroalga *Ascophyllum nodosum* and the gastropod *Osilinus lineatus*).

c) Salinity

Salinity gradients occur in estuaries and coastal areas near riverine inputs. This factor affects particularly the species pool, as only few species can tolerate low or variable salinities. Salinity can affect the vertical distribution of species: in the supralittoral zone salinity can increase considerably in crevices and rock pools (Raffaelli and Hawkins, 1996).

d) Physical disturbance

In rocky intertidal and subtidal assemblages, physical disturbances associated with partial or total loss of biomass have been recognised as primary mechanisms that generate mosaics of patches at different stages of recovery, and control abundance and diversity of species (Dayton, 1971; Menge, 1976; Sousa, 1979; 2001; Paine and Levin, 1981; Airoldi, 2000 a, 2003). Waves, excessive heat, scour from sediment and other debris are examples of natural disturbances that cause mortality of organisms and open discrete patches of open space (Dayton, 1971; Hawkins and Hartnoll, 1983; Airoldi and Virgilio, 1998).

14.1.4. Local scale – Biological interactions and behaviour

On rocky shores, the following biological interactions and processes are extremely important in influencing species distribution at small spatial scales:

a) Grazing/predation
b) Competition for space
c) Biologically mediated disturbance (algal sweeping, bioturbation)
d) Facilitation (positive interactions, sheltering etc.)
e) Biodeposition and sediment trapping
f) Larval and adult behaviour

Local biodiversity reflects the direct and indirect interactions among and within species. Trophic interactions are particularly strong on hard substrata, for example limpet grazing on algae on rocky shores (Hawkins, 1981; Hawkins *et al.*, 1992). Competition for space or resources often reduces the diversity of species assemblages but diversity can often be higher at intermediate levels of physical and biological disturbance (Caswell, 1978). Examples are biologically mediated disturbances like algal canopy sweeping on rocky shores and bioturbation in sediments (Rhoads, 1974). Certain species can also improve conditions for other species and so increase the local biodiversity. Such «facilitation» effects includes several mechanisms, e.g. sheltering from canopy-forming macro-algae or mussel beds promoting recruitment of polychaetes and small crustaceans. Some species build 3-dimensional structures that alter the physical conditions leading to changes in the species assemblage. Examples include reef-building polychates consolidating sand beds, encrusting algae creating complex secondary substrata, and meadow-forming seagrass attenuating wave energy. Organisms changing the hydrodynamic regime by wave attenuation or flow reduction will often promote sediment trapping offering new habitats for sediment-living organisms or exclude species sensitive to high sediment load. Finally, spatial heterogeneity of abiotic and biotic factors may interact with behaviour during all life stages. Gregarious responses during the settlement phase in barnacles are one example that leads to aggregated distribution patterns.

14.1.4.1. Interactions between physical and biological factors

The upper limits of vertical distribution of species are generally set by physical factors whilst the lower limits are set by competition, predation and grazing. However, there are some exceptions, especially lower on the shore, where algal upper limits can be set by grazing (Hawkins and Jones, 1992; Boaventura *et al.*, 2002) or competition (Hawkins and Hartnoll, 1985). On wave exposure gradients both direct physical effects and indirect biological interactions can set the distribution patterns of species. For example, limpets prevent establishment of algae on wave beaten shores (Hawkins and Hartnoll, 1983; Moschella *et al.*, 2005; Jonsson *et al.*, 2006) whilst algal persistence is probably controlled by wave action (Jonsson *et al.*, 2006).

14.1.5. Micro scale – Complexity

On even smaller scales (< 10 cm), factors such as heterogeneity in surface topography (roughness) affect the availability of refuge from hydrodynamics and grazing (Fretter and Manly, 1977; Underwood and Chapman, 1998). In sediments, small-scale gradients in grain size and compaction (both horizontally and vertically in the sediment column) may lead to changes in porous flow and chemical composition with strong effects on infauna assemblages.

14.1.6. Human activities

Human activities alter the marine environment at various scales from global (e.g., climate change) to the local (point source pollution). Major factors likely to interact with natural processes in the coastal zone are outlined below. These factors need to be considered when predicting the impacts of LCS construction:

a) Global changes

Anthropogenic release of greenhouse gases is now widely accepted to be influencing the climate of the planet. Various predictive scenarios have been made. In short, air and sea temperatures will increase, as will sea level (IPCC, 2001a,b). The Atlantic Ocean and adjacent seas will become stormier in part due to greater frequency of NAO positive winter values. Thus, in addition to rise in average temperature and wave height, the incidence of extreme events will be more likely. Southern species will migrate towards the poles. Increased likelihood of extreme events will lead to an increasing number of LCS being built along the coast. This in turn will have marked effects in the distribution of species. There is evidence from the Delos project and climate change programmes (e.g., the MarClim project coordinated by the MBA – www.marclim.mba.ac.uk) of species extending their ranges using artificial structures as stepping stones between areas of natural hard substrates or in their absence extending their distribution (Herbert *et al.*, 2003). A good example is the southern snail, *Gibbula umbilicalis*, which has been found at Elmer 60 km east of its previous limit. Southern fish species such as anchovies (*Engraulis* sp.) and sardines (pilchards, *Sardina pilchardus*) have also been found around the breakwater at Elmer.

b) Spread of exotic species

The arrival of new species from different biogeographic provinces has increased in recent years. The main vectors are ships and aquaculture. Thus new highly competitive species in Europe such as seaweeds *Undaria* and *Sargassum* (from Japan) can arrive in an area and markedly change the ecology of an LCS (Floc'h *et al.*, 1996; Staehr *et al.*, 2000). Coupled with global environmental change, escapes of non-native species from aquaculture become more likely (e.g. *Crassostrea*, an oyster of far eastern origin).

c) Disturbance due to maintenance and food harvesting of LCS

Frequent maintenance of LCS, such as replacement or relocation of boulders within a structure, can cause severe disturbance to epibiotic assemblages. Maintenance of LCS reduces effectively species diversity by keeping the assemblages at an early successional stage, thus dominated by opportunistic species such as ephemeral algae (*Ulva* spp., *Porphyra* sp.). As a consequence, frequent maintenance, while increasing the availability of uncolonised space (bare rock), will have profound effects on the species richness and on the biomass supported by LCS.

d) Broad-scale eutrophication

Eutrophication (anthropogenic nutrient enrichment) is a common phenomenon in enclosed bays and estuaries due to a combination of agricultural run-off and human and agricultural wastes (Correggiari *et al.*, 1992). It can also scale up to larger areas such as the northern Adriatic, parts of the Baltic and the southern North Sea and possibly the Irish Sea, resulting in eutrophic seas (Allen *et al.*, 1998). On a large scale, atmospheric input of nitrogen can also be important.

Eutrophication causes several effects in the marine ecosystem. Higher concentration of nutrients will lead to an increase in the abundance of phytoplankton and consequently greater food resources for filter-feeders such as mussels. However, the likelihood of harmful algal blooms (e.g. red tides) will also increase causing anoxia and thus killing macroalgae and marine invertebrates (Southgate *et al.*, 1984). Macroalgal growth, for example ephemeral green algae, will also be faster in eutrophic conditions, in many instances being able to outpace grazing activities.On LCS, eutrophic waters coupled with high levels of disturbance will create optimal conditions for proliferation of slippery green algae.Sediments, in turn, will tend to become muddy and compact, leading to substantial changes in the chemical gradients in the sediment (e.g., anoxia) which will, in turn, modify the infaunal composition (i.e., reduction of diversity,and proliferation of opportunistic species). Impacts of eutrophication will be worse on the landward side of LCS, where water movement is significantly reduced, particularly if the structures are connected to the shore by groynes.

e) Localised acute and chronic pollution

Acute pollution incidents (e.g., oil spills) and chronic point source pollution (e.g., heavy metals, persistent organics including leachates from antifouling paints) will affect the species composition and successional processes of benthic assemblages. On rocky shores acute incidents such as oil spills (e.g., Torrey Canyon) generally lead to mass-mortality of organisms, in particular more sensitive species such as limpets (Southward and Southward, 1978). Following deaths of these grazers, early successional, opportunistic species such as ephemeral algae will flourish. Other macroalgae such as fucoids will follow but marine invertebrates such as barnacles and limpets will take longer to recolonise. Epibiotic assemblages on LCSs will be similarly affected by such pollution incidents. Chronic pollution can severely affect the epibiotic species. For example, predatory whelks, which are commonly found on LCSs, have been shown to be particularly sensitive to TBT pollution from antifouling paints which can induce «imposex» (females become masculinised) leading to sterility (Gibbs and Bryan, 1986; Bryan *et al.*, 1986; Spence *et al.*, 1990). This problem is still evident near marinas and commercial ports, despite the ban of TBT on small boats throughout Europe.

Under certain conditions, however, the effects on benthic communities caused by both acute and chronic pollution generally tend to reverse once the pollution source is eliminated or reduced. For example, after the clean-up of the river Mersey (near Liverpool, UK) limpets (*Patella vulgata*) and dogwhelks (*Nucella lapillus*) have been found recolonising LCSs on Merseyside in recent years.

f) Overexploitation of natural living resources

Overfishing and the proliferation of coastal infrastructures such as marinas and sea defences have significantly reduced the fish stocks, particularly for species that tend to settle in shallow coastal waters. LCSs, however, seem to create suitable habitats (particularly the sheltered landward side) for settlement of juveniles of commercial fish such as sea bass, sole and plaice, and crustaceans, such as lobster and crabs. LCSs therefore could represent new nursery grounds for fish, contributing to enhance the local fishery.

g) Effects of recreational use of LCS

Shellfish harvesting and recreational use of LCSs can lead to disturbance through collection of a range of organisms for food, bait, or aquaria, and trampling, particularly during summer (Durán and Castilla, 1989; Kingsford *et al.*, 1991; Dye, 1992; Keough and Quinn, 1998; Fraschetti *et al.*, 2001; Moreno, 2001). These activities are likely to affect the persistence, growth and abundance of more vulnerable species, thus leading to changes in diversity and dynamics of the whole assemblage, as largely documented on rocky shores (reviewed in Thompson *et al.*, 2002). For example, on LCSs along the North Adriatic Sea mussels are subject to intensive harvesting, creating patches of bare space and increasing the abundance of pioneer species such as ephemeral algae. Intensive fishing removes top level predators and may alter the food webs leading to an increase in lower trophic levels such as limpets and an associated reduction, in algal abundance, especially ephemerals (Bulleri *et al.*, 2000). Similar effects could occur if predatory birds such as oystercatchers are scared away by human activities (Coleman *et al.*, 2003). Scaring away birds will also reduce guano deposition that will reduce green algal bloom such as *Prasiola*, on the top of structure (Wootton, 1991).

14.2. TOOLS FOR ASSESSMENT OF IMPACTS

14.2.1. Rapid field assessment protocol for evaluation of ecological conditions of the proposed LCS

As part of the scoping study (see Section 6.10) a rapid field assessment of local ecological features should be carried out to characterise the physical and biological features of the site and enable prediction of impacts of the planned LCS. Much of the information will also be gathered as part of site characterization for engineering purposes and so it may be possible to make savings by combining these surveys.

Below is a checklist of information to be collected in a preliminary site visit. This is based on the work that can be done by a team of experienced coastal ecologists. The time necessary to accomplish the field survey will vary depending on the site where the LCS will be built. In general, more time is required for field surveys in the subtidal and microtidal shores due to technical difficulties in accessing the sites.

The site and at least two adjacent beaches 10 km to either side should be visited. In macrotidal shores, it is essential to carry out the field visit at low tide and also high tide, ideally on spring tides, whereas in microtidal habitats the visit should include a scuba diving survey. The area visited should also be defined by GPS coordinates.

At each site a sketch of the beach profile at low tide (or by diving, for subtidal systems), on 3-4 transects should be drawn.

Biotopes at various shore levels (e.g. HWS, MHWN, MTL, MLWN, MLWS on macrotidal shores or depth intervals on microtidal shores) should be described using standard classification schemes (e.g. Connor *et al.*, 1995; Garrabou *et al.*, 1998). Some digging and sieving along with photographs of the area may be required to help identification of biotopes and characterisation of sediment characteristics (grain size, oxic layer).

Visits to adjacent rocky shores or any artificial structures (seaside piers, groynes, harbour walls, moles, jetties, existing sea walls etc.) should be made, carrying out a rapid assessment of rocky shore biotopes present (using BioMar classification). Particularly, evidence of scouring around any hard substrates should be noted. In the assessment, the presence of the following key species should be recorded: mussels, as they both play an important role in filtration (Wilkinson *et al.*, 1996), but they can also interfere with performance of LCS if very abundant (by reducing porosity of structures); *Sabellaria*, a reef forming worm that can reduce porosity as do mussels; limpets, winkles & topshells, which are important for controlling algal growth (Jenkins *et al.*, 1999, 2001; Thompson *et al.*, 2000; Boaventura *et al.*, 2002); green algae, that can represent a nuisance for recreational use of LCS and may indicate disturbance; fucoids, as they can provide an indication of wave exposure (e.g. for Atlantic: *Ascophyllum* is an indicator of sheltered shores whilst *Fucus* is an indicator of more exposed sites, Raffaelli and Hawkins, 1996); proportion of dead and live barnacles, as an index of scouring on the structures; presence of starfish and gastropod *Nucella*, which feed on mussels and can control their abundance (Minchin and Dugan, 1989); *Cystoseira* species, as they could provide information on the environmental quality in the Mediterranean (Benedetti-Cecchi *et al.*, 2001), as well as seagrasses that could also contribute too stabilize the coastline; *Capitella* and other indicators of organic enrichment in soft bottoms (Airas and Rapp, 2003). It also important to search for presence of alien species (*Sargassum, Undaria, Caulerpa, Rapana, Occulina*, non native oysters such as *Crassostrea gigas* in the UK). In the absence of hard structures navigation buoys can be a good indicator of the likelihood of local subtidal epifaunal assemblages.

Accumulation of algal and seagrass detritus on the beach should be quantified, as the presence of LCS is likely to increase the accumulation rate, which could have both negative and positive effects (Alongi and Tenore, 1985). From a recreational viewpoint, the accumulation of detritus is seen as a negative impact, while they may contribute to stabilizing the coastline.

Algal and seagrass detritus in the strandline should be examined to assess the pool of algal species in the region, as well as the dead shell assemblages that could provide information on the mollusc diversity of the region (Hily *et al.*, 1992).

The boundary between terrestrial and marine habitats should be surveyed, noting whether they are artificial or natural, or have physical or biological features of scientific or natural interest such as vegetated shingle banks, sand dunes, coastal lagoons). In addition, photographs should be taken.

14.2.2. Baseline ecological surveys

As part of the Environmental Impact Assessment procedure, a detailed survey of the selected site for the LCSs and the relevant coastal cell should be carried out to assess both local and large-scale effects.

14.2.2.1. Local effects (near field)

Survey profiles of the beach (at least 3 transects) to run at right angles and across the proposed site of the structure(s). If possible undertake the survey at the end of summer (August/September) and at the end of the winter (February/March).

Along these transects take at least 3 to (preferably) 5 sediment cores (the size depend on the grain size but at least 15 cm diameter, 20 cm deep, 40 x 40 cm sediment boxes or 600 cm^2 grabs) at vertical intervals along each transect (at least 5 but no more than 10 intervals per transect). Spacing depends on the shore communities present (on the basis of rapid assessment). The default option is uniform spacing. At each sampling station, take at least 2 samples for analysis of sediment granulometry, organic matter and chlorophyll a (using standard methods: see HMSO, 1983; Holme and McIntyre, 1971).

14.2.2.2. Far field effects and broader context

In addition to replicated transects at the site, at least two reference or control locations should be surveyed ideally on either side of the construction site, using the same survey and sampling protocol described above.

The reference sites should be selected to be as similar as possible in terms of wave exposure and geomorphology. This survey should be carried out in the same period as that used to assess local effects.

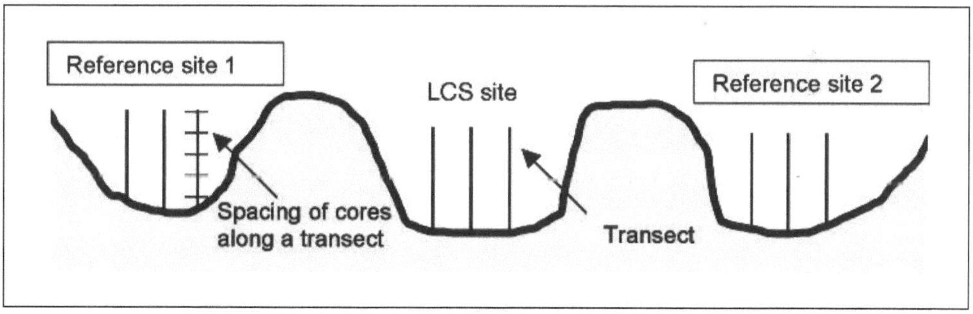

Figure 14.1. Diagram showing sampling design to be carried out in the pre-construction phase.

14.2.2.3. Methods

Each parameter should be measured using the following standard methods.

a. Organic matter. Organic matter can be estimated by oxidation methods: the simplest method is by burning organic material and taking the differences between dry weight and ash free dry weight. Wet oxidation using potassium permanganate can also be used (see in Holme and McIntyre, 1971).

b. Granulometry. Standard methods using nested sets of sieves or automated system should be used (see in Holme and McIntyre, 1971). A sample pre-treatment with hydrogen peroxide should be used to eliminate residual organic detritus such as shell fragments and algal debris. However, care should be taken when sediment consist of carbonate particles. The basic parameters that should be measured are percentage of silt and clay and very coarse sand, mean grain size, median and phi.

c. Chlorophyll a. Chlorophyll can be extracted using solvents (methanol, ethanol or acetone) and quantified using spectrophotometric of fluorimetric techniques. Standard methods for soft sediments (HMSO, 1982) should be followed. Presence of pheopygments should be estimated through acidification of extracted chlorophyll.

d. Macrofauna. Sediment samples should be sieved on a 0.5 mm sieve and the biota retained and preserved in formalin (see in Holme and McIntyre, 1971). Samples should be sorted and, when possible, organisms should be identified and quantified to species level.

Data should be analysed using a combination of multivariate (e.g. MDS, PCA, BIOENV, Clarke and Warwick, 2001) and univariate (e.g. ANOVA,Underwood, 1997) analysis, preferably on the basis of a beyond BACI experimental design (Underwood, 1992, 1994).

14.2.3. A biotope model for prediction of impacts on soft-bottoms

14.2.3.1. Introduction

Within the framework of the DELOS project, a methodology was developed that can be used to predict the environmental effects of adding an LCS to a coastline area. The method is based on a combination of predictive modelling of physical changes in the environment and analysis of these changes from the viewpoint of effects on species habitats. This approach is particularly suitable for sandy beaches where the macrofauna communities are controlled almost entirely by physical processes (McArdle and McLachlan, 1992) i.e. each population is structured by its response to the physical environment rather than by biological interactions (McLachlan *et al.,* 1995).

14.2.3.2. Methodology

The methodology is based on a three-step approach, namely: predictive modelling, selection of biotopes, collection of baseline data and analysis of impacts.

14.2.3.2.1. Predictive modelling

The DELFT3D package, developed by WL Delft Hydraulics and the MIKE 21 suite, developed by DHI Water & Environment, can be utilised amongst others to describe wave action hydrodynamics, and sediment transport in the midfield and farfield of a study area. Both model suites consist of a number of integrated modules which together allow the simulation of hydrodynamic flow (under the shallow water assumption), computation of the transport of water-borne constituents such as salinity and heat, short wave generation and propagation, sediment transport and morphological changes, and the modelling of ecological processes and water quality parameters (see Lesser *et al.*, 2003).

14.2.3.2.2. Biotope selection

The second stage involves finding a way of linking the physical changes to effects on the ecology and this is done, for instance, by using the BioMar Classification developed for the UK and Ireland by Connor et al. (Connor *et al.,* 1997). A biotope is defined as «the habitat together with its recurring associated community of species, operating together at a particular scale» (Connor *et al.*, 1997). The classification provides a link between the physical environment and its associated biological community, which is exploited in this methodology in order to predict changes in the latter as a result of changes in the former. All the output produced by the physical model (such as current velocity, bed shear stress, height zone) are subsequently converted to classifications to match the BioMar physical parameters definitions. Other parameters used as part of the BioMar classification (salinity and substratum type) were input directly rather than produced as a result of the model.

14.2.3.2.3. Baseline data collection

In order to prepare an impact study, baseline data need to be collected for a study site.

a. Physical data. Bathymetric, tidal range and wave data measurements from the area are necessary as inputs for the model. For waves, typical stormy weather conditions should be included as these conditions could be most structuring for local biotopes distributions. In addition, a map of substratum types is needed. The substrate definitions given in the BioMar system are most suitable. On the basis of above data, the mathematical model can produce values for maximum bed shear stress and maximum current velocities for each cell based on combinations of waves and currents and pre-design locations of LCS and/or other structures.

b. Biological data. Fieldwork should be carried out in order to produce an accurate map of biotopes for the real situation for comparison with the situation predicted by the model. Biotopes should be mapped using GPS to mark the boundaries. Infaunal cores should also be collected to confirm the biotope designations.

14.2.3.3. Results

The environmental impact of any amount of cases (various breakwater layouts in combination with various environmental forcing conditions) can be predicted by numerical modelling. The result for each case is a set of BioMar class values for physical parameters being designated for each cell. A procedure is then applied that selects the biotopes that can occur within the predicted set of parameter class values for each cell. Biotopes recorded in the field during baseline data survey can be compared with those predicted by the model. This enables calibration of the model to the present situation and allows evaluation of the type and magnitude of changes for each computed case in a straightforward fashion.

For the field situation at the Elmer study site, a total of six biotopes were mapped. The predictive accuracy of the model (Delft 3D was used in this case) for the situation of a breakwater with no waves was 65%. For the situation of a breakwater present with waves, the model accurately predicted 69% of the biotopes that had been recorded. As expected, for the control situation without breakwater, with relatively few biotopes, the model achieved a high accuracy rate of 97% although this dropped to 76% if the situation with waves was modelled. The hierarchical nature of the BioMar classification means that the model can also be used to predict biotope complexes, the next level up in the hierarchy. These initial trials with the model are encouraging and the model is still being refined in order to develop a tool for more accurately predicting change in the identity and extent of biotopes as a result of the addition of breakwaters.

CHAPTER 15

Design tools related to socio-economics

15.1. GENERAL DESCRIPTION OF COST BENEFIT ANALYSIS

(Polomé, UTW)

This section summarises the relevant information from Hanley *et al.* (1993), Ridell and Green (1999), U.S. Environmental Protection Agency (2000), Lipton *et al.* (1995), Bateman and Willis (1999), and Polomé *et al.* (2001).

Although there are several techniques for appraising policies and projects which impact the environment, the DELOS project concentrates on **Cost-Benefit Analysis (CBA)**. Only CBA can in itself decide whether it is worth implementing a policy or not in the sense that the sum of all the positive impacts of that policy outweighs or not the sum of its negative impacts. In any CBA, several steps must be conducted, they are briefly described in this chapter.

When benefits are complex to estimate and/or their estimation is liable to large errors, it is common to assume that all the projects under consideration have roughly the same benefits. To choose among different projects, one then may resort to **Cost-Effectiveness Analysis** (CEA). In essence the same steps as in CBA apply, but only the costs, and not the benefits, of the project are taken into account. Later on we will define costs and benefits differently, but for CEA only construction and financial costs matter, because intangible or non-market costs are outside of the realm of CEA.

Step 1: **Definition of the project.** This step includes the reallocation of resources being proposed; and the population of gainers and losers to be considered.

Step 2: **Identification of project impacts.** Draws a qualitative and exhaustive list of the impacts resulting from the project implementation. *Additionality* refers to the net impact of the project, for example, the impact on beach erosion of a coastal defence must be computed net of other changes in beach erosion that would have occurred without this policy change. *Displacement* refers to shifting a problem somewhere else, for example when a defence structure at one point of the coast causes erosion downdrift. When perfect displacement occurs within the population defined in the previous step, then the project has no value.

Step 3: **Relevant economic impacts.** We assume that society is interested in maximising the weighted sum of utilities across its members. These utilities depend upon, among other things, consumption levels of marketed goods (e.g. fish) and non-marketed ones (e.g. fine views, clean beaches, risk of inundation). We term positive impacts on that sum benefits, and negative impacts costs. For example, a sea defence project could affect the landscape and have adverse effects on fish spawning grounds. The former is relevant to CBA if at least one person is not indifferent to the landscape change, the latter is relevant if at least one fisherman or one angler captures fewer fishes. The fact that there is no market for landscape is irrelevant, all that matters is that an impact on production or on utility can be recorded.

Step 4: **Physical quantification of relevant impacts**. The physical amounts of benefits and costs flows for a project are determined, and the time at which they will occur is identified.

Step 5: **Monetary valuation of relevant effects**. The essential idea behind monetary valuation is to express all the relevant impacts in a common unit. At this step, the analyst in a CBA has to predict prices for value flows extending into the future, correct market prices when necessary, and calculate prices where none exists.

Step 6: **Discounting**. Once all the costs and benefits have been expressed in monetary terms, we convert them into present value terms using the real interest rate. A value of 6% is often advised in practice, but 3% has been used in coastal defence.

Step 7: **Applying the Net Present Value (NPV) test**. The main purpose for applying CBA is to select projects which are efficient in terms of their use of resources. This is achieved if the project sum of discounted benefits exceeds the sum of discounted costs, that is the Net Present Value test. There are a number of alternative tests, but they all refer to the same idea.

Step 8: **Sensitivity Analysis**. It is instructive to recalculate the NPV when the value of key parameters are changed (interest rate, physical quantities or qualities, prices, project life span).

15.2. CLASSIFICATION OF COSTS AND BENEFITS AND INVENTORY OF COASTAL ASSETS

(Polomé, UTW)

15.2.1. Principle of economic value and typology of values

The concept of economic value that we will use in these guidelines is the Willingness To Pay (WTP) defined as the maximum amount of money a person is willing to exchange to acquire a good or service that he considers desirable. The economic value does not refer to an exchange of money or to a price, the goal is to convert «utility» or «well-being» into money to match it against monetary costs such as those of building a coastal defence scheme. The WTP is used, and not prices, because of the presence of non-marketed goods such as a coastal defence. A government provides the defence scheme, but cannot charge the consumers for it. Economics addresses this issue by converting the change of well-being into money, and compares it to the actual money that has been spent on providing the good.

Several methods exist to estimate the sum of WTP for different classes of public goods. Defining economic value is important because it makes clear that a broad class of benefits should be considered in a CBA, not only those benefits generated by a monetary transaction. Yet, economic value is not the only criterion for deciding on public projects and projects should be restricted by equity considerations, precautionary environmental standards, and regional economic constraints.

The value of the coastal defence scheme is composed of the sum of the values of the consequences of that scheme on the seafront and on its residents, provided it is possible to avoid double-counting. Often different types of values will require different valuation

Table 15.1. Coastal Defence Values. Adapted from Bower and Turner (1998).

Value name	*Example*	*Valuation Method*
	Use	
Direct Use	– Construction & maintenance costs – Fishing – Agriculture – Transport, navigation	Market pricing (possibly adjusted)
	- Recreation	Travel cost Stated preferences
Indirect Use	– Flood control – Storm protection – Sedimentation – Habitat loss reduction – Landscape – Human health	Market pricing Hedonic pricing Stated preferences
	Non-use and Option use	
Option	– Insurance value of preserving options for use	Stated preferences
Quasi-option	– Value of increased information in the future (biodiversity)	Stated preferences
Existence and Bequest	– Knowing that a species or system is conserved – Passing on natural assets intact to future generations – Moral resource/Non-human rights	Stated preferences

methods. Classical typologies of values following Turner *et al.* (1992) and Bower and Turner (1998) are presented in Table 15.1. This table is best interpreted as «motives for valuing» the assets given in the examples. The third column indicates the valuation methods that would be most suitable for estimating each value. This is not an indication that it has been estimated. An overview of the valuation methods is given in the next section.

15.2.2. Overview of the valuation techniques

Haab and McConnell (2002) provide an excellent technical reading for this section. The valuation techniques are divided into stated and revealed preferences. **Revealed preferences** methods rely on market information and have several steps. First, estimate the demand curve of a market good. Second, based on that estimate, forecast the change in demand caused by the change that we want to value and compute the new market equilibrium. The change in consumer surplus is the change in area below the demand curve and the price line. The price of a market good is sometimes equivalent to the marginal social cost and marginal social benefit of a unit of that good; as an approximation, and if the market can be said to be competitive, the social benefit of a project that increases marginally the output of such a good can be taken as the product of price times quantity.

For some goods, there is normally no observable demand but there is a complementary or substitute market good that can be used instead. The **travel cost method** is concerned with changes in the quality of a recreational site. The basic idea is that the consumer surplus of the demand for travel to that site is equivalent to the consumer surplus for that site. **Hedonic pricing** captures the WTP associated with variations in property values that result from the presence or absence of specific environmental attributes. The **production function approaches** link environmental changes to changes in production relationships. This may relate to firms producing goods and services, or to households producing services that generate utility. The main idea of the approaches in this group is that changes in expenditures are due to the need to substitute other inputs for changes in environmental quality. One such approach is called avoided cost (or defensive expenditure): the value of an environmental improvement can be inferred directly from the reduction in expenditures on defensive activities. The dose-response function is another such approach (also known as factor income method), it links environmental quality and the output level of a marketed commodity, such as water pollution impacts on fisheries.

Stated preferences methods are used for changes in non marketed good with no complementary or substitute market good demand (landscape, natural or cultural heritage. In that case, one can only resort to directly asking individuals (in a survey) how much they are willing to pay to obtain that change. The precise way to ask that question is the subject of much debate and has given rise in practice to several methods. The ones that have been most used are contingent valuation and choice experiment. The **contingent valuation** is the most developed stated preferences method and is very well documented. It consists in directly asking individuals to state their WTP for some previously described change in a non-marketed good. There are several ways of asking such a valuation question and design of such question is the key issue in contingent valuation. The **choice experiment** method strives to place the respondent in a natural choice situation: two to four options are carefully described using attribute levels and pictures (for example, different kinds of defence structure may be pictured, along with levels of biodiversity such as number of birds, and some measure of recreation, e.g. expected fish catch), the cost to the respondent of each option is simply another attribute. The respondent is then asked to indicate which option he prefers. Statistical techniques are used to estimate trade-offs between attributes, which result in monetary values when the costs is used in the trade-off.

15.2.3. Typologies of coastal assets

The purpose of this section is to present types of assets the supply of which may be modified by a coastal defence scheme (see Bower and Turner, 1998; Fankhauser, 1995; Penning-Rowsell *et al.*, 1992). For a detailed list, see Polomé, (2002).

Mitigation benefits or costs

- Reducing damage (including preventing complete destruction) to coastal properties from coastal storms and eroding shorelines.
- Reducing salinity intrusion.
- Reducing sedimentation in navigation channels and in harbour areas.
- Reducing sedimentation on spawning beds and coral reefs.
- Restoration or preservation of habitats.
- Restoration of recreational opportunities, e.g. sand beach.
- Human health in the sense that defence reduces the risk of accident (e.g. storm impact).
- Reducing damages to cultural and heritage assets. Note: buildings can be valued in

two ways-erosion can cause complete loss, in that case we seek the discounted value flow of the whole building as in Yohe, Neumann and Marshall (1999) or Fankhauser (1995); but erosion may simply mean that the probability of temporary flooding increases, that is only an inconvenience not a complete loss, that would be valued through hedonic pricing.

Enhancement benefits or costs

- Increased output of the seafront caused by the defence scheme, e.g. creation of recreational opportunities. In general, an LCS can be seen as a type of artificial reef, and thus may increase fish output.
- Deepening of navigation channels (as a result of the scheme).
- Finfish and shellfish yield declines.
- Water quality that is affected by changes in marine currents or sewage system caused by the defence scheme; can be positive (improved sewage systems) and negative (eutrophication, red tides).
- Conflicts among different types of recreation users of beach areas caused by the defence scheme.

Preservation benefits or costs. This refers to natural areas that are preserved, directly or indirectly, by the defence scheme. One example is the Aldeburgh British scheme in which inland and seafront marshes were indirectly protected by a sea wall. The benefits stemming from the preservation of a natural ecosystem are generally recreational use and non-use. An in-depth case is described in Goodman *et al.* (1996). Offshore sand and gravel mining (e.g. to find the sand for beach nourishment) may affect fisheries and habitats.

Indirect economic benefits or costs. These are «second round» effects, e.g. assume a defence scheme improves recreational opportunities by allowing scuba diving (maybe

Table 15.2. Reported values for direct consumptive use.

Asset.	*Benefit/cost*
Land of all types including land for residential, commercial and industrial activities *and* agriculture	Loss of land
Yohe, Neumann and Marshall, 1999. In the absence of threat, land prices follow the equation $d[\ln(P_t)] = \alpha + \lambda L + \psi Y + \beta d[\ln(P_{t-1})]$ where P_t is the real price at t, L is the population growth rate, and Y is the per capita income growth rate. The symbol $d[\,]$ indicates a growth rate. This equation is estimated for each of the 30 sites in their sample. Land values continue to follow the equation and drop to zero when inundation occurs. The authors estimated the equation with US data, but do not indicate any value directly. For an application, it is necessary to collect local prices and estimate the equation.	
Fankhauser (1995). Average land value is set to \$2 M/km^2 for open coasts and beaches and \$5 M/km^2 for wetlands (non-built lands only).	
Fisheries	Yield changes
Farber (2001). M \$ 0.25-0.36 expected over 100 years for 170 km Louisiana barrier islands system through protection from storms.	

Table 15.3. Reported values type for direct non-consumptive use.

Asset	*Benefit/cost*
Bird viewing	Preservation, enhancement
Loomis and Crespi (1999). Value per day of viewing (1992, US $) 29.91 for one viewer in the USA. Other data have shown that a 1% change in the number of birds seen per trip results in a change of 0.173% bird viewing trips. It is assumed that a reduction of 1% of wetland area results in an equal reduction of bird population, which in turn results in an equal reduction of birds seen per trip. Transferring to a particular site still requires to know the number of visitors.	
Waterfowl hunting	Preservation, enhancement
Loomis and Crespi (1999). Value per day of hunting (1992, US $) 30.45 for one hunter in the USA, a 1% change in wetland acres results in a 0.275% change in hunter days. Transferring to a particular site still requires to know the number of visitors. Waterfowl hunting is much more practiced in the USA than in Europe, it is not expected that this value can be transferred to a European context.	
Beach visitation (informal recreation)	Preservation, enhancement
Loomis and Crespi (1999). Value per day of visit (1992, US $) 16.3 for one visitor in the USA. A 1% change in the length of shoreline (in meters) results in a change of .425% change in the number of visits in North-eastern US, of 0.096% in Southern US, and of 0.147% in Western US.	
Silberman and Klock (1988); Ruijgrok (1999); Whitmarsh et al. (1999); King (1995); Green (personal communication); Hanemann (personal communication): this is the data used in the next section.	
Penning-Rowsell et al. Yellow Manual (1992). UK£ 7.55 VOE per visit for generic beach. See also the section on benefit transfer.	
Fouquet et al. (1991) in Green (2001). UK£ 7.15 VOE per visit for generic shingle bank.	
Costa et al. (1992) in Green (2001). UK£ 8.75 VOE per visit for generic promenade.	
NOAA (1995) (personal communication). US$ 11 WTP for use of generic beach per visit.	
All recreational seafront activities	Preservation, enhancement
Farber (2001). M $ 1.12-1.33 expected over 100 year for 170 km Lousiana barrier islands system through protection from storms.	

because interesting species have settled in). The «first round» benefits come directly from the increased recreational activity (in as much as it a net increase). A «second round» benefit may be the establishment of a specialised shop for scuba diving.

Another example is constructions in hazardous areas in relation to coastal storms that are built because of the protection granted by the defence scheme (resulting possibly in a stronger scheme being necessary in the future, see Cordes *et al.*, 1998 and 2001).

15.2.4. Indicative values per type of coastal asset

In this section, we present references to actual figures of values for some of the above types of benefits. The literature does not cover all the potential benefit and costs of coastal defence. There is only one type of value for which there is a substantial number of estimates, this is

Table 15.4.Reported values for indirect use

Asset	*Benefit/cost*
Residential, commercial and industrial non-heritage **buildings**	Inundation (complete loss)
Yohe, Neumann and Marshall (1999). Building prices follow $d[\ln(P_t)] = \alpha + \lambda L + \psi Y + \beta d[\ln(P_{t-1})]$ with the same symbols as in Table 15.2. This equation is estimated for each of the 30 sites in their sample. Buildings start depreciating 30 years before inundation in an efficient market and reach zero at T at which time they are abandoned. If the market is not efficient or if abandonment is uncertain then the market has less than 30 years to react and properties do not have a value of zero at time of abandonment, they investigate a scenario of no foresight at all, as if SLR would occur instantly, and the equation applies until T. The authors estimated the equation with US data, but do not indicate any value directly. For an application, it is necessary to collect local prices and estimate the equation.	
Fankhauser (1995). Average value set to \$ 200 M/km^2 for cities and harbour.	
Farber (2001). M \$ 15.3 (M \$ 21.5) expected over 100 years for 170 km Louisiana barrier islands system through protection from 90.5º W (91.5º W) storms. 1 km of barrier protects 30 km^2 of land.	
Dorfman et al. (1996). Given a probability P of loss, an increase of 1% of the risk of inundation causes a decrease of .2 P% of the house price.	

Table 15.5. Reported values for non-use values.

Asset	*Benefit/cost*
Ecosystem and natural heritage, beach	Preservation for motives of Option, Quasi-option, Existence or Bequest non-use values
Silberman and Klock (1988). US \$ 16.3 as a one-time contribution/visitor.	
Ecosystem and natural heritage, global (large areas including all coastal types of natural assets)	Preservation for motives of Option, Quasi-option, Existence or Bequest non-use values
Goodman et al. (1996). UK£ 48.36 for maintenance, annual for 30 years, for an English or Welsh household for the whole length of the English and Welsh coast.	

informal beach recreation that is studied in detail in the next two sections of this report. For some classes of benefits (land protection, bird viewing, waterfowl hunting), benefit transfer results are available in the literature, although their applicability in the context of the DELOS project is limited.

In the context of the DELOS project, it is possible that in some circumstances not all economic values are acceptable, but only those that lead to a measurable flow of money generated by the use of resources. These are financial values, a subset of the economic values. English Nature Research Reports No. 182 is dedicated to marine and coastal wildlife areas in England and details the methodology of collecting data on the financial values of a given site.

15.3. TRANSFER OF EMPIRICAL VALUES

(Polomé, UTW)

The objective on this section is to present an example of benefit transfer for coastal defence. Enough data to attempt a transfer exercise are available only for informal beach recreation.

15.3.1. Data sets

The data set comes from three sources. The first one is a library search of published and unpublished papers, including reports and theses. This list of references can be found in the report of the DELOS WP 4.1 (Polomé, 2002). It is important not to restrict the search to published papers.

The second source of data comes from Professor Colin Green (Flood Hazard Research Centre, Middlesex University) who gave us several unpublished results. The data are very scarce regarding the description of each site being valued and the socio-economic characteristics of the local or visiting populations. A second problem comes from the valuation procedure used to acquire these data, following the Penning-Rowsell *et al.* (1992), comparatively with the international standards applied in valuation. The Value Of Enjoyment (VOE, detailed in the next section) has been used instead of the internationally used WTP. VOE is to be seen more as an average of the prices of experiences similar to a visit to the beach; WTP is the maximum amount a person would pay to visit the beach. Those values are quite different. Another difficulty with the VOE is that it does not seem to take substitute sites into account. The literature on valuation has solved this problem by resorting to what is known as Multiple Site Travel Cost Models (see e.g. Herriges and Kling, 1999), but this methodology is scarcely applied for beach recreation.

The third source of data comes from Professor Michael Hanemann (University of California at Berkeley). The data originate from studies by the US National Oceanic and Atmospheric Administration (NOAA) with the purpose of issuing recommended values for informal beach recreation. The NOAA currently recommends a rough value of 11$ per beach day per visitor, but Professor Michael Hanemann, after carefully reconsidering each study, recommends values ranging from 11 to 23$, with an average of 15$ for Florida beaches (personal communication). This reconsideration was admitted in a court of law. Professor Michael Hanemann's data are also very scarce regarding the physical description of the beach and the socio-economic characteristics of the visitors. On the other hand, they are based on more conventional valuation concepts.

Apart from those data problems, another general shortcoming of benefit transfer relates to the number of visits to the beach. All the available values are per visit to the beach. To estimate the value of the beach itself, it is still necessary to know the total amount of visitors to the beach and their number of visits. That information was not available. Counting the visitors to a beach is not easy and is prone to errors. Professor Colin Green considers that the main problem in valuation of beach recreation is counting the visitors. Another problem related to counting is estimating the number of visits per person. Another problem is on-site sample bias. This bias is due to the fact that when we randomly select visitors on-site at a beach, it is more likely that we will encounter a person who visits often than a person who visits rarely. This will bias the estimate of the count, see Shaw (1988) on this issue.

The final data set that has been used as a starting point for the regressions had 106 observations, but only 38 different sites. Some sites have been observed during more than one year, and for some sites there were hypothetical behaviour questions such as «how much would you value this beach if it was eroded». Only three countries provide data: the UK, with 79 observations, the US with 22, and the Netherlands with 5.

A first category of variables, *X*, is the site characteristics. Sites are classified according to 3 types: Coastal resort (74 observations), Beach (5) and Dune (2). There are 25 observations for which the site type is not known, but there are reasons to believe that they are coastal resorts, and this is what is assumed from here on. Another variable that is available per site is a rough measure of quality. A site can be in its current state (64 observations), eroded (20 observations) or defended (24 observations).

A second category of variables, *Y*, is the socio-economic variables. They are equally very sparse. There are 4 categories of respondents: the local visitors (16 observations), the non-local visitors among which those who stay a single day (15) and those who stay more time (15), and those observations for which this distinction is not made. This last category is a kind of average of the other three. For some sites under some circumstances, there was a value for each category. In this case, the average value (the last category) has been excluded from the regressions (15 observations removed).

The last category of variables, *Z*, relates to the study itself. A first variable in this category is the year the study took place, ranging from 1975 to 1995, with the most studies in the early nineties. The following *Z* variables are available:

Table 15.6. Study characteristics.

Value concept	*Count*	*Valuation method*	*Count*
VOE	78	Open-ended CV	89
WTP for use	13	Bidding game CV	2
CS	15	TC	15

The value itself is expressed per visit per person in € of 2001, adjusted by the consumer retail price index of the relevant countries up to 2001 and then converted to € using the average rate for 2001. The average of the values is nearly 17€, with standard deviation around 14, minimum 1, maximum nearly 92. Table 15.7 compares the data used in this report with the three other known references in which a value for transfer is presented.

Table 15.7. Value per visit to a generic beach (€ 2001).

Source	*Country*	*Current state*	*Eroded*	*Defended*	*Value concept*
Average of data available for this report	UK	17.7	9.1	20.6	VOE
	US	23.1	–	–	WTP for use or Consumer surplus
Yellow manual (1992)	UK	15.6	8.2	18.7	VOE
NOAA (1995)	US	13.9	–	–	WTP for use
Loomis and Crespi (1999)	US	22.4	–	–	WTP for use

15.3.2. Regression models and transfer

Given the previous provisions, in this section we show an example of benefit transfer in the case of informal beach recreation. A benefit transfer function is usually linear, at least in the sense of first degree approximation. To formalise the model, start with the prototype model from Brouwer (2000):

$$V_i = \alpha + \beta X_i + \gamma Y_i + \delta Z_i + \varepsilon_i \qquad (15.1)$$

where $\alpha\ \beta\ \gamma\ \delta$ are parameters, V is the value per site per visit for a given policy, X, Y and Z have been defined above and i indexes the studies. Because we have no data on several variables that could explain the value, such as beach width and length or respondents' income, Ordinary Least Squares (OLS) estimation of the Brouwer's linear model will be biased. This is a standard result with OLS: missing regressors lead to bias.

Since in the current dataset, there is often more than one observation for a single site, the model can be written as:

$$V_{it} = \alpha_i + \beta X_{it} + \gamma Y_{it} + \delta Z_{it} + \varepsilon_{it} \qquad (15.2)$$

where V_{it} the value for site i under the circumstance t. The circumstance can refer to a different point in time (a different year), or to some hypothetical situation (for example, the site is eroded). This is a panel data model, the main difference with Brouwer's linear model is that the intercept term a is now specific to each site because it is indexed by i. This is critical because the site-specific intercept term will account for all the differences in values across sites not accounted for in the regressors, and thus avoid the bias problem referred to above.

When the goal of the study is to predict the value of one site given some characteristics, bias in the estimated coefficients is not important. Therefore Brouwer's linear model can be estimated using OLS. When the goal of the study is to estimate the marginal effect of some characteristic of the beach, it is critical to estimate the coefficients without bias and then the panel data model is best. This is illustrated below.

The date (T) of the study is inserted in the regressions as a natural trend starting in 1975 (normalised to 1). The 4 categories of visitors (local residents, day visitors, stay visitors and unspecified type) are represented using three dichotomous variables (Local, Day, Stay), with the omitted category being the unspecified type. The 3 remaining categories of quality of the site (eroded, current quality, defended) are represented using two dichotomous variables (Eroded, Defended), the omitted category is the current quality.

The concept of value has three categories (VOE, WTP for use, Consumer Surplus). The 3 categories have been represented by 2 dichotomous variables (WTP, CS), the omitted category being VOE. In the panel data model, it turns out that the sum of these 2 variables is a vector of zeros and ones identical to the sum of certain site-specific constants. Therefore, one of these 2 variables had to be removed to enable estimation. Since the decision to remove is arbitrary, we present the 2 sets of results: in the first one (Table 15.8.a) the variable removed is the dummy indicating the Consumer Surplus, in the second one (Table 15.8.b) it is the dummy indicating the WTP for use.

The tables are quite similar with the exception of the intercept term, this is reasonable because of the two different dummies (WTP or CS). Neither the effect of time (T) nor of the

Table 15.8. Panel data estimates.

a) Variable	*Coefficient*	*P-value*
T	0.218	0.4933
DAY	4.700	0.2224
LOCAL	1.547	0.6873
STAY	4.116	0.2853
WTP	– 15.671	0
ERODED	– 8.369	0
DEFEND	3.295	0.0158
Intercept	19.383	0.0019

b) Variable	*Coefficient*	*P-value*
T	0.222	0.4845
DAY	6.256	0.1054
LOCAL	3.121	0.4183
STAY	5.673	0.142
CS	15.902	0
ERODED	– 8.316	0
DEFEND	3.482	0.0108
Intercept	10.216	0.0834

type of respondents (Local residents, Day visitors, Stay visitors or Unspecified) are statistically significant.

The quality of the site (Current, Defended, Eroded) is very significant. «Current» refers to the beach as it is at the moment of the study; it denotes a coastal site that is enjoyable under normal conditions. «Eroded» indicates a state, usually hypothetical, in which only a narrow range of the beach remains in place, if any. «Defended» indicates that a coastal defence scheme, also usually hypothetical, is implemented that partially modifies the aspect of the beach and may enlarge it.

Finally, the high significance of the concept of value used (VOE, WTP for use, Consumer surplus) is worrisome. It is acceptable that different concepts of value yield different values, but the problem is that different survey design (Open-ended CV or Travel cost model) have been used for the different concepts. Therefore, we cannot tell whether the differences in value are genuine or are led by the method used. If it is the former, we would still have to decide which concept of value is more appropriate. If it is the latter, then benefit transfer of informal beach recreation is flawed since a different method leads to a different value for the same beach. These are the conclusions of the panel data models regarding the effect of invidual characteristics on the site value.

The results of estimating Brouwer's model directly by OLS are shown in Table 15.9. Since the OLS estimates are biased, they are not interpreted.

Table 15.9. OLS (biased) estimates.

Variable	*Coefficient*	*P-Value*	*Variable*	*Coefficient*	*P-Value*
Constant	– 9.35	0.22	WTP	– 22.66	0.08
U.S.	23.56	0.11	CS	– 12.44	0.42
NL	1.39	0.94	ERODED	– 9.27	0.04
BEACH	– 10.94	0.32	Unspecified defence	2.95	0.53
DUNE	– 10.47	0.51	Defended by nourishment	– 1.47	0.85
DAY	– 7.82	0.14	Defended by nourishment plus groynes	3.13	0.69
LOCAL	– 9.78	0.06	T	1.87	0.00
STAY	– 8.00	0.13			

To run a transfer exercise on the basis of the regressions above, for each site run the above regressions (the 2 panel data regressions and the OLS) without this site's observation(s) and predict its value using the level of the regressors specific to this site. Then, to measure the

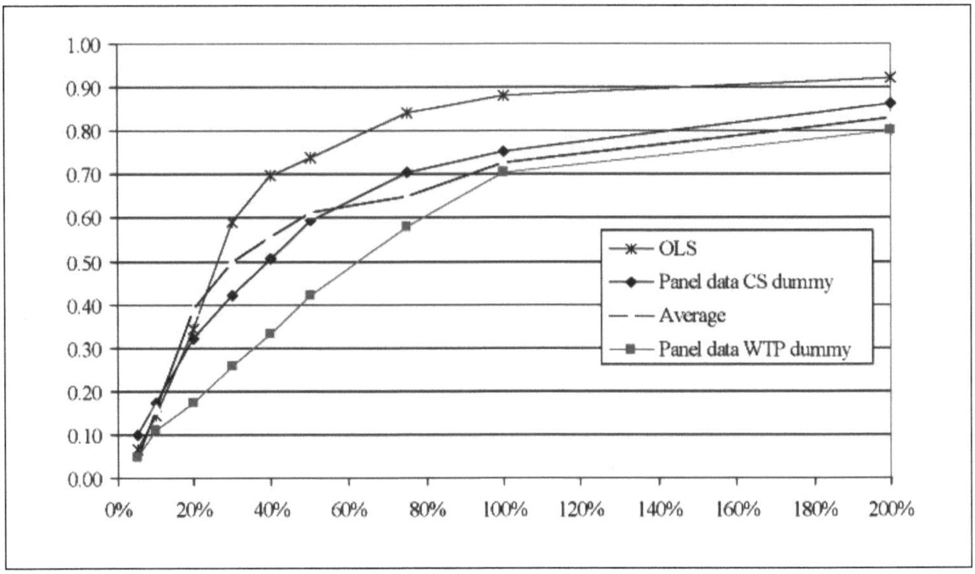

Figure 15.1. Benefit transfer cumulative distribution of prediction errors.

gain of precision obtained by carrying a new study, compare the predicted value with the one obtained from the original study. The measure of prediction error is the proportion of deviation from the value(s) reported for the site in absolute term. We also present the simple value transfer prediction which consists in predicting for one site the average value of the other sites.

Figure 15.1 reports the proportion (vertical axis) of predictions that falls below the error level indicated on the horizontal axis. We call that the cumulative distribution of prediction errors. For example, the proportion of predictions of less than a 40% error is about 70% for OLS and 55% when the prediction is the average of the values of the other sites. We say that model A predicts better than model B when the cumulative distribution of prediction errors of model A is above that of model B. In that sense, the panel data models are worse than a simple average of values (but that does not undermine their qualities for an unbiased estimation of regression coefficients). For prediction purposes, our best model is the OLS.

In summary, we have shown that to transfer benefit Brouwer's equation could be estimated by OLS. Figure 15.1 reports the risk of error in doing so. To find out about the marginal effect of some characteristic, panel data models could be used.

15.4. NON-MARKETABLE RECREATIONAL USE VALUE OF A BEACH

(Marzetti, UB)

15.4.1. Introduction

Within the Cost Benefit Analysis (CBA) framework (see Sections 15.1 and 15.2), different methods exist for evaluating the non-marketable use (present informal recreational use) of a beach in different scenarios (status quo, erosion and expansion), and a wide economic literature on this topic is available (Polomé *et al.*, 2001).

This section does not describe in detail how to estimate the non-marketable beach use, but focuses on the contingent valuation method (CVM) in the Value of Enjoyment (VOE) version (Penning-Rowsell *et al.*, 1992) which, within the DELOS Project, was applied to the following Italian case-studies: Lido di Dante, on-site survey of 600 interviews (Sub-section 11.4.7); Trieste (Barcola seafront), resident survey of 600 interviews (Marzetti, 2003a; Marzetti and Lamberti, 2004); Pellestrina, on-site and resident surveys of a total of 150 interviews (Sub-section 11.3.7); and Ostia, on-site survey of 100 interviews (Sub-section 11.5.5).

After a brief description of this e valuation method, we focus on two main issues: i) the estimate of the recreational use value in different seasons, and ii) the extension of the market (or the aggregation level) which is not only national but international where the site is visited by foreigners.

15.4.2. Methodology used for the Italian case-studies: the questionnaire

The CVM is based on the well-known economic consumer theory: *individual values* reflect individual preferences - or enjoyment, or welfare - according to the constraints perceived by the consumer (visitor). By means of a survey, the CVM aims to create a hypothetical market which permits respondents to express the non-marketable use value for a beach change. The sample of the relevant population is random.

Every respondent expresses a value which is contingent to the hypothetical scenario created within the survey. Different beach scenarios are considered (Marzetti, 2003a). When a beach changes due to erosion or expansion, the consequent VOE change of a daily beach visit represents a benefit or a loss, depending on whether the beach change is considered an improvement or a worsening of the status quo respectively.

A CVM survey consists of different steps: i) survey design (questionnaire), ii) pilot survey, iii) sampling design, iv) main survey. At the heart of the CV approach is the questionnaire, which attempts to develop plausible scenarios in which evaluations can be made. The basic VOE questionnaires used for the Italian case-studies are those published in the *Yellow Manual* (Penning-Rowsell *et al.*,1992, Appendices 4.2 (a) and (b)). They were adapted to the specific characteristics of the Italian case-studies.

In its wording a questionnaire is generally divided into parts: i) to collect information about respondent's residence; more specifically if s/he is resident (people who live at the site considered), or day-visitor (non-residents who visit the site, but return home the same day) or tourist (non-residents who visit the site and stay the night at that site); ii) to collect information about the type of beach recreational use, and number of visits; iii) to evaluate the enjoyment of a daily visit to the seafront in its current condition; iv) to evaluate the change of enjoyment after the possible beach change (erosion or artificial expansion) and, if the respondent would go to another beach, to find out the VOE and cost of transport of the alternative beach; v) to collect data about the social characteristics of respondents; vi) to obtain information from the interviewers about respondents' understanding of the questionnaire.

The structure of the valuation question is as follows (Penning-Rowsell *et al.*,1992): «We are trying to find out how much value you, as an individual, put on your enjoyment of this visit to this seafront today. Now this is an unusual question to ask so let me explain it to you in this way: Think of a visit or activity you have done in the past which gave you the same amount of enjoyment as your visit to this seafront today (a show card with a list of possibilities is shown). Now think about how much that visit (or other activities) cost you. Remember that the cost of a visit may include petrol and parking costs or bus or train fares as well as admission charges and any costs. You can use the costs of that visit (or other activities) as a guide to the value of your enjoyment of today's visit to this seafront. So, now, what value do you put on your individual enjoyment of this visit to this seafront?».

This elicitation question is asked about each different scenario. The format is Open-Ended (OE), because respondents are free to state any amount. In addition, because the CVM survey results depend on the information given to respondents about the beach changes being evaluated, in order to limit the risk of respondents giving an incorrect interpretation of a hypothetical change to the beach, a photograph or a photomontage is shown and carefully explained.

15.4.3. The use value according to seasons

At many coastal sites, weather and temperature conditions are very different according to the season: very hot and sunny in summer, and cold in winter. At these sites it is useful to distinguish the beach use and its value according to the different seasons. This distinction permits a more accurate description of the recreational beach use.

For the Italian case-studies of Lido di Dante, Trieste and Pellestrina it was possible to organise only one-time surveys in spring/summer 2002, therefore in the VOE questionnaire respondents were asked if they also visit the beach in autumn/winter (Marzetti, 2003a). If the reply is yes, they were also asked to elicit the beach use value in autumn/winter.

Day-visitors and tourists in Lido di Dante and Pellestrina visit the beach mainly in spring/summer (high season), while residents who visit the beach in autumn/winter (low season) are 60% in Lido di Dante, 73.5% in Trieste and 48.8% in Pellestrina. In Table 15.10, the mean

Table 15.10. Daily beach use values per person. (*: whole sample; **: people who visit the beach in autumn/winter only).

Mean value (€)	*Spring/summer*			*Autumn/winter*	
	Status quo	*Erosion*	*Expansion*	*Status quo*	*Expansion*
Lido di Dante	27.67	13.26	28.37	4.10*	
Developed beach area	25.41	11.47	27.43	16.38**	
Semi-developed beach area	27.21	9.94	26.35	17.60**	
Natural beach area	32.44	21.49	33.39	19.62**	
Trieste (residents)	5.24		8.32	5.25*	6.45*
Pellestrina	9.23			3.54*	
Residents	9.69			11.04**	
Non-residents	8.72			6.95**	

use values in spring/summer are computed considering the whole sample, while as regards Lido di Dante and Pellestrina, the mean use values in autumn/winter are computed in respect of the number of respondents who visit the beach (**) as well as in respect of the whole sample – i.e. including those who visit and those who do not visit the beach (*).

The daily beach use value in autumn/winter may differ considerably from that in spring/summer. It also changes according to the different characteristics of the beach and the kind of visitor. As regards the status quo, considering the whole sample, the mean use value in Lido di Dante and Pellestrina in the low season is lower than the mean value in the high season, while it is slightly higher in Trieste. In Pellestrina residents who visit the beach in autumn/winter give a much higher mean value than non-residents. As regards Lido di Dante, the seasonal use value is also computed for three different beach areas («developed beach» means «sunbathing establishment on the beach»). Respondents who visit the different beach areas in the low season give lower values than for the high season.

15.4.4. Use value for foreigners and aggregation level

In the CBA in general it is recommended that the aggregation level is national economy and not merely local economy (Penning-Rowsell *et al.*, 1992, p. 64). Nevertheless, when foreign tourists visit the site, this phenomenon cannot be neglected (see also Daniel, 2001; Marzetti, 2003a). The existence of international tourism – typical of a number of Italian beaches – means that preservation of the beach is also of international importance. The presence of foreign tourists characterises a situation in which the recreational value is not only relevant to the national community who pay for the conservation project. Foreigners use the free beach because it is a public good, but they pay nothing. Thus, at international tourist sites, as regards the relevant population, foreign visitors should be interviewed to avoid «losing» the «foreign use value», which could be an important part of the total recreational value of the beach.

Foreigners were interviewed at the tourist site of Lido di Dante. They were 32.1% of tourists and 17.7% of the whole sample. Table 15.11 shows that at this resort foreign visitors (excluding Dutch tourists) elicited higher use values (spring/summer) than Italian visitors.

If every respondent elicits how much enjoyment s/he would obtain from the use of a beach, it is also appropriate to compute the *aggregate* value or total recreational net benefit per year of the beach change considered. We need to test whether the beach aggregate value

Table 15.11. Foreigners' daily beach use value in Lido di Dante.

Mean value (€)	*Spring/summer*		
	Status quo	*Erosion*	*Expansion*
Nationals	26.45	12.49	17.99
Foreigners:			
German	30.93	16.45	28.65
French	30.00	14.04	33.36
Swiss	53.33	28.70	36.38
Dutch	22.50	5.50	25.00
Other nationalities	30.33	14.08	31.73

per year could be increased by the implementation of a LCS project. The unit of measure for the valuation is the *recreation day* on the beach, and the number of visits is considered as the quantity consumed of beach recreational services. Including foreigners, beach visitors are divided into those who continue to visit the site and those who would visit an alternative site if the beach changed (Penning-Rowsell *et al.*, 1992). If people continue to visit the beach after the project implementation, the individual gain (loss) per visit (D) is the difference between the VOE of a visit after the implementation of the project (Vp) and the VOE of a visit in the current condition (Vs). For each individual it is:

$$D = VP - Vs. \tag{15.3}$$

If, after the implementation of a project, individuals visit another site because they dislike the change, the gain or loss per visit is the difference between the VOE at the site in the status quo and the VOE at the alternative site plus the possible increase in the cost of the visit to the new site. In this case, for each individual it is:

$$Da = (Vs - Va) + (Ca - C) \tag{15.4}$$

where Da is the gain, or loss, Va the VOE at the other site, Ca the cost per visit to the alternative site, and C the cost per visit to the status quo. As regards the Lido di Dante and Trieste case-studies, Table 15.12 shows the mean daily gain (loss) for a beach change.

Finally, the aggregate gain (loss) is estimated for each season as follows:

$$B = Nq_m D_m \tag{15.5}$$

Table 15.12. Daily mean gain (loss) in Euros per person according to seasons and scenarios.

	Erosion loss	*Expansion gain*
Lido di Dante		
Spring/summer	12.29	1.29
Trieste		
Spring/summer		3.07
Autumn/winter		1.39

where B is the total gain (loss), D_m the mean gain (loss) per adult visit – obtained by computing the mean of the individual gains (losses) of those who continue to visit the beach and of those who decide to visit an alternative beach according to equations 15.3 and 15.4 respectively – Nq_m the total number of beach use days obtained by multiplying the total relevant population of the site N by the individual mean number of visits q_m. The total aggregated gain (loss) per year is the sum of the aggregated gain (loss) for the different seasons.

Individual mean gains and losses should be estimated for residents, day-visitors and tourists, and data about the total number of visits of locals, day-visitors and tourists are needed in order to compute the total recreational benefits per annum. The number of tourist visits – both national and foreign – are usually available; arrivals and night stays in a site can usually be obtained from local records. Data about residents' and day-visitors' visits are not always available. The CVM enables data to be obtained about residents and day-visitors interviewed by asking them how often they visit the beach each year in the different seasons. In particular, the Lido di Dante CV survey shows that in spring/summer 44.8% of respondents are day-visitors and visit the beach on average just under 23 days, while residents visit the beach on average about 47 days. In Trieste, as regards spring/summer, the mean number of residents' daily visits is about 15, and as regards autumn/winter about 13 days.

15.4.5. Conclusions

Within the DELOS Project, the Italian CV surveys showed that visitors are sensitive to the protection of coastal sites from erosion and flooding and that the great majority of them are in favour of defence projects. The mean use values are from 5 to 28 € per beach visit. As shown in Polomé *et al.* (2005), the mean value of a recreational visit to beaches in the status quo in the United States and United Kingdom (20 € with reference to 2001) is within the bounds of the Italian case studies. In Italy the VOE may also vary considerably accordingly with the season (spring/summer or autumn/winter). The distinction of the use value and number of visits according to different seasons can better describe the recreational beach use, and permit a more accurate computation of the aggregate use value of a beach change. In addition, as regards the relevant population, the inclusion of foreign visitors also refines the aggregate value computation, mainly for sites where foreigners are numerous.

15.5. THE BENEFIT OF PROTECTION OF LAND/HINTERLAND

(van der Veen, UTW)

This section discusses mitigating benefits as presented in Section 6.2.c. Preventing damage is a benefit that should be counted in a CBA (see Section 6.1). In Section 6.2.d. we show how damage to buildings due to inundation should be handled. However, we want to comment a little bit on this, because there are several methodological problems in defining damage. We mainly refer to a recent report by the EU (van der Veen, Vetere Arellano and Nordvik, 2003) on «A common methodology for damage estimation».

The problem of protection of the hinterland is one of the primary triggers of building protective measures along the coast. The question behind for economists is the following: «What is it we are protecting?».

A first and quick answer to this question is the value to society of the damage after an inundation. Probability times effect then is an indicator of risk to society. However, the current measures of risk to society mainly focus on direct economic effects and do not cover indirect economic damage. Secondly, by concentrating on risk we refrain from the resilience of society after a disaster and the ability of society to adapt. Otherwise stated, the question is «How vulnerable are we for disasters?» Our idea what is vulnerability is lead by the following quotation:

> «...*Moreover, with sea level changes occurring slowly throughout the century, economically rational foresight will make sure that protection will be afforded only to property that is worth more than the protection costs and settlements will be avoided were costs will outweigh benefits....*»
>
> (Lomborg 1998).

15.5.1. Risk and Vulnerability

Risk and vulnerability are words that have gone through a certain process changing its meaning and connotation. See also (Blaikie *et al.*, 1994). It is in our view the more engineering mode of dealing with the question how vulnerable society is for disasters.

$$S = \sum_{i=1}^{n} \alpha_i m_i S_i \tag{15.3}$$

S = Total damage
α = Damage factor
m = Number of entities in damage class
S_i = Damage value
n = Number of damage classes i

Common practice in the flooding (engineering) literature is to visualize risk and thus the underlying effect by counting unit losses (Parker *et al.*, 1987). With different flood-depths, depth-damage data is used to asses flood losses. The current state of this type of models (Vrisou van Eck and Kok, 2001) is that data on land cover is collected and downloaded into a GIS environment. Damage assessment then counts the number of units of a certain type in the affected area and multiplies this with a damage factor. The latter is basically a relationship that is empirically derived from surveys, in which a relationship is established between depth and damage. The damage factor is the heart of the method and thus plays an important role in estimating damage. In standard research on flood management the value of damage is based on a replacement value. As discussed by (van der Veen *et al.* 2003) this might not reflect the economic value of the goods at risk, see also Cole, 1998; Rose and Lim, 2002; Cochrane, 1997; Rose and Benavides, 1998; MAFF, 2000; Freeman *et al.*, 2002. This annoying matter is caused by a few misunderstandings:

1. There is no agreement on the economic points of departure. Financial appraisals are mixed up with cost-benefit analyses (CBA). In the latter, the usual concept is economic costs, which relates to opportunity costs in welfare economics, whereas a financial appraisal is often a base for investigating the sum of money to be recovered from insurance companies.
2. There is confusion on time and spatial scales: Financial appraisal limits itself to a single organisation, whereas CBA requires well-defined borders, like a region, a nation, or the European Union.
3. Stock concepts are confused with flow concepts.
4. The borderline between direct and indirect costs is not well defined.

The distinction between stocks and flows relate to the difference between direct and indirect costs (Cochrane 1997). If factory B is flooded suppliers of goods and services are hit, as well as firms that purchase goods B (Figure 15.2). In the end final demand of

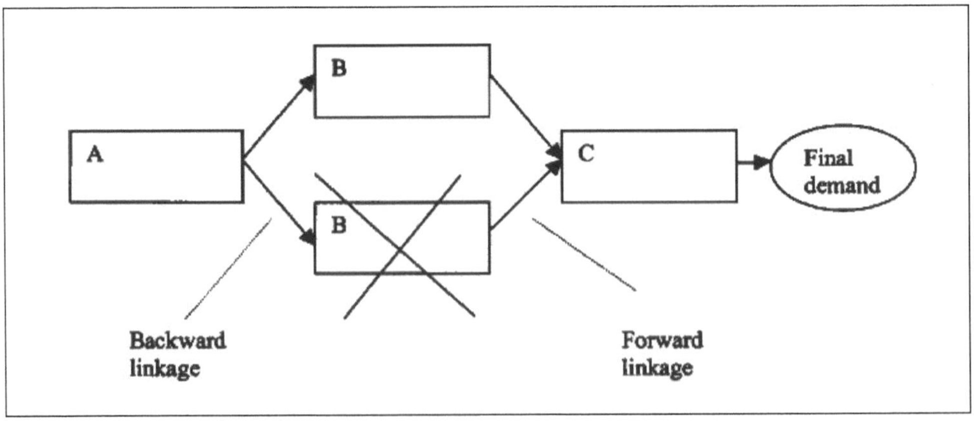

Figure 15.2. Forward and Backward Linkages in an Economy, when Factory B is Damaged.

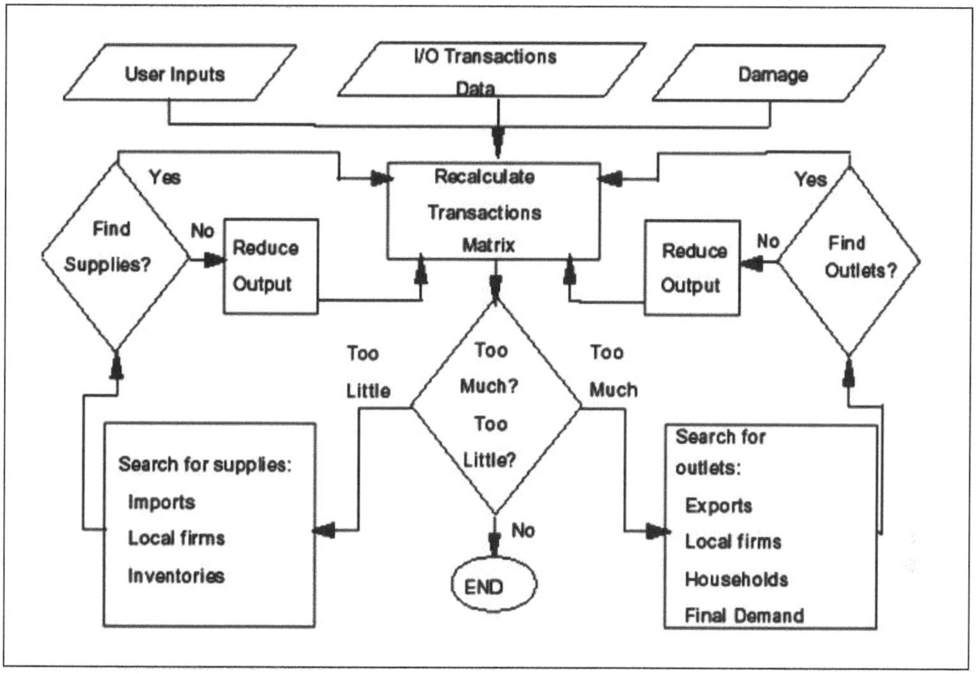

Figure 15.3. Determining redundancy in an economy (FEMA, 1999).

consumption, investment, export and government spending is touched. Part of a risk concept thus implies taking into account forward and backward linkages in a regional or national economy. However, this risk concept does not allow for redundancy in an economy: if there is a second firm *B* that is able to take over the production, an economy is less vulnerable.

By extending the concept of risk to a vulnerability concept we have to include the coping capacity of a region/nation to deal with floods. What is this coping capacity of society after a disaster? As a point of departure we take the concept of vulnerability as introduced by (Parker *et al.*, 1987). Vulnerability V is introduced with the following formula:

$$V = f(S, D, T) \tag{15.4}$$

where S = susceptibility, defined by the probability and extent to which the physical presence of water will affect inputs or outputs of an activity; D = dependence, reflecting the degree to which an activity requires a particular good as an input to function normally; T = transferability, the ability of an activity to respond to a disruptive threat by deferring or using substitutes or relocating.

Susceptibility refers to the geo-location of a site that is under investigation. Some sites are more prone to flooding and may encounter more often flooding. Susceptibility therefore relates to the geo-concept of damage. Dependency and transferability relate to the characteristics of the economic system. Dependency and transferability are concepts that are thus best understood when representing the economic system as a network of interrelated activities. Within such a network, there are certain functions and sectors that are important for the functioning of the network as a whole. To assess how important such functions are, we can distinguish two characteristics. The first refers to how dependent we are upon output produced at a site and the latter refers to the local redundancy in the network. Both concepts are highly interrelated.

Note that introducing concepts like dependency and transferability we relate to the concept of economic costs in Cost-Benefit Analysis as discussed in (EPA, 2000). The concept of economic costs is a dynamic one accounting for adaptations in an economic structure.

The choice between alternatives in order to cope with the consequences of a disaster is elaborated in (FEMA, 1999), see Figure 15.3.

We recommend as a guideline to give more attention to the notion of vulnerability as an alternative to the conventional concept of risk in order to reckon with the dynamics in an economy.

15.6. THE VALUE OF HABITAT DISRUPTION

(Polomé, UTW)

This section presents a case study as an illustration of the methodology for estimating the value of habitat disruption.

The object of valuation is a small (2 ha) restored natural area called Normerven, situated in the Dutch Waddenzee. It was restored using a system of two low crested structures that are overtopped on some high winter tides. This is done on purpose to maintain a mudflat that is adequate for bird breeding. After a first failed attempt, the restoration appears to work well as revealed by a dramatic increase in the number of breeding birds and stability of the structure over the last 5 years.

Access is forbidden to Normerven to avoid disturbing the nesting birds and the site is in a relatively remote area; the greatest part of the value of the site should be non-use. Normerven was actually cheap to build, yet significant for some bird species in the South Waddenzee. Since the restoration of Normerven has had no market impact, only «stated preferences» methods of valuation could be used. That means designing a survey.

Value was elicited through a dichotomous choice question. The respondents were asked to choose between an alternative plan (1 to 10 new sites at a certain cost) and the classical «do-nothing» plan, that is not building any more site (that has a cost of zero). Each respondent was shown 1 out of 14 possible alternatives and had to choose between this alternative and the classical «do-nothing» option, that is 2 cards (visual aids). Before arriving to that question, the respondents were thoroughly described the site of Normerven and its history. The respondents were indicated the cost of each alternative, as well as the geographical location of each site and the expected number of breeding pairs of birds. There were 14 choice situations in which the number of new sites could be 1, 3, 5 or 10, and the cost could range from 6 to 150 € per year . The «cost» of building more sites is called the bid in this context because the interest is to find out the respondents' value for the alternative shown, as if the interviewer was «selling» it. The choice situation was repeated 3 times to increase the available information per respondent. The payment vehicle must be feasible. We chose the real estate tax.

Following the NOAA panel recommendations (1993), in a contingent valuation, one should always use a referendum context for credibility. In our case, that means telling the respondents that there is a referendum on whether or not to build new sites similar to

Table 15.13. Empirical estimates of the coefficients of Eq (15.5).

Regressors	*Coefficient*	*P-value*
Constant	– 0.353	0.217
ln(bid)	– 0.387	0.000
# sites – 7	– 0.063	0.001
(# sites – 7)^2	– 0.008	0.072
Context	*Reference: No context and Donation*	
Opinion poll dummy	0.355	0.001
Consultation dummy	0.487	0.000
Referendum dummy	0.325	0.002

Normerven. However that seemed strange for a country in which referenda are exceptional and we feared that it could distort the image of the good to value. To answer this concern thoroughly, we split our sample in 5 and each subsample was given a different context: Referendum, Opinion poll, Consultative referendum, Donation, and No context. In each case, the wording of the whole survey was identical but for a few sentence that described the context.

The sample was selected randomly from the census file of the North region of the North-Holland province. Each potential respondent received a letter informing them that an interviewer from the University of Twente would pay them a visit about a survey on the environment of this region. Each potential respondent was followed-up as much as possible. The actual survey was run sequentially to find the best bids, that is the survey was administered in rounds of about 100 questionnaires (see e.g. Hanemann and Kanninen, 1999, for a survey). After each round, a brief analysis of the answers to the bids made it possible to update them. We obtained 600 observations.

We tried several econometric models to analyse those data. The one that was finally selected is the following.

$$\Pr\{No\} = \Phi\left(\alpha + \alpha_i (Site - 7)^2 + \sum_j \alpha_j Context_j + \sum_k \alpha_k X_k + \beta \ln(Bid) \right) \quad (15.5)$$

The effect of the bid is very significant and in the expected direction. There is a very significant effect of the normalised number of sites and a weakly significant effect of the squared number of sites. Jointly, these two variables imply that there can be «too many new sites», that is, when the normalised number of sites is close to the zero the probability of a Yes answer is maximal. Regarding the decision contexts, there is no significant difference between the donation context and the absence of a context. Likelihood ratio tests can be used to show that the three other contexts can be pooled together without significant difference, but that they cannot be dropped from the regression, neither individually nor jointly. Therefore, globally the contexts are very significant, but there is in fact only 2 groups: No context and Donation on the one side, Opinion poll, Consultation and Referendum on the other. There are other significant regressors but they are not presented here because they are not relevant to this analysis.

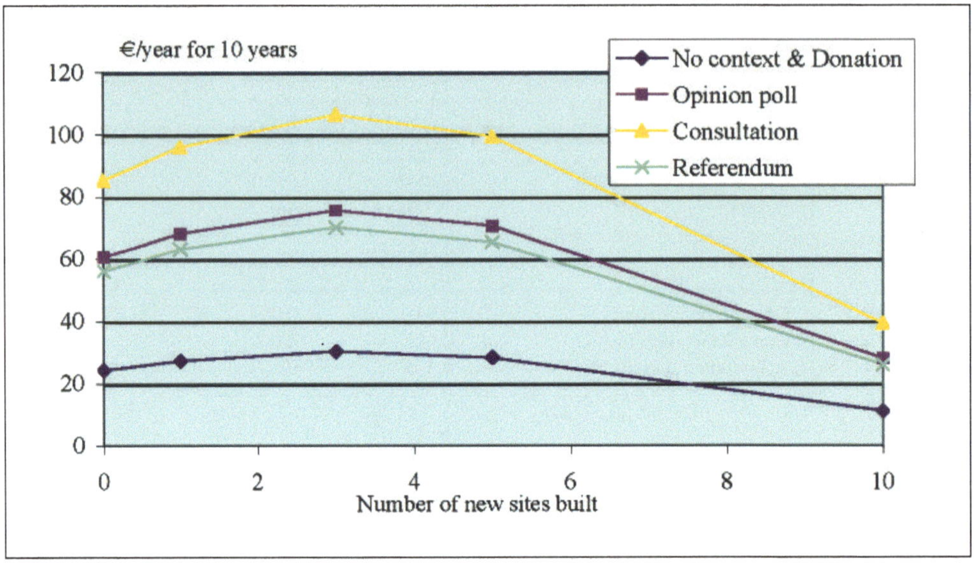

Figure 15.4. Median WTP over the sample, including income (see Table 15.13).

The model that has been defined above is a RUM (Random Utility Model). It is compatible with economic theory and can be used to extract a welfare measure as shown by Hanemann (1984). The relevant welfare measure in this case is the WTP. We computed the median WTP for each individual in the sample for each decision context and for 0, 1, 3, 5 and 10 new sites on top of Normerven. Then we took the median over the sample. The results are presented in Figure 15.4.

The decision contexts which had the largest positive coefficients coincide with the largest value. The respondents do not distinguish between no context and donation. Although this is not apparent from the picture, there is no significant differences between the Opinion poll, Consultation and Referendum contexts. Therefore there is essentially only two groups of contexts: with and without government intervention, with welfare being higher in the former case. Also, quite in contrast to the NOAA Panel expectation, the referendum context does not produce the most conservative welfare estimate.

The value of the original Normerven itself can be extrapolated as shown in Figure 15.4. It is apparent that it is this first site that generated most value. From there, the WTP follows a quadratic curve that culminates at 3 new sites than starts decreasing (5 sites are still worth more than one site). One might expect that when the number of sites increases, the value should also increase. That could be the general economic intuition, but that is not true in general. In the case of a natural area, when it becomes bigger, it starts competing with other uses, there is some sort of congestion. Therefore, it is indeed possible that the utility of 10 additional sites is actually lower than that of 5 new sites. In other words, the last 5 sites have a negative utility.

This case-study has shown several things that may be important in the design of coastal defence in general and of LCSs in particular. First it has been shown that it is possible to value LCSs even when they do not have any market impact. Second, that the context in which a defence is provided is important. Third, that there can be «too much of a good thing», that is, it is not because one defence site has been highly valued that replication of it will have the same value. It is even possible that excess defence causes congestion and that adding more defence sites decreases the value of the whole. The latter is of course a critical argument against the transfer of benefit for constructions such as a coastal defence.

15.7. OPTIONS USE AND NON-USE VALUES OF A COASTAL CULTURAL HERITAGE

(Marzetti, UB)

15.7.1. Introduction

This section deals with the CVM in the WTP version for evaluating option use value and non-use values (bequest and existence values) about heritage sites which was applied within the DELOS Project to Venice as World Heritage Site (UNESCO) in summer 2002 (Marzetti, 2003b; Marzetti and Lamberti, 2003).

For its architectural and historical characteristics, Venice attracts about ten million visitors per year (tourists and day-visitors), but is affected by floods and high water phenomena which may take the nature of extreme flooding events. Its coastal defence program consists of different kinds of interventions. We mention the defence of buildings, the defence and rebalance of the morphological and hydrodynamic system of the lagoon, the defence of the natural barriers of Lido and Pellestrina islands by the building of artificial beaches protected by low crested structures, and the temporary closure of the three inlets with mobile floodgates built inside the lagoon across each inlet (MO.S.E.). Its sustainable management (involving a considerable amount of public funds) requires policy-makers to have a clear understanding of all benefits and costs (see Sections 15.1 and 15.2). Here we focus on option use and non-use values, because they are not established by the market. Option use value means that a person may be willing to pay for the option of visiting Venice in the future; bequest value measures the amount a person would pay for the preservation for use by future generations; while existence value represents the amount the person who makes the valuation would pay only for knowing that Venice as a cultural heritage exists.

Our aim is not to describe in detail how to estimate in monetary terms these non-marketable values because a wide economic literature on the topic is available (in particular, see Arrow *et al.*, 1993), but we focus on two aspects of the CVM in the WTP version: i) the relevant population which, at international tourist heritage sites, is also made up of foreigners, and ii) respondent's probability of paying the amount elicited. Finally, results of the Venice case-study are presented.

15.7.2. Aggregation level: the international community

In the CBA the aggregation level is usually that at the national economy. Nevertheless, in the case of heritage sites of international or world interest the relevant population cannot be made up of nationals only, but consists of the world community or a part of it (see King, 1995). As regards option value and non-use values, not only national and foreign users (residents, day-visitors and tourists), but also national and foreign non-users (people who have never visited and will never visit the site in question) should be interviewed. In particular, foreigners should be interviewed to avoid «losing» the foreign economic value, which may be a very important part of the Total Economic Value (TEV). In Venice foreigners are very numerous and come from all the world. In 1996, they were more than 50% of day-visitors (not staying overnight in Venice), and 80% of tourists (Cellerino, 1998).

An international or world CVM survey is complex and expensive. For this reason, as regards Venice, given the available funds, an on-site survey of 1000 face-to-face interviews (10-15 minutes each) to visitors – tourists and day-visitors, nationals and foreigners – aged 18 plus in its most crowded streets was carried out (random sample), and a pilot survey was performed to test the questionnaire. In this case the option use and non-use values can only be ascribed to the population sampled.

15.7.3. The CVM questionnaire: the probability of paying

When the quantity of the good considered is fixed, as in the case of heritage sites, the WTP is the amount respondents are willing to pay for maintaining or improving the existing quality level of the site. The payment vehicle used for the evaluation of option value and non-use values about Venice is an extra payment to a non-profit agency.

In its final wording the questionnaire is divided into six sections. The first section aims to select people for the interview (visitors only). Residents were excluded, as well as commuters to Venice for work or study and non-residents who are staying in Venice more than one year. The second section seeks information about respondent's recreational activities in Venice, while the third section investigates respondent's attitudes towards the cultural goods in general.

The fourth section is the heart of the questionnaire since it includes the elicitation questions. Different formats exist for eliciting the WTP, and we refer the reader to the existing literature (see, for example, Hausman, 1993; Bateman *et al.*, 1999). As regards the Venice case-study, the modified double referendum format (double dichotomous choice plus an open-ended question) was chosen (see, in particular, Silberman and Klock, 1988; Silberman *et al.*, 1992; Seip and Strand, 1992; Arrow *et al.*, 1993; Goodman *et al.*, 1996; Shechter *et al.*, 1998; and Scarpa *et al.*, 1999). First of all respondents are presented with a detailed description of the Venice defence programme for the high water phenomenon through the description of Photomontage 11.20, asked if they are favourable or contrary to the project, and reminded that there are many other worthy causes to contribute to. Then they are asked i) whether they are willing to pay one Euro per year to a non-profit agency for that programme; if the reply is yes, ii) they are asked whether they are willing to pay more; if the reply is still yes, iii) the maximum willingness to pay is asked. In addition, respondents willing to pay are also asked to specify their donation motives, while respondents unwilling to pay are asked the non-donation motives.

Given the hypothetical nature of a contingent market, the elicited WTP could be different from the true WTP or actual donation. Respondents may be uncertain to different degrees about their actual WTP (see Champ *et al.*,1997; Ready *et al.*, 2001). Therefore, respondents willing to pay are also asked how certain they are to pay on a scale from 0 to 100 if the sum elicited is actually asked.

Finally, the fifth section asks respondents' socio-economic characteristics, while the last section is addressed to the interviewer, mainly to collect information about respondents' understanding of the questionnaire.

15.7.4. The option use and non-use values of visitors in Venice

In Venice at the time of the survey, the randomly chosen visitors included tourists (55.7%) and day-visitors (44.3%). Foreign respondents (European and non-European) were 75.8% of the whole sample.

The great majority of respondents think that cultural heritage sites in general have to be protected, as first choice, because «they are our future» (47.5%) and as second choice because they «represent our past» (36.8%). In particular, 93% of respondents are in favour of the implementation of the protection programme of Venice. The main visitors' activity in Venice is walking around the streets, and the second is to visit museums.

As regards the elicitation questions, 71.1% of interviewees would be willing to pay at least 1 Euro to cover the cost of the flood and coastal defence programme, in particular 77.7% of Italians and 69% of foreigners. Moreover 40.9% of respondents would be willing to pay more than 1 € in order to protect Venice. We highlight that, in the case of option value and non-use-values of heritage sites, particularly interested people could be willing to pay high sums, so extreme values were also considered in the computation of the mean WTP. Considering the whole sample, the elicited mean WTP for the defence of Venice per year is 4.85 € (median 1 €, std. dev. 11.16). In particular, on average, tourists are willing to donate more (5.56 €) than day-visitors (3.95 €).

As regards the distinction between the elicited WTP and the true WTP, as shown in figure 15.2, 64.4% of respondents claiming to be willing to pay at least 1 € for the defence programme are 100% sure that, if actually asked to pay, they would pay the amount elicited. The rest of respondents are unsure in different degrees, and of these respondents 1.3% claim to be very uncertain.

As regards donation motives, the most important motive, as first choice, is to preserve Venice for future generations (53.7% of respondents willing to pay), while the second most important motive is to preserve the option of visiting Venice in the future (17.4%); 12.2% of interviewees would be willing to pay to allow other people to enjoy Venice and 10.5% just to know that Venice exists, no matter whether they will ever visit it again. As second choice, the most important motive of donation is giving money to a good cause (21.8%), and the second most important is to preserve the option of visiting Venice in the future (18.8%). We highlight that the WTP is asked as a lump sum, and it is not split into option value and non-use values.

As regards non-donation motives, 28.9% of respondents are not willing to donate to the

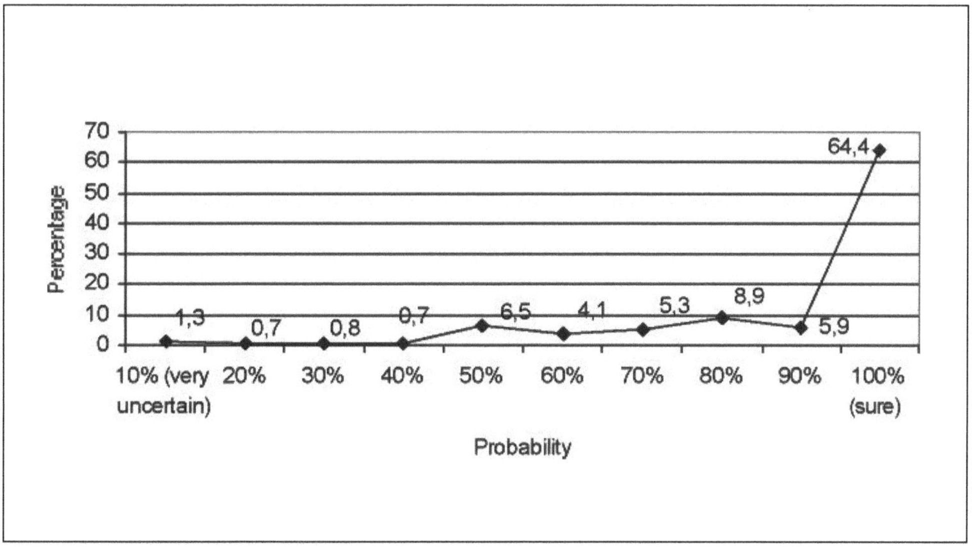

Figure 15.5. Probability of paying the amount declared; percentages of respondents.

protection programme for the following main reasons: 37.7% of these respondents think that paying for the Venice defence project is the state's duty; 18.3% says that protection is not their problem because they do not live in Venice (in particular 20.4% of foreigners unwilling to pay); 11.8% think that money should be spent on some other project; 11.4% claim that non-profit foundations waste money.

15.7.5. Conclusion

The Venice CVM survey results highlight that day-visitors and tourists seem very sensitive to the defence of heritage sites, that it is important also to interview foreign visitors, because at international heritage sites these may be the majority of visitors, and that data about the subjective probability of paying also has to be collected in order to estimate the true WTP.

15.8. VISITORS PREFERENCES ABOUT BEACH DEFENCE TECHNIQUES AND BEACH MATERIALS

(*Marzetti, UB*)

15.8.1. Introduction

This section describes an approach for investigating preferences about different kinds of beach defence techniques and beach materials which was applied to the DELOS case-studies of Lido di Dante, Pellestrina and Ostia (see Sections 11.3, 11.4, 11.5 and 12.4.8). We found no specific bibliography on this topic.

To save time and money, a CVM questionnaire is a good opportunity to collect information other than the economic data. Therefore, in order to design LCS which meet the preferences of beach visitors, here we present some questions to find out respondents' opinions regarding project characteristics and the motive of preference.

15.8.2. Questions about kinds of defence structures and beach materials

The following questions can be asked to beach visitors (Marzetti *et al.*, 2003):

i) *The beach can be protected from erosion with different techniques. Which of these techniques do you prefer?* A photomontage of different kinds of LCS, such as those in Figure 15.5 (1. parallel breakwaters, 2. nourishment, 3. groynes, and 4. composite intervention with submerged breakwaters), should be created and shown to respondents.

ii) *Why did you choose this technique?*
iii) *How do you rate (on a scale from 0 to 10) the presence of groynes on a beach?*
iv) *Do you prefer a beach of fine sand, coarse sand or gravel?*

Comparing the preferences about different defence techniques in the three Italian case-studies considered, Table 15.14 shows that, as regards question i), the composite intervention is preferred in Lido di Dante and Pellestrina, while nourishment is preferred in Ostia.

As regards question ii), Table 15.15 highlights the two main motives of preference (in order of importance) according to the different defence structures. Aesthetic motives prevail in all the case-studies. The second motive differs according to the different sites: water quality is given in Lido di Dante for all the techniques, while in Ostia and Pellestrina it is the second preferred in two out of four techniques. In particular, the most preferred technique for aesthetic motives is the composite intervention in Lido di Dante, and nourishment in

Figure 15.6. Photomantage 1 (1. parallel breakwaters, 2. nourishment, 3. groynes, and 4. composite intervention with submerged breakwatters).

Table 15.14. Preferences about four defence techniques: percentage of respondents.

Defence techniques	*Lido di Dante*	*Ostia*	*Pellestrina*
E/S* parallel breakwaters	23.7%	36%	15%
Nourishment	19.8%	53%	20%
Groynes	21.2%	6%	24%
Composite intervention	32.5%	5%	35%
(* E/S means emerged/submerged)			

Table 15.15. Defence structures – the two main motives of preferences (in order of importance).

Defence techniques	*Lido di Dante*	*Ostia*	*Pellestrina*
E/S parallel breakwaters	Aesthetic motives Water quality	Water quality Aesthetic motives	Aesthetic motives Water quality
Nourishment	Aesthetic motives Water quality	Aesthetic motives Suitable for beach activities	Aesthetic motives Water quality
Groynes	Aesthetic motives Water quality	Aesthetic motives Water quality	Aesthetic motives Suitable for beach activities
Composite intervention	Aesthetic motives Water quality	Aesthetic motives Water quality	Suitable for beach activities Aesthetic motives

Ostia; the composite intervention is the most preferred in Pellestrina for suitability for beach activities.

As regards question iii), on a scale from 0 to 10, a medium-high level of preference is assigned to groynes in all the three considered sites. Finally, as regards question iv), asked only to Ostia and Pellestrina respondents, the majority of them prefer fine sand as first choice, while coarse sand is the second preferred beach material.

15.8.3. Conclusion

These results cannot be generalised to represent visitors' preferences on other sites, unless beaches and visitors are very similar to those considered in DELOS. If data from very similar beaches and population are not available, a specific survey is recommended. Within the DELOS Project, the data here presented highlight the sensitivity of beach visitors to aesthetic characteristics and suitability of beach defence structures for recreational activities.

LCS design guidelines

References

Abbott M.B., A. Damgaard and G.S. Rodenhuis (1973). System 21, Jupiter. A design system for two-dimensional nearly-horizontal flows. *J. Hydraul. Res.*, **1**, 1-28.

Ahrens J.P. and J. Cox (1990). Design and performance of reef breakwaters. *Journal of Coastal Research*, Special issue, **7**, Spring, 61-75.

Ahrens J.P. (1987). *Characteristics of Reef Breakwaters*, Technical report CERC-87-17, Waterways Experiment Station, US Army Corps of Engineers.

Airas S., H.T. Rapp (2003). *Correlation between environmental gradients and the abundance of selected marine invertebrates*. Institutt for Fiskeri- og Marinbiologi Rapport. **7**, 28 pp.

Albani M. and D. Romano (1999). *Total Economic Value and Evaluation techniques,* in Bishop R.C. and D. Romano, Environmental Resource Valuation. Applications of the Contingent Valuation Method in Italy, Kluwer Academic Publishers, Boston, 47-71.

Allen J.R., D.J. Slinn, T.M. Shammon, R.G. Hartnoll, S.J. Hawkins (1998). Evidence for eutrophicationof the Irish Sea over four decades. *Limn. Ocean.*, **43** (8), 1970-1974.

Allsop N.W.H. and A.R. Channel (1989). Wave reflections in harbours: reflection performance of rock armoured slopes in random waves. Hydraulic Research, Wallingford, Report OD **102**.

Allsop N.W.H., P. Besley and L. Madurini (1995). Overtopping performance of vertical and composite breakwaters, seawalls and low reflection alternatives. Paper 4, 6 in *Final Proceedings of MCS Project*, University of Hanover, Hanover, Germany.

Alongi D.M., K.R. Tenore (1985). Effect of detritus supply on trophic relationships within experimental benthic food webs. 1. Meiofauna-polychaete (Capitella capitata (Type I) Fabricius) interactions. *Journal of Experimental Marine Biology and Ecology*, **88**, (2), 153-166.

Alsina J.M., Cáceres I., Sánchez-Arcilla A., González-Marco D., Sierra J.P. and Montoya F. (2003). Morphodynamics in the neighbourhood of a submerged breakwater. *3rd Symposium on River, Coastal and Estuarine Morphodynamics* (Barcelona, Spain), 1018-1028.

Alsina J.M. (2005). *Development of a morphodynamic numerical model. Application to LCS impact assessment*. Ph. D. Thesis, Universitat Politècnica de Catalunya, Barcelona, Spain, 233 p.

Andersen O.H. and J. Fredsoe (1983). *Transport of suspended sediment along the coast*. Progress Report n. **54**, Inst. of Hydrodynamics and Hydraulic Eng. ISVA. Tech. Univ. of Denmark, 33-46.

Andersen O.H., Hedegaard I.B., Ronberg J.K. and Deigaard R. (1991). Model for morphological changes in the coastal zone. In: *Proc. hit. Symp. on the Transport of Suspended Sediment and Its Mathematical Modelling*, Florence, 327-338.

Archetti R., M. Tirindelli, G. Gamberini and A. Lamberti (2003). *Analysis of currents around a low crested barrier: comparison between field and numerical results.* Proc. MEDCOAST 2003. 1731-1740.

ARPA (2001). *Stato del litorale emiliano-romagnolo all'anno 2000.* Regione Emilia-Romagna-ARPA, Bologna, 2001.

Arrow K., R. Solow, P.R. Portney, E.E. Learner, R. Radner and H. Shuman (1993). *Report of the NOAA Panel on Contingent Valuation.* Report To the General Counsel of the US National Oceanic and

Atmospheric Administration. US Department of Commerce. Natural resource damage assessments under the Oil Pollution Act of 1990, Federal Register, **58**, (10), 4601-4614.

ASCE (1994). Coastal groins and Nearshore breakwaters. ASCE Thechnical engineering and design guides as adapted from the US Army Corps of Engineers, Nº **6**.

Bacchiocchi F., L. Airoldi, V.U. Ceccherelli, A. Lamberti and E. Drei (1999). Analysis of the benthic community in two areas protected by stone reef barriers. In *Proceedings of the 7th International Conference on Artificial Reefs and Related Aquatic Habitats* (G. Relini, G. Ferrara and E. Massaro eds). San Remo, Italy, 599-604.

Bacchiocchi F., L. Airoldi (2003). Structure, distribution, and dynamics of epibiota on different typologies of coastal defence works. *Estuar. Coast. Shelf Sci.*, **56**, 1157-1166.

Bagnold R.A. (1966). *An approach to the sediment transport problem from general physics*. U.S. Geological Survey Professional Paper. 422-I, 11-137.

Bakker W.T. (1969). The dynamics of a coast with a groyne system, *11th Coastal Engineering Conference* (ASCE), 1001-1020.

Bailard J.A. (1981). An Energetics Total Load Sediment Transport Model for a Plane Sloping Beach: Local Transport. *Journal of Geophysical Research*, **86**, C3, 10938-10954.

Banyard L. and D.M. Herbert (1995) *The effect of wave angle in the overtopping of seawalls*. SR.396, in: Wave Overtopping of Seawalls, Design and Assessment Manual, HR Wallingford, 1999, TR. W**178**, available at http://www.environment-agency.gov.uk/commondata/105385/w178.pdf

Baquerizo A. (1995). *Reflexión del oleaje en Playas. Métodos de evaluación y de predicción*. Ph. D. Dissertation, Universidad de Cantabria, Santander, Spain, 180 p.

Barbanti C. (2002). *Modellazione morfodinamica e valutazione dei benefici di una difesa costiera: il caso di Pellestrina (VE)*. Degree Thesis, in Italian, University of Bologna.

Bateman I.J., I.H. Langford, J. Rasbash (1999). *Willingness-to-Pay Question Format Effect in Contingent Valuation Studies*, in I.J. Bateman, K.G. Willis (1999), Valuing Environmental Preferences: Theory and Practice of the Contingent Valuation Method in the US, EU and Developing Countries, Oxford University Press, Oxford, 511-39.

Bateman I.J. and K.G. Willis (1999). *Valuing Environmental Preferences: Theory and Practice of the Contingent Valuation Method in the US, EU and Developing Countries*. Oxford: Oxford University Press.

Battjes J.A., and J.P.F.M. Janssen (1978). Energy loss and set-up due to breaking of random waves, *Proc. 16th Int. Conference on Coastal Eng. ASCE*, **1**, 569-587.

Battjes J.A., and H.W. Groenendijk (2000). Wave height distributions on shallow foreshores. *J. of Coastal Eng.*, **40**, 161-182.

Beach management manual, Simm D.(ed), A.H. Brampton, N.W. Beech and J.S. Brook, CIRIA 1996, Reprint 2001, London.

Bellotti G. and M. Brocchini (2001). On the shoreline boundary conditions for boussinesq type models. *International Journal for Numerical Methods in Fluids*, **37** (4): 479-500.

Bellotti G. and M. Brocchini (2002). On using Boussinesq type equations near the shoreline: a note of caution. *Ocean Engineering*, **29** (12), 1569-1575.

Bellotti G. (2004). A simplified model of rip currents systems around discontinuous submerged barriers. *Coastal Engineering*, **51**, 4, 323-335.

Benedetti-Cecchi L., F. Pannacciulli, F. Bulleri, P.S. Moschella, L. Airoldi, G. Relini, F. Cinelli (2001). Predicting the consequences of anthropogenic disturbance: Large-scale effects of loss of canopy algae on rocky shores. *Marine Ecology Progress Series*, **214**: 137-150.

Berkhoff J.C.W. (1972). Computation of Combined Refraction-Diffraction. *Proc. 13th Int. Conf. Coastal Eng.*, Vancouver 1972, ASCE. New York, **1**, Chap. 24, 471-490.

Bijker E.A. (1971): Longshore Transport Computations. *Journal of Waterway, Harbours and Coastal Eng. Division,* ASCE, **97** (WW4), 687-701.

Birkemeier W.A. (1985). Field Data on Seaward Limit of Profile Change, *J. of Waterways, Port, Coast. and Ocean Eng.,* ASCE, **111** (3), 598-602.

Black K.P. and C. Andrews (2001a). Sandy shoreline response to offshore obstacles. Part I: Salient and tombolo geometry and shape. *Journal of Coastal Research*, 29/82.

Black K.P. and C. Andrews (2001b). Sandy shoreline response to offshore obstacles. Part II: Discussion of formative mechanisms. *Journal of Coastal Research*, 29/94.

Blaikie P., T. Cannon, I. Davis and B. Wisner (1994). *At Risk: natural hazards, people's vulnerability, and disasters*. London, Routledge.

Boaventura D., M. Alexander, P.D. Santina, N.D. Smith, P. Re, da L.C. Fonseca, S.J. Hawkins (2002). The effects of grazing on the distribution and composition of low-shore algal communities on the central coast of Portugal and on the southern coast of Britain. *Journal of Experimental Marine Biology and Ecology*, **267** (2), 185-206.

Bombace G., L. Castriota and A. Spagnolo (1995). Benthic communities on concrete and coal-ash blocks submerged in an artificial reef in the central Adriatic Sea. *Proc. 30th European Marine Biological Symposium* (Hawkins, L. E. & Hutchinson, S., eds), 281-290.

Borsboom M., N. Doorn, J. Groeneweg, and M. Van Gent (2000). A Boussinesq-type wave model that conserves both mass and momentum. *Proc. 27th ICCE Conference*, Sydney, 148-161.

Borsboom M., N. Doorn, J. Groeneweg and M. Van Gent (2001). Nearshore wave simulations with a Boussinesq-type model including breaking. *Proc. Coastal Dynamics 2001 Conference*, Lund, 759-768.

Bower Blair T. and R. Kerry Turner (1998). Characterising and analysis benefits from integrated coastal management (ICM). *Ocean & Coastal Management*, **38**, 41-66.

Briand M.H.G. and Kamphuis J.W. (1993). Sediment transport in the surf zone: a quasi 3-D numerical model. *Coastal Eng.*, **20**, 135-156.

Briganti R., J.W. van der Meer, M. Buccino and M. Calabrese (2003). Wave transmission behind low crested structures. ASCE, *Proc. Coastal Structures*, Portland, Oregon.

Briganti R., R.E. Musumeci, G. Bellotti, M. Brocchini and E. Foti (2004). Boussinesq Modelling Of Breaking Waves: Description Of Turbulence. *Journal of Geophysical Research-Oceans*; **109**.

Brookshire D.S., L.S. Eubanks and Randall A. (1983). Estimating option Prices and Existence Values for Wildlife Resources, *Land Economics*, **59** (1), 1-15.

Brotto M.T., M. La Terza (1996). Aspetti esecutivi del ripascimento artificiale del litorale di Pellestrina realizzato nell'ambito della salvaguardia di Veneziane della sua laguna. *Proc. IV AIOM Congress*, 469-480, in Italian.

Brouwer R. (2000). Environmental value transfer: state of the art and future prospects. *Ecological Economics*, **32** (1), 137-52.

Bruun P. (1954). *Coast erosion and the development of beach profiles*. Beach Erosion Board, Technical Memorandum, n. **44**.

Bulleri F., M. Menconi, F. Cinelli and L. Benedetti-Cecchi (2000). Grazing by two species of limpets on artificial reefs in the northwest Mediterranean. *Journal of Experimental Marine Biology and Ecology*, **255**, 1-19.

Burcharth H.F. and Christensen C. (1991). *On stationary and non-stationary porous flow in coarse granular materials*. MAST G6-S, Project I, Aalborg University, Department of Civil Engineering, Report.

Burcharth H.F. and O.H. Andersen (1995). On the one-dimensional steady and unsteady porous flow equations, *Coastal Engineering*, **24**, 233-257.

Burcharth H.F. Kramer M., Lamberti A. and B. Zanuttigh (2006). Stability at low-crested detached breakwaters, *Coastal Engineering*, **53**(4), 381-394.

Burger G. (1995). *Stability of low-crested breakwaters, MSc*. Thesis Delft University, report H1878/H2415 and Final Proceedings, EU research project Rubble mound breakwater failure modes, MAST 2 contract MAS2-CT92-0042.

Bustamante R.H. and G.M. Branch (1996). The dependence of intertidal consumers on kelp-derived organic matter on the west coast of South Africa. *Journal of Experimental Marine Biology and Ecology*, **196**, 1-28.

Cáceres I. (2004). *Q-3D near-shore circulation, development and utilities.* PhD Thesis, Universitat Politècnica de Catalunya, Barcelona, Spain, 147 pp.

Calabrese M., V. Vicinanza, M. Buccino (2002). Large scale experiments on the behaviour of low crested and submerged breakwaters in presence of broken waves. *Proc. 28th Int. Conf. on Coastal Engineering*, ASCE, 1900-1912.

Calabrese M., D. Vicinanza and M. Buccino (2003). 2D wave set up behind low-crested and submerged breakwaters. *Proc. of the 13th Int. Offshore and Polar Eng. Conf.*, 831-836.

Calabrese M., Vicinanza D. and Buccino, M. (2005). Verification and recalibration of an engineering method for predicting 2D wave setup behind submerged breakwaters, *Proc. Int. Coastal Symp.'05*, Hofn, Iceland, available at www.itv.is/ics2005/Data/B8.4/Calabrese_PA.pdf.

Callow M.E., R.L. Fletcher (1994). The influence of low surface-energy materials on bioadhesion – a review. *International Biodeterioration and Biodegradation*, **34** (3-4), 333-348.

Cappietti L. Martinelli L. and B. Zanuttigh (2004). Experimental analysis of filtration and set-up in presence of low-crested structures, *Atti del XXIX Convegno Nazionale di Idraulica e Costruzioni Idrauliche* (Trento, 7-10 Settembre), **3**, 713-720, BIOS ed., Cosenza.

Cayocca F. (1997). Long-term morphological modeling of a tidal inlet: the Arcachon Basin, France, *Coastal Eng.*, **42**, 115-142.

Carson R.T. (1999). *Contingent Valuation: A User's Guide.* Discussion Paper, 99-26, University of California at San Diego.

Casadei C., D. Ceccaroni, E. Drei and A. Lamberti (1998). Individuazione delle correnti nella zona protetta di Lido di Dante (Ra). *Atti XXVI Convegno Nazionale di Idraulica e Costruzioni Idrauliche*, **III**, 233-244, in Italian.

Ceccherelli V.U. and R. Rossi (1984). Settlement, growth and production of the mussel Mytilus galloprovincialis. *Marine Ecology Progress Series* **16**, 173-184.

Cellerino R. (1998). *Venezia Atlantide. L'impatto economico delle acque alte*, Franco Angeli ed., Milan, in Italian.

CEM (2001). *Coastal Engineering Manual*, «Fundamentals of Design» Part VI, Chapter 5, US Army Corps of Engineers.
http://users.coastal.ufl.edu/~shepp/loc6430/Coastal_Engineering_Manual.htm oppure: http://chl.erdc.usace.army.mil/CHL.aspx?p=s&q=Publications;8

Champ P.A., R.C. Bishop, T.C. Brown and D.W. McCollun (1997). Using Donation Mechanisms to Value Nonuse Benefits from Public Good, *Journal of Environmental Economics and Management*, **33**, 156-62.

Chatham C.E. (1972). Movable-bed model studies of perched beach concept, *Proc. 13th Int. Conf. Coastal Eng.* ASCE, 1197-1216.

Chow V.T. (1959). *Open-Channel Hydraulics.* McGraw-Hill, 680.

Christensen E.D. and R. Deigaard (2001). Large eddy simulation of breaking waves, *Coastal Engineering*, **42**, 53-86.

Christensen E.D., D-J. Waltra and N. Emerat (2002). Vertical variation of the flow across the surf zone, *Coastal Engineering*, **45**, (3-4), 169-198.

Christensen E.D., B. Zanuttigh and J.A. Zyserman (2003). Validation of Numerical Models against Laboratory Measurements of Waves and Currents around Low Crested Structures. *Coastal Structures*, Portland, Oregon, 876-888.

Clarke, K. R. and Warwick, R. M., (2001). *Change in Marine Communities: An Approach to Statistical Analysis and Interpretation.* 2nd edition: PRIMER-E, Plymouth, UK, 172 pp.

Clementi E., Cappietti L. and Martinelli L., (2006). Analisi sperimentale di tracimazione, piling up e filtrazione per scogliere a cresta bassa, *Proc. XXX Convegno di Idraulica e Costruzioni Idrauliche*, 15 pp, in Italian.

Cochrane H.C. (1997). *Forecasting the Economic Impact of a Midwest Earthquake. Economic Consequences of Earthquakes: Preparing for the Unexpected.* B.G. Jones. Buffalo, USA, National Center for Earthquake Engineering Research.

Coeffe Y. and Pechon P. (1982). Modelling of sea-bed evolution under waves action. In: *Proc. 18th Int. Coastal Eng. Conf.*, Cape Town, **I**, 149-1160.

Colantoni P., G. Gabbianelli, F. Mancini and W. Bertoni (1997). Coastal defence by breakwaters and sea-level rise: the case of the Italian Northern Adriatic Sea. *Bulletin de l'Institut Oceanographique*, **18**, 133-150.

Cole S. (1998). *Decision support for calamity preparedness: socioeconomic and interregional impacts*. Engineering and socioeconomic impacts of earthquakes. M. Shinozuka, A. Rose and R. T. Eguchi. Buffalo, MCEER.

Collins K.J., A.C. Jensen, A.P.M. Lockwoodand S.J. Lockwood (1994). Coastal structures, waste materials and fishery enhancement. *Bulletin of Marine Science*, **55**, 1240-1250.

Connell S.D., T.M. Glasby (1999). Do urban structures influence the local abundance and diversity of subtidal epibiota? A case study from Sydney harbour, Australia. *Marine Environmental Research*, **47**, 373-387.

Connell S.D. (2000). Floating pontoons create novel habitats for subtidal epibiota. *Journal of Experimental Marine Biology and Ecology*, **247**, 183-194.

Connell S.D. (2001). Urban structures as marine habitats: an experimental comparison of the composition and abundance of subtidal epibiota among pilings, pontoons and rocky reefs. *Marine Environmental Research*, **52**, 115-125.

Connor D.W., K. Hiscock, R.L. Foster-Smith, R. Covey (1995). *A classification system for benthic marine biotopes*. In: Biology and ecology of shallow coastal waters. Olsen & Olsen, Fredensborg (Denmark), 155-166.

Cooper N.J., D.M. King, J.M. Hooke (1996). Collaborative research studies at Elmer beach, West Sussex, UK. Proceedings of EUROCOAST Littoral '96. In: Taussik and Mitchell (eds.) Partnership in coastal engineering, Samara, Cardigan, UK 369-376.

Cordes J.J. and A.M.J. Yezer (1998). In Harm's Way: Does Federal Spending on Beach Enhancement and Protection Induce Excessive Development in Coastal Areas? *Land Economics*, **74:1**, 128-45.

Cordes J.J., D.H. Gatzlaff and A.M.J. Yezer (2001). To the Water's Edge, and Beyond: Effects of Shore Protec-tion Projects on Beach Development. *The Journal of Real Estate Finance and Economics*, **22** (2/3), 287-302.

Correggiari A., F. Frascari, S. Miserocchi and D. Fontana (1992). *Breakwaters and eutrophication along the Emila-Romagna coast*. In: Marine coastal eutrophication. The response of marine transitional systems to human impact: problems and perspectives for restoration (Vollenweider, R. A., Marchetti, R. & Viviani, R., eds). Elsevier, Amsterdam, 277-290.

Crisp D.J. (1974). *Factors influencing the settlement of marine invertebrate larvae*. In: Grant PT, Mackie AM (eds) Chemoreception in Marine Organisms. Academic Press, London, 177-265.

CUR/CIRIA (1991). *Rock Manual. Manual on the use of rock in coastal and shoreline engineering*. CUR Report 154, The Netherlands; CIRIA Special Publication 83, UK.

d'Angremond K., J.W. van der Meer and R.J. de Jong (1996). Wave transmission at low-crested structures. ASCE, *Proc. ICCE*, Orlando, Florida, 3305-3318.

Dally W.R., Dean R.B.G. and Dalrymple R.A. (1984). A Model for Breaker Decay on Beaches. *Proceedings of the 19th International Conference on Coastal Engineering*, ASCE, 82-98.

Dally W., R. Dean and R. Dalrymple (1985). Wave height variation across beaches of arbitrary profile. *J. Geophys. Res.*, **90**, (C6).

Dally W.R. and J. Pope (1986). *Detached breakwaters for shore protection*. Technical report CERC-**86-1**, U.S. Army Engineer WES, Vicksburg, MS.

Dally W.R. and Brown C.A. (1995). A modelling investigation of the breaking wave roller with application to cross-shore currents. *Journal of Geophysical Research*, **100**, C12, 24873-24883.

Damgaard J., Dodd N., Hall L. and T. Chesher (2002). Morphodynamic modelling of rip channel growth, *Coastal Eng.*, **45**, 199-221.

Daniel H. (2001), Replenishment versus retreat: the cost of maintaining Delaware's beaches, *Ocean*

and Coastal Management, **44**, 87-104.
Davis N., G.R. Van Blaricom, P.K. Dayton (1982). Man-mad structures on marine sediments: Effect on adjacent communities. *Marine Biology*, **70**(3), 295-303.
Davis J.L.D., L.A. Levin and S.M. Walther (2002). Artificial armored shorelines: sites for open-coast species in a southern California bay. *Marine Biology*, **140**, 1249-1262.
De Vriend H.J. and Stive M.J.F. (1987). Quasi-3D modelling of nearshore currents. *Coastal Engineering*, **11**, 565-601.
De Vriend H.J., J. Zyserman, J. Nicholson, J.A. Roelvink, P.H. Péchon, and H.N. Southgate (1993). Medium term 2DH coastal area modelling. *Coastal Engineering*, **21**(1-3), 193-224.
De Vriend H.J. (1996). *Mathematical modelling of meso-tidal barrier island coasts, Part I: empirical and semi-empirical models*, Adv. in Coastal and Ocean Eng., **2**, Liu ed. World Scientific, 115-149.
Dean R.G. (1977). *Equilibrium beach profiles: U.S. Atlantic and Gulf coasts*. Department of Civil Engineering, Ocean Engineering Report n. **12**, University of Delaware, Newark, Delaware.
Dean R.G. and R.A. Dalrymple (1992). *Water Waves mechanics for Engineers and Scientists*. Advanced Series on Ocean Engineering, **2**, World Scientific.
Dean R.G., R. Chen and A.E. Brower (1997). Full scale monitoring study of a submerged breakwater, Palm Beach, Florida, USA. *Coastal Engineering*, **29**, 291-315.
Debski D. & Loveless J. (1997). *The design and performance of submerged breakwaters*. Report for Ministry of Agriculture Fisheries and Food, University of Bristol.
Defra (2001). *Guidance on the role of flood and coastal defence in nature Conservation in England.*
Deigaard R., J. Fredsøe and I.B. Hedegaard (1986a). Suspended Sediment in the Surf Zone. *Journal of Waterway, Port, Coastal and Ocean Eng.*, ASCE, **112**, (1), 115-128.
Deigaard R., J. Fredsøe and I.B. Hedegaard (1986b). Mathematical Model for Littoral Drift. *Journal of Waterway, Port, Coastal and Ocean Eng.*, ASCE, **112**, (3), 351-369.
Deigaard, R., I. Brøker and J. Fredsøe (1988). A Model for Off/Onshore sediment Transport. *Proceedings of IAHR Symposium on Mathematical Modelling of Sediment Transport in the Coastal Zone*, Copenhagen, Denmark.
Deigaard R. (1993). A Note on the Three-dimensional Shear Stress Distribution in a Surf Zone, *Coastal Engineering*, **20**, 157-171.
Deigaard R., N. Dronen, J. Fredsoe, J.H. Jensen and M.P. Jorgensen (1999). A morphological stability analysis for a long straight barred coast, *Coastal Eng.*, **36**, 171-195.
Del Valle R., R. Medina and M.A. Losada (1993). Dependence of Coefficient *K* on Grain Size, *J. of Waterways, Port, Coast. and Ocean Eng.*, **119** (5), 568-574.
Desprez M. (2000). Physical and biological impact of marine aggregate extraction along the French coast of the eastern English Channel: short- and long-term post-dredging restoration. *ICES Journal of Marine Science*, **57** (5), 1428-1438.
Dibajnia M. and A. Watanabe (1996). A Transport Rate Formula for Mixed Sands, *Proceedings of 25th Coastal Eng. Conference*, Orlando, USA, 3791-3804.
Diskin M.H., M.L. Vajda and I. Amir (1970). Piling up behind low and submerged permeable breakwaters. *J. of Waterways, Harbour and Coastal division*, **96**, 353-372.
Doorn N. and M.R.A. van Gent (2003). Pressures by breaking waves on a slope computed with a VOF-model. *Proc. Conf. on Coastal Structures*, J. Melby (ed.), ASCE, 728-739.
Dorfman J.H., A.G. Keeler and W. Kriesel (1996). Valuing Risk-Reducing Interventions with Hedonic Models: The Case of Erosion Protection. *Journal of Agricultural and Resource Economics*, **21** (1), 109-19.
Drei E. (1996). *Analisi dell' intervento di ripascimento nella zona di Lido di Dante e di alcuni dei suoi aspetti economici*, Degree thesis, University of Bologna, in Italian, 1-210.
Drei E., S. Turchetto, R. Archetti and A. Lamberti (2001). Wave and Current Field Measurement Around Low Crested Structure. *Proc. Ocean Waves Measurement and Analysis*, 115-124.
Durán L.R. and J.R. Castilla (1989). Variation and persistence of the middle rocky intertidal

community of central Chile, with and without human harvesting. *Marine Biology*, **103**, 555-562.

Dye A.H. (1992). Experimental studies of succession and stability in rocky intertidal communities subject to artisanal shellfish gathering. *Netherlands Journal of Sea Research*, **30**, 209-217.

Ebersole B.A. (1985). Refraction-diffraction model for linear water waves. *Journal of Waterways, Port, Coastal and Ocean Engineering*, **111**, 989-953.

Elfrink B., I. Brøker and R. Deigaard (2000). Beach Profile Evolution due to Oblique Wave Attack. Proceedings of 27th Coastal Eng. Conference, Sydney, Australia, **4**, 3021-3034.

Elias E.P.L., Cleveringa J., Buijsman M.C., Roelvink J.A. and Stive M.J.F. (2006). Field and model data analysis of sand transport patterns in Texel Tidal inlet (the Netherlands), *Coastal Eng.*, **53** (5-6), 505-529.

EPA (2000). *Guidelines for preparing economic analyses*. Washington, U.S. Environmental Protection Agency.

Ergun S. (1952). Fluid flow through packed columns, *Chem. Eng. Prog.*, **48** (2), 89-94.

Fankhauser S. (1995). Protection versus Retreat: The Economic Costs of Sea-Level Rise. *Environment and Planning A*, **27** (2), 299-319.

Farber S. (2001). *The Economic Value of Coastal Barrier Islands: A Case Study of the Louisiana Barrier Island System.* University of Pittsburgh, **26**, Pittsburgh.

FEMA (1999). *Hazus: Technical Manual, Volume 3*. Washington, USA, FEMA.

Ferrante A., L. Franco, S. Boer (1993). Modelling and monitoring of a perched beach at Lido di Ostia. *Proc. 23rd ICCE Int. Conf. Coastal Engineering*, **3**, 3305-3318, Venice, October 1992, Ed. B. Edge, ASCE, NY.

Fleming CA. and Hunt J.N. (1976). Application of a sediment transport model. *Proc. 15th Int. Coastal Eng. Conf.*, Honolulu, **I**, 184-1202.

Forteath G.N.R., G.B. Picken, R. Ralph and J. Willimas (1982). *Marine growth studies on the North Sea Oil Platform Montrose Alpha Marine Ecology Progress Series*, **8**, 61-68.

Franco C. and L. Franco (1999). Overtopping Formulas for Caisson Breakwaters with Nonbreaking 3D Waves. *Journal of Waterway, Port, Coastal, and Ocean Engineering*, **125**, (2), 98-108.

Fredsøe J., Deigaard R. (1992). *Mechanics of Coastal Sediment Transport*. Advanced Series in Ocean Engineering, (3). World Scientific Publishing, Singapore, 369 pp.

Freeman P.K., L.A. Martin, R. Mechler, K. Warner and P. Hausmann (2002). *Catastrophes and Development; Integrating natural catastrophes into development planning*. Working papers series n. **4**. Washington, The World Bank.

Garcia N., J.L. Lara and I.J. Losada (2004). *2-D Numerical analysis of near-field flow at low-crested rubble-mound breakwaters. Coastal Engineering*, **51**(10), 991-1020.

Garrabou J., J. Riera, M. Zabala (1998). Landscape pattern indices applied to Mediterranean subtidal rocky benthic communities. *Landscape Ecology*, **13** (4), 225-247.

Gislason K., J. Fredsøe and B.M. Sumer (2003). Morphodynamic modeling in front of vertical and sloping walls. *Proc. Coastal Structures 2003*, Portland, Oregon, U.S.A.

Givler L.D., and R.M. Sorensen (1986). *An Investigation of the Stability of Submerged Homogeneous Rubble-Mound Structures Under Wave Attack.* Report IHL 110-86, H.R. IMBT Hydraulics, Lehigh University, Philadelphia, PA.

Glasby T.M. and S.D. Connell (1999). Urban structures as marine habitats. *Ambio*, **28**, 595-598.

Glasby T.M. (2000). Surface composition and orientation interact to affect subtidal epibiota. *Journal of Experimental Marine Biology and Ecology*, **248**, 177-190.

Glasby T.M. and S.D. Connell (2001). Orientation and position of substrata have large effects on epibiotic assemblages. *Marine Ecology Progress Series*, **214**, 127-135.

Glukovskiy B.Kh. (1966). Investigation of sea wind waves (in Russian), Leningrad, Gidrometeo-Izdat (reference obtained from E. Bouws (1979). Spectra of extreme wave conditions in the Southern North sea considering the influence of water depth), *Proc. of Sea Climatology Conf.*, Paris, 51-71.

Goda Y. (2000). *Random Seas and Design of Maritime Structures.* Advanced Series on Ocean

Engineering, World Scientific.

Gomez-Pina G. (1995). *Análisis de perfiles de playa en las fachadas cantábrica y atlántica de la costa española y su aplicación a proyectos de regeneración.* Universidad de Cantabria E.T.S.I.C.C.P. Dpto. de Ciencias y Técnicas del Agua y Medio Ambiente.

González M. (1995). *Morfologia de playas en equilibrio: planta y perfil.* Ph.D. Thesis, Departamento de Ciencias y Tecnicas del Agua y del Medio Ambiente, Universidad de Cantabria. Santander, Spain.

González M., R. Medina and M.A. Losada (1997). Equilibrium beach profiles: effect of refraction, *Proc. Coastal Dynamics*, **97**, 933-942.

González M., R. Medina, M. Losada (1999). Equilibrium beach profile model for perched beaches, *Coastal Engineering*, **36**, (4), 1999, 343-357.

González M., R. Medina (1999). Equilibrium shoreline response behind a single offshore breakwater, *Proc. Coastal Sediment '99.*

González M., R. Medina (2001). On the application of static equilibrium bay formulations to natural and man-made beaches. *Coastal Engineering*, **43**, 209-225.

González M., R. Medina (2002). A new methodology for the design of static equilibrium beaches and the application in nourishment projects. *Proc. 28th Int. Conf. Coastal Eng.*, ASCE, **3**, 3753-3764.

González-Marco D. (2005). *Modelado numérico de la propagación del oleaje. Una herramienta para la Ingeniería Marítima y la Predicción Operativa.* PhD Thesis, Universitat Politècnica de Catalunya, Barcelona, Spain, 304 pp. (in spanish).

Goodman S.L., W. Seabrooke, H.M. Daniel, S.A. Jaffry and H. James (1993). *Results of a Contingent Valuation Study of Non-use Values of Coastal Resources.* Research Report to MAFF Flood and Coastal Defence Division.

Goodman S.L., W. Seabrooke, H.M. Daniel, S.A. Jaffry, and H. James (1996). *Results of a Contingent Valuation Study of Non-use Values of Coastal Resources.* MAFF.

Gourlay M.R. (1980). Beach processes in the vicinity of offshore breakwaters. *Proc. 17th Int. Coastal Engineering Conference*, Sydney, Australia, ASCE, 1320-1339.

Gourlay M.R. (1994). Wave transformation on a coral reef. *Coastal Eng.*, **23**, 17-42.

Grant W.D. and Madsen O.S. (1979). Combined wave and current interaction with a rough bottom. *Journal of Geophysical Research*, **85**, C4, pp. 1797-1808.

Grant W.D. and Madsen O.S. (1986). The continental shelf bottom boundary layer. *In M. Van Dyke (ed.), Annual Review of Fluid Mechanics*, **18**, 265-305.

Grasmeijer B.T. and Van Rijn L.C. (1998). Breaker bar formation and migration. *Proceedings of the 26th International Conference on Coastal Engineering*, ASCE.

Gravens M.B., N.C. Kraus, and H. Hanson (1991). *GENESIS: Generalized model for simulating shoreline change, report 2, workbook and user's manual.* Technical Report CERC-89-19, Report 2 of a series, U.S. Army Engr. Wtrways. Experiment Station, Vicksburg, Miss.

Green C. (2001). *The Environment: How can it be valued? Ought it to be Valued? How does Valuation Assist Coastal Sustainability?* Flood Hazard Research Centre, Middlesex University: 26.

Greiner R., M.D. Young, A.D. McDonald and M. Brooks (2000). Incentive instruments for the sustainable use of marine resources, *Ocean & Coastal Management*, **43**, 29-50.

Grilli S.T., M.A. Losada and F.L. Martín (1994). Characteristics of solitary wave breaking induced by breakwaters, *Journal of Waterway, Port, Coastal and Ocean Eng.*, **120**, (1), 74-92.

Kamphuis J.W. (1991). Wave transformation, *J. of Coastal Eng.*, **15**, 173-184.

Haab T., and K.E. McConnell (2002). Valuing Environmental and Natural Resources: The Econometrics of Non-Market Valuation Cheltenham, UK: Edward Elgar.

Habitat Directive 1992/43/EC of the European Parliament of the Council (1992).

Hallermeier R.J. (1978). Uses for a Calculated Limit Depth to Beach Erosion, *Proc. of the 16th Int. Conf. on Coastal Eng.*, ASCE, Hamburg, 1493-1512.

Hallermeier R.J. (1981). A Profile Zonation for Seasonal Sand Beaches from Wave Climate, *Coastal*

Engineering, **4**, 253-277.

Hanley, Nick and Clive L. Spash (1993). *Cost-Benefit Analysis and the Environment*. Aldershot: Edward Elgar.

Hanson H., M.B. Gravens and N.C. Kraus (1988). Prototype applications of a generalized shoreline change numerical model. *Proc. 21st Coastal Engineering Conference* (ASCE), 1265-1279.

Hanson H. and Kraus N.C. (1989). *GENESIS: Generalised model for simulating shoreline change*. Report 1. Technical Report CERC-89-19, Vicksburg, Mississii: U.S. Army Corps of Engineers, 185 p.

Hanson H. and Kraus N.C. (1990). Shoreline response to a single transmissive detached breakwater. *Proc. 22nd Int. Conf. on Coastal Eng.*, Delft, The Netherlands, 2034-2046.

Hanson H. and M. Larson (1999). Extension of GENESIS into the cross-shore dimension-from 1-line to N-line. *5th Conference on Coastal and Port Engineering in Developing Countries*, 312-323.

Hanson H. (2003). Overview of beach nourishment in the European Union. Project, practices and objectives. *Shore & Beach*, **71**, (2), 9-19.

Hanson H., S. Aarninkhof, M. Capobianco, J.A. Jiménez, M. Larson, R.J. Nicholls, N.G. Plant, H.N. Southgate, H.J. Steetzel, M.J.F. Stive and H.J. Vriend (2003). Modelling of coastal evolution on yearly to decadal time scales. *Journal of Coastal Research*, **19** (4), 790-811.

Hashimoto H. (1974). A model to predict the influence or erosion to Neighboring beaches with a detached breakwaters. *Proc. 21st Conf. on Coastal Eng. in Japan* (in Japanese).

Hasselmann K. (1973). *Measurements of wind-wave growth and swell decay during the Joint North Sea Wave Project (JONSWAP)*. Deutsche Hydr. Zeit., Reihe A (8°), No. 12, 1973.

Hausman J.A. (ed.) (1993). *Contingent Valuation. A Critical Assessment*. North Holland, Amsterdam.

Hawkins S.J. (1981). *The influence of Patella grazing on the fucoid-barnacle mosaic on moderately exposed rocky shores*. Kieler Meeresforsch, 537-543.

Hawkins S.J. and R.G. Hartnoll (1983). Grazing of intertidal algae by marine invertebrates. *Oceanography and Marine Biology Annual Review*, **21**, 195-282.

Hawkins S.J., A.J. Southward and R.L. Barrett (1983). Population structure of *Patella vulgata* L. during succession on rocky shores in southwest England. *Oceanologica Acta*, Special Volume: 103-107.

Hawkins, S. J., and R. G. Hartnoll (1985). Factors determining the upper limits of intertidal canopy-forming algae, *Marine Ecology Progress Series*, **20**, 265-272.

Hawkins S.J., R.G. Hartnoll, J.M. Kain and T.A. Norton (1992). *Plant-animal interactions on hard substrata in the Northeast Atlantic*. In: D.M. John, H.S. J. and P.J. H. (Editors), Plant-Animal Interactions in the Marine Benthos. Oxford University Press, Oxford, UK, 1-32.

Hawkins S.J. and H.D. Jones (1992). *Marine Conservation Society Marine field course guide 1*. Rocky Shores, 144 pp. Immel, London.

Hedges T.S. and Reis M.T. (1998). Random waves overtopping of simple sea walls: a new regression model. *Proceedings of the Institution of Civil Engineers. Water, Maritime and Energy*, **130**, 1-10.

Heideman J.C., George R.J. (1981). *Biological and engineering parameters for macrofouling growth on platforms offshore Lousiana. The Ocean – An international workplace*. Oceans 81 Conference Record, Boston, Massachussets, September 16-18, 1981, **1**: 550-557.

Herbich J.B. (1989). Shoreline Change Due to Offshore Breakwaters. *Proceeding XXIII Congress, IAHR*. Ottawa, Canada, **C**., 317-321.

Herbich J.B. (2000). *Handbook of Dredging Engineering*. 2nd ed., McGraw-Hill, New York.

Herriges J.A. and C.L. Kling, eds. (1999). *Valuing Recreation and the Environment*. Edward Elgar, Cheltenham.

Hily C., P. Potin, J.-Y. Floch (1992). Structure of subtidal algal assemblages on soft-bottom sediments: Fauna/flora interactions and role of disturbances in the Bay of Brest, France. *Marine ecology progress series*, **85**, 1-2:115-130.

Hirose N., A. Watanuki and M. Saito (2002). New Type Units for Artificial Reef Development of

Ecofriendly Artificial Reefs and the Effectiveness Thereof. *Proc. 30th International navigation congress*, PIANC.

Ho S.K. (1971). *Crenulate shaped bays*. Thesis n. 346, Assoc. Institute of Technology at Bangkok, Thailand.

Holme N.A., A.D. McIntyre (1971). *Methods for the study of marine benthos*. Blackwell Scientific Publications, Oxford and Edimburgh, pp. 334.

Holmes S.P., C.G. Sturgess, M.S. Davies (1997). The effect of rock-type on the settlement of Balanus balanoides (L.) Cyprids. *Biofouling*, **11** (2), 137-147.

Horikawa K. and C.T. Kuo (1966). A study on wave transformation inside surf zone. *Proc. 10th Coastal Eng. Conf. Tokyo*, 217-233.

Hsu J.R.C., R. Silvester and Y.M. Xia (1987). New characteristics of equilibrium shape bays. *Proc. 8th Austral. Conf. on Coastal & Ocean Eng.*, 140-144.

Hsu J.R.C. and C. Evans (1989). Parabolic bay shapes and applications. *Proc. Institution of Civil Engineers,* London, **87** (Part 2), 556-570.

Hsu J.R.C., R. Silvester and Y.M. Xia (1989a). Static equilibrium bays: New relationships. *J. Waterway, Port, Coastal and Ocean Eng.*, ASCE, **115**(3), 285-298.

Hsu J.R.C., R. Silvester and Y.M. Xia (1989b). Generality on static equilibrium bays. *Coastal Eng.*, **12**, 353-369.

Hsu J.R.C., R. Silvester and Y.M. Xia (1989c). Applications of headland control. *J. of Waterway, Port, Coast and Oc. Eng.*, ASCE, **115** (3), 299-310.

Hsu J.R.C. and R. Silvester (1990). Accretion behind single offshore breakwater. *J. Wtrway., Port, Coast. and Oc. Engrg.*, ASCE, **116** (3), 367-380.

Hsu J.R.C., T. Uda and R. Silvester (1993). Beaches downcoast of harbours in bays. *Coastal Eng.*, **19**: 163-181.

Hsu T.-J., T. Sakakiyama, P.L.-F. Liu (2002). A numerical model for wave motions and turbulence flows in front of a composite breakwater. *Coastal Engineering*, **46**, 25-50.

Hughes S.A. (1993). Physical models and laboratory techniques in coastal engineering. *Advanced Series on Ocean Engineering*, **7**, World Scientific Publishing, Singapore.

Hur D., N. Kawashima and K. Iwata (2003). Experimental study of the breaking limit of multi-directional random waves passing over an impermeable submerged breakwater, *Ocean Engineering*, **30**, (15), 1923-1940.

Ibragimov I.A. and R.Z. Has'minskii (1981). *Statistical Estimation, Asymptotic Theory.* Springer verlag, New York.

IDROSER (1996). *Progetto di Piano per la Difesa dal Mare e la Riqualificazione Ambientale del Litorale della Regione Emilia-Romagna.* Regione Emilia-Romagna-Idroser, Bologna, 1996.

Irish J.L., W.J. Lollycrop, L.E. Parson (1997). Accuracy of sand volumes as a function of survey density. 3736-3749. *Proc. of the 25th Int. Conf. on Coastal Eng.*, **3**, Orlando, Florida (ed. B.L. Edge), New York.

Isobe M. and Horikawa K. (1982). Study on water particle velocities of shoaling and breaking waves. *Coastal Engineering in Japan*, **25**, 109-123.

Italian Ministry of Public Works, National Research Council (1996). *Technical instructions for the design of maritime dikes.* GNDCI, n. 1450.

Jackson J.B.C. (1986). Modes of dispersal of clonal benthic invertebrates: consequences for species' distribution and genetic structure of local populations. *Bulletin of Marine Science*, **39**, 588-606.

Jenkins S.R., S.J. Hawkins, T.A. Norton (1999). Direct and indirect effects of a macroalgal canopy and limpet grazing in structuring a sheltered inter-tidal community. *Marine Ecology Progress*, 81-92.

Jenkins S.R., F. Arenas, J. Arrontes, J. Bussell, J. Castro, R.A. Coleman, S.J. Hawkins, S. Kay, B. Martinez, J. Oliveros, M.F. Roberts, S. Sousa, R.C. Thompson and R.G. Hartnoll (2001). European-scale analysis of seasonal variability in limpet grazing activity and microalgal abundance. *Marine Ecology Progress Series*, **211**, 193-203.

Jiménez J.A. and Sánchez-Arcilla A. (2002). Preliminary analysis of shoreline evolution in the leeward of low-crested breakwaters using one-line models. *Proc.of the 1st year meeting of Delos Project* (Barcelona, Spain), www.delos.unibo.it.

Jones C.A. and F. Lupi. (1999). The Effect of Modeling Substitute Activities on Recreational Benefit Estimates. *Marine Resource Economics*, **14** (4), 357-74.

Jonsson P., L. Granhag, P.S. Moschella, P. Åberg, S.J. Hawkins and R.C. Thompson. Interaction between wave action and grazing control the distribution of intertidal macroalgae. *Ecology*, **87** (5), 1169-1178.

Kaminsky G.M. and N.C. Kraus (1993). Evaluation of depth-limited wave breaking criteria. *Proc. Ocean Wave Measurement and Anal., Waves 93*. ASCE, 180-193.

Kamphuis J.W. (1991). Alongshore Sediment Transport Rate. *J. of Waterway, Port, Coastal and Ocean Eng.*, ASCE, **117**, 624-640.

Kamphuis J.W. (2002). Alongshore Transport Rate of Sand. *Proceedings of 28th Coastal Eng. Conference*, Cardiff, Wales, 2478-2490.

Kana T.W., C.J. Andrassy (1995). *Beach profile spacing: practical guidance for monitoring nourishment projects. Proc. of the 24th Int. Conf. on Coastal Eng.*, **2**, 2100-2114.

Katopodi I. and J.S. Ribberink (1992). Quasi-3D modelling of suspended sediment transport by currents and waves. *Coastal Engineering*, **18** (1-2), 83-110.

King O.H. (1995). Estimating the value of marine resources: a marine recreation case, *Ocean and Coastal Management*, **27**, (1-2), 129-41.

King D.M., N.J. Cooper, J.C. Morfett and D.J. Pope (2000). Application of offshore breakwaters to the UK: a case study at Elmer beach. *Journal of Coastal Research*, **16**, 172-187.

King O.H. (1995). Estimating the value of marine resources: a marine recreation case. *Ocean & Coastal Management*, **27** (1-2), 129-41.

Kirby J.T. and R.A. Dalrymple (1983). A parabolic equation for the combined refraction-diffraction of Stokes waves by mildly varying topography. *J. Fluid Mechanics*, **136**, 543-566.

Kirby J.T. (1986a). Higher order approximations in the parabolic equation method for water waves. *Journal of Geophysical Research*, **91** (C1), 933-952.

Kirby J.T. (1986b). Rational approximations in the parabolic equation method for water waves. *Coastal Engineering*, Elsevier, **10**, 355-378.

Klopman G. (1996). *Extreme wave heights in shallow water*. Report H2486, WL/Delft Hydraulics, The Netherlands.

Koerner R. (1990). *Designing with geosynthetics*. Prentice-Hall, New York.

Kofoed J.P. and H.F. Burcharth (2002). Estimation of overtopping rates on slopes in wave power deviced and other low crested structures. *Proc. 28th. ICCE*, **2**, 2191-2202

Komar P. and D.L. Inman (1970). Longshore Sand Transport on Beaches. *Journal of Geophysical Research*, **75**, (30), 5914-5927.

Komar P. (1975). Nearshore currents: generation by obliquely incident waves and longshore variations in breaker height. *Proc. of the Symposium on Nearshore Sediment Dynamics*, Ed. J.R. Hails and A. Carr, Wiley, London, 17-45.

Kottegoda N. and R. Rosso (1997). *Statistics, Probabilità, and reliability for civil and environmental engineers*. Mc Graw Hill Companies.

Kramer M. and H.F. Burcharth (2003). Stability of Low-Crested Breakwaters in Shallow Water Short Crested Waves. *Proc. Coastal Structures 2003*, ASCE, 137-149.

Kramer M., B. Zanuttigh, W. Baoxing, J.W. Van der Meer, A. Lamberti, and H.F. Burcharth (2003), *Wave basin experiments, internal report*. DELOS deliverable D31, available from the Internet www.delos.unibo.it.

Kramer M., B. Zanuttigh, J.W. Van der Meer, C. Vidal, F.X. Gironella (2005). Laboratory experiments on low-crested breakwaters, *Coastal Eng*. **52** (10-11), 867-885.

Kraus N.C. (2001). On equilibrium properties in predictive modeling of coastal morphology change. *Proc. Coastal Dynamics*, ASCE, 1-15.

Kraus N.C., M. Larson, and D.L. Kriebel (1991). Evaluation of Beach Erosion and Accretion predictors. *Proc. Coastal Sediments '91*, ASCE, 572-587.

Kriebel D.L. and R.G. Dean (1985). Numerical simulation of time-dependent beach and dune erosion. *Coastal Engineering*, **9**, 221-245.

Lamberti A. (1995). *Preliminary Results on Main Armour*. Toe Berm Interaction, Final Proceedings, MAST II Project, Rubble Mound Breakwater Failure Modes, vol. **2**, Aalborg University, ed.

Lamberti A., B., Zanuttigh, M. Kramer (2003). Waves and currents flows around low crested structures. *Proc. Coastal Structures, 2003*. ASCE, 860-862.

Lamberti A., B. Zanuttigh, M. Tirindelli (2003). 3D Hydrodynamic tests with low crested structures: analysis of overtopping and velocity fields. *Proc. ISOPE 2003*, EA, 562-569.

Lamberti A. and B. Zanuttigh (2005). An integrated approach to beach management in Lido di Dante, Italy., Estaurine *Coastal and Shelf Science*, **62**, n. 3, 441-451, Pacyna, J., Barrett, K., Namiesnik J. ed., Elsevier.

Lamberti A., Archetti R., Kramer M., Paphitis D., Mosso C. and Di Risio M. (2005). Prototype experience regarding low-crested structures. *Coastal Engineering*, **52** (10-11), 841-866.

Lara J.L., I.J. Losada, P.L.-F. Liu (2004). Time average characteristics of breaking waves on a gravel slope. *Journal of Fluid Mechanics,* under revision.

Larson M. and N.C. Kraus (1989). *SBEACH: Numerical model for simulating storm-induced beach change. Report 1: Empirical foundation and model development*. Technical Report CERC-89-9, Coastal Engineering Research Center, U.S. Army Corps of Engineers.

Larson M., H. Hanson, and N.C. Kraus (1997). Analytical solutions of the one line model of shoreline change near coastal structures. *J. Waterway, Port, Coastal and Ocean Engineering*, **123**(4), 180-191.

Latteux B. (1980). Harbour design including sedimentological problems using mainly numerical technics. *Proc. 16th Int. Coastal Eng. Conf.*, Sydney, 2213-2219.

Laustrup C., H.T. Madsen (1994). Design of Breakwaters and Beach Nourishment. *24th Proc. Int. Conference on Coastal Engineering*, Kobe, Japan, ASCE.

Leont'yev I.O. (1999). Modelling of morphological changes due to coastal structures, *Coastal Eng.*, **38**, 143-166.

Lesser G.R., J.H. de Vroeg, J.A. Roelvink, M. de Gerloni, V. Ardone (2003). Modelling the morphological impact of submerged offshore breakwaters. *Proc. Coastal Sediments '03 Conference*, Clearwater, ASCE, New York.

Lesser G.R., Roelvink J.A., van Kester J.A.T.M. and Stelling G.S. (2004). Development and validation of a three-dimensional morphological model, *Coastal Eng.*, **51**, 883-915.

Lewis J.R. (1964). *The Ecology of Rocky Shores*. English Universities Press, London.

Lin P., P.L.-F. Liu (1998). A numerical study of breaking waves in the surf zone. *Journal of Fluid Mechanics*, **359**, 239-264.

Lippmann T.C. and R.O. Holman (1990). The spatial and temporal variability of sand bar morphology. *Journal of Geophysical Research,* **95** (C7), 11, 575-11, 590.

Lipton D.W., K. Wellman, I.C. Sheifer and R.L. Weiher (1995). *Economic Evaluation of Natural Resources*. A Handbook for Coastal Policymakers.

Liu P. L.-F. (1990). *Wave transformation*. In The Sea, Wiley-Interscience Publication, vol. 9, 27-63.

Liu P.L.-F., P. Lin, K.A. Chang, T. Sakakiyama (1999). Numerical modeling of wave interaction with porous structures. *J. Waterway, Port, Coastal and Ocean Engineering*, **125**, 322-330.

Liu P.L.-F., P. Lin, T. Hsu, K. Chang, I.J. Losada, C. Vidal, T. Sakakiyama T. (2000). A Reynolds averaged Navier-Stokes equation model for nonlinear water wave and structure interactions. *Proc. Coastal Structures '99*, 169-174.

Liu P.L.-F. (2004). Solitary wave runup and force on a vertical barrier. *Journal of Fluid Mechanics*, **505**, 225-233.

Lomborg B. (1998). *The Skeptical Environmentalist*. Cambridge, Cambridge University Press.

Longuet-Higgins M.S. (1970). Longshore Currents Generated by Obliquely Incident Waves. *Journal of Geophysical Research*, **75**, (33), 6678-6801.

Loomis J. and J. Crespi (1999). *Estimated effects of climate change on selected outdoor activities in the United States*. In The impact of climate change on the United States economy. Robert Mendelsohn and James- E. Neumann eds. Cambridge; New York and Melbourne: Cambridge University Press, pp. xi, 331.

Losada I.J., M.A. Losada and A.J. Roldán (1992). Oblique incident wave past vertical thin barriers. *Applied Ocean Research*, **14**, (3), 191-200.

Losada I.J., J.L. Lara and N. Garcia (2003). 2-D experimental and numerical analysis of wave interaction with low-crested breakwaters including breaking and flow recirculation. *Proc. Coastal Structures 2003*, ASCE.

Losada I.J., J.L. Lara, E. Damgaard, N. García (2005). Modelling of velocity and turbulence fields around and within low-crested rubble-mound breakwaters, *Coastal Eng.*, **53** (10-11), 887-913.

Loveless J. and D. Debski (1997). *The Design and Performance of Submerged Breakwaters*. Ministry of Agriculture Fisheries and Food, contract CSA 2606, Department of Civil Engineering, University of Bristol.

Loveless J.H., D. Debski, A.B. MacLeod (1998). Sea level set-up behind detached breakwaters. *Proc. Int. Conf. Coastal Eng.*, ASCE, 1665-1678.

Loveless J. (1999). *The Design and performance of Detached Rubble Mound Breakwaters*. Ministry of Agriculture, Fisheries and Food (MAFF), December 1999.

Madsen O.S. and W.D. Grant (1976). *Sediment Transport in the Coastal Environment*. MIT Ralph M. Parsons Laboratory, Report 209.

Madsen P.A. (1983). Wave Reflection from a Vertical Permeable Wave Absorber. *Coastal Eng.*, **7**, 381-396.

Madsen O.S., Poon Y.K. and Graber H.C. (1988). Spectral wave attenuation by bottom friction: theory. *Proc. of the 21st Int. Conference on Coastal Engineering*, ASCE, 492-504.

Madsen P.A., R. Murray and O.R. Sørensen (1991). A New Form of the Boussinesq Equations with Improved Linear Dispersion Characteristics (Part 1). *Coastal Eng.*, **15**, 371-388.

Madsen P A. and O.R. Sørensen (1992). A New Form of the Boussinesq Equations with Improved Linear Dispersion Characteristics. Part 2: A Slowly-varying Bathymetry, *Coastal Eng.*, **18**, 183-204.

Madsen O.S. (1994). Spectral wave-current bottom boundary layer flows. *Proc. of the 24th Int. Conference on Coastal Engineering*, ASCE, 384-398.

Madsen P.A., O.R. Sørensen and H.A. Schäffer (1997a). Surf zone dynamics simulated by a Boussinesq type model. Part I: Model description and cross-shore motion of regular waves. *Coastal Eng.*, **32**, 255-288.

Madsen P.A., O.R. Sørensen and H.A. Schäffer (1997b). Surf zone dynamics simulated by a Boussinesq type model. Part II: Surf beat and swash zone oscillations for wave groups and irregular waves. *Coastal Eng.*, **32**, 289-320.

MAFF (2000). *Flood and coastal defence project appraisal guidance: economic appraisal*. Ministry of Agriculture, Fisheries and Food, London, UK.

Mangor K. (2001). *Shoreline Management Guidelines*. DHI Water & Environment, Hørsholm, Denmark, 232 pp.

Martin D., Bertasi F., Colangelo M.A., Frost M., Hawkins S.J., Macpherson E., Moschella P.S., Satta M.P., Thompson R.C., deVries M. and Ceccherelli V.U. (2005). Ecological impacts of low crested structures on soft bottoms and mobile infauna: how to evaluate and forecast the consequences of an unavoidable modification of the native habitats. *Coastal Engineering*, **52** (10-11), 1027-1051, Elsevier.

Martinelli L. (2003). *Modello fisico presso il LIC di Bari, 24 Marzo-17 Aprile 2003* (in Italian). Internal Report, University of Bologna, Italy. Electronic format avaliable under request (luca.martinelli@mail.ing.unibo.it).

Martinelli L., B. Zanuttigh and A. Lamberti (2006). Hydrodynamic and morphodynamic response of

isolated and multiple low crested structures: Experiments and simulations, *Coastal Engineering*, **53** (4), 363-379.
Marzetti Dall'Aste Brandolini S. and B. Zanuttigh (2003). Socio-economic valuation of beach protected from erosion in Lido di Dante, Italy. *Proc. MEDCOAST 2003*, **1**, 319-330.
Marzetti Dall'Aste Brandolini S. (2003). *Economic and Social Valuation about European Coastal Sites*. Deliverable 28/A, www.delos.unibo.it.
Marzetti S. (2003). *Willingness to Pay for the Defence of Venice (Italy) as World Heritage Site*. In Assessment of Option Value and Non-Use Values – The Case-Studies of Venice (Italy) and Normerven (The Netherlands), Deliverable 28B, www.delos.unibo.it.
Marzetti S. and L. Franco (2003). Assessment of the daily loss of enjoyment of the Ostia beach because of erosion. *Delos workshop*, Santander 2003.
Marzetti S. and A. Lamberti (2003). Economic and social valuation of the defence system of Venice and its lagoons (Italy). *Proc. MEDCOAST 2003*, **1**, 307-318.
Marzetti Dall'Aste Brandolini S., L. Franco, A. Lamberti and B. Zanuttigh (2004). *Preferences about different kinds of defence structures and beach materials: the Italian case-studies of Lido di Dante, Ostia and Pellestrina island*. Deliverable 28/C, www.delos.unibo.it.
Marzetti S. and A. Lamberti (2004). Evaluation of the recreational use of Barcola beach in Trieste (Italy), *Proc. Sustainable tourism*, Pineda, Brebbia and Mugic ed.s, WIT Press.
Massie P.E., (1986). Coastal Engineering. Volume I Introduction and Volume II Harbour and Beach problems. Lecture notes Delft University of Technology, the Netherlands.
Mayer S., A. Garapon and L.S. Sørensen (1998). A fractional step method for unsteady free-surface flow with application to non-linear wave dynamics. *Intl. Journal for Numerical Methods in Fluids*, **28**, (2), 293-315.
Mc Cormick M.E. (1993). Equilibrium shoreline response to breakwaters. *Journal of Waterway, Port, Coastal and Ocean Engineering*, vol. **119**, n. 6, November/Dec., 657-670.
McCowan J. (1894). On the highest wave of permanent type, *Philosophical Magazine*, **38**, 351-358.
Medina R., A.M. Bernabeu, C. Vidal and M. Gonzalez (2000). Relationship between beach morphodynamics and equilibrium profiles. *Proc. 27th International Conf. Coastal Eng.* ASCE.
Mei C.C. (1989). The Applied dynamics of Ocean surface waves. World Scientific Publishing, Singapore.
Melito I. and J.A. Melby (2002). Wave runup, trasmission, and reflection for structures armoured with CORE-LOC. *Coastal Engineering*, Elsevier, **45**, 33-52.
Miche R. (1944). *Mouvements ondulatoirs des mers en profondeur constante on decroissant*. Annals des Ponts et Chausses.
Minchin D., C.B. Duggan (1989). Biological control of the mussel in shellfish culture. *Aquaculture*, **81** (1), 97-100.
Ming D.H. and Y. Chiew (2000). Shoreline Changes behind Detached Breakwater. *Journal of Waterway, Port, Coast and Oc. Eng.*, ASCE, **126** (2), 63-70.
Ministry of Public Works (1991). *Technical instructions for design and construction of coastal defence systems.*
Ministry of Public Works (1996). *Technical instructions for breakwater design.*
Mitchell R.C. and R.T. Carson (1989). *Using Surveys to Value Public Goods: the Contingent Valuation Method, Resources for the Future*. Washington, D.C.
Mitsuyasu H., Tasai F., Suhara T., Mizuno S., Ohkusu M., Honda T. and Rikiishi K. (1975). Observations of the directional spectrum of ocean waves using a cloverleaf buoy. *Journal of Physical Oceanography*, **5**, 750-760.
Mohan R.K., T.W. Kana (1997). The effect of grain sorting on profile stability of nourished beaches. *Proc. of the 25th Int. Conf. of Coastal Eng.*, **4**, 4756-4769.
Moore B. (1982). *Beach profile evolution in response to changes in water level and wave height*. M.S. Thesis, University of Delaware.
Moreno C.A. (2001). Community patterns generated by human harvesting on Chilean shores: a review.

Aquatic Conservation: Marine and Freshwater Ecosystem **11**, 19-30.

Moschella P., M. Abbiati, P. Åberg, L. Airoldi, J.M. Anderson, F. Bacchiocchi, G.E. Dinesen, E. Gacia, L. Granhag, P. Jonsson, M.P. Satta, A. Sundelöf, R.C. Thompson and S.J. Hawkins. (2005) Low-crested coastal defence structures as artificial habitats for marine life: using ecological criteria in design. *Coastal Engineering*, **52** (10-11), 1053-1071, Elsevier.

Mösso C. (2004). *Estudio numérico-experimental de la dinámica de la zona cercana a la costa*. Ph D Thesis, Universitat Politècnica de Catalunya, Barcelona, Spain (in spanish).

Muñóz-Pérez J.J., L. Tejedor and R. Medina (1998). Equilibrium beach profile model for reef-protected beaches. *Journal of Coastal Research*, **15**, (4), 950-957.

Nicholson J., Broker I., Roelvink J.A., Price D., Tanguy J.M. and Moreno L. (1997). Intercomparison of coastal area morphodynamic models, *Coastal Eng.*, **31**, 97-123.

Nielsen P. (1985). *A Short Manual of Coastal Bottom Boundary Layers and Sediment Transport*. Public Works Dept. of New South Wales.

Nielsen P. (1992). *Coastal Bottom Boundary Layers and Sediment Transport*. Advances Series on Ocean Engineering, **4**, World Scientific Publishing, Singapore.

Nir J. (1976). *Detached breakwaters, groynes and artificial structures on the Mediterranean shore and their influence on the structure of the Israeli SHORE*. Rep. n. 3, 76/2, Jerusalem, March 1976 Ministry of Industry and Commerce, Geological Institute, Marine Geology Section, (in Hebrew).

Nir Y. (1982). Offshore artificial structures and their influences on the Israel and Sinai Mediterranean beaches. *Proc. 18th Int. Coastal Engineering Conference*, Cape Town, ASCE, 1837-1856.

NOAA Panel report: See Arrow (1993).

Noble R.M. (1978). Coastal structures' effects on shoreline. *Proc. 16th Int. Coastal Engineering Conference*, Hamburg, ASCE, 2069-2085.

O'Connor B.A. and Nicholson J. (1989). *Modelling changes in coastal morphology*. In: S.S.Y. Wang (Editor), Sediment Transport Modeling, ASCE, 160-165.

O'Connor M. and C. Spash, eds. (1999). *Valuation and the environment: Theory, method and practice*. Cheltenham, U.K. and Northampton, Mass.: Elgar.

Osiecki D.A. and Dally W.R. (1996). The influence of rollers on longshore currents. *Proc. of the 25th Int. Conference on Coastal Engineering*, ASCE, 3419-3430.

Oumeraci H., H. Scüttrumpf and M. Bleck (1999). *Wave overtopping at seadikes. Comparison of Physical model tests and numerical computations*. OPTICREST Task 3.5 and 5. MAS3-CT97-0116, Dept. Hydr. and Coastal Eng., Tech. Univ. Braunsweigh, Germany.

Owen M.W. (1980). *Design of sea walls allowing for wave overtopping*. Report EX 924, Hydraulics Research, Wallingford, U.K.

Ozasa H. and A.H. Brampton (1980). Mathematical modelling of beaches backed by seawalls. *Coastal Eng*. **4**, 47-63.

Page H.M., J.E. Dugan, D.S. Dugan, J.B. Richards and D.M. Hubbard (1999). Effects of an offshore oil platform on the distribution and abundance of commercially important crab species. *Marine Ecology Progress Series*, **185**, 47-57.

Palmquist R.B. (1988). Welfare Measurement for Environmnetal Improvements Using the Hedonic Model: The Case of Nonparametric Marginal Prices. *Journal of Environmental Economics and Mangement*, **15**, 297-312.

Panizzo A., R. Briganti, J.W. van der Meer and L. Franco (2003). Analysis of wave transmission behind low-crested structures using neural networks. ASCE, *Proc. Coastal Structures 2003*, Portland, Oregon.

Parker D.J., C.H. Green and P.M. Thompson (1987). *Urban flood protection benefits, a project appraisal guide*. Aldershot, Gower.

Pearse P.H. (1968): A New Approach to the Evaluation of Non-Priced Recreational Resources', *Land Economics*, February, 87-99.

Penning-Rowsell E., C.H. Green, P.M. Thompson, A.M. Coker, S.M. Tunstall, C. Richards, and D.J.

Parker (1992). *The Economics of Coastal Management: a Manual of Benefit Assessment Techniques (Yellow Manual).* London, Belhaven Press.

Pergent-Martini C., V. Rico-Raimondo, G. Pergent (1996). Impact of nutrients on Posidonia oceanica seagrass beds: Preliminary data. Proceedings of the coastal and lagoonal management. *Journal de Recherche Oceanographique*, **21**, (1-2), 35-39.

Perlin M. and R.G. Dean (1979). Prediction of beach planforms with littoral controls. *16th Int. Conf. on Coastal Engineering Conference (ASCE)*, 1818-1838.

Perlin M. (1979). Predicting beach planforms in the lee of breakwater. *Symp. on Coastal structures '79*, ASCE, **2**, 792-808.

Petit H.A.H., P. Tönjes, M.R.A. van Gent and P. van den Bosch (1994). Numerical simulation and validation of plunging breakers using a 2D Navier-Stokes model. *ASCE, Proc. ICCE'94*, **1**, 511-524, Kobe.

Phillps J.D. (1985). Estimation of optimal beach profile sample intervals, *Journal of Coastal Research*, **1**(2), 187-191.

PIANC (1992). *Guidelines for the design and construction of flexible revetments incorporating geotextiles in marine environment.* PIANC, Supplement to bulletin n. 78/79, Brussels, Belgium.

Pierson W.J. Jr. and L. Moskowitz (1964). A proposed spectral form for fully developed wind seas based on the similarity law of S.A. Kitaigorodski. *Journal of Geophysical Research*, **64**, (24), 1964, 5181-5190.

Pylarczyk K.W. (1990). Coastal Protection. *Proc. of the Short Course on Coastal Protection,* Delft University of Technology, Balkema, The Netherlands.

Pilarczyk K.W. (2000). *Geosynthetics and Geosystems in Hydraulic and Coastal Engineering.* Published by A.A. Balkema, Rotterdam.

Pilarczyk K.W. (2003). Design of low-crested (submerged) structures. An overview. *6th Proc. International Conference on Coastal and Port Engineering in Developing Countries (COPEDEC)*, Colombo, Sri Lanka; www.tawinfo.nl.

Polomé P., A. van der Veen and S. Marzetti (2001). *DELOS WP 1.3 Final Report (Literature Review)*, www.delos.unibo.it.

Polomé P. (2002). *DELOS WP 4.1 Final Report (Benefit Transfer),* www.delos.unibo.it.

Polomé P., S. Marzetti and A. van der Veen (2005). Social effects of coastal defence: economics and societal demands, *Coastal Engineering,* **52** (10-11), 819-840.

Powell K.A. and N.W.H. Allsop (1985). *Low-crest breakwaters, hydraulic performance and stability.* Report SR 57, HR Wallingford.

Preti M., F. Bonsignore, M. Guerrero, L. Martinelli, L. Grandi and N. De Nigris (2002). *Stato del litorale emiliano-romagnolo all'anno 2000.* I quaderni di ARPA, Bologna, 2002, 1-125.

Putrevu U. and I.A. Svendsen (1999). Three-dimensional dispersion of momentum in wave-induced nearshore currents. *Eur. J. Mech. B/Fluids*, 83-101.

Raffaelli D.G. & S.J. Hawkins (1996). *Intertidal Ecology*. Chapman and Hall, London, 356 pp. Re-issued by Kluwer 1999.

Ranasinghe R., Symondsa, R., Black K. and Holman, R. (2004). Morphodynamics of intermediate beaches: a video imaging and numerical modelling study, *Coastal Eng*., **51**, 629-655.

Raven H.C. (1996). *A Solution Method for the Nonlinear Ship Wave Resistance Problem.* PhD Thesis, Technical University of Delft.

Reish D.J. (1964). Discussion of the Mytilus californianus community on newly constructed rock jetties in southern California (Mollusca: Bivalvia). *The Veliger*, **7**, 95-101.

Relini G., F. Tixi, M. Relini and G. Torchia (1998). The macrofouling on offshore platforms at Ravenna. *International Biodeterioration & Biodegradation*, **41**, 41-55.

Reniers A.J.H.M., J.A. Roelvink and A.R. van Dongeren (2000). Morphodynamic response to wave group forcing. *Proc. of the 27th Int. Conference on Coastal Engineering*, 3218-3228, Am. Soc. of Civ. Eng., Sidney.

Reniers A.J.H.M., J.A. Roelvink and E.B. Thornton (2003). Morphodynamic modeling of an embayed

beach under wave group forcing, accepted for publication, *Journal of Geophysical Research*.
Rerdy R. C., Navrud S. and W.R. Horton (2001). How do respondents with uncertain willingness to pay answer contingent valuation questions?, *Land Economics*, **77**(3), 315-326.
Rhoads (1974). Organism-sediment relations on the muddysea floor. *Oceanogr. Mar. Biol. Ann. Rev.*, **12**, 263-300
Riddell K. and C. Green (1999). *Flood and coastal defence project appraisal guidance: economic appraisal. VII, 104*. Ministry of Agriculture, Fisheries and Food, MAFF Publication: London.
Roelvink J.A. and I. Baker (1993). Cross-shore profile models. *Coastal Engineering*, **21**, 163-191.
Roelvink J.A., M.J.P. Duin, M.B. de van Vries (2002). *DELOS, Improved morphodynamic modelling and application to cases*. Deliverable 27, www.delos.unibo.it.
Roelvink J.A., G. Lesser, C. Van Ha, A. Luijendijk (2002). Morphological pilot experiment in a 3D wave-current basin. *Proc. of ICCE*, J.M. Smith (ed.), ASCE, New York, 2664-2676.
Rojanakamthorn S., M. Isobe, A. Watanabe (1990). *Modeling of wave transformation on submerged breakwater*. Proc. 22nd Coastal Engineering Conference, ASCE, 1060-1073.
Rosati J.D. (1990). *Functional design of breakwaters for shore protection: Empirical methods*. Technical Report CERC-90-15, U.S. Army Engr. Wtways. Experiment Station, Coast. Engrg. Res. Ctr., Vicksburg, Miss.
Rose A. and J. Benavides (1998). *Regional economic impacts. Engineering and socioeconomic impacts of earthquakes*. M. Shinozuka, A. Rose and R.T. Eguchi. Buffalo, MCEER.
Rose A. and D. Lim (2002). Business interruption losses from natural hazards: conceptual and methodological issues in the case of the Northridge earthquake. *Environmental hazards*, **4**, 1-14.
Rosen D.S., and M. Vajda (1982). Sedimentological influences of detached breakwaters. *Proc. 18th Int. Conf. Coastal Eng.*, ASCE, **3**, 1930-1949.
Ruijgrok E.C.M. (1999). *Valuation of Nature in Coastal Zones*. Vrije Universiteit, Amsterdam, 236.
Ruol P. and A. Faedo (2002). Physical model study on low-crested structures under breaking wave conditions. *Proc. Medcoast 2002*, 83-96.
Ruol P., A. Faedo and A. Paris (2004). Prove sperimentali sul comportamento di una scogliera a cresta bassa e sul fenomeno del piling-up a tergo di essa. *Studi Costieri*, **6**, in Italian.
Ruol P., L. Martinelli, B. Zanuttigh, L. Cappietti, D. Vicinanza and Faedo, A. (2006). Sui fenomeni di tracimazione, filtrazione e piling up per opere di difesa a cresta bassa. *Studi Costieri*, **9**, 51-74, in Italian.
Sánchez-Arcilla A., J.M. Alsina, I. Cáceres, D. González-Marco, J. P. Sierra and C. Peña (2004). Morphodynamics on a beach with a submerged detached breakwater. *Proc. of the 29th Int. Conf. on Coastal Eng*. Lisbon, Portugal, 2836-2848.
Sánchez-Arcilla A., J.P. Sierra, I. Cáceres, D. González-Marco, J.M. Alsina, F. Montoya y J. Galofré (2005). Beach dynamics in the presence of a low crested structure. The Altafulla case. *Journal of Coastal Research*, SI **39**, in press.
Santelices B. (1990). Patterns of reproduction, dispersal and recruitment in seaweeds. *Oceanography and Marine Biology Annual Review*, **28**, 177-276.
Sasaki T., (1975). Simulation of shoreline and nearshore current, *Proc. of Civil Eng. in the Oceans*, **III,** ASCE, Univ. of Delaware, 179-196.
Sato K., Shuto N., and Tanaka H. (1995). *Numerical simulation of the sand spit flushing at a river mouth*. In: Chinese Hydraulic Engineering Society and International Research and Training Center on Erosion and Sedimentation (Editors), Adv. in Hydro-Science and Eng., Vol. II, Part B. Tsinghua University Press, Beijing, 1399-1406.
Sawamoto M. and T. Yamashita (1986). Sediment Transport Rate due to Wave Action. *Journal of Hydrosc. and Hydr. Eng.*, **4**, (1).
Scarpa R., G. Sirchia and M. Bravi (1998). *Kernel vs. Logit Modelling of Single Bounded CV Responses: Valuing Access to Architectural and Visual Arts Heritage in Italy*. In Bishop R.C. and D. Romano (1998), Environmental Resource Valuation. Applications of the CVM in Italy,

Kluwer Academic Publishers, Boston.

Schoonees J.S. and A.K. Theron (1995). Evaluation of 10 cross-shore sediment transport/morphological models. *Coastal Engineering*, **25**, 1-41.

Schüttrumpf H., H. Oumeraci (2001). *Prediction of wave overtopping flow parameters on the crest and landward slope of sea dikes*. OPTICREST Research Report on Subtask 3.5.

Seabrook S.R. and K.R. Hall (1998). Wave transmission at submerged rubble mound breakwaters. *Proc. 26th Int. Conf. on Coastal Engineering*, ASCE, 2000-2013.

Seiji M., T. Uda, and S. Tanaka (1987). Statistical study on the effect and stability of detached breakwaters. *Coastal Engineering in Japan* **30** (1).

Seip K. and J. Strand (1992). Willingness to Pay for Environmental Goods in Norway: A Contingent valuation Study with Real Payment. *Environmental and Resource Economics*, **2**, 91-106.

Serra J., J.R. Medina (1996). Beach monitoring program of Valencia (Spain). *Proc. Coastal Eng. Conf.* Orlando, Florida (ed. B.L.Edge), ASCE, **3**, 2871-2884.

Shaw D. (1988). On-site Samples' Regressions, *Journal of Econometrics*, **37**, 211-23.

Shechter M., B. Reiser and N. Zaitsev (1998). Measuring Passive Use Value. *Environmental and Resources Economics*, **12**, 457-78.

Shibayama T. and K. Horikawa (1980). Bed Load Measurement and Prediction of Two-Dimensional Beach Transformation due to Waves. *Coastal Engineering in Japan*, **23**, 179-190.

Shih R.W.K., (1990). Permeability characteristics of rubble material - new formulae. *Proc. Int. Conf. Coastal Eng.*, ASCE, New York., 1827-1846.

Shinohara K., T. Tsubaki (1966). Model Study on the Change of Shoreline of Sandy Beach by Offshore Breakwater. *Proc. 10th Conf. Coastal Eng. Tokio*, 1, 550-563.

Silberman J. and M. Klock (1988). The Recreation Benefits of Beach Renourishment. *Ocean and Shoreline Management*, **11**, 73-90.

Silberman J., D.A. Gerlowski and N.A. Williams (1992). Estimating Existence Value for Users and Nonusers of New Jersey Beaches. *Land Economics*, **68**, (2), 225-36.

Silvester R. (1970a). Coastal defense. *Proc. Inst. Civ. Eng.*, **45**, 677-682.

Silvester R. (1970b). Growth of crenulate shaped bays to equilibrium. *J. Waterway and Harbor Div.*, ASCE, **96** (ww4), 275-287.

Silvester R., Y. Tsuchiya and T. Shibano (1980). Zeta bays, pocket beaches and headland control. *Proc. 17th Int. Conf. Coastal Eng.*, ASCE, **2**, 1306-1319.

Silvester R. and J.R.C. Hsu (1997). *Coastal Stabilization*. Advanced series on Ocean Engineering, **14**, Ed. World Scientific.

Simms J.D., A.H. Brampton and J.K. Brooke (1996). *Beach Management manual*. Report Number 153. CIRIA London.

Sleath J.F.A. (1978). Measurements of Bed Load in Oscillatory Flow. *Journal of Waterway, Port, Coastal and Ocean Eng.*, ASCE, **104**, (WW4), 291-307.

Sørensen O.R. and L.S. Sørensen (2001). Boussinesq type modelling using unstructured finite element technique. *Proc. 27th Coastal Eng. Conf.*, 190-202.

Sørensen O.R., H.A. Schäffer and L.S. Sørensen (2004). Boussinesq type modelling using an unstructured finite element technique. *Coastal Eng.*, **50**, 181-198.

Sorensen R.M. and N.J. Beil (1988). Perched beach profile response to wave action. *Proc. 21st. International Conference on Coastal Eng.* ASCE, 1482-1491.

Soulsby R. (1997). *Dynamics of Marine Sands*. Thomas Telford, London.

Southward A.J. and J.H. Orton (1954). The effects of wave-action on the distribution and numbers of the commoner plants and animals living on the Plymouth Breakwater. *Journal of the Marine Biological Association of the United Kingdom*, **33**, 1-19.

SPM 1984. *Shore Protection Manual*, 4th edition. US Army Corps of Engrs, Coastal Engng. Res. Center, US Govt Printing Office, Washington DC.

Stær P.A., M.F. Pedersen, M.S. Thomsen, T. Wernberg, D. Krause-Jensen (2000). Invasion of Sargassum muticum in Limfjorden (Denmark) and its possible impact on the indigenous macroalgal community. *Mar. Ecol. Progr. Ser.*, **207**, 79-88.

Steetzel H.J. (1993). Cross-shore transport during storm surges. Thesis, Technical University of Delft.

Steetzel H.J. and J.H. de Vroeg (1999). Application of a multi-layer approach for morphological modelling. *Coastal Sediments, '99*, 2206-2218.

Stephenson and Stephenson (1972). *Life between tidemarks on rocky shores.* WH Freeman, San Francisco, 425 pp.

Stive M.F.F. and J.A. Battjes (1984). A Model for Offshore Sediment Transport. *Proc. of the 19th Coastal Eng. Conference*, Houston, USA, 1420-1436.

Suh K. and R.A. Dalrymple (1987). Offshore breakwaters in laboratory and field. Journal of Waterway, Port, Coastal and Ocean Engineering, ASCE, **113**, (2), 105-121.

Sumer B.M. and Fredsøe J. (2000). Experimental study of 2D scour and its protection at a rubble mound breakwater, *Coastal Engineering*, **40** (1), 59-87.

Sumer B.M. and J. Fredsøe (2002). *The Mechanics of Scour in the Marine Environment.* World Scientific, 552 p.

Sumer M., Fredsøe J., Lamberti A. & B. Zanuttigh, Dixen M., Gislason K., Di Penta A., (2005). Local scour and erosion around low-crested structures, *Coastal Engineering*, **52** (10-11), 995-1025, Elsevier.

Svendsen I.A. and U. Putrevu (1994). Nearshore mixing and dispersion. *Proc. Roy. Soc. Lond. A* **445**, 561-576.

Svendsen I.A., J. Veeramony, J. Bakunin, and J. Kirby (2000). The flow in weak turbulent hydraulic jumps. *J. Fluid. Mech.*, **418**, 25-57.

Swart D.H. (1975). A schematisation of onshore-offshore transport. *Proc. 14th Int. Coastal Engineering Conference* ASCE, 884-900.

Tan S. and Y. Chiew (1994). Analysis of Bayed Beaches in Static Equlibrium. *J. Waterway, Port, and Ocean Eng.*, ASCE, **120**, (2), 145-153.

Tanguy J.M. and Zhang B. (1994). Morphological modeling of the evolution of a trench connecting a lagoon to the sea at Keta - Ghana. *Proc. Int. Symp. on Waves - Physical and Numerical Modelling*, Vancouver.

Tolman H.L. (1992). An evaluation of expressions for wave energy dissipation due to bottom friction in the presence of currents. *Coastal Engineering*, **16**, 165-179.

Thompson R., M. Roberts, T. Norton, S. Hawkins (2000). Feast or famine for intertidal grazing molluscs: a mis-match between seasonal variations in grazing intensity and the abundance of microbial resources. *Hydrobiologia*, **440**, 1-3: 357-367.

Toyoshima O. (1974). *Design of a detached breakwater system.* Proceedings, 14th International Conference on Coastal Engineering, Copenhagen Denmark, 1419-1431.

Trowbridge J. and D. Young (1989). Sand Transport by Unbroken Water Waves under Sheet Flow Conditions. *Journal of Geophysical Research*, **94**, (C8), 10971-10991.

Turner C.H., E.E. Ebert and R.R. Given (1969). Man-made reef ecology. California Department of Fish and Game, *Fish Bulletin*, **146**, 1-221.

Turner R.K., I. Bateman and J.S. Brooke (1992). *Valuing the Benefits of Coastal Defence: a Case Study of the Adelburgh Sea-defence Scheme.* In Valuing the Environment: Economic Approaches to Environmental Valuation. A. Coker and C. Richards eds. New York, J. Wiley & Sons, U.S. Environmental Protection Agency, 2000, Guidelines for Preparing Economic Analyses, X, 205. U.S. Environmental Protection Agency.

Turner R.K. (2000). Integrating natural and socio-economic science in coastal management. *Journal of Marine Systems*, **25**, 447-60.

Underwood, A. J., (1992). Beyond Baci - The detection of environmental impacts on populations in the real, but variable, world, *Journal of Experimental Marine Biology and Ecology*, **161** (2), 145-178.

Underwood, A. J., (1994). On beyond baci - sampling designs that might reliably detect environmental disturbances, *Ecological applications*, **4** (1), 3-15.

Underwood, A. J., (1997). *Experiments in ecology. Their logical design and interpretation using analysis of variance,* Cambridge University Press, Cambridge, UK, 522 pp.

Van der Meer J.W. (1988). *Rock slopes and gravel beaches under wave attack.* PhD-thesis Delft University of Technology, the Netherlands.

Van der Meer J.W. (1990). *Low-Crested and Reef Breakwaters.* Report H198/Q638, Delft Hydraulics Laboratory, The Netherlands.

Van der Meer J.W. (1990a). *Extreme shallow water wave conditions.* Report H198, Delft Hydraulics Laboratory, The Netherlands.

Van der Meer J.W. (1990b). *ENDEC verification on slope 1:10.* Report H986, Delft Hydraulics Laboratory, The Netherlands.

Van der Meer J.W. (1991). *Stability and Transmission at Low-Crested Structures.* Delft Hydraulics Publication n. 453, Delft Hydraulics Laboratory, The Netherlands.

Van der Meer J.W. (1992). Conceptual Design of rubble mound breakwaters. *Proc. of the Short Course on Design and Reliability of Coastal Structures*, 447-510.

Van der Meer J.W. (1993) *Wave run-up and overtopping on dikes* (in Dutch), Technical Advisory Committee on Water Defences, Delft, The Netherlands.

Van der Meer J.W. and I.F.R. Daemen (1994). Stability and wave transmission at low crested rubble-mound structures, *J. of Waterway, Port Coastal and Ocean Engineering*, **1**, 1-19.

Van der Meer J.W. and Janssen J.P.F.M. (1995). *Wave run-up and wave overtopping at dikes.* In: Kobayashi, N., Demirbilek, Z. (Eds.), Wave Forces on Inclined and Vertical Structures, 1-27, Ch. 1.

Van der Meer J.W. and J.P.F.M. Janssen (1995). *Wave run-up and wave overtopping at dikes.* Task Committee Reports, ASCE, 1995.

Van der Meer J.W., K. d'Angremond, and E. Gerding (1995). Toe Structure Stability of Rubble Mound Breakwaters. *Proceedings of the Advances in Coastal Structures and Breakwaters Conference, Institution of Civil Engineers*, Thomas Telford Publishing, London, UK, 308-321.

Van der Meer J.W., W.H. Tutuarima and G. Burger (1996). Influence of rock shape and grading on stability of low crested structures. *Proc. 25th Int. Conf. on Coastal Eng.*, ASCE, 1957-1970.

Van der Meer J.W., E. Regeling, J.P. de Waal (1998). A code for dike height design and examination. Coastlines, Structures and Breakwaters. *Proc. ICE*, Editor N.W.H. Allsop, London, 5-19.

Van der Meer J.W., H.J. Regeling and J.P. de Waal (2000). Wave transmission: spectral changes and its effect on run-up and overtopping. ASCE, *Proc. ICCE*, Sydney, Australia, 2156-2168.

Van der Meer J.W., B. Wang, A. Wolters, B. Zanuttigh and M. Kramer (2003). Oblique wave transmission over low-crested structures. ASCE, *Proc. Coastal Structures,* Portland, Oregon.

Van der Veen A., A.E. Steenge, M. Bockarjova and C.J.J. Logtmeijer (2003). *Structural economic effects of large-scale inundation: A simulation of the Krimpen dike breakage.* Proceedings: Joint NEDIES and University of Twente Workshop – In search of a common methodology for damage estimation. Report EUR XXXXX EN (2003), Office for Official Publications of the European Communities. A. van der Veen, A.L. Vetere Arellano and J.P. Nordvik. Bruxelles, European Union.

Van Gent M.R.A. (1993). *Stationary and oscillatory flow through coarse porous media,* Communications on Hydraulic and Geotechnical Engineering, Report N. 93-9, Delft University of Technology 1993.

Van Gent M.R.A. (1995-a). *Wave interaction with permeable coastal structures.* Ph.D.-thesis, Delft University of Technology, ISBN 90-407-1182-8, Delft University Press, Delft.

Van Gent M.R.A. (1995-b). Porous flow through rubble mound material. *J. of Waterway, Port, Coastal and Ocean Engineering*, ASCE, **121**, (3), 176-181, New York.

Van Rijn L.C. (1993). *Principles af Sediment Transport in Rivers, Estuaries and Coastal Seas.* Amsterdam: Aqua Publishing, 673 p.

Van Rijn L.C. (2000). *General View on Sand Transport by Currents and Waves.* Delft Hydraulics. Report Z2899.30, Delft, The Netherlands.

Van Rijn L.C., Walstra D.J.R., Grasmeijer B., Sutherland J., Pan S. and Sierra, J. P. (2003). The predictability of cross-shore bed evolution of sandy beaches at the time scale of storms and seasons using process-based Profile models. *Coastal Engineering*, **47**, 295-327.

Veeramony J. and I.A. Svendsen (1999). *Modelling the flow in surf zone waves.* Research Report CACR-99-04, Center for Applied Coastal Research, University of Delaware.

Veeramony J. and I.A. Svendsen (2000). The flow in surf zone waves. *Coast. Eng.* **39**.

Vichetpan N. (1969). *Equilibrium shapes of coastline in plan.* Thesis, No. 280, Asian Ins. of Technology, Bangkok, Thailand.

Vidal C., M.A. Losada, R. Medina, E.P.D. Mansard and G. Gomez-Pina (1992). A Universal Analysis for the Stability of both Low-Crested and Submerged Breakwaters. *Proc. 23rd Int. Conference on Coastal Engineering*, 1679-1692, Italy.

Vidal C., M.A. Losada and E.P.D. Mansard (1995). Stability of Low-Crested Rubble-Mound Breakwater heads. *Journal of Waterway, port, coastal and Ocean Engineering*, **121**, (2), 114-122.

Vidal C., R. Medina, and F.L. Martín (2000). A Methodology to Assess the Armour Unit Stability of Low-Crested and Submerged Rubble-Mound Breakwaters. *Proceedings of Coastal Structures '99 Conference*, Balkema, Rotterdam, 721-725.

Vrisou van Eck N. and M. Kok (2001). *Standard Methode Schade en Slachtofers als gevolg van overstromingen.* Lelystad & Delft, HKV & Dienst weg en waterbouwkunde.

Wang H., Miao G. and Lin L.H. (1992). A time-dependent nearshore morphological response model. *Proc. 23rd Int. Coastal Eng. Conf.*, Venice. ASCE, 2513-2527.

Watanabe A. (1985). Three-dimensional predictive model of beach evolution around a structure. *Proc. Water Wave Research Conf.*, Hannover, 121-141.

Watanabe A., Maruyama K., Shimizu T. and Sakakiyama T. (1986). Numerical prediction model of three-dimensional beach deformation around a structure. *Coastal Engineering in Japan*, **29**, 179-194.

Water Framework Directive 2000/60/EC of the European Parliament and of the Council (2000).

Weggel J.R. (1972). Maximum breaker height, *J. Way, Harbour and Coastal Eng. Div.*, ASCE, **98**, (4), 529-548.

Whitmarsh D., Northen J. and Jaffry S. (1999). Recreational Benefits of Coastal Protection: a Case Study. *Marine Policy*, **23**, (4-5), 453-63.

Wilkinson S.B., Zheng, Weizhong, J.R. Allen, N.J. Fielding, V.C. Wanstall, G. Russel, S.J. Hawkins (1996). Water quality improvements in Liverpool docks: the role of filter feeders in algal and nutrient dynamics, The role of episodic events. *Marine Ecology*, **17** (1-3), 197-211.

Wright L.D. and A.D. Short (1984). Morphodynamic variability of surf zones and beaches: A synthesis. *Marine Geology*, **56**, 93-11.

Yamaguchi M. and Nishioka Y. (1984). Numerical simulation on the change of bottom topography by the presence of coastal structures. *Proc. 19th Int. Coastal Eng. Conf.*, Houston, TX. ASCE, 1732-1748.

Yelland M. and Taylor P.K. (1996). Wind Stress Measurements from the Open Ocean. *Journal of Physical Oceanography*, **26**, 541-558.

Yohe G., J.E. Neumann and P. Marshall (1999). *The economic damage induced by sea level rise in the United States.* In The impact of climate change on the United States economy. Robert Mendelsohn and James- E. Neumann eds. Cambridge, New York and Melbourne, Cambridge University Press, pp. xi, 331.

Zanuttigh B., L. Martinelli, A. Lamberti (2004): Hydrodynamics induced by low-crested rubble-mound structures: experimental analysis and 2DH numerical simulations. *XXIX Convegno di Idraulica e Costruzioni Idrauliche*, Trento, Italy, **3**, 645-652.

Zyserman J.A., Jorgensen K and Christensen E.D. (1999). *Sediment transport and morphology in the vicinity of shore parallel breakwaters. Proc. Coastal Structures'99*, 857-863.

Zyserman J., Johnson H. K., Zanuttigh B. and L. Martinelli, (2005). Far field erosion and morphological effects, *Coastal Engineering*, **52** (10-11), 977-994, Elsevier.

LCS design guidelines

Index

Kolibri 360 GmbH
Ferdinand-Jühlke-Straße 7
99095 Erfurt, Deutschland
produktsicherheit@kolibri360.de